The Alcohol Hangover

The Alcohol Hangover: Causes, Consequences, and Treatment

Editors

Joris C Verster
Lizanne Arnoldy
Sarah Benson
Andrew Scholey
Ann-Kathrin Stock

MDPI • Basel • Beijing • Wuhan • Barcelona • Belgrade • Manchester • Tokyo • Cluj • Tianjin

Editors

Joris C Verster
Utrecht University
The Netherlands

Lizanne Arnoldy
Swinburne University
Australia

Sarah Benson
Swinburne University
Australia

Andrew Scholey
Swinburne University
Australia

Ann-Kathrin Stock
TU Dresden
Germany

Editorial Office
MDPI
St. Alban-Anlage 66
4052 Basel, Switzerland

This is a reprint of articles from the Special Issue published online in the open access journal *Journal of Clinical Medicine* (ISSN 2077-0383) (available at: https://www.mdpi.com/journal/jcm/special_issues/Alcohol_Hangover).

For citation purposes, cite each article independently as indicated on the article page online and as indicated below:

LastName, A.A.; LastName, B.B.; LastName, C.C. Article Title. *Journal Name* **Year**, *Volume Number*, Page Range.

ISBN 978-3-0365-0356-1 (Hbk)
ISBN 978-3-0365-0357-8 (PDF)

Cover image courtesy of depositphotos.com under the extended license.

© 2021 by the authors. Articles in this book are Open Access and distributed under the Creative Commons Attribution (CC BY) license, which allows users to download, copy and build upon published articles, as long as the author and publisher are properly credited, which ensures maximum dissemination and a wider impact of our publications.

The book as a whole is distributed by MDPI under the terms and conditions of the Creative Commons license CC BY-NC-ND.

Contents

About the Editors . . . . **ix**

Joris C. Verster, Lizanne Arnoldy, Sarah Benson, Andrew Scholey and Ann-Kathrin Stock
The Alcohol Hangover Research Group: Ten Years of Progress in Research on the Causes, Consequences, and Treatment of the Alcohol Hangover
Reprinted from: *J. Clin. Med.* **2020**, *9*, 3670, doi:10.3390/jcm9113670 . . . . **1**

Joris C. Verster, Andrew Scholey, Aurora J.A.E. van de Loo, Sarah Benson and Ann-Kathrin Stock
Updating the Definition of the Alcohol Hangover
Reprinted from: *J. Clin. Med.* **2020**, *9*, 823, doi:10.3390/jcm9030823 . . . . **9**

Joris C. Verster, L. Darren Kruisselbrink, Karin A. Slot, Aikaterini Anogeianaki, Sally Adams, Chris Alford, Lizanne Arnoldy, Elisabeth Ayre, Stephanie Balikji, Sarah Benson, Gillian Bruce, Lydia E. Devenney, Michael R. Frone, Craig Gunn, Thomas Heffernan, Kai O. Hensel, Anna Hogewoning, Sean J. Johnson, Albertine E. van Lawick van Pabst, Aurora J.A.E. van de Loo, Marlou Mackus, Agnese Merlo, René J.L. Murphy, Lauren Owen, Emily O.C. Palmer, Charmaine J.I. van Rossum, Andrew Scholey, Chantal Terpstra, Vatsalya Vatsalya, Sterre A. Vermeulen, Michelle van Wijk and Ann-Kathrin Stock
Sensitivity to Experiencing Alcohol Hangovers: Reconsideration of the 0.11% Blood Alcohol Concentration (BAC) Threshold for Having a Hangover
Reprinted from: *J. Clin. Med.* **2020**, *9*, 179, doi:10.3390/jcm9010179 . . . . **15**

Joris C. Verster, Aurora J. A. E. van de Loo, Sally Adams, Ann-Kathrin Stock, Sarah Benson, Andrew Scholey, Chris Alford and Gillian Bruce
Advantages and Limitations of Naturalistic Study Designs and Their Implementation in Alcohol Hangover Research
Reprinted from: *J. Clin. Med.* **2019**, *8*, 2160, doi:10.3390/jcm8122160 . . . . **23**

Joris C Verster, Aurora J.A.E. van de Loo, Sarah Benson, Andrew Scholey and Ann-Kathrin Stock
The Assessment of Overall Hangover Severity
Reprinted from: *J. Clin. Med.* **2020**, *9*, 786, doi:10.3390/jcm9030786 . . . . **33**

Chantal Terpstra, Andrew Scholey, Joris C. Verster and Sarah Benson
Prevalence of Hangover Resistance According to Two Methods for Calculating Estimated Blood Alcohol Concentration (eBAC)
Reprinted from: *J. Clin. Med.* **2020**, *9*, 2823, doi:10.3390/jcm9092823 . . . . **47**

Marlou Mackus, Aurora JAE van de Loo, Johan Garssen, Aletta D. Kraneveld, Andrew Scholey and Joris C. Verster
The Role of Alcohol Metabolism in the Pathology of Alcohol Hangover
Reprinted from: *J. Clin. Med.* **2020**, *9*, 3421, doi:10.3390/jcm9113421 . . . . **59**

Aurora J.A.E. van de Loo, Marlou Mackus, Oran Kwon, Illathu Madhavamenon Krishnakumar, Johan Garssen, Aletta D. Kraneveld, Andrew Scholey and Joris C. Verster
The Inflammatory Response to Alcohol Consumption and Its Role in the Pathology of Alcohol Hangover
Reprinted from: *J. Clin. Med.* **2020**, *9*, 2081, doi:10.3390/jcm9072081 . . . . **73**

Joris C. Verster, Sterre A. Vermeulen, Aurora J. A. E. van de Loo, Stephanie Balikji, Aletta D. Kraneveld, Johan Garssen and Andrew Scholey
Dietary Nutrient Intake, Alcohol Metabolism, and Hangover Severity
Reprinted from: *J. Clin. Med.* **2019**, *8*, 1316, doi:10.3390/jcm8091316 85

Vatsalya Vatsalya, Hamza Z. Hassan, Maiying Kong, Bethany L. Stangl, Melanie L. Schwandt, Veronica Y. Schmidt-Teron, Joris C. Verster, Vijay A. Ramchandani and Craig J. McClain
Exacerbation of Hangover Symptomology Significantly Corresponds with Heavy and Chronic Alcohol Drinking: A Pilot Study
Reprinted from: *J. Clin. Med.* **2019**, *8*, 1943, doi:10.3390/jcm8111943 103

Joris C. Verster, Karin A. Slot, Lizanne Arnoldy, Albertine E. van Lawick van Pabst, Aurora J. A. E. van de Loo, Sarah Benson and Andrew Scholey
The Association between Alcohol Hangover Frequency and Severity: Evidence for Reverse Tolerance?
Reprinted from: *J. Clin. Med.* **2019**, *8*, 1520, doi:10.3390/jcm8101520 117

Sam Royle, Lauren Owen, David Roberts and Lynne Marrow
Pain Catastrophising Predicts Alcohol Hangover Severity and Symptoms
Reprinted from: *J. Clin. Med.* **2020**, , 280, doi:10.3390/jcm9010280 131

Albertine E. van Lawick van Pabst, Lydia E. Devenney and Joris C. Verster
Sex Differences in the Presence and Severity of Alcohol Hangover Symptoms
Reprinted from: *J. Clin. Med.* **2019**, *8*, 867, doi:10.3390/jcm8060867 147

Joris C. Verster, Lizanne Arnoldy, Aurora J.A.E. van de Loo, Sarah Benson, Andrew Scholey and Ann-Kathrin Stock
The Impact of Mood and Subjective Intoxication on Hangover Severity
Reprinted from: *J. Clin. Med.* **2020**, *9*, 2462, doi:10.3390/jcm9082462 165

Chris Alford, Zuzana Martinkova, Brian Tiplady, Rebecca Reece and Joris C. Verster
The Effects of Alcohol Hangover on Mood and Performance Assessed at Home
Reprinted from: *J. Clin. Med.* **2020**, *9*, 1068, doi:10.3390/jcm9041068 179

Andrew Scholey, Sarah Benson, Jordy Kaufman, Chantal Terpstra, Elizabeth Ayre, Joris C. Verster, Cory Allen and Grant J. Devilly
Effects of Alcohol Hangover on Cognitive Performance: Findings from a Field/Internet Mixed Methodology Study
Reprinted from: *J. Clin. Med.* **2019**, *8*, 440, doi:10.3390/jcm8040440 193

Chris Alford, Callum Broom, Harriet Carver, Sean J. Johnson, Sam Lands, Rebecca Reece and Joris C. Verster
The Impact of Alcohol Hangover on Simulated Driving Performance during a 'Commute to Work'—Zero and Residual Alcohol Effects Compared
Reprinted from: *J. Clin. Med.* **2020**, *9*, 1435, doi:10.3390/jcm9051435 205

Sarah Benson, Elizabeth Ayre, Harriet Garrisson, Mark A Wetherell, Joris C Verster and Andrew Scholey
Alcohol Hangover and Multitasking: Effects on Mood, Cognitive Performance, Stress Reactivity, and Perceived Effort
Reprinted from: *J. Clin. Med.* **2020**, *9*, 1154, doi:10.3390/jcm9041154 219

Craig Gunn, Graeme Fairchild, Joris C. Verster and Sally Adams
The Effects of Alcohol Hangover on Executive Functions
Reprinted from: *J. Clin. Med.* **2020**, *9*, 1148, doi:10.3390/jcm9041148 231

Antje Opitz, Christian Beste and Ann-Kathrin Stock
Alcohol Hangover Differentially Modulates the Processing of Relevant and Irrelevant Information
Reprinted from: *J. Clin. Med.* **2020**, *9*, 778, doi:10.3390/jcm9030778 247

Antje Opitz, Jan Hubert, Christian Beste and Ann-Kathrin Stock
Alcohol Hangover Slightly Impairs Response Selection but not Response Inhibition
Reprinted from: *J. Clin. Med.* **2019**, *8*, 1317, doi:10.3390/jcm8091317 265

Julia Berghäuser, Wiebke Bensmann, Nicolas Zink, Tanja Endrass, Christian Beste and Ann-Kathrin Stock
Alcohol Hangover Does Not Alter the Application of Model-Based and Model-Free Learning Strategies
Reprinted from: *J. Clin. Med.* **2020**, *9*, 1453, doi:10.3390/jcm9051453 281

Lydia E. Devenney, Kieran B. Coyle, Thomas Roth and Joris C. Verster
Sleep after Heavy Alcohol Consumption and Physical Activity Levels during Alcohol Hangover
Reprinted from: *J. Clin. Med.* **2019**, *8*, 752, doi:10.3390/jcm8050752 301

Joris C Verster, Aikaterini Anogeianaki, Darren Kruisselbrink, Chris Alford and Ann-Kathrin Stock
Relationship between Alcohol Hangover and Physical Endurance Performance: Walking the Samaria Gorge
Reprinted from: *J. Clin. Med.* **2020**, *9*, 114, doi:10.3390/jcm9010114 315

Joris C Verster, Thomas A Dahl, Andrew Scholey and Jacqueline M Iversen
The Effects of SJP-001 on Alcohol Hangover Severity: A Pilot Study
Reprinted from: *J. Clin. Med.* **2020**, *9*, 932, doi:10.3390/jcm9040932 331

Andrew Scholey, Elizabeth Ayre, Ann-Kathrin Stock, Joris C Verster and Sarah Benson
Effects of Rapid Recovery on Alcohol Hangover Severity: A Double-Blind, Placebo-Controlled, Randomized, Balanced Crossover Trial
Reprinted from: *J. Clin. Med.* **2020**, *9*, 2175, doi:10.3390/jcm9072175 343

About the Editors

Joris C Verster (1970) studied psychology and obtained his Ph.D. at the Utrecht Institute for Pharmaceutical Sciences in 2002. Verster is Associate Professor at the Division of Pharmacology at Utrecht University, The Netherlands, and Adjunct Professor of Human Psychopharmacology at the Centre for Human Psychopharmacology of Swinburne University of Technology, Melbourne, Australia. Dr. Verster investigates the impact of exposome pressure (e.g., lifestyle factors) on health, immune fitness, and behavior. The aim of this multidisciplinary research is to develop and implement tools and methods (biomarkers, clinical assessments, and questionnaires), and to collect real world evidence on how a wide range of environmental factors may impact health outcomes and quality of life. Dr. Verster has a track record of clinical trials examining the effects of CNS drugs and psychoactive substances on cognitive and psychomotor functioning, mood, sleep, and daily activities, such as driving a car. He conducts extensive research into the effects of alcohol use as risk factor in health and disease, and the causes, consequences, and treatments of an alcohol hangover. Verster is the founder of the Alcohol Hangover Research Group, and his research is regularly covered by international media. Joris serves as scientific advisor for both industry and governmental organizations.

Lizanne Arnoldy is a Ph.D. candidate from the Centre for Human Psychopharmacology at Swinburne University, where she explores the role of diet on aging brains, as indexed by diffusion weighted imaging. Arnoldy has a background in pharmacy, and previously worked for Utrecht University where she conducted research for the World Health Organization.

Sarah Benson is a Postdoctoral Research Fellow at the Centre for Human Psychopharmacology, Swinburne University. She completed her Ph.D. in 2016, where she explored the cognitive and behavioral effects of alcohol and caffeine. Her research portfolio has focused primarily on cognitive impairment and enhancement, alcohol hangover, and drug use and functional outcomes. Sarah aims to disentangle the neurocognitive effects of licit and illicit substances in order to identify relationships between their behavioral effects and pharmacological properties, as well as to reduce dangerous behaviors and harm.

Andrew Scholey is Professor of Human Psychopharmacology based in Melbourne, Australia. He is a leading international researcher for the neurocognitive effects of nutritional interventions, recreational drugs, supplements, and food components. He has published >250 peer-reviewed journal articles, as well as over 25 books and book chapters. He has attracted around $25 million in research funding. Andrew has been a lead investigator on a series of studies regarding human biobehavioral effects of nutritional interventions, focusing on potential neurocognition-enhancing and anti-stress/anxiolytic properties (including first-into-human neurocognitive assessment of ginseng, sage, curcumin, and lemon balm amongst others). His current research focuses on neuroimaging and biomarker techniques to better understand the mechanisms of cognitive enhancement. Andrew works closely with the industry to allow for the rapid translation of research into evidence-based end-user health benefits.

Ann-Kathrin Stock (1987) studied psychology and obtained her doctorate at the University of Bochum in 2014, where she focused on dopaminergic modulation of action control. She is currently a senior researcher at the Faculty of Medicine/Chair of Cognitive Neurophysiology and at the Faculty of Psychology/Chair of Biopsychology at TU Dresden, Germany. Dr. Stock's research focuses on neurobiochemical modulators of action control and executive functions. With respect to drugs of abuse, her research focuses on both short- and long-term effects of alcohol and stimulant (ab)use.

Editorial

The Alcohol Hangover Research Group: Ten Years of Progress in Research on the Causes, Consequences, and Treatment of the Alcohol Hangover

Joris C. Verster [1,2,*], Lizanne Arnoldy [1,2], Sarah Benson [2], Andrew Scholey [2] and Ann-Kathrin Stock [3,4]

1. Division of Pharmacology, Utrecht Institute for Pharmaceutical Sciences (UIPS), Utrecht University, 3584CG Utrecht, The Netherlands; larnoldy@swin.edu.au
2. Centre for Human Psychopharmacology, Swinburne University, Melbourne VIC 3122, Australia; sarahmichellebenson@gmail.com (S.B.); andrew@scholeylab.com (A.S.)
3. Cognitive Neurophysiology, Department of Child and Adolescent Psychiatry, Faculty of Medicine, TU Dresden, Fetscherstr. 74, 01,307 Dresden, Germany; Ann-Kathrin.Stock@uniklinikum-dresden.de
4. Biopsychology, Department of Psychology, School of Science, TU Dresden, Zellescher Weg 19, 01,069 Dresden, Germany
* Correspondence: j.c.verster@uu.nl

Received: 2 November 2020; Accepted: 2 November 2020; Published: 16 November 2020

Abstract: The alcohol hangover is defined as the combination of negative mental and physical symptoms, which can be experienced after a single episode of alcohol consumption, starting when blood alcohol concentration (BAC) approaches zero. Here, we present the book "The alcohol hangover: causes, consequences, and treatment", written to celebrate the 10th anniversary of the Alcohol Hangover Research Group (AHRG), summarizing recent advances in the field of alcohol hangover research.

Keywords: alcohol; hangover; causes; consequences; treatments

---

The alcohol hangover is defined as the combination of negative mental and physical symptoms, which can be experienced after a single episode of alcohol consumption, starting when blood alcohol concentration (BAC) approaches zero [1,2]. Despite the fact that the alcohol hangover is the most commonly reported negative consequence of alcohol consumption [3], a relatively small amount of research has been devoted to this topic. The latter is surprising as the alcohol hangover is associated with negative mood, cognitive impairment, and physical effects [4]. Here, we present the book "The alcohol hangover: causes, consequences, and treatment", written to celebrate the 10th anniversary of the Alcohol Hangover Research Group (AHRG).

In 2010, the AHRG was founded to raise the profile of alcohol hangover research [5]. The AHRG scientific meetings aim to bring together active and internationally diverse alcohol hangover researchers to generate discussion on recent developments in hangover research. The objectives of these meetings are to discuss recent findings and future research directions, to raise the profile of alcohol hangover research, and to start new research collaborations. Over the past 10 years, 11 successful AHRG meetings have been held across the world [5–11]. In 2010, the first AHRG meeting was held as a satellite symposium of the Research Society on Alcoholism conference in San Antonio, Texas, USA. Subsequent AHRG meetings were held in Paris in 2010 (France), Utrecht in 2011 (The Netherlands), Wolfville in 2012 (Canada), Keele in 2013 (UK), Bellevue in 2014 (USA), Perth in 2015 (Australia), New Orleans in 2016 (USA), Utrecht in 2017 (The Netherlands), Utrecht in 2018 (The Netherlands), and Wailoaloa Beach, Nadi in 2019 (Fiji). Proceedings of most of the AHRG meetings have been published [5–11].

In the decade since its inception, the AHRG has moved the field forward significantly. The inaugural meeting resulted in the publication of a consensus paper on best practice in hangover research, and an evaluation of the gaps in knowledge that should be addressed by future research [12]. Among the key accomplishments was the development of a definition for the alcohol hangover [1]. In addition, international research collaborations resulted in a significant increase in the number of published articles on the alcohol hangover (see Figure 1).

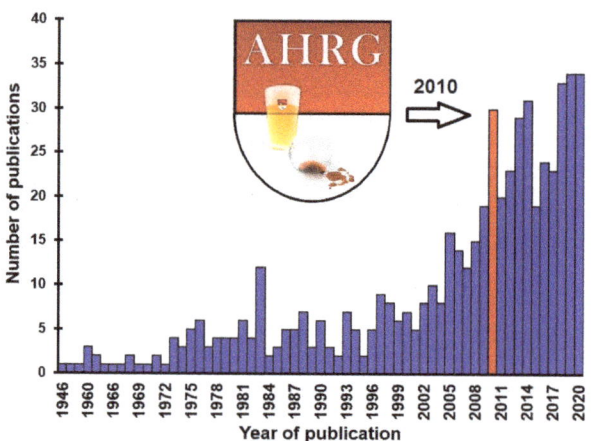

Figure 1. Published articles on the alcohol hangover. Number of publications taken from PubMed (https://pubmed.ncbi.nlm.nih.gov, assessed on 13 October 2020), searching for "alcohol hangover". In 2010 (red bar), the Alcohol Hangover Research Group (AHRG) was founded.

In conjunction with its 10th anniversary, a Special Issue of the Journal of Clinical Medicine on the alcohol hangover was curated by members of the AHRG. After peer-review, twenty-five articles were accepted for the Special Issue, and this collection is combined in this book, entitled "The alcohol hangover: causes, consequences, and treatment".

The first five articles discuss significant methodological advances. In the first article, an update of the definition of the alcohol hangover is discussed [2]. The update of the definition was necessary, as current thinking about the traditional threshold for experiencing hangovers (a BAC of at least 0.11%) had changed. The new consensus, which is discussed in article two, reflects observations that hangovers can be experienced at any BAC [13].

Article three discusses the advantages and limitations of naturalistic study designs and their implementation in alcohol hangover research [14]. In contrast to traditional, controlled clinical trials, hangover research often applies a naturalistic study design in which investigators do not interfere with the drinking session. The article explains why this is important in terms of ecological validity (i.e., a real-life drinking session at a venue of choice, with corresponding behaviors and real-life alcohol consumption levels), and to what extent the naturalistic design has an impact on reliability and validity of study outcomes in comparison to highly controlled clinical trials. Article four discusses the assessment of overall hangover severity [15]. Traditional research has used composite symptom scales to assess hangover severity. The advantage of this approach is that information is gathered about the presence and severity of individual hangover symptoms. However, the research discussed in article four demonstrates that there are several disadvantages to this approach (e.g., the choice of included individual symptoms in a scale determines the overall hangover severity score, which therefore differs between currently used scales). The findings suggest that a one-item hangover severity assessment has advantages over composite symptom scale scores. In the final part of this section, article five discusses the prevalence of hangover resistance according to two methods for calculating estimated

BAC [16]. The findings discussed show that different equations used to calculate estimated BAC yield different outcomes. The latter is an important finding, and future consensus is warranted among AHRG members to ensure harmonization in reporting estimated BAC to allow direct comparisons of research from different groups.

The next two articles discuss the "causes" of alcohol hangover, and articles review the current knowledge on the pathology of the alcohol hangover. Whereas previous reviews on the causes of the alcohol hangover relied heavily on research data from the 1970s by the Finnish group Ylikhari et al., [17–19], the articles in this book provide major advances in the understanding of the pathology of the alcohol hangover. Article six reviews the role of alcohol metabolism in the pathology of the alcohol hangover [20], and article seven presents new data on the inflammatory response to alcohol consumption and its contribution to the alcohol hangover [21]. The data reveal that the rate of ethanol metabolism is an important predictor of next day hangover severity. In addition, the impact of oxidative stress and the balance between free radicals and antioxidants is discussed, as well as the role of acetaldehyde in eliciting an inflammatory response to alcohol (e.g., the release of cytokines), which ultimately elicits the alcohol hangover.

The following six articles discussing a variety of factors ('correlates') that may exacerbate or attenuate hangover symptoms. Article eight presents data on the effect of dietary nutrient intake on alcohol hangover severity [22]. Dietary nutrients are frequently included as ingredients in hangover treatments. Therefore, it is of interest to verify which of these, taken as part of daily diet, are associated with experiencing less severe hangovers. The results indicate that drinkers who consume food rich in zinc and nicotinic acid report less severe hangovers. Both nutrients are involved in the breakdown of ethanol and acetaldehyde, which may explain these findings. Article nine discusses the fact that different drinking levels are associated with experiencing differential levels of hangover severity [23]. The data confirm previous findings that hangover symptom severity is most severe among heavy and chronic drinkers. Article 10 discusses the interesting finding that when individuals experience hangovers more frequently, their severity increases [24]. Contrary to the common notion that drinkers get used to the amount of alcohol they consume and become "immune" to the adverse effects of drinking at this level, this observation suggests that reverse tolerance develops. Article 11 discusses the finding that hangover symptom severity is to some extent determined by the level of pain catastrophizing of drinkers [25]. Reporting higher levels of pain catastrophizing, in particular rumination, was associated with experiencing more severe hangovers. This finding is important, as it may have implications for the percentage of drinkers reporting being hangover resistant, and illustrates that the psychological perception of "what is pain?" and "what is mild, moderate, or severe?" differs between individuals, and thus impacts the reporting of the presence and rating of the severity of hangover symptoms. The latter is important as to date, no objective assessments for alcohol hangover (symptom) severity are available, and researchers have to rely on subjective reporting. Article 12 reviews possible sex differences in the presence and severity of hangover symptoms [26]. In contrast to acute alcohol effects (e.g., greater ratings of subjective intoxication in women), sex differences in the next-day effects of alcohol consumption appear to be limited. Finally, article 13 discusses the impact of mood and subjective intoxication on hangover severity [27]. Whereas baseline mood and mood while drinking had no relevant impact on next-day hangover severity, subjective intoxication (i.e., the level of drunkenness) showed to be a strong determinant of hangover severity.

Eleven subsequent articles discuss various aspects of cognitive, psychomotor, and physical performance during the hangover state ('consequences'). Article 14 describes the results of a study that assessed cognitive functioning and mood, applying a naturalistic study design [28]. The study demonstrates that participants can be tested at home using mobile technology to collect data. This methodology has clear advantages for participants (they do not have to come to the research center) and logistics for researchers (no lab space needed). Article 15 describes an investigation in which participants were approached on premise after consuming alcohol [29]. Both objective (breathalyzer) and subjective assessments (interview) of alcohol consumption and intoxication were made, and subjects were invited to complete online assessments of hangover severity and cognitive performance the next morning. Articles 14 and 15 demonstrate the utility of mobile/online assessments for hangover research. Future direct comparisons should investigate if validity and reliability of at home testing are equal to that of testing in controlled laboratory environments. Driving a car is one of the common daily activities that are potentially dangerous, as the use of alcohol can significantly impair driving performance and increase the chances of having an accident. Numerous studies have demonstrated that driving a car or bicycle while intoxicated is unsafe [30,31]. Previous research has also shown that simulated highway driving whilst experiencing a hangover is significantly impaired [32]. Despite this knowledge, a substantial number of both private and professional drivers continue to drive a car while experiencing a hangover [33]. Article 16 discusses the impact of alcohol hangover on simulated driving performance during a "commute to work" [34]. The study also revealed that during a relatively short drive, driving performance was significantly impaired while hungover. Driving is also a clear example of multitasking. In article 17, Benson et al. investigate alcohol hangover effects on another behavior, which can be translated to everyday workload, namely multitasking [35]. A hangover was associated with worse mood (reduced alertness and contentment, and increased anxiety and mental fatigue), and poorer multitasking performance, with greater effort needed to complete the tasks compared to the non-hangover condition. Interestingly, stress reactivity was not differentially affected by the hangover. The effects of alcohol hangover on executive functions are discussed in article 18. The investigation by Gunn et al. found that the alcohol hangover impairs core executive function processes that are important for everyday behaviors, such as decision-making and planning [36].

Affected daily behaviors and cognitive functioning during the hangover may ultimately be related to impaired information processing during the hangover. Three articles from Stock and colleagues investigated this in more detail. In article 19, they demonstrate that the alcohol hangover differentially modulates the processing of relevant and irrelevant information [37], and article 20 discusses findings showing that the alcohol hangover slightly impairs response selection but not response inhibition [38]. Finally, article 21 shows that the alcohol hangover does not alter the application of model-based and model-free learning strategies [39]. Together, these three articles provide further insights into the nature of slowed and impaired information processing during a hangover.

Whereas much research is devoted to cognitive aspects of the alcohol hangover, our understanding of the effects on physical state and sports performance has been limited. In article 22, Devenney et al. [40] report on physical activity level assessments comparing hangover and alcohol-free days. Using mobile technology, the continuous assessments of activity levels showed that during hangover subjects performed at lower activity levels, and vigorous activities were absent. Additionally, the assessments revealed that sleep quality was significantly poorer after the evening of alcohol consumption. Article 23 discusses the results of a study investigating the effects of alcohol consumption and hangover on endurance performance. Subjects on holiday walking the 18 km Samariá Gorge on the island of Crete in Greece were surveyed before and after they completed the walk. The analysis revealed that a variety of factors may predict walking performance and effort required to perform the walk, ranging from baseline physical state, immune fitness, to past night sleep quality, and also alcohol consumption and hangover severity [41].

Two articles discuss the outcomes of recent clinical trials that evaluated potential new hangover treatments. Despite a clear demand from drinkers who experience hangovers [42], currently there are no hangover treatments where the effectiveness has been demonstrated in independent double-blind, placebo-controlled clinical trials [43–45]. The increasing knowledge on the pathology of the alcohol hangover has resulted in focusing treatment development on products that aim to reduce the inflammatory response to alcohol and/or to enhance alcohol metabolism. In article 24, results of a pilot study are presented examining the effectiveness of SJP-005, a combination product of naproxen and fexofenadine, aiming to prevent hangovers by reducing the inflammatory response to alcohol consumption [46]. In article 25, the effects of Rapid Recovery are discussed—a hangover treatment aiming to reduce oxidative stress—and thereby preventing hangovers or reducing their severity [47].

Taken together, "The alcohol hangover: causes, consequences, and treatment" provides a comprehensive overview of current insights and research into many aspects of the alcohol hangover. The book highlights the advances in the field over the past decade, fueled by successful collaborations of researchers of the Alcohol Hangover Research Group and others investigating the interesting yet sometimes puzzling phenomenon of the alcohol hangover.

Funding: This research received no external funding.

Acknowledgments: This Special Issue and forthcoming book was composed to celebrate the 10th anniversary of the Alcohol Hangover Research Group. We wish to thank all members of the AHRG for their support and input.

Conflicts of Interest: S.B. has received funding from Red Bull GmbH, Kemin Foods, Sanofi Aventis, Phoenix Pharmaceutical and GlaxoSmithKline. Over the past 36 months, A.S. has held research grants from Abbott Nutrition, Arla Foods, Bayer, BioRevive, DuPont, Fonterra, Kemin Foods, Nestlé, Nutricia-Danone, Verdure Sciences. He has acted as a consultant/expert advisor to Bayer, Danone, Naturex, Nestlé, Pfizer, Sanofi, Sen-Jam Pharmaceutical, and has received travel/hospitality/speaker fees from Bayer, Sanofi and Verdure Sciences. Over the past 36 months, J.C.V. has held grants from Janssen and Sequential Medicine, and acted as a consultant/expert advisor to More Labs, Red Bull, Sen-Jam Pharmaceutical, Toast!, Tomo, and ZBiotics. A.K.S. has received funding from the Daimler and Benz Foundation. L.A. has no conflicts of interest to declare.

References

1. Van Schrojenstein Lantman, M.; van de Loo, A.J.; Mackus, M.; Verster, J.C. Development of a definition for the alcohol hangover: Consumer descriptions and expert consensus. *Curr. Drug Abuse Rev.* **2016**, *9*, 148–154. [CrossRef] [PubMed]
2. Verster, J.C.; Scholey, A.; van de Loo, A.J.A.E.; Benson, S.; Stock, A.-K. Updating the definition of the alcohol hangover. *J. Clin. Med.* **2020**, *9*, 823. [CrossRef] [PubMed]
3. Verster, J.C.; van Herwijnen, J.; Olivier, B.; Kahler, C.W. Validation of the Dutch Brief Young Adult Alcohol Consequences Questionnaire (B-YAACQ). *Addict. Behav.* **2009**, *34*, 411–414. [CrossRef] [PubMed]
4. Van Schrojenstein Lantman, M.; Mackus, M.; van de Loo, A.J.A.E.; Verster, J.C. The impact of alcohol hangover symptoms on cognitive and physical functioning, and mood. *Hum. Psychopharmacol.* **2017**, *32*, e2623. [CrossRef]
5. Verster, J.C.; Stephens, R. The importance of raising the profile of alcohol hangover research. *Curr. Drug Abuse Rev.* **2010**, *3*, 64–67.
6. Howland, J.; Rohsenow, D.J.; McGeary, J.E.; Streeter, C.; Verster, J.C. Proceedings of the 2010 symposium on hangover and other residual alcohol effects: Predictors and consequences. *Open Addict. J.* **2010**, *3*, 131–132. [CrossRef]
7. Verster, J.C.; Alford, C.; Bervoets, A.C.; de Klerk, S.; Grange, J.; Hogewoning, A.; Jones, K.; Kruisselbrink, D.; Owen, L.; Piasecki, T.M.; et al. The Alcohol Hangover Research Group. Hangover research needs: Proceedings of the 5th Alcohol Hangover Research Group meeting. *Curr. Drug Abuse Rev.* **2013**, *6*, 245–251. [CrossRef]
8. Mackus, M.; Adams, S.; Barzilay, A.; Benson, S.; Blau, L.; Iversen, J.; Johnson, S.J.; Keshavarzian, A.; Scholey, A.; Smith, G.S.; et al. Proceeding of the 8th Alcohol Hangover Research Group meeting. *Curr. Drug Abuse Rev.* **2016**, *9*, 106–112. [CrossRef]
9. Merlo, A.; Adams, S.; Benson, S.; Devenney, L.; Gunn, C.; Iversen, J.; Johnson, S.J.; Mackus, M.; Scholey, A.; Stock, A.K.; et al. Proceedings of the 9th Alcohol Hangover Research Group Meeting. *Curr. Drug Abuse Rev.* **2017**, *10*, 68–75. [CrossRef]
10. Merlo, A.; Abbott, Z.; Alford, C.; Balikji, S.; Bruce, G.; Gunn, C.; Iversen, J.; Iversen, J.; Johnson, S.J.; Kruisselbrink, D.L.; et al. Proceedings of the 10th Alcohol Hangover Research Group meeting in Utrecht, The Netherlands. *Proceedings* **2020**, *43*, 4. [CrossRef]
11. Palmer, E.; Arnoldy, L.; Ayre, E.; Benson, S.; Balikji, S.; Bruce, G.; Chen, F.; van Lawick van Pabst, A.E.; van de Loo, A.J.A.E.; O'Neill, S.; et al. Proceedings of the 11th Alcohol Hangover Research Group meeting in Nadi, Fiji. *Proceedings* **2020**, *43*, 1. [CrossRef]
12. Verster, J.C.; Stephens, R.; Penning, R.; Rohsenow, D.; McGeary, J.; Levy, D.; McKinney, A.; Finnigan, F.; Piasecki, T.M.; Adan, A.; et al. The Alcohol Hangover Research Group consensus statement on best practice in alcohol hangover research. *Curr. Drug Abuse Rev.* **2010**, *3*, 116–127. [CrossRef] [PubMed]

13. Verster, J.C.; Kruisselbrink, L.D.; Slot, K.A.; Anogeianaki, A.; Adams, S.; Alford, C.; Arnoldy, L.; Ayre, E.; Balikji, S.; Benson, S.; et al. Sensitivity to experiencing alcohol hangovers: Reconsideration of the 0.11% blood alcohol concentration (BAC) threshold for having a hangover. *J. Clin. Med.* **2020**, *9*, 179. [CrossRef] [PubMed]
14. Verster, J.C.; van de Loo, A.J.A.E.; Adams, S.; Stock, A.-K.; Benson, S.; Alford, C.; Scholey, A.; Bruce, G. Advantages and limitations of naturalistic study designs and their implementation in alcohol hangover research. *J. Clin. Med.* **2019**, *8*, 2160. [CrossRef] [PubMed]
15. Verster, J.C.; van de Loo, A.J.A.E.; Benson, S.; Scholey, A.; Stock, A.-K. The assessment of overall hangover severity. *J. Clin. Med.* **2020**, *9*, 786. [CrossRef]
16. Terpstra, C.; Benson, S.; Verster, J.C.; Scholey, A. Prevalence of hangover resistance according to two methods for calculating estimated blood alcohol concentration (eBAC). *J. Clin. Med.* **2020**, *9*, 2923. [CrossRef]
17. Swift, R.; Davidson, D. Alcohol hangover: Mechanisms and mediators. *Alcohol. Health Res. World* **1998**, *22*, 54–60.
18. Prat, G.; Adan, A.; Sánchez-Turet, M. Alcohol hangover: A critical review of explanatory factors. *Hum. Psychopharmacol.* **2009**, *24*, 259–267. [CrossRef]
19. Penning, R.; van Nuland, M.; Fliervoet, L.A.L.; Olivier, B.; Verster, J.C. The pathology of alcohol hangover. *Curr. Drug Abuse Rev.* **2010**, *3*, 68–75. [CrossRef]
20. Mackus, M.; van de Loo, A.J.E.A.; Garssen, J.; Kraneveld, A.D.; Scholey, A.D.; Verster, J.C. The role of alcohol metabolism in the pathology of alcohol hangover. *J. Clin. Med.* **2020**, *9*, 3421. [CrossRef]
21. Van de Loo, A.J.A.E.; Mackus, M.; Kwon, O.; Krishnakumar, I.; Garssen, J.; Kraneveld, A.D.; Scholey, A.; Verster, J.C. The inflammatory response to alcohol consumption and its role in the pathology of alcohol hangover. *J. Clin. Med.* **2020**, *9*, 2081. [CrossRef] [PubMed]
22. Verster, J.C.; Vermeulen, S.; van de Loo, A.J.A.E.; Balikji, S.; Kraneveld, A.D.; Garssen, J.; Scholey, A. Dietary nutrient intake, alcohol metabolism, and hangover severity. *J. Clin. Med.* **2019**, *8*, 1316. [CrossRef] [PubMed]
23. Vatsalya, V.; Hassan, H.; Kong, M.; Stangl, B.; Schwandt, M.; Schmidt-Teron, V.; Verster, J.C.; Ramchandani, V.; McClain, C. Exacerbation of hangover symptomology significantly corresponds with heavy and chronic alcohol drinking: A pilot study. *J. Clin. Med.* **2019**, *8*, 1943. [CrossRef]
24. Verster, J.C.; Slot, K.; Arnoldy, L.; van Lawick van Pabst, A.; van de Loo, A.J.A.E.; Benson, S.; Scholey, A. The Association between alcohol hangover frequency and severity: Evidence for reverse tolerance? *J. Clin. Med.* **2019**, *8*, 1520. [CrossRef] [PubMed]
25. Royle, S.; Owen, L.; Roberts, D.; Marrow, L. Pain Catastrophising predicts alcohol hangover severity and symptoms. *J. Clin. Med.* **2020**, *9*, 280. [CrossRef]
26. Van Lawick van Pabst, A.E.; Devenney, L.E.; Verster, J.C. Sex differences in the presence and severity of alcohol hangover symptoms. *J. Clin. Med.* **2019**, *8*, 867, Correction in **2019**, *8*, 1308. [CrossRef]
27. Verster, J.C.; Arnoldy, L.; van de Loo, A.J.A.E.; Benson, S.; Scholey, A.; Stock, A. The impact of mood and subjective intoxication on hangover severity. *J. Clin. Med.* **2020**, *9*, 2462. [CrossRef]
28. Alford, C.; Martinkova, Z.; Tiplady, B.; Reece, R.; Verster, J.C. The Effects of Alcohol Hangover on Mood and Performance Assessed at Home. *J. Clin. Med.* **2020**, *9*, 1068. [CrossRef]
29. Scholey, A.; Benson, S.; Kaufman, J.; Terpstra, C.; Ayre, E.; Verster, J.C.; Allen, C.; Devilly, G. Effects of Alcohol Hangover on Cognitive Performance: Findings from a Field/Internet Mixed Methodology Study. *J. Clin. Med.* **2019**, *8*, 440. [CrossRef]
30. Taylor, B.; Irving, H.M.; Kanteres, F.; Room, R.; Borges, G.; Cherpitel, C.; Greenfield, T.; Rehm, J. The more you drink, the harder you fall: A systematic review and meta-analysis of how acute alcohol consumption and injury or collision risk increase together. *Drug Alcohol. Depend.* **2010**, *110*, 108–116. [CrossRef]
31. Verster, J.C.; van Herwijnen, J.; Olivier, B.; Volkerts, E.R. Nonfatal bicycle accident risk after an evening of binge drinking. *Open. Addict. J.* **2009**, *2*, 1–5. [CrossRef]
32. Verster, J.C.; Bervoets, A.C.; de Klerk, S.; Vreman, R.A.; Olivier, B.; Roth, T.; Brookhuis, K.A. Effects of alcohol hangover on simulated highway driving performance. *Psychopharmacology* **2014**, *231*, 2999–3008. [CrossRef] [PubMed]
33. Verster, J.C.; van der Maarel, M.; McKinney, A.; Olivier, B.; de Haan, L. Driving during alcohol hangover among Dutch professional truck drivers. *Traffic Inj. Prev.* **2014**, *15*, 434–438. [CrossRef] [PubMed]

34. Alford, C.; Broom, C.; Carver, H.; Johnson, S.; Lands, S.; Reece, R.; Verster, J.C. The impact of alcohol hangover on simulated driving performance during a 'commute to work'—Zero and residual alcohol effects compared. *J. Clin. Med.* **2020**, *9*, 1435. [CrossRef] [PubMed]
35. Benson, S.; Ayre, E.; Garrisson, H.; Wetherell, M.; Verster, J.C.; Scholey, A. Alcohol hangover and multitasking: Effects on mood, cognitive performance, stress reactivity, and perceived effort. *J. Clin. Med.* **2020**, *9*, 1154. [CrossRef] [PubMed]
36. Gunn, C.; Fairchild, G.; Verster, J.C.; Adams, S. The effects of alcohol hangover on executive functions. *J. Clin. Med.* **2020**, *9*, 1148. [CrossRef]
37. Opitz, A.; Beste, C.; Stock, A. Alcohol hangover differentially modulates the processing of relevant and irrelevant information. *J. Clin. Med.* **2020**, *9*, 778. [CrossRef]
38. Opitz, A.; Hubert, J.; Beste, C.; Stock, A. Alcohol hangover slightly impairs response selection but not response inhibition. *J. Clin. Med.* **2019**, *8*, 1317. [CrossRef]
39. Berghäuser, J.; Bensmann, W.; Zink, N.; Endrass, T.; Beste, C.; Stock, A. Alcohol hangover does not alter the application of model-based and model-free learning strategies. *J. Clin. Med.* **2020**, *9*, 1453. [CrossRef]
40. Devenney, L.; Coyle, K.; Roth, T.; Verster, J.C. Sleep after heavy alcohol consumption and physical activity levels during alcohol hangover. *J. Clin. Med.* **2019**, *8*, 752. [CrossRef]
41. Verster, J.C.; Anogeianaki, A.; Kruisselbrink, D.; Alford, C.; Stock, A. Relationship between alcohol hangover and physical endurance performance: Walking the Samaria Gorge. *J. Clin. Med.* **2020**, *9*, 114. [CrossRef] [PubMed]
42. Mackus, M.; van Schrojenstein Lantman, M.; van de Loo, A.J.A.E.; Nutt, D.J.; Verster, J.C. An effective hangover treatment: Friend or foe? *Drug Sci. Policy Law* **2017**. [CrossRef]
43. Pittler, M.H.; Verster, J.C.; Ernst, E. Interventions for preventing or treating alcohol hangover: Systematic review of randomized trials. *Br. Med. J.* **2005**, *331*, 1515–1518. [CrossRef] [PubMed]
44. Verster, J.C.; Penning, R. Treatment and prevention of alcohol hangover. *Curr. Drug Abuse Rev.* **2010**, *3*, 103–109. [CrossRef]
45. Jayawardena, R.; Thejani, T.; Ranasinghe, P.; Fernando, D.; Verster, J.C. Interventions for treatment and/or prevention of alcohol hangover: Systematic review. *Hum. Psychopharmacol.* **2017**, *32*, e2600. [CrossRef]
46. Verster, J.C.; Dahl, T.; Scholey, A.; Iversen, J. The Effects of SJP-001 on Alcohol Hangover Severity: A Pilot Study. *J. Clin. Med.* **2020**, *9*, 932. [CrossRef]
47. Scholey, A.; Ayre, E.; Stock, A.; Verster, J.C.; Benson, S. Effects of Rapid Recovery on Alcohol Hangover Severity: A Double-Blind, Placebo-Controlled, Randomized, Balanced Crossover Trial. *J. Clin. Med.* **2020**, *9*, 2175. [CrossRef]

Publisher's Note: MDPI stays neutral with regard to jurisdictional claims in published maps and institutional affiliations.

© 2020 by the authors. Licensee MDPI, Basel, Switzerland. This article is an open access article distributed under the terms and conditions of the Creative Commons Attribution (CC BY) license (http://creativecommons.org/licenses/by/4.0/).

Communication

Updating the Definition of the Alcohol Hangover

Joris C. Verster [1,2,3], Andrew Scholey [3], Aurora J. A. E. van de Loo [1,2], Sarah Benson [3] and Ann-Kathrin Stock [4,*]

1. Division of Pharmacology, Utrecht Institute for Pharmaceutical Sciences (UIPS), Utrecht University, 3584CG Utrecht, The Netherlands; j.c.verster@uu.nl (J.C.V.); a.j.a.e.vandeloo@uu.nl (A.J.A.E.v.d.L.)
2. Institute for Risk Assessment Sciences (IRAS), Utrecht University, 3584CM Utrecht, The Netherlands
3. Centre for Human Psychopharmacology, Swinburne University, Melbourne, VIC 3122, Australia; andrew@scholeylab.com (A.S.); Sarahmichellebenson@gmail.com (S.B.)
4. Cognitive Neurophysiology, Department of Child and Adolescent Psychiatry, Faculty of Medicine, TU Dresden, Fetscherstr. 74, 01307 Dresden, Germany
* Correspondence: Ann-Kathrin.Stock@uniklinikum-dresden.de

Received: 26 February 2020; Accepted: 14 March 2020; Published: 18 March 2020

Abstract: In 2016, the Alcohol Hangover Research Group defined the alcohol hangover as "the combination of mental and physical symptoms experienced the day after a single episode of heavy drinking, starting when blood alcohol concentration (BAC) approaches zero". In the light of new findings and evidence, we carefully reviewed the different components of that definition. Several studies demonstrated that alcohol hangovers are not limited to heavy drinking occasions. Instead, data from both student and non-student samples revealed that at a group level, alcohol hangover may occur at much lower BAC levels than previously thought. Regression analysis further revealed that for individual drinkers, the occurrence of hangovers is more likely when subjects consume more alcohol than they usually do. However, hangovers may also occur at a drinker's usual BAC, and in some cases even at lower BAC (e.g. in case of illness). We also carefully reviewed and modified other parts of the definition. Finally, hangovers are not necessarily limited to the 'next day'. They can start at any time of day or night, whenever BAC approaches zero after a single dinking occasion. This may also be on the same day as the drinking occasion (e.g. when drinking in, or until the morning and subsequently having a hangover in the afternoon or evening). To better reflect the new insights and sharpen the description of the concept, we hereby propose to update the definition of the alcohol hangover as follows: "The alcohol hangover refers to the combination of negative mental and physical symptoms which can be experienced after a single episode of alcohol consumption, starting when blood alcohol concentration (BAC) approaches zero", and recommend to use this new definition in future hangover research.

Keywords: alcohol; hangover; definition

1. Introduction

In 2016, the Alcohol Hangover Research Group defined the alcohol hangover as "the combination of mental and physical symptoms experienced the day after a single episode of heavy drinking, starting when blood alcohol concentration (BAC) approaches zero" [1]. The development of this definition was a welcome and necessary addition to the substance abuse and addiction research field. Since then, ongoing research has generated new insights and there have been continuous discussions among researchers about how to further improve the definition of the alcohol hangover. Updating the current definition is necessary to describe the alcohol hangover more precisely against the background of new findings in the field. These specifications address recently discussed issues and further remove ambiguity from the previous wording.

2. Heavy Drinking

The most important discussion pertains to the amount of alcohol consumption that is required to elicit a hangover. Given this discussion, there is controversy about the word 'heavy' in the definition of alcohol hangover. First, the word 'heavy' is unspecific, as it does not define what exact amount of alcohol should actually be consumed to elicit a hangover. Second, it suggests that hangovers occur only when large amounts of alcohol are consumed. However, the Alcohol Hangover Research Group recently reached a consensus to abandon the criterion that a BAC of 0.11% or higher is needed to provoke a hangover [2]. This conclusion was drawn based on an increasing body of evidence showing that drinkers also report hangovers at BACs that are much lower than both the suggested threshold of 0.11%, as well as the binge drinking threshold of 0.08% issued by the National Institute on Alcohol Abuse and Alcoholism (NIAAA) [3].

For example, Verster et al. [4] found that non-student subjects ($N = 176$) who consumed a mean (SD) of 3.0 (1.8) alcoholic drinks (10 g ethanol each) reported considerable next-morning hangover severity, i.e., a mean (SD) overall hangover severity score of 4.6 (2.4) on a 0–10 scale. Despite this, their peak estimated BAC was 0.03%. In another student sample, Kruisselbrink et al. [5] found significant hangover symptoms in subjects having consumed as few as two beers, with a mean maximum BAC of 0.036%. These observations are neither consistent with 'heavy drinking' in the definition of the alcohol hangover, nor with the binge drinking threshold suggested by the NIAAA. Surveys completed by large student samples from Canada ($N = 5540$) and The Netherlands ($N = 6002$) further confirmed that alcohol hangovers are reported across all BAC levels [6,7]. Thus, hangovers may occur at any reasonable BAC level, and are not limited to 'heavy' drinking only. Given this, we need to modify the current definition of the alcohol hangover and omit the referral to heavy drinking.

3. The Concept of Alcohol Hangover Versus Risk Factors and Possible Causes

When developing a definition, it is vital to accurately describe the concept (i.e., alcohol hangover). Furthermore, a proper definition of a phenomenon should not contain potential risk factors for its occurrence. There are many risk factors for hangover including, but not limited to, the amount of consumed alcohol (compared to normal), peak BAC, congener content of drinks, smoking, activities during drinking (e.g. dancing or sitting in a bar), or the emotional state during drinking. While these risk factors are of course important to investigate and mention in relation to alcohol hangover, they should not be included in a definition of the concept itself, as the observation/diagnosis of a condition should be separate from the risk/likelihood that it will occur. It is however important to still refer to 'alcohol consumption' in the definition of alcohol hangover as this behavior is mandatory to elicit the condition (rather than a mere risk factor).

Evaluating hangover experiences from individual drinkers has shown that developing a useful, short, and accurate description of the relevant amount of alcohol intake to elicit a hangover is not straightforward. First of all, the presence and severity of alcohol hangovers may vary from day to day [8], even when the same amount of alcohol is consumed and the same BAC is reached. In line with this, regression analyses revealed that neither the amount of consumed alcohol, nor BAC, were strong predictors of hangover severity [2]. Instead, the relative increase in alcohol consumption, as compared to what subjects normally consume on a typical drinking occasion, was the best predictor of overall hangover severity [2]. Thus, the chances of having a hangover are significantly increased when drinking more alcohol than usual, whatever the usual amount consumed. However, including the phrase 'relatively elevated amounts of alcohol consumption' into the definition would exclude a substantial amount of drinkers who also experience hangovers, but do not fulfill this criterion. For example, there are drinkers who almost always experience a hangover, also when only drinking their usual amount of alcohol. One of the many potential reasons for this could be deficient metabolization of alcohol and/or its metabolite acetaldehyde in the liver. Genetic variation in alleles for alcohol dehydrogenase (ADH) and aldehyde dehydrogenase (ALDH), which are the enzymes necessary for metabolizing ethanol into acetaldehyde and further into acetate, may account for this. Twin studies showed that heritability of this genetic variation is related to about 45% of the reported hangover severity [9,10]. In this context, in populations of Asian descent, subjects with ALDH2*2 alleles, i.e., those who breakdown acetaldehyde more slowly, usually report significantly worse hangovers [11,12], and are more likely to experience hangovers at lower alcohol consumption levels than others. In such hangover-sensitive drinkers, the amount of alcohol does not need to be elevated to elicit a hangover. Aside from this, hangovers may also occur when drinking less alcohol than usual on a given occasion. This might for example be the case if subjects experience illness or reduced immune fitness [13,14], or in case of elevated negative mood while drinking [15,16]. It is hence challenging to encompass the different scenarios that may result in a hangover in a modified definition, without making it very lengthy. We therefore propose to substitute 'heavy drinking' with 'alcohol consumption' without any further reference to the amount consumed. Further, to more accurately reflect the day-to-day variability in the likelihood of developing a hangover despite more or less equal circumstances [8], we further propose to change 'experienced' into 'which can be experienced'. This also acknowledges the fact that about 10% to 20% of drinkers report not having a hangover, even after consuming large amounts of alcohol [6,7].

4. Alcohol Hangover Symptoms

The definition refers to a 'combination of mental and physical symptoms'. Hangover symptoms are generally perceived to be negative, but the original wording does not specify whether these symptoms are expected to be negative or positive. Therefore, we suggest to modify this as a 'combination of negative mental and physical symptoms'. In line with previous discussions [1], hangover symptoms are not listed as part of the definition. Symptoms vary between drinkers and between drinking occasions, even when same amounts of alcohol are consumed [8,17]. Including specific symptoms instead of the general description 'combination of negative mental and physical symptoms' would thus significantly limit the applicability of the definition.

5. Timing of Drinking

The previous definition states that hangover is 'experienced the day after ... ', which was included to clearly differentiate the intoxication phase from the hangover phase. In the vast majority of cases, the hangover starts when waking up after an afternoon, evening, or night of drinking, followed by a period of sleep. Yet, this definition would not properly match cases where an individual drinks past midnight, in the morning, or during the day [18]. For example, a UK study revealed that almost 20% of all 'drinking occasions' took place before 5:00 p.m. [19]. In these instances, drinkers may experience a hangover in the afternoon or evening of the same day. We therefore decided to omit the next day criterion and changed the definition to 'experienced after'.

6. Differentiating between Alcohol Hangover and Withdrawal

The definition refers to 'a single episode ... '. This was included to differentiate hangovers in social drinkers from withdrawal symptoms experienced by individuals with alcohol use disorders (i.e., alcoholism), who tend to not only engage in alcohol binges but also maintain a rather steady baseline level of alcohol consumption with continuous drinking for several days, or even longer. This leads to extensive homeostatic adaptations in the regulation of many vital parameters as well as neuroadaptive processes [20], which foster the development of alcohol tolerance. These counter-regulatory mechanisms require clinical treatment as they may cause life-threatening complications when the BAC approaches zero. In contrast to individuals with alcohol use disorder, social drinkers lack such extensive tolerance, as they do by definition not engage in such continuous drinking. Given the functional differences in the symptom-associated drinking patterns as well as the underlying physiological mechanisms, we therefore decided not to alter this part of the definition.

7. Differentiating between Alcohol Hangover and Intoxication

The definition states that hangover is 'starting when blood alcohol concentration (BAC) approaches zero'. This is crucial in order to clearly distinguish between alcohol intoxication and alcohol hangover on the basis of timing. We did therefore not alter this part of the definition.

8. Conclusions

We hereby propose to update the definition of alcohol hangover as follows: "The alcohol hangover refers to the combination of negative mental and physical symptoms which can be experienced after a single episode of alcohol consumption, starting when blood alcohol concentration (BAC) approaches zero".

Author Contributions: Conceptualization, J.C.V., A.S., A.-K.S. writing—original draft preparation, J.C.V.; writing—review and editing, J.C.V., A.S., A.-K.S., A.J.A.E.v.d.L. and S.B. All authors have read and agreed to the published version of the manuscript.

Conflicts of Interest: S.B. has received funding from Red Bull GmbH, Kemin Foods, Sanofi Aventis, Phoenix Pharmaceutical and GlaxoSmithKline. Over the past 36 months, A.S. has held research grants from Abbott Nutrition, Arla Foods, Bayer, BioRevive, DuPont, Kemin Foods, Nestlé, Nutricia-Danone, Verdure Sciences. He has acted as a consultant/expert advisor to Bayer, Danone, Naturex, Nestlé, Pfizer, Sanofi, Sen-Jam Pharmaceutical, and has received travel/hospitality/speaker fees Bayer, Sanofi and Verdure Sciences. Over the past 36 months, J.C.V. has held grants from the Dutch Ministry of Infrastructure and the Environment, Janssen, Nutricia, and Sequential, and acted as a consultant/expert advisor to Clinilabs, Morelabs, Red Bull, Sen-Jam Pharmaceutical, Toast!, and ZBiotics. A.K.S. has received funding from Daimler and Benz.A.J.A.E.V.D.L. has no conflicts of interest to declare.

References

1. Van Schrojenstein Lantman, M.; van de Loo, A.J.; Mackus, M.; Verster, J.C. Development of a definition for the alcohol hangover: Consumer descriptions and expert consensus. *Curr. Drug Abuse Rev.* **2016**, *9*, 148–154. [CrossRef]
2. Verster, J.C.; Kruisselbrink, L.D.; Slot, K.A.; Anogeianaki, A.; Adams, S.; Alford, C.; Arnoldy, L.; Ayre, E.; Balikji, S.; Benson, S.; et al. Sensitivity to experiencing alcohol hangovers: Reconsideration of the 0.11% blood alcohol concentration (BAC) threshold for having a hangover. *J. Clin. Med.* **2020**, *9*, 179. [CrossRef]
3. NIAAA. *NIAAA Council Approves Definition of Binge Drinking*; NIAAA Newsletter; NIAAA: Washington, DC, USA, 2004; Volume 3.
4. Verster, J.C.; Kruisselbrink, L.D.; Anogeianaki, A.; Alford, C.; Stock, A.K. Relationship of alcohol hangover and physical endurance performance: Walking the Samaria Gorge. *J. Clin. Med.* **2020**, *9*, 114. [CrossRef] [PubMed]
5. Kruisselbrink, L.D.; Martin, K.L.; Megeney, M.; Fowles, J.R.; Murphy, R.J.L. Physical and psychomotor functioning of females the morning after consuming low to moderate quantities of beer. *J. Stud. Alcohol Drugs* **2006**, *67*, 416–420. [CrossRef] [PubMed]
6. Kruisselbrink, L.D.; Bervoets, A.C.; de Klerk, S.; van de Loo, A.J.A.E.; Verster, J.C. Hangover resistance in a Canadian university student population. *Addict. Behav. Rep.* **2017**, *5*, 14–18. [CrossRef] [PubMed]
7. Verster, J.C.; de Klerk, S.; Bervoets, A.C.; Kruisselbrink, L.D. Can hangover immunity really be claimed? *Curr. Drug Abuse Rev.* **2013**, *6*, 253–254. [CrossRef] [PubMed]
8. Hensel, K.O.; Longmire, M.R.; Köchling, J. Should population-based research steer individual health decisions? *Aging* **2019**, *11*, 9231–9233. [CrossRef] [PubMed]
9. Slutske, W.S.; Piasecki, T.M.; Nathanson, L.; Statham, D.J.; Martin, N.G. Genetic influences on alcohol-related hangover. *Addiction* **2014**, *109*, 2027–2034. [CrossRef] [PubMed]
10. Wu, S.H.; Guo, Q.; Viken, R.J.; Reed, T.; Dai, J. Heritability of usual alcohol intoxication and hangover in male twins: The NAS-NRC Twin Registry. *Alcohol Clin. Exp. Res.* **2014**, *38*, 2307–2313. [CrossRef] [PubMed]
11. Wall, T.L.; Horn, S.M.; Johnson, M.L.; Smith, T.L.; Carr, L.G. Hangover symptoms in Asian Americans with variations in the aldehyde dehydrogenase (ALDH2) gene. *J. Stud. Alcohol* **2000**, *61*, 13–17. [CrossRef] [PubMed]

12. Yokoyama, M.; Yokoyama, A.; Yokoyama, T.; Funazu, K.; Hamana, G.; Kondo, S.; Yamashita, T.; Nakamura, H. Hangover susceptibility in relation to aldehyde dehydrogenase-2 genotype, alcohol flushing, and mean corpuscular volume in Japanese workers. *Alcohol Clin. Exp. Res.* **2005**, *29*, 1165–1171. [CrossRef] [PubMed]
13. Van de Loo, A.J.A.E.; van Schrojenstein Lantman, M.; Mackus, M.; Scholey, A.; Verster, J.C. Impact of mental resilience and perceived immune functioning on the severity of alcohol hangover. *BMC Res. Notes* **2018**, *11*, 526. [CrossRef] [PubMed]
14. Van de Loo, A.J.A.E.; Mackus, M.; van Schrojenstein Lantman, M.; Kraneveld, A.D.; Garssen, J.; Scholey, A.; Verster, J.C. Susceptibility to alcohol hangovers: The association with self-reported immune status. *Int. J. Environ. Res. Public Health* **2018**, *15*, 1286. [CrossRef] [PubMed]
15. Harburg, E.; Davis, D.; Cummings, K.M.; Gunn, R. Negative affect, alcohol consumption and hangover symptoms among normal drinkers in a small community. *J. Stud. Alcohol* **1981**, *42*, 998–1012. [CrossRef] [PubMed]
16. Harburg, E.; Gunn, R.; Gleiberman, L.; DiFranceisco, W.; Schork, A. Psychosocial factors, alcohol use, and hangover signs among social drinkers: A reappraisal. *J. Clin. Epidemiol.* **1993**, *46*, 413–422. [CrossRef]
17. Verster, J.C.; van de Loo, A.J.A.E.; Benson, S.; Scholey, A.; Stock, A.-K. The assessment of overall hangover severity. *J. Clin. Med.* **2020**, *9*, 786.
18. Thompson, C.; Milton, S.; Egan, M.; Lock, K. Down the local: A qualitative case study of daytime drinking spaces in the London Borough of Islington. *Int. J. Drug Policy* **2018**, *52*, 1–8. [CrossRef] [PubMed]
19. Ally, A.K.; Lovatt, M.; Meier, P.S.; Brennan, A.; Holmes, J. Developing a social practice-based typology of British drinking culture in 2009–2011: Implications for alcohol policy analysis. *Addiction* **2016**, *111*, 1568–1579. [CrossRef] [PubMed]
20. Roberto, M.; Varodayan, F.P. Synaptic targets: Chronic alcohol actions. *Neuropharmacology* **2017**, *122*, 85–99. [CrossRef] [PubMed]

© 2020 by the authors. Licensee MDPI, Basel, Switzerland. This article is an open access article distributed under the terms and conditions of the Creative Commons Attribution (CC BY) license (http://creativecommons.org/licenses/by/4.0/).

Communication

Sensitivity to Experiencing Alcohol Hangovers: Reconsideration of the 0.11% Blood Alcohol Concentration (BAC) Threshold for Having a Hangover

Joris C. Verster [1,2,3], L. Darren Kruisselbrink [4], Karin A. Slot [1], Aikaterini Anogeianaki [1], Sally Adams [5], Chris Alford [6], Lizanne Arnoldy [1], Elisabeth Ayre [3], Stephanie Balikji [1], Sarah Benson [3], Gillian Bruce [7], Lydia E. Devenney [8], Michael R. Frone [9], Craig Gunn [5], Thomas Heffernan [10], Kai O. Hensel [11,12], Anna Hogewoning [1], Sean J. Johnson [6,13], Albertine E. van Lawick van Pabst [1], Aurora J.A.E. van de Loo [1], Marlou Mackus [1], Agnese Merlo [7], René J.L. Murphy [4], Lauren Owen [14], Emily O.C. Palmer [15], Charmaine J.I. van Rossum [1], Andrew Scholey [3], Chantal Terpstra [3], Vatsalya Vatsalya [16,17,18,19,20], Sterre A. Vermeulen [1], Michelle van Wijk [1] and Ann-Kathrin Stock [21,*] on behalf of the Alcohol Hangover Research Group

1 Division of Pharmacology, Utrecht Institute for Pharmaceutical Sciences (UIPS), Utrecht University, 3584CG Utrecht, The Netherlands; j.c.verster@uu.nl (J.C.V.); k.a.slot@uu.nl (K.A.S.); kanogeianaki@gmail.com (A.A.); l.arnoldy@uu.nl (L.A.); stephaniebalikji@hotmail.com (S.B.); annahogewoning@gmail.com (A.H.); albertinevanlawick@live.nl (A.E.v.L.v.P.); a.j.a.e.vandeloo@uu.nl (A.J.A.E.v.d.L.); marloumackus@gmail.com (M.M.); c.j.i.vanrossum2@students.uu.nl (C.J.I.v.R.); s.a.vermeulen@students.uu.nl (S.A.V.); m.wijk@students.uu.nl (M.v.W.)
2 Institute for Risk Assessment Sciences (IRAS), Utrecht University, 3584CM Utrecht, The Netherlands
3 Centre for Human Psychopharmacology, Swinburne University, Melbourne, VIC 3122, Australia; eayre@swin.edu.au (E.A.); sarahmichellebenson@gmail.com (S.B.); andrew@scholeylab.com (A.S.); chantalterpstra92@gmail.com (C.T.)
4 Centre of Lifestyle Studies, School of Kinesiology, Acadia University, Wolfville, NS B4P 2R6, Canada; darren.kruisselbrink@acadiau.ca (L.D.K.); rene.murphy@acadiau.ca (R.J.L.M.)
5 Addiction and Mental Health Group, Department of Psychology, University of Bath, Bath BA2 7AY, UK; sa221@bath.ac.uk (S.A.); cag35@bath.ac.uk (C.G.)
6 Psychological Sciences Research Group, University of the West of England, Bristol BS16 1QY, UK; chris.alford@uwe.ac.uk (C.A.); JohnsonS11@cardiff.ac.uk (S.J.J.)
7 Education and Social Sciences, University of the West of Scotland, Paisley PA1 2BE, UK; gillian.bruce@uws.ac.uk (G.B.); agnese.merlo@gmail.com (A.M.)
8 School of Psychology, Life and Health Sciences, Ulster University, Coleraine, Co. Londonderry BT52 1SA, UK; lydiadevenney@gmail.com
9 Department of Psychology, University at Buffalo, The State University of New York, Buffalo, NY 14203, USA; mrf@buffalo.edu
10 Department of Psychology, Faculty of Health and Life Sciences, Northumbria University, Newcastle upon Tyne NE1 8ST, UK; tom.heffernan@northumbria.ac.uk
11 Cambridge Biomedical Campus, Department of Paediatrics, Addenbrooke's Hospital, Cambridge University Hospitals NHS Foundation Trust, Cambridge CB2 0QQ, UK; kai.hensel@gmail.com
12 Faculty of Health, Department of Paediatrics, Center for Clinical & Translational Research (CCTR), Witten/Herdecke University, 58455 Witten, Germany
13 Centre for Trials Research, Cardiff University, Cardiff CF14 4YS, UK
14 Department of Psychology, School of Health and Society, University of Salford, Salford 5 M6 6PU, UK; L.J.Owen2@salford.ac.uk
15 Department of Medicine, Imperial College London, London W12 0NN, UK; e.palmer@imperial.ac.uk
16 Department of Medicine, University of Louisville, Louisville, KY 40202, USA; vatsalya.vatsalya@louisville.edu
17 Alcohol Research Center, University of Louisville, Louisville, KY 40202, USA

18 Hepatobiology & Toxicology Center, University of Louisville, Louisville, KY 40202, USA
19 National Institute on Alcohol Abuse and Alcoholism, NIH, Bethesda, MD 20892, USA
20 Robley Rex Louisville VAMC, Louisville, KY 40206, USA
21 Cognitive Neurophysiology Department of Child and Adolescent Psychiatry, Faculty of Medicine of the TU Dresden, University of Dresden, D-01307 Dresden, Germany
*   Correspondence: Ann-Kathrin.Stock@uniklinikum-dresden.de

Received: 7 December 2019; Accepted: 7 January 2020; Published: 9 January 2020

Abstract: The 2010 Alcohol Hangover Research Group consensus paper defined a cutoff blood alcohol concentration (BAC) of 0.11% as a toxicological threshold indicating that sufficient alcohol had been consumed to develop a hangover. The cutoff was based on previous research and applied mostly in studies comprising student samples. Previously, we showed that sensitivity to hangovers depends on (estimated) BAC during acute intoxication, with a greater percentage of drinkers reporting hangovers at higher BAC levels. However, a substantial number of participants also reported hangovers at comparatively lower BAC levels. This calls the suitability of the 0.11% threshold into question. Recent research has shown that subjective intoxication, i.e., the level of severity of reported drunkenness, and not BAC, is the most important determinant of hangover severity. Non-student samples often have a much lower alcohol intake compared to student samples, and overall BACs often remain below 0.11%. Despite these lower BACs, many non-student participants report having a hangover, especially when their subjective intoxication levels are high. This may be the case when alcohol consumption on the drinking occasion that results in a hangover significantly exceeds their "normal" drinking level, irrespective of whether they meet the 0.11% threshold in any of these conditions. Whereas consumers may have relative tolerance to the adverse effects at their "regular" drinking level, considerably higher alcohol intake—irrespective of the absolute amount—may consequentially result in a next-day hangover. Taken together, these findings suggest that the 0.11% threshold value as a criterion for having a hangover should be abandoned.

Keywords: alcohol; hangover; sensitivity; subjective intoxication; blood alcohol concentration

---

Alcohol hangover is defined as the combination of mental and physical symptoms experienced the day after a single episode of heavy drinking, starting when blood alcohol concentration (BAC) approaches zero [1]. The hangover state can comprise a variety of symptoms which differ in presence and severity among drinkers [2,3]. These symptoms include, but are not limited to, nausea, sleepiness, concentration problems, and headache. In the 2010 consensus paper of the Alcohol Hangover Research Group [4], it was stated that in order to experience a hangover per se, a minimum BAC of 0.11% should be reached. In the current consensus paper, we discuss why the 0.11% threshold value as a criterion for having a hangover should be abandoned.

BAC varies depending on the combination of the amount of alcohol consumed and drinking duration. A smaller impact is also evident for other factors such as sex and body weight. For example, a BAC of 0.11% roughly equates to consuming about 6 US standard drinks (14 g of alcohol each) or 8.4 European standard drinks (10 g of alcohol each) over a period of 2 hours [5]. This threshold was based on a study by Chapman et al. [6] in which participants experienced hangovers at this BAC level. At first glance, observing drinking levels of student samples and corresponding average BACs [3,7–9], the threshold seems well selected. However, a closer look at the data revealed that this threshold could well be an arbitrary one. Research on large Dutch and Canadian student samples [7,8] revealed that a substantial number of drinkers who did not reach the consensus BAC level of 0.11% still reported having a hangover. Other studies also confirmed this observation. For example, the data of van Schrojenstein Lantman et al. [3] identified that 19.4% of N = 1833 students who had a hangover after their past month's heaviest drinking occasion had an estimated BAC below 0.11%. Data from another survey [9] revealed that 22.5% of N = 989 students had an estimated BAC below 0.11% at their past month's heaviest drinking occasion that resulted in a hangover. Aggregating the data of these two studies [3,9] revealed that 20.5% of N = 2822 students who reported a hangover had an estimated BAC well below 0.11% the night before (see Figure 1). In each of these studies [3,7–9], BAC was (retrospectively) estimated using a modified Widmark formula [10] based on self-reported alcohol consumption, and taking into account sex and body weight.

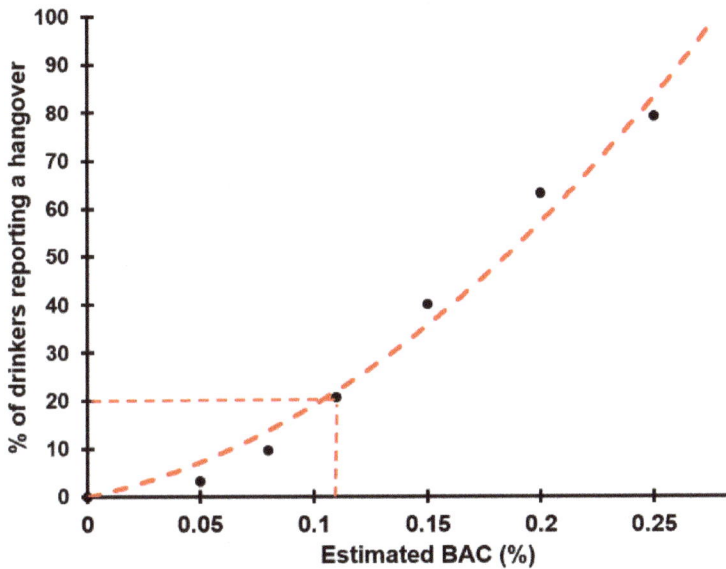

Figure 1. Percentage of students reporting a hangover at different blood alcohol concentrations. Aggregated data from N = 2822 students who reported a hangover after their past month's heaviest drinking occasion. Data from references [3,9]. Abbreviation: BAC = blood alcohol concentration.

In non-student samples, alcohol consumption levels are often considerably lower. Nevertheless, these drinkers report having hangovers as well. An illustrative example for this was provided by a recent study conducted among N = 307 adults in Crete, Greece [11]. Among them, N = 176 reported having had a hangover. These individuals were on average 39.0 (10.3) years old (59.7% men) and had consumed a mean (SD) of 3.0 (1.8) alcoholic drinks the previous evening over a drinking period from 17:40 (1.8 h) to 20:13 (1.9 h). Their mean (SD) BAC, estimated via a modified Widmark formula [10], equaled 0.03% (0.03). While the amount of alcohol consumed was low in comparison to student samples, it is still likely that they consumed significantly more alcohol than they usually do at home (i.e., as compared to their usual weekly alcohol intake of 5.9 alcoholic drinks). As a consequence, they reported a mean (SD) being drunk/ intoxicated score of 4.7 (2.6) rated on a scale ranging from 0 (absent) to 10 (extreme) [12,13]. Their mean (SD) overall hangover severity, rated on an 11-point scale ranging from 0 (absent) to 10 (extreme) [14], was 4.6 (2.1). In line with other studies [15,16], both subjective intoxication and estimated BAC correlated significantly with overall hangover severity. The correlations between hangover severity and past evening's drinking behavior (see Figure 2) revealed that subjective intoxication yielded the strongest correlation with overall hangover severity (Figure 2B), followed by the number of alcoholic drinks consumed (Figure 2A). Although significant, the correlations between hangover severity and estimated BAC (Figure 2C) and drinking duration (Figure 2D) were smaller in magnitude. Figure 2C further shows that participants also reported having hangovers of moderate to high severity at lower BAC levels. In fact, the estimated BAC level of almost all participants (98.3%) fell below 0.11%.

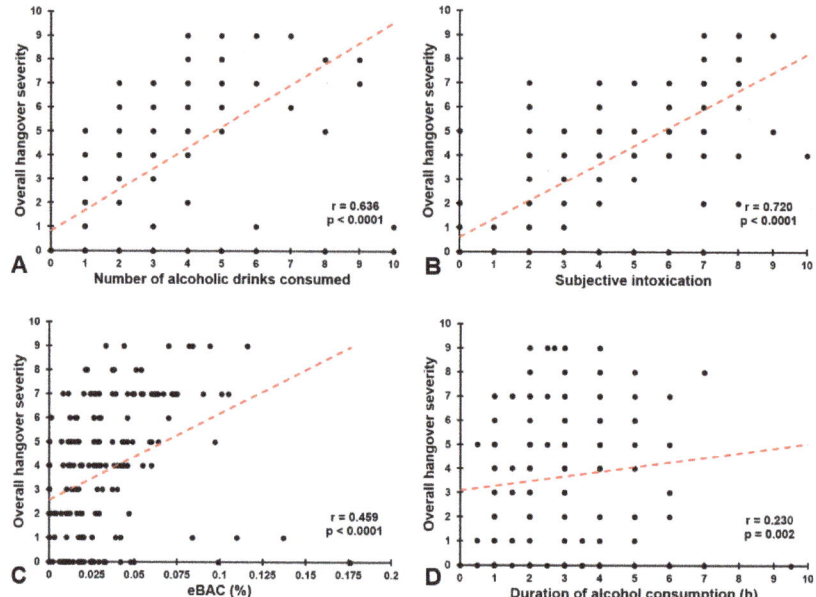

Figure 2. Correlations between overall hangover severity and drinking variables. Depicted are the correlations between overall hangover severity and (**A**) number of alcoholic drinks consumed the previous evening, (**B**) subjective intoxication while drinking, (**C**) estimated blood alcohol concentration (eBAC) on the previous evening, and (**D**) hours of drinking alcohol on the previous evening. Dotted lines represent Spearman's rho correlations. Data from Reference [11].

A stepwise regression analysis revealed that four variables accounted for 58% of the variance in overall hangover severity. When looking at their unique contributions to the variance explained, the strongest predictor was subjective intoxication (48.5%), followed by sleep quality (7.2%), estimated BAC (1.2%), and body mass index (BMI; 1.1%). These findings are in line with two other recent regression analyses [17,18]. Both studies showed that subjective intoxication (perceived drunkenness), and not BAC, was the strongest predictor of hangover severity.

It is important to have an understanding as to why people experience hangovers at low BAC levels. An explanation could be found through a closer examination of the participants' weekly alcohol consumption. If participants usually consume one or two alcoholic drinks per drinking occasion, and then consume two or three times as much while on holiday, this "higher than usual" drinking level may cause a hangover. This could potentially occur even if the absolute number of alcoholic drinks is still low as compared to some student samples [14]. The increase in the number of consumed alcoholic drinks, as compared to "regular" drinking occasions, and the corresponding increase in subjective intoxication, highly correlates with experiencing a hangover the next day. Importantly, this increase is independent of the absolute BAC levels. In other words, hangovers may occur at any BAC level, and their occurrence is more likely if individuals drink substantially more alcohol than they usually do on occasions that do not result in a hangover.

The impact of an increase in alcohol consumption relative to a regular drinking occasion was also demonstrated in a recent study [19]. This naturalistic study comprised an alcohol test day resulting in a hangover, and an alcohol-free control day. Various demographic data (e.g., age, sex, height, and weight) and data on drinking variables (including number of drinks, drinking duration, and the number of additional drinks they had consumed on the hangover drinking occasion as compared to a regular non-hangover drinking occasion) were collected in students aged 18 to 30 years. BAC was estimated using a modified Widmark equation [10]. Overall hangover severity was rated on an 11-point scale ranging from 0 (absent) to 10 (extreme) [14]. The number of hangover episodes that participants had experienced during the past year was also assessed. Dancing frequency during the drinking occasion was rated as "none", "sometimes", "often", or "almost all the time" and the number of cigarettes smoked, drug use, and total sleep time were also recorded. The Five-Shot questionnaire alcohol screening test was used to analyze general drinking behavior [20]. Personality (i.e., somatization, obsession-compulsion, interpersonal sensitivity, depression, anxiety, hostility, phobic anxiety, paranoid ideation, and psychoticism) was assessed with the Brief Symptom Inventory (BSI) [21]. Risk taking was assessed with the RT18 questionnaire [22]. N = 93 participants were included in this study with a mean (SD) age of 21.0 (2.9) years old, and 41.9% were male. On the alcohol test day, participants consumed 9.2 (4.6) alcoholic drinks over a time period of 6.3 (2.2) hours. They reported consuming 6.5 (4.2) more alcoholic drinks than they would normally consume on a regular non-hangover drinking occasion. Although not assessed in this study, assuming the alcohol was consumed within a similar time frame on both occasions, the increase in alcohol consumption likely corresponded to a significant rise in BAC relative to a regular drinking occasion. Mean (SD) next-day hangover severity was 3.5 (2.5). Although the average estimated BAC was relatively high, i.e., 0.16% (0.09), about one-third of the hungover participants (30.4%) had had an estimated BAC below the 0.11% cutoff level. A stepwise linear regression analysis including all the assessed variables revealed that four variables accounted for 31.7% of the variance in overall hangover severity (See Table 1). The analysis showed that with regard to the unique contribution to variance explained of individual variables, the increase in alcohol consumption relative to a regular drinking occasion was the strongest predictor of hangover severity.

Table 1. Summary of the regression analysis.

Variables	Model	Contribution
Increase in alcohol consumption relative to a "regular" drinking occasion	17.8%	17.8%
Body mass index (kg/m^2)	24.0%	6.2%
Dancing frequency on the drinking occasion	28.5%	4.5%
Number of past year's hangovers	31.7%	3.2%

Variables were included if they significantly ($p < 0.05$) contributed to the model. Significant Spearman's rho correlations were found between hangover severity and increase in alcohol consumption relative to a regular drinking occasion ($r = 0.435$, $p < 0.0001$), dancing frequency on the drinking occasion ($r = 0.288$, $p = 0.005$), and the number of hangovers in the past year ($r = 0.529$, $p < 0.0001$). The correlation between hangover severity and body mass index did not reach statistical significance ($r = -0.144$, $p = 0.168$). Data from Reference [18].

The findings discussed above do not imply that calculating estimated BAC serves no relevant purpose in future hangover research. Quite the opposite, BAC is a valuable measure that must be implemented in experimental studies. Calculating the estimated BAC enables researchers to administer individual amounts of alcohol that have been adjusted for sex and body weight, in order to achieve comparable BAC levels across study participants who undergo experimentally induced intoxication. BAC assessment also serves an important purpose during the process of recruitment of participants. Whenever a certain dose of alcohol is administered in an experimental study, researchers require a prior estimate of whether a participant will experience a next-day hangover at the designated BAC level. For this purpose, the estimated BAC can be calculated for a regular drinking occasion that usually results in a hangover. Preferably, this estimated BAC should be evaluated for more than one drinking occasion, as a recent analysis of data from an experimental study showed that there is a subset of approximately 20% of study participants for whom there was a substantial intra-individual hangover severity difference between the test days, even when the administered amount of alcohol and achieved BAC where the same [23]. With this prior understanding, the researcher can identify a group of drinkers who are resistant to developing hangovers at the (estimated) BAC level that will be achieved in their experimental study, and exclude these individuals. One could also consider excluding participants who report great intra-individual differences. Alternatively, one could identify individuals who are very sensitive to acute alcohol effects, or already develop hangovers at much lower BAC levels than the designated study BAC. It would be ethically inappropriate to include these individuals and they should be excluded in order to reduce drop-out rate due to anticipated adverse events such as vomiting. Finally, future experiments could also use actual BAC measures to avoid recall bias which may emerge with retrospective recall [24]. Similarly, subjective intoxication ratings could be measured in real time rather than retrospectively. However, research has shown that subjective intoxication, either assessed in real time (while drinking) [25] or the next morning (retrospectively, as in the presented survey data discussed in this paper) both highly correlate with hangover severity. Research on predictors of hangover severity and possible tolerance to hangovers is relatively new [19], and future research should explore the spectrum of additional factors, such as genetics, environment, drinking behaviors, and alcohol metabolism to further understand how variations in BAC (e.g., as a result of drinking more alcohol than usual) influence the presence and severity of alcohol hangovers.

Taken together, the research reviewed here suggests that the level of subjective intoxication and the increase in alcohol consumption relative to a regular drinking occasion are stronger predictors of next-day hangover severity than (estimated) BAC. Furthermore, a substantial number of alcohol drinkers experience hangovers at BAC levels well below 0.11%. Therefore, we argue that the current consensus regarding the BAC 0.11% threshold value as a criterion for having a hangover should be abandoned.

Author Contributions: Conceptualization and design, J.C.V., L.D.K., and A.-K.S., data acquisition, K.A.S., L.A., A.A.; data analysis, J.C.V.; writing original draft, J.C.V.; all authors critically reviewed the manuscript for important intellectual content and approved the final version. All authors have read and agreed to the published version of the manuscript.

Funding: This research received no external funding.

Conflicts of Interest: C.A. has undertaken sponsored research, or provided consultancy, for a number of companies and organizations including Airbus Group Industries, Astra, British Aerospace/Bae Systems, Civil Aviation Authority, Duphar, Farm Italia Carlo Erba, Ford Motor Company, ICI, Innovate UK, Janssen, LERS Synthélabo, Lilly, Lorex/Searle, UK Ministry of Defense, Quest International, Red Bull GmbH, Rhone-Poulenc Rorer, and Sanofi Aventis. S.B. has received funding from Red Bull GmbH, Kemin Foods, Sanofi Aventis, Phoenix Pharmaceutical and GlaxoSmithKline. S.J. has undertaken sponsored research for Pfizer, AstraZeneca, Merck, Gilead, Novartis, Roche, Red Bull GmbH, the Department for Transport, and Road Safety Trust. A.S. has held research grants from Abbott Nutrition, Arla Foods, Bayer Healthcare, Cognis, Cyvex, GlaxoSmithKline, Naturex, Nestle, Martek, Masterfoods, Wrigley, and has acted as a consultant/expert advisor to Abbott Nutrition, Barilla, Bayer Healthcare, Danone, Flordis, GlaxoSmithKline Healthcare, Masterfoods, Martek, Novartis, Unilever, and Wrigley. L.O. has received research funding from Cultech Ltd., FIT-Bioceuticals, PT Academy, Danone, Nutria Research, GlaxoSmithKline, Kraft Foods, Naturex, and The European Union (FP7). Over the past 3 years, J.C.V. has received grants/research support from the Dutch Ministry of Infrastructure and the Environment, Janssen Research and Development, and Sequential, and has acted as a consultant/advisor for Clinilabs, More Labs, Red Bull, Sen-Jam Pharmaceutical, Toast!, and ZBiotics. The other authors have no potential conflicts of interest to disclose.

References

1. Van Schrojenstein Lantman, M.; van de Loo, A.J.; Mackus, M.; Verster, J.C. Development of a definition for the alcohol hangover: Consumer descriptions and expert consensus. *Curr. Drug Abuse Rev.* **2016**, *9*, 148–154. [CrossRef] [PubMed]
2. Penning, R.; McKinney, A.; Verster, J.C. Alcohol hangover symptoms and their contribution to overall hangover severity. *Alcohol Alcohol.* **2012**, *47*, 248–252. [CrossRef] [PubMed]
3. Van Schrojenstein Lantman, M.; Mackus, M.; van de Loo, A.J.A.E.; Verster, J.C. The impact of alcohol hangover symptoms on cognitive and physical functioning, and mood. *Hum. Psychopharmacol.* **2017**, *32*. [CrossRef] [PubMed]
4. Verster, J.C.; Stephens, R.; Penning, R.; Rohsenow, D.; McGeary, J.; Levy, D.; McKinney, A.; Finnigan, F.; Piasecki, T.M.; Adan, A.; et al. The Alcohol Hangover Research Group consensus statement on best practice in alcohol hangover research. *Curr. Drug Abuse Rev.* **2010**, *3*, 116–127. [CrossRef] [PubMed]
5. National Institute of Alcohol Abuse and Alcoholism. Rethinking Drinking. Alcohol and Your Health. Calculators. Available online: https://www.rethinkingdrinking.niaaa.nih.gov/tools/calculators/Default.aspx (accessed on 1 December 2019).

6. Chapman, L.F. Experimental induction of hangover. *Q. J. Stud. Alcohol* **1970**, *5*, 67–86.
7. Verster, J.C.; de Klerk, S.; Bervoets, A.C.; Kruisselbrink, L.D. Can hangover immunity really be claimed? *Curr. Drug Abuse Rev.* **2013**, *6*, 253–254. [CrossRef]
8. Kruisselbrink, L.D.; Bervoets, A.C.; de Klerk, S.; van de Loo, A.J.A.E.; Verster, J.C. Hangover resistance in a Canadian university student population. *Addict. Behav. Rep.* **2017**, *5*, 14–18. [CrossRef]
9. Van de Loo, A.J.A.E.; Mackus, M.; van Schrojenstein Lantman, M.; Kraneveld, A.D.; Garssen, J.; Scholey, A.; Verster, J.C. Susceptibility to alcohol hangovers: The association with self-reported immune status. *Int. J. Environ. Res. Public Health* **2018**, *15*, 1286. [CrossRef]
10. Watson, P.E.; Watson, I.D.; Batt, R.D. Prediction of blood alcohol concentrations in human subjects. Updating the Widmark Equation. *J. Stud. Alcohol Drugs* **1981**, *42*, 547–556. [CrossRef]
11. Verster, J.C.; Kruisselbrink, L.D.; Anogeianaki, A.; Alford, C.; Stock, A.K. Relationship of alcohol hangover and physical endurance performance: Walking the Samaria Gorge. *J. Clin. Med.* **2020**, *9*, 114. [CrossRef]
12. Verster, J.C.; Benjaminsen, J.M.E.; van Lanen, J.H.M.; van Stavel, N.M.D.; Olivier, B. Effects of mixing alcohol with energy drink on objective and subjective intoxication: Results from a Dutch on-premise study. *Psychopharmacology* **2015**, *232*, 835–842. [CrossRef]
13. Van de Loo, A.J.A.E.; van Andel, N.; van Gelder, C.A.G.H.; Janssen, B.S.G.; Titulaer, J.; Jansen, J.; Verster, J.C. The effects of alcohol mixed with energy drink (AMED) on subjective intoxication and alertness: Results from a double-blind placebo-controlled clinical trial. *Hum. Psychopharmacol.* **2016**, *31*, 200–205. [CrossRef]
14. Hogewoning, A.; van de Loo, A.J.A.E.; Mackus, M.; Raasveld, S.J.; de Zeeuw, R.; Bosma, E.R.; Bouwmeester, N.H.; Brookhuis, K.A.; Garssen, J.; Verster, J.C. Characteristics of social drinkers with and without a hangover after heavy alcohol consumption. *Subst. Abuse Rehab.* **2016**, *7*, 161–167. [CrossRef]
15. Rohsenow, D.J.; Howland, J.; Winter, M.; Bliss, C.A.; Littlefield, C.A.; Heeren, T.C.; Calise, T.V. Hangover sensitivity after controlled alcohol administration as predictor of post-college drinking. *J. Abnorm. Psychol.* **2012**, *121*, 270–275. [CrossRef]
16. Piasecki, T.M.; Alley, K.J.; Slutske, W.S.; Wood, P.K.; Sher, K.J.; Shiffman, S.; Heath, A.C. Low sensitivity to alcohol: Relations with hangover occurrence and susceptibility in an ecological momentary assessment investigation. *J. Stud. Alcohol. Drugs* **2012**, *73*, 925–932. [CrossRef]
17. Köchling, J.; Geis, B.; Wirth, S.; Hensel, K.O. Grape or grain but never the twain? A randomized controlled multiarm matched-triplet crossover trial of beer and wine. *Am. J. Clin. Nutr.* **2019**, *109*, 345–352. [CrossRef]
18. Arnoldy, L.; Benson, S.; Scholey, A.; Verster, J.C. Psychological factors affecting hangover severity. *Drug Alcohol. Rev.* **2019**, *38* (Suppl. S1), S23.
19. Verster, J.C.; Slot, K.A.; Arnoldy, L.; Van Lawick van Pabst, A.E.; van de Loo, A.J.A.E.; Benson, S.; Scholey, A. The association between alcohol hangover frequency and severity: Evidence for reverse tolerance? *J. Clin. Med.* **2019**, *8*, 1520. [CrossRef]
20. Seppä, K.; Lepistö, J.; Sillanaukee, P. Five-Shot Questionnaire on Heavy Drinking. *Alcohol. Clin. Exp. Res.* **1998**, *22*, 1788–1791. [CrossRef]
21. Derogatis, L.R. *Brief Symptom Inventory*; Clinical Psychometric Research: Baltimore, MD, USA, 1975.
22. De Haan, L.; Kuipers, E.; Kuerten, Y.; Van Laar, M.; Olivier, B.; Verster, J.C. The RT-18: A new screening tool to assess young adult risk-taking behavior. *Int. J. Gen. Med.* **2011**, *4*, 575–584.
23. Hensel, K.O.; Longmire, M.R.; Köchling, J. Should population-based research steer individual health decisions? *Aging* **2019**, *11*, 9231–9233. [CrossRef]
24. Verster, J.C.; van de Loo, A.J.A.E.; Adams, S.; Stock, A.K.; Benson, S.; Alford, C.; Scholey, A.; Bruce, G. Naturalistic study design in alcohol hangover research: Advantages, limitations, and solutions. *J. Clin. Med.* **2019**, *8*, 2160. [CrossRef]
25. Scholey, A.; Benson, S.; Kaufman, J.; Terpstra, C.; Ayre, E.; Verster, J.C.; Allen, C.; Devilly, G. Effects of alcohol hangover on cognitive performance: A field/internet mixed methodology approach. *J. Clin. Med.* **2019**, *8*, 440. [CrossRef]

 © 2020 by the authors. Licensee MDPI, Basel, Switzerland. This article is an open access article distributed under the terms and conditions of the Creative Commons Attribution (CC BY) license (http://creativecommons.org/licenses/by/4.0/).

Perspective

Advantages and Limitations of Naturalistic Study Designs and Their Implementation in Alcohol Hangover Research

Joris C. Verster [1,2,3,*], Aurora J. A. E. van de Loo [1,2], Sally Adams [4], Ann-Kathrin Stock [5], Sarah Benson [3], Andrew Scholey [3], Chris Alford [6] and Gillian Bruce [7]

1. Utrecht Institute for Pharmaceutical Sciences (UIPS), Faculty of Science, Division of Pharmacology, Utrecht University, 3584 CG Utrecht, The Netherlands; a.j.a.e.vandeloo@uu.nl
2. Institute for Risk Assessment Sciences (IRAS), Faculty of Veterinary Medicine, Utrecht University, 3584 CM Utrecht, The Netherlands
3. Centre for Human Psychopharmacology, Swinburne University, Melbourne VIC 3122, Australia; sarahmichellebenson@gmail.com (S.B.); andrew@scholeylab.com (A.S.)
4. Addiction and Mental Health Group, Department of Psychology, University of Bath, Bath BA2 7AY, UK; sa221@bath.ac.uk
5. Cognitive Neurophysiology, Department of Child and Adolescent Psychiatry, Faculty of Medicine, TU Dresden, 01307 Dresden, Germany; Ann-Kathrin.Stock@uniklinikum-dresden.de
6. Psychological Sciences Research Group, University of the West of England, Bristol BS16 1QY, UK; chris.alford@uwe.ac.uk
7. Education and Social Sciences, University of the West of Scotland, Paisley PA1 2BE, UK; Gillianbruce@gmail.com
* Correspondence: J.C.Verster@uu.nl; Tel.: +31-30-253-6909

Received: 2 November 2019; Accepted: 5 December 2019; Published: 6 December 2019

Abstract: In alcohol hangover research, both naturalistic designs and randomized controlled trials (RCTs) are successfully employed to study the causes, consequences, and treatments of hangovers. Although increasingly applied in both social sciences and medical research, the suitability of naturalistic study designs remains a topic of debate. In both types of study design, screening participants and conducting assessments on-site (e.g., psychometric tests, questionnaires, and biomarker assessments) are usually equally rigorous and follow the same standard operating procedures. However, they differ in the levels of monitoring and restrictions imposed on behaviors of participants before the assessments are conducted (e.g., drinking behaviors resulting in the next day hangover). These behaviors are highly controlled in RCTs and uncontrolled in naturalistic studies. As a result, the largest difference between naturalistic studies and RCTs is their ecological validity, which is usually significantly lower for RCTs and (related to that) the degree of standardization of experimental intervention, which is usually significantly higher for RCTs. In this paper, we specifically discuss the application of naturalistic study designs and RCTs in hangover research. It is debated whether it is necessary to control certain behaviors that precede the hangover state when the aim of a study is to examine the effects of the hangover state itself. If the preceding factors and behaviors are not in the focus of the research question, a naturalistic study design should be preferred whenever one aims to better mimic or understand real-life situations in experimental/intervention studies. Furthermore, to improve the level of control in naturalistic studies, mobile technology can be applied to provide more continuous and objective real-time data, without investigators interfering with participant behaviors or the lab environment impacting on the subjective state. However, for other studies, it may be essential that certain behaviors are strictly controlled. It is, for example, vital that both test days are comparable in terms of consumed alcohol and achieved hangover severity levels when comparing the efficacy and safety of a hangover treatment with a placebo treatment day. This is best accomplished with the help of a highly controlled RCT design.

Keywords: study design; naturalistic study; randomized controlled trial; alcohol; hangover; blinding; mobile technology

---

1. Introduction

The alcohol hangover is defined as a combination of mental and physical symptoms, experienced the day after a single episode of heavy drinking, starting when the blood alcohol concentration approaches 0 [1]. Studies in this research area examine the causes, functional consequences, and potential treatments of the next day (i.e., post-intoxication) effects of alcohol consumption. The alcohol hangover is associated with cognitive and psychomotor impairment [2] and mood changes [3], and may negatively affect daily activities, such as driving a car [4,5] or job performance [6]. The World Health Organization (WHO) estimates that 5.1% of the global burden of disease and injury is attributable to alcohol use and its consequences [7], and a recent UK study rated the economic costs of having hangovers in terms of absenteeism and presenteeism at 4 billion GBP per year [8]. Despite this, the pathology of the alcohol hangover is poorly understood [9,10], and although there is great market demand [11], there are currently no effective hangover treatments available [12].

Both randomized controlled trials (RCTs) and naturalistic study designs are commonly applied in hangover research. Although increasingly applied in social sciences and medical research, the suitability of using naturalistic study designs remains a topic of debate. To examine this, our paper compares the naturalistic study design with the traditional controlled experimental design, in particular RCTs. It discusses the advantages and disadvantages of both designs and suggests solutions for issues of concern.

Traditionally, medical science has been based on clinical observations of patients and control samples. In the fields of psychiatry and psychology, for example, participants either self-report their mood or an investigator observes their behavior. This was common practice before the introduction of RCTs. However, since their introduction, the quality, methodology, and reporting of medical science has been continuously optimized [13], and the RCT is, therefore, currently often viewed as the gold standard that allows for the most precise and systematic investigations. RCTs are, for example, commonly used to investigate the efficacy and safety of a medicinal drug in a specific patient population. The RCT design is characterized by having several inclusions, exclusion, and discontinuation criteria that apply to participants, including lifestyle rules with regard to, for example, alcohol and drug use and smoking. RCTs are ideally double or triple blind to avoid influencing the study outcome, and participants are randomly allocated to treatment conditions. The treatment order is varied (cross-over) to account for any learning or order effects. All study-related activities are highly standardized and conducted per protocol, with the aim to have all test days as identical to each other as possible. In theory, the only methodological difference between the test days is the administered treatment or intervention. This way, it is thought that the study gathers 'clean' data about the effect of the treatment or intervention. However, this level of control comes at the cost of RCTs creating highly artificial situations, which lack ecological validity and/or potentially differ from the effects observed in the participants' everyday life.

On the other hand, the aim of the naturalistic study design is to mimic real-life as closely as possible, and as such is characterized by a minimum of lifestyle rules for participants, in which the investigators do not (actively) interfere with their activities. Hence, several behaviors and activities of the participants are not standardized and not regulated by a study protocol. Participants continue their normal lives and may visit the testing site for assessments or bio-sample collection or may even be able to undertake these assessments whilst remaining in their usual environment. Commonly, the only instruction is to behave normally (e.g., take their medication as prescribed or drink alcohol as they would on a normal night out), complete scheduled assessments (e.g., a sleep diary or online scales), and visit the testing site at set times.

The naturalistic design is increasingly utilized in various research areas and has been successfully applied in phase III studies and pharmacovigilance research, e.g., to investigate the efficacy of antipsychotics in schizophrenia patients [14] or breast cancer patients [15]. The following sections will discuss the commonalities and differences between RCTs and naturalistic study designs, advantages and disadvantages, and possible solutions to common pitfalls.

2. Recruitment, Screening, and Test Days

Both RCTs and naturalistic studies have highly controlled data collection on test days. This includes conducting standardized and validated tests according to good clinical practice (GCP) and utilizing standard operating procedures at pre-set times specified in the study protocol. Furthermore, both study designs can have various lifestyle rules (e.g., no alcohol or drug use, no smoking), which can be verified by objective assessments on the test day. In this respect, naturalistic studies do usually not differ from RCTs.

Recruitment, screening, selecting, and training of participants can also be equally rigorous in RCTs and naturalistic studies. Both study designs can apply the same inclusion and exclusion criteria. Objective assessments can be conducted to verify the criteria (e.g., blood chemistry, urinalysis, and electrocardiography), and participants can be familiarized with and trained in completing psychometric tests, treatment administration, and completing mood scales. The main reason that rigorous screening and selection of study participants in RCTs is common is that it ensures a more homogenous study sample. It is expected that there will be more variability between study participants in responsiveness to the administered treatments when the eligibility criteria are loosened. Loosening eligibility criteria may then decrease the chances of successfully demonstrating efficacy or safety. To demonstrate the true drug effect, assessments should not be obscured by various external uncontrolled factors. Unfortunately, applying a large number of eligibility criteria usually results in a considerable number of screening failures (i.e., participants not meeting all criteria for participation) or drop-outs and compliance failures (i.e., participants discontinuing or failing to adhere to the study protocol). This is commonly seen in RCTs [16–18]. In addition, a number of people may not participate in the first place when they are informed about the strict lifestyle rules and the hassle of screening procedures (e.g., blood drawings and medical examinations). Unfortunately, this may induce a (self-)selection bias in the study sample.

The extent to which RCT participants in drug development are representative for the patient population can therefore be questioned [17,18]. While some 'safety-related' eligibility criteria are obviously necessary, other eligibility criteria (e.g., cut off values for body weight ranges) are often not strongly justified by supporting scientific evidence [16]. Not applying or loosening unjustified eligibility criteria will increase recruitment speed and result in a study sample that better reflects the entire patient population. Some recent RCTs have, therefore, included a 'real life' arm in their study, including participants who did not meet the stringent eligibility criteria of the RCT [19]. As naturalistic studies aim to mimic real life, eligibility criteria are often less strict than those applied in RCTs. This may significantly increase the ecological validity of the study, which is usually low in RCTs [14].

3. Level of Control, Supervision, and Monitoring

All RCT study-related activities are closely monitored at the testing site (e.g., clinic or lab). However, this is not always the case in naturalistic study designs, in which researchers are not necessarily present.

One issue is not reporting behavior. As participation in research studies is typically confidential, and sometimes anonymous, there should be no objective reason for participants not to report certain behaviors. However, if these behaviors are restricted by discontinuation criteria, participants may decide not to report them in order to prevent themselves from being excluded from further study participation. Another reason could be social desirability, as participants may be less likely to report behaviors or incidents that they either perceived to be detrimental to their self-image or that they fear may result in negative judgement from others. Another issue may be misreporting. Participants may not report certain behaviors simply because they were not asked about them (e.g., a researcher refrains from questioning participants about drug use, because an inclusion criterion to participate in the study was not using drugs), or they view these behaviors as irrelevant to the study (e.g., a participant being unaware that drinking a cup of coffee can improve subsequent cognitive test performance). Fortunately, there are several ways to retrospectively and objectively verify the occurrence of study-relevant behaviors, including assessments for residual alcohol use (breathalyzer), drug use (urine tests), and recent smoking (exhaled carbon monoxide), or monitor activity and sleep episodes (actigraphy).

In both naturalistic studies and RCTs, it is also increasingly common to implement ambulatory assessments in the study design, for example cognitive tests or questionnaires completed online/at home. The advantage of not having to schedule visits to the testing site makes it easier to participate in the study and thus reduces chances of dropouts. It also allows for repeated testing at fixed time intervals, which may help to reduce the risk of study-relevant events not being recalled correctly. At home, testing has been successfully implemented in numerous phase III studies, using the same tests that would have been conducted in the clinic (e.g., online cognitive tests, blood pressure assessment, or self-administered blood glucose tests). In short, the use of mobile technologies enable compliance monitoring. Furthermore, mobile technology, home testing, and the internet provide various ways to ensure valid and reliable real-time assessments of cognitive and physical functioning, mood, and biomarkers [20–23].

However, in naturalistic studies, assessments are often limited to retrospective and subjective self-reports. When relying entirely on self-reports, recall bias and memory loss may have a significant impact on the accuracy of the collected data. For example, research has shown that people under- or over-estimate the amount of alcohol consumed [22,23] and that subjective and objective assessment of sleep parameters are not always in concordance with each other [23]. The latter should be taken into account when interpreting the data obtained in naturalistic studies.

To prevent the presence of observers/researchers from influencing the behaviors of study participants, one could consider monitoring the subject's behaviors in real time via video streaming, without the awareness of study participants that they are being filmed. However, this approach would raise ethical, privacy, and data security concerns. A better alternative to this would be to apply mobile technology to objectively measure behaviors, including parallel objective measures to help triangulate data obtained from other measures.

Activity, sleep, and physiological parameters, such as heartrate and body temperature, can, for example, all be measured in real time using activity watches or 'wearables'. Behavioral and mood data can be collected by real-time self-reports via smartphone apps (e.g., entering every drink they consumed). Alternatively, wearable technology (watches) that may record transdermal alcohol concentrations are currently being developed. In the future, these devices could be used to complement or partly replace self-reports. Moreover, they could help to reduce drop-out rates as a number of "passive" measurements could be conducted without requiring any effort from the participants. Importantly, this would also help to obtain a more complete picture in studies that investigate aversive effects, such as a hangover, which might lead to systematic drop-outs on the more severe end of the symptom scale. Taken together, mobile technology would not only reduce the strain on study participants, but potentially also make the measurements more objective. In addition, test batteries used in RCTs are often administered as single assessments or, at best infrequently. These can therefore easily miss critical events or periods. Mobile data collection can include participant actioned recording of events and more regular testing, or continuous psychophysiological assessments, including wearable devices, which can all provide a better picture of participant behavior and subjective state.

As part of mobile testing, conducting an online survey is another common way to collect data from participants. This is effective if the subject sample is large or if it is not necessary or possible for participants to visit the research facility (e.g., due to obstacles, such as bad weather, large distances, or physiological constraints). While online methodologies are an easy way to collect data, there are several disadvantages. For example, the researcher cannot be certain whether the scheduled participant is completing the survey or whether someone else is doing it in their place. Furthermore, the condition of the participant cannot be verified by the researcher (e.g., they might be drunk or drugged while completing the survey or may not be giving the assessment their full attention), which may reduce the accuracy and validity of the resulting data. Further enhancing this methodology can increase reliability of the collected data, for example by video streaming. Video streaming can confirm if the scheduled participant is actually present and can verify how the participant conducts a test or completes questionnaires. It further enables the researcher to observe the general health and makes it possible to record real-time observer-rated adverse effects.

4. Level of Standardization of Tests and Procedures

While the scrutiny of recruitment, screening, and test day assessments can have comparable levels of control and standardization in RCTs and naturalistic studies, the designs differ significantly with regard to the standardization and activities of participants during the intervention phase. In RCTs, every activity of the participant takes place in the testing facility. Activities are scheduled at pre-set times and conducted according to standard procedures. This includes treatment administration, meals, activities, time going to bed, or the environment where participants spend time (i.e., the testing site). Moreover, all assessments and activities are standardized and precisely monitored and recorded by the researchers. The rationale to conduct an RCT in this way is clear: By minimizing the non-intervention-related variability (i.e., the uncontrolled "noise") in all potentially study-relevant parameters, the chance of observing a true treatment effect increases.

In contrast, in naturalistic studies, participants continue with their usual activities and researchers do not observe them or provide instructions on how to behave. Thus, the researchers do not interfere with the participants' activities. Consequently, behaviors are unstandardized and self-initiated. The rationale for this approach is to closely mimic real life, i.e., to maximize ecological validity. This ecological validity is important because it best reflects the way in which phenomena, such as hangovers, emerge, and medicinal treatments will be actually used when marketed. Additionally, eligibility criteria in naturalistic studies may be less strict compared to those of RCTs to ensure the study sample better reflects the heterogeneous population who will use a treatment or intervention in clinical practice and provide a better picture of efficacy. Thus, rather than a limitation, the lack of standardization can be considered to be a benefit of the naturalistic study design.

A related discussion is the use of subjective versus objective assessments and the quest for the inclusion of biomarker assessments in a study. Cytokine concentrations, for example, can vary in cases of depression [24] or during the hangover state [25]. It can thus be interesting to assess cytokine changes in blood or saliva. The alcohol hangover state is a subjective experience which, up till now, cannot be objectively measured. Although this can be viewed as a significant limitation of this research area, it should be underlined that biomarkers are per definition (at best) proxy-measures if one aims to measure mood or how the participant feels. Clinical observations may be an alternative, but these usually do not substitute for subjective assessments of the severity or nature of mood states. To date, the best way to rate mood levels is by asking participants to report how they feel [26]. Interestingly, in this regard, the outcome of these subjective assessments is not always in correspondence with the outcome of objective biomarker assessments. Participants can, for example, report feeling perfectly fine while having a clinically relevant increase in blood pressure. Alternatively, participants can report sleep complaints and poor sleep quality while their polysomnographic outcomes are within normal ranges. Together, these findings advocate to include both subjective and objective assessments in future studies, irrespective of whether the study design is RCT or naturalistic.

5. Implications for Hangover Research

To provoke the hangover state, an evening of supervised alcohol consumption is typically scheduled in RCTs. The amount and type of alcoholic drink (and placebo) and the pace of drinking are usually pre-defined, and drinking is conducted within a pre-set time frame. This is typically conducted in a clinical setting, often accompanied by other participants who do not know each other. Food and other beverage intake (e.g., water) are prohibited or controlled, as are the cognitive and physical activities of the participants. All activities are closely monitored and recorded by the researchers, including blood alcohol concentration (BAC) assessments to verify alcohol consumption levels and adverse event recording. The evening activities are often concluded by a night of supervised sleep in the clinic, with a pre-set bed-time and wake-up time. Sleep quality and duration can be monitored with polysomnography or study personnel.

In contrast, in naturalistic studies, participants drink in a familiar setting (e.g., a bar or at home) with people they know, engaging in their usual activities. These normally differ from activities employed in RCTs (e.g., dancing in a club versus reading a magazine in the laboratory). In naturalistic studies, participants can eat food when they feel hungry and smoke and are exposed to external stimuli which are not replicated in the RCT setting (e.g., visiting multiple bars, walking outside in the rain, waiting for a bus to travel home). They can go to bed when feeling sleepy without being restricted by study procedures, which often dictate a much earlier time-to-bed than people have in real life after an evening out. As they sleep in their own beds, they will not experience the sleep problems that are common in RCTs, in which participants sleep in a new and unknown clinical environment (e.g., the first night effect) [27,28]. In addition, participants can apply their personal sleep habits, sleep hygiene activities, and wake-up rituals in naturalistic studies. Finally, socializing, expectancies, and motives for alcohol consumption most likely differ between real-life situations and RCTs and may impact assessment outcomes. Thus, in naturalistic studies, participants can either drink alone or have an evening with friends in a setting of their own choice. Bedtime is self-initiated, and participants sleep at home in their own bed. The next morning, participants come to the testing site for the assessments on the test day. Past evening behaviors are recorded retrospectively (e.g., via questionnaires or an interview), and in case of mobile technology use, objective data read-outs are obtained from the devices.

Whether or not it is important to monitor the drinking session depends entirely on the aim of individual research projects. For some studies, it may be essential that certain behaviors are strictly controlled. For example, when comparing the efficacy and safety of a hangover treatment with a placebo treatment, it is vital that both test days are comparable, in terms of consumed alcohol and achieved hangover severity levels. In this case, a strictly controlled RCT design would be favorable. If one chooses to use a naturalistic study design in efficacy studies, a statistical analysis should account for differences between the test days (e.g., in the form of co-variates or propensity scores). However, it is not always possible to accurately account for all variables. This could, for example, be because they depend on subjective self-reports (e.g., alcohol intake), because certain information is lacking (e.g., congener content of drinks), or because a certain factor has not (yet) been recognized as relevant (e.g., a certain genotype or developmental factors). In summary, several important factors that differ between test days (e.g., certain behaviors) that may bias the comparison between treatment and placebo will likely remain unknown or unrecognized and, therefore, not properly accounted for.

On the other hand, if one is primarily interested in the effects of the (subjective) alcohol hangover itself on cognitive performance, mood, or other variables, then the behaviors that provoked the hangover state are of limited importance. In this case, there is no clear need to monitor the amount and type of alcohol consumed, estimated peak BAC, and the setting and behaviors during the drinking session. In extremis, participants could then be recruited in the morning after an evening out and allocated to a hangover or control group, or groups that consumed alcohol or not. This would be the ultimate way of not interfering with participant drinking behavior, as participants were unaware that they were going to participate in a research study at the time they displayed the study-relevant behavior (e.g., drinking or staying sober). This design was successfully applied by Devenney et al. [29], who recruited participants at university venues in the morning, i.e., on the day following the drinking session. However, if one is interested in how drinking variables and behaviors during the drinking session cause or relate to hangover variables, it is essential that these are accurately measured. Statistical analysis can then take into account the observed interindividual differences in naturalistic studies.

There are obvious advantages of applying a naturalistic study design in alcohol hangover research, as the drinking session reflects what people do in normal life. In contrast to RCTs, they are not forced to adapt to a drinking regime, including consuming alcoholic beverages that are not their regular choice during a pre-set drinking time period that may differ from a normal night out. In fact, research consistently shows that in real life situations, most people consume much larger quantities of alcohol over a longer period of time, as compared to the pre-set dosages of alcohol that are administered in clinical studies to provoke a hangover. This results in significantly higher (and more realistic) BAC levels in naturalistic studies, as compared to many RCTs [30].

Assessments during the hangover state can then take place in the clinic, following a highly standardized and controlled protocol, similar to RCTs. Alternatively, Scholey et al. [31] utilized online cognitive testing in a naturalistic hangover study and demonstrated that this was an effective way to collect objective data in real time during the hangover state. This study also addressed the issue of participant drop out. It has been argued that participants who experience severe events may not continue participation in naturalistic studies. This would of course bias the study outcome in favor of a treatment. Scholey et al. [31] compared their study participants with their dropouts. For both groups, peak BAC was assessed in real time the evening before the (hangover) test day, and no significant BAC difference was observed between participants who did and did not complete the test day assessments. Hence, there are presumably other reasons than mere degree of intoxication that determine whether participants discontinue study participation or not. A different approach has been the use of mobile technology, including screen-based tests, to enable participants to be assessed within the privacy and safety of their own homes, without the need to travel to the test center when hungover, avoiding dropouts [32].

Finally, studies comprising alcohol administration to humans usually require ethics approval. For many ethics committees, it appears that a noteworthy difference is made based on whether the alcohol is actually administered to participants by the experimenters (RCTs) or whether they administer it themselves in an unsupervised setting (naturalistic studies). Ethics committees often limit the amount of alcohol researchers are allowed to administer to participants of RCTs to a blood alcohol concentration (BAC) below 0.12%, while in study protocols for naturalistic studies it is unknown how much alcohol participants will consume. Naturalistic studies consistently demonstrate that actual drinking levels are associated with much higher BACs. For example, Hogewoning et al. [30] reported an estimated BAC of 0.2%. When interviewing naturalistic study participants, they attest that they had a 'normal' night out, including their usual drinking behavior. This is an odd situation considering that, in RCTs, alcohol consumption is closely monitored with a physician and study personnel present, while participants can drink alcohol freely and unsupervised in naturalistic studies. Monitoring the level of alcohol consumed will also aid in evaluating hangover treatments. True symptom levels may not be assessed in the laboratory due to alcohol dosing restrictions, where effectively only 'sub-clinical' hangover symptom levels are evaluated.

Of note, viewpoints and safety concerns of ethics committee members are not always in agreement with those of study participants. For example, Petrie et al. [33] investigated the stress and pressure/imposition experienced by RCT participants for a variety of study-related handlings (e.g., blood pressure assessment, blood drawing) and compared their ratings to those of ethics committee members. The study revealed that several commonly applied procedures, such as taking a saliva sample or completing a questionnaire or mood scale, were rated as significantly less stressful by RCT participants compared to the ratings anticipated by ethics board members.

Petrie et al. [33] also compared the experienced stress levels in RCT procedures with those experienced in daily life and found that many relatively harmless experiences (e.g., stress when 'asked to donate to a charity in the street' or being 'caught in the rain') were rated as more stressful by study participants than completing a mood scale or delivering a saliva or urine sample. The overall conclusion of the study was that study-related stress and the impact of procedures in the standardized data collection may be overestimated by some ethics committees. Unfortunately, the restrictions that ethics committees feel inclined to impose upon proposed research projects (especially RCTs) can have a significant impact on the ecological validity of these studies and the consequential validity of the findings.

6. Concluding Remarks

The commonalities and differences between RCT designs and naturalistic studies are summarized in Table 1.

Table 1. Commonalities and differences between randomized controlled trial (RCT) and naturalistic study designs.

Study Design	RCT	Naturalistic Study Design
Ecological validity	Low to medium	High
External validity	Low to medium	High
Internal validity	High	Low to medium
Criterion validity	High	High
Construct validity	High	High
Screening	**Controlled, per protocol**	**Controlled, per protocol**
Inclusion, exclusion criteria	Yes	Yes
Familiarize with procedures and tests	Yes	Yes
Drinking session	**Per protocol, supervised**	**Uncontrolled, not supervised**
Instructions	Per protocol	None, or minimal
Amount, type of drink	Pre-set, standardized	Self-initiated
Start and stop time of drinking	Pre-set, per protocol	Self-initiated
Behaviors during drinking	Restricted, per protocol	Free (as normal)
Drinking environment	Research unit	Free (e.g., bar(s), at home)
Social aspects of drinking (if not alone)	With strangers	With friends or strangers
Real time assessments (e.g., BAC)	Yes	Possible via mobile technology
Food and water intake	Restricted, per protocol	Free (as normal)
Smoking	Restricted, per protocol	Free (as normal)
Sleep	**Controlled and supervised**	**Uncontrolled, not supervised**
Time to bed, wake up time	Restricted, per protocol	Self-initiated
Sleep hygiene and related behaviors	Restricted, per protocol	Self-initiated
Monitoring, real time assessments	Yes	Possible via mobile technology
Sleep environment	Sleep unit	At home or elsewhere
Test days	**Controlled and standardized**	**Controlled and standardized**
Psychometric tests, mood assessments	Standardized and validated	Standardized and validated
Time of testing	Per protocol at pre-set time	Per protocol at pre-set time
Conductance of study procedures	Per protocol	Per protocol
Supervision, monitoring	Yes	Yes

Description of validity types: Ecological validity = to what extend the study reflects a realistic hangover drinking occasion; external validity = to what extent can findings be generalized to the population as a whole; internal validity = to what extent can the design demonstrate causal effects; criterion validity = to what extent are measures related to study outcomes; construct validity = the degree to which the administered tests measure what they claim or purport to be measuring. Abbreviation: Blood alcohol concentration (BAC). Please note that this table is intended to contrast the RCT and naturalistic study design. Some studies might incorporate features of both designs (e.g., supervised and standardized alcohol administration, but unsupervised sleep at home). Additionally, studies with the same design type may differ significantly in the levels of control, standardization, and quality.

RCT designs are preferred for studies that require strictly-controlled study procedures. Treatment efficacy and safety studies, for example, require controlled treatment administration and the variability in participants' behaviors (e.g., alcohol intake, physical activity, food intake, and sleep) should be kept to a minimum. However, RCTs, per definition, modify and structure participant behaviors in a standardized and, therefore, often "unnatural" way. Therefore, a naturalistic study design is preferred if one aims to better understand or mimic real-life interventions. The lack of standardization of naturalistic studies should, therefore, be considered as a benefit of the study design.

Additionally, free drinking in naturalistic studies often exceeds the intoxication limits deemed safe and ethically approved for RCT studies, which further increases the ecological validity of naturalistic hangover studies compared to RCT hangover studies. To improve the level of control in naturalistic studies, mobile technology can be used to assess objective real-time data and control the quality of assessment, without investigators interfering with participant behaviors.

Author Contributions: Conceptualization: All authors; draft of first version of the manuscript: J.C.V. and G.B.; all authors approved the final version.

Conflicts of Interest: S.B. has received funding from Red Bull GmbH, Kemin Foods, Sanofi Aventis, Phoenix Pharmaceutical, and GlaxoSmithKline. A.S. has held research grants from Abbott Nutrition, Arla Foods, Bayer Healthcare, Cognis, Cyvex, GlaxoSmithKline, Naturex, Nestle, Martek, Masterfoods, and Wrigley and has acted as a consultant/expert advisor to Abbott Nutrition, Barilla, Bayer Healthcare, Danone, Flordis, GlaxoSmithKline Healthcare, Masterfoods, Martek, Novartis, Unilever, and Wrigley. Over the past three years, J.C.V. has received grants/research support from the Dutch Ministry of Infrastructure and the Environment, Janssen, Nutricia, and Sequential and acted as a consultant/advisor for Clinilabs, More Labs, Red Bull, Sen-Jam Pharmaceutical, Toast!, and ZBiotics. C.A. has undertaken sponsored research, or provided consultancy, for a number of companies and organizations, including Airbus Group Industries, Astra, British Aerospace/BAeSystems, Civil Aviation Authority, Duphar, FarmItalia Carlo Erba, Ford Motor Company, ICI, Innovate UK, Janssen, LERS Synthélabo, Lilly, Lorex/Searle, UK Ministry of Defense, Quest International, Red Bull GmbH, Rhone-Poulenc Rorer, and Sanofi Aventis. The other authors have no potential conflicts of interest to disclose.

References

1. Van Schrojenstein Lantman, M.; van de Loo, A.J.A.E.; Mackus, M.; Verster, J.C. Development of a definition for the alcohol hangover: Consumer descriptions and expert consensus. *Curr. Drug Abuse Rev.* **2016**, *9*, 148–154. [CrossRef] [PubMed]
2. Gunn, C.; Mackus, M.; Griffin, C.; Munafò, M.R.; Adams, S. A systematic review of the next-day effects of heavy alcohol consumption on cognitive performance. *Addiction* **2018**, *113*, 2182–2193. [CrossRef] [PubMed]
3. McKinney, A. A review of the next day effects of alcohol on subjective mood ratings. *Curr. Drug Abuse Rev.* **2010**, *3*, 88–91. [CrossRef] [PubMed]
4. Verster, J.C. Alcohol hangover effects on driving and flying. *Int. J. Disabil. Hum. Dev.* **2007**, *6*, 361–367. [CrossRef]
5. Verster, J.C.; Bervoets, A.C.; de Klerk, S.; Vreman, R.A.; Olivier, B.; Roth, T.; Brookhuis, K.A. Effects of alcohol hangover on simulated highway driving performance. *Psychopharmacology* **2014**, *231*, 2999–3008. [CrossRef]
6. Frone, M.R.; Verster, J.C. Alcohol hangover and the workplace: A need for research. *Curr. Drug Abuse Rev.* **2013**, *6*, 177–179. [CrossRef]
7. World Health Organization. *Global Status Report on Alcohol and Health 2018*; WHO: Geneva, Switzerland, 2018.
8. Bhattacharya, A. *Financial Headache. The Cost of Workplace Hangovers and Intoxication to the UK Economy*; Institute of Alcohol Studies: London, UK, 2019.
9. Penning, R.; van Nuland, M.; Fliervoet, L.A.; Olivier, B.; Verster, J.C. The pathology of alcohol hangover. *Curr. Drug Abuse Rev.* **2010**, *3*, 68–75. [CrossRef]
10. Tipple, C.T.; Benson, S.; Scholey, A. A Review of the Physiological Factors Associated with Alcohol Hangover. *Curr. Drug Abuse Rev.* **2016**, *9*, 93–98. [CrossRef]
11. Mackus, M.; Van De Loo, A.J.; Nutt, D.; Verster, J.C.; Lantman, M.V.S. An effective hangover treatment: Friend or foe? *Drug Sci. Policy Law* **2017**, *3*. [CrossRef]
12. Jayawardena, R.; Thejani, T.; Ranasinghe, P.; Fernando, D.; Verster, J.C. Interventions for treatment and/or prevention of alcohol hangover: Systematic review. *Hum. Psychopharmacol. Clin. Exp.* **2017**, *32*, e2600. [CrossRef]
13. Brunoni, A.R.; Tadini, L.; Fregni, F. Changes in clinical trials methodology over time: A systematic review of six decades of research in psychopharmacology. *PLoS ONE* **2010**, *5*, e9479. [CrossRef] [PubMed]
14. Fagiolini, A.; Rocca, P.; De Giorgi, S.; Spina, E.; Amodeo, G.; Amore, M. Clinical trial methodology to assess the efficacy/effectiveness of long-acting antipsychotics: Randomized controlled trials vs naturalistic studies. *Psychiatry Res.* **2017**, *247*, 257–264. [CrossRef] [PubMed]
15. Fleming, L.; Randell, K.; Stewart, E.; Espie, C.A.; Morrison, D.S.; Lawless, C.; Paul, J. Insomnia in breast cancer: A prospective observational study. *Sleep* **2019**, *42*, zsy245. [CrossRef] [PubMed]
16. Van Spall, S.H.G.; Toren, A.; Kiss, A. Eligibility criteria of randomized controlled trials published in high-impact general medical journals. *JAMA* **2007**, *297*, 1233–1244. [CrossRef] [PubMed]
17. Roehrs, T.; Verster, J.C.; Koshorek, G.; Withrow, D.; Tancer, M.; Roth, T. How representative are insomnia clinical trials? *Sleep Med.* **2018**, *51*, 118–123. [CrossRef]
18. Huls, H.; Abdulahad, S.; van de Loo, A.J.A.E.; Mackus, M.; Roehrs, T.; Roth, T.; Verster, J.C. Inclusion and exclusion criteria of clinical trials for insomnia. *J. Clin. Med.* **2018**, *7*, 206. [CrossRef]

19. Björkholm, C.; Gannedahl, A.; Cars, T.; Rive, B.; Merclin, N.; Umuhire, D.; Milz, R.; van Sanden, S.; Lööv, S.-A.; Hellner, C.; et al. Constructing a real-world evidence arm to quantify the placebo effect in a randomized control trial utilizing electronic-medical record depression scales. *Eur. Neuropsychopharmacol.* **2020**, in press.
20. Verster, J.C.; Tiplady, B.; McKinney, A. Mobile technology and naturalistic study designs in addiction research. *Curr. Drug Abuse Rev.* **2012**, *5*, 169–171. [CrossRef]
21. Adams, Z.; McClure, E.A.; Gray, K.M.; Danielson, C.K.; Treiber, F.A.; Ruggiero, K.J. Mobile devices for the remote acquisition of physiological and behavioral biomarkers in psychiatric clinical research. *J. Psychiatr. Res.* **2017**, *85*, 1–14. [CrossRef]
22. Monk, R.L.; Heim, D.; Qureshi, A.; Price, A. "I have no clue what I drunk last night" using Smartphone technology to compare in-vivo and retrospective self-reports of alcohol consumption. *PLoS ONE* **2015**, *10*, e0126209. [CrossRef]
23. Devenney, L.E.; Coyle, K.B.; Roth, T.; Verster, J.C. Sleep after heavy alcohol consumption and physical activity levels during alcohol hangover. *J. Clin. Med.* **2019**, *8*, 752. [CrossRef]
24. Janssen, D.G.A.; Verster, J.C.; Baune, B.T. Depression, immunity and cytokine function: A psychoneuroimmunological review on depression and pharmacological response. *Hum. Psychopharmacol.* **2010**, *25*, 201–215. [CrossRef]
25. Van de Loo, A.J.A.E.; Mackus, M.; Knipping, K.; Brookhuis, K.A.; Kraneveld, A.D.; Garssen, J.; Verster, J.C. Changes over time in saliva cytokine concentrations the day following heavy alcohol consumption. *Allergy* **2017**, *72*, 482.
26. Baldwin, W. Information no one else knows: The value of self-report. In *The Science of Self-Report*; Psychology Press: Hove, UK, 1999; pp. 15–20.
27. Webb, W.B.; Campbell, S.S. The first night effect revisited with age as variable. *Waking Sleep.* **1979**, *3*, 319–324.
28. Le Bon, O.; Staner, L.; Hoffmann, G.; Dramaix, M.; San Sebastian, I.; Murphy, J.R.; Kentos, M.; Pelc, I.; Linkowski, P. The first-night effect may last more than one night. *J. Psychiatr. Res.* **2001**, *35*, 165–172. [CrossRef]
29. Devenney, L.E.; Coyle, K.B.; Verster, J.C. Memory and attention during an alcohol hangover. *Hum. Psychopharmacol.* **2019**, *34*, e2701. [CrossRef]
30. Hogewoning, A.; van de Loo, A.J.A.E.; Mackus, M.; Raasveld, S.J.; De Zeeuw, R.; Bosma, E.R.; Bouwmeester, N.H.; Brookhuis, K.A.; Garssen, J.; Verster, J.C. Characteristics of social drinkers with and without a hangover after heavy alcohol consumption. *Subst. Abuse Rehabil.* **2016**, *7*, 161–167. [CrossRef]
31. Scholey, A.; Benson, S.; Kaufman, J.; Terpstra, C.; Ayre, E.; Verster, J.C.; Allen, C.; Devilly, G. Effects of alcohol hangover on cognitive performance: A field/internet mixed methodology approach. *J. Clin. Med.* **2019**, *8*, 440. [CrossRef]
32. Alford, C.; Martinkova, Z.; Johnson, S.; Tiplady, B.; Verster, J.C. The effects of alcohol hangover on performance and mood including risk taking when assessed at home. *Alcohol. Clin. Exp. Res.* **2015**, *39*, 28A.
33. Petrie, K.J.; Faasse, K.; Notman, T.A.; O'Carroll, R. How distressing is it to participate in medical research? A calibration study using an everyday events questionnaire. *J. R. Soc. Med. Short Rep.* **2013**, *4*, 1–7. [CrossRef]

© 2019 by the authors. Licensee MDPI, Basel, Switzerland. This article is an open access article distributed under the terms and conditions of the Creative Commons Attribution (CC BY) license (http://creativecommons.org/licenses/by/4.0/).

Review

The Assessment of Overall Hangover Severity

Joris C Verster [1,2,3], Aurora J.A.E. van de Loo [1,2], Sarah Benson [3], Andrew Scholey [3] and Ann-Kathrin Stock [4,*]

1. Division of Pharmacology, Utrecht Institute for Pharmaceutical Sciences (UIPS), Utrecht University, 3584CG Utrecht, The Netherlands; j.c.verster@uu.nl (J.C.V.); a.j.a.e.vandeloo@uu.nl (A.J.A.E.v.d.L.)
2. Institute for Risk Assessment Sciences (IRAS), Utrecht University, 3584CM Utrecht, The Netherlands
3. Centre for Human Psychopharmacology, Swinburne University, Melbourne, VIC 3122, Australia; sarahmichellebenson@gmail.com (S.B.) andrew@scholeylab.com (A.S.)
4. Cognitive Neurophysiology, Department of Child and Adolescent Psychiatry, Faculty of Medicine, TU Dresden, Fetscherstr. 74, 01307 Dresden, Germany
* Correspondence: Ann-Kathrin.Stock@uniklinikum-dresden.de

Received: 25 February 2020; Accepted: 11 March 2020; Published: 13 March 2020

Abstract: The aim of this study was to critically evaluate and compare the different methods to assess overall hangover severity. Currently, there are three multi-item hangover scales that are commonly used for this purpose. All of them comprise a number of hangover symptoms for which an average score is calculated. These scales were compared to a single, 1-item scale assessing overall hangover severity. The results showed that the hangover symptom scales significantly underestimate (subjective) hangover severity, as assessed with a 1-item overall hangover severity scale. A possible reason for this could be that overall hangover severity varies, depending on the frequency of occurrence of individual symptoms included in the respective scale. In contrast, it can be assumed that, when completing a 1-item overall hangover scale, the rating includes all possible hangover symptoms and their impact on cognitive and physical functioning and mood, thus better reflecting the actually experienced hangover severity. On the other hand, solely relying on hangover symptom scales may yield false positives in subjects who report not having a hangover. When the average symptom score is greater than zero, this may lead to non-hungover subjects being categorized as having a hangover, as many of the somatic and psychological hangover symptoms may also be experienced without consuming alcohol (e.g., having a headache). Taken together, the current analyses suggest that a 1-item overall hangover score is superior to hangover symptom scales in accurately assessing overall hangover severity. We therefore recommend using a 1-item overall hangover rating as primary endpoint in future hangover studies that aim to assess overall hangover severity.

Keywords: alcohol; hangover; symptoms; severity; measurement; scale; single item assessment

1. Introduction

The alcohol hangover is defined as the combination of negative mental and physical symptoms which may be experienced the day after a single episode of alcohol consumption, starting when blood alcohol concentration (BAC) approaches zero [1,2]. Alcohol hangovers are typically characterized by a combination of symptoms affecting subjective mood, cognition and physical functioning [3–6]. These symptoms have been shown to negatively impact daily activities such as job performance [7] and driving [8–10]. The annual economic costs of alcohol hangover in terms of absenteeism and presenteeism have been estimated to be 173 billion USD for the USA [11] and 4 billion GBP for the UK [12]. Since the foundation of the Alcohol Hangover Research Group in 2010, the amount of research on the causes, consequences, treatment, and prevention of hangovers has been growing rapidly. Accurate measurement tools are essential to assess hangover severity, for example in experimental and naturalistic studies, in intervention studies examining treatment efficacy, and in survey research. Generally, they may be filled in by any drinker of all (adult) ages and in any drinking-related (research) context. There is, however, ongoing debate about which measure of hangover severity is most suitable. In this paper, we compare the three most widely used hangover symptom scales with a 1-item overall hangover severity rating, and discusses why the latter is a more reliable and useful measure to be included in future hangover research.

2. Characteristics of an Effective Patient-Reported Outcome Measure (PROM)

Currently, there are no biomarkers that accurately and objectively assess hangover severity. Therefore, one has to rely on Patient-Reported Outcome Measures (PROMs). A PRO instrument is often a questionnaire/scale or single item that is directly answered by the patient, capturing the patient's experience without interpretation of the patient's response by a clinician or anyone else [13]. To the ensure the validity of a PRO, it is fundamental that it reliably measures the concept it is intended to measure (i.e., hangover). In the case of a multi-item scale, it is furthermore important that all of the individual items adequately contribute to the final conceptual framework of the instrument [13]. Any given scale can only be considered to have sufficient validity as a measuring tool when these conditions are fulfilled. The complex nature of alcohol hangover severity, which includes multidomain facets associated with the presence and severity of variable symptoms, and their impact on cognitive and physical functioning and mood, makes it quite challenging to develop multi-item scales that accurately assess the concept of hangover severity. Nevertheless, there are currently three hangover symptom scales available for this purpose, and they are commonly used in hangover research [14–16].

Slutske et al. [14] developed the Hangover Symptoms Scale (HSS) to assess the frequency with which drinkers experienced hangover symptoms in the last year. The scale consists of 13 items including "felt extremely thirsty or dehydrated", "felt more tired than usual", "experienced a headache", "felt very nauseous", "vomited", "felt very weak", "had difficulty concentrating", "more sensitive to light and sound than usual", "sweated more than usual", "had a lot of trouble sleeping", "was anxious", "felt depressed", and "experienced trembling or shaking". Items can be scored either dichotomously (experienced the symptom, or not), or on a 5-point scale from "never", "2 times or less" (once or twice per year), "3–11 times" (more than once or twice, but less than once per week), "12–51 times" (more than once a month, but not every week), and "52 times or more" (once per week or more frequently). It is important to underline that the original HSS outcome is a frequency measure. The scale has, however, been modified and used to assess hangover severity, for example by using the same items but changing the item scoring into a symptom rating ranging from 0 (absent) to 10 (extreme) [3].

Rohsenow et al. [15] developed the Acute Hangover Scale (AHS) to assess hangover severity. The scale consists of nine items including "hangover", "thirsty", "tired", "headache", "dizziness/faintness", "loss of appetite", "stomachache", "nausea", and "heart racing", which are rated on a scale ranging from 0 to 7. The anchors of the scale are "none" (score of 0), "mild" (score of 1), "moderate" (score of 4), and "incapacitating" (score of 7). Overall hangover severity is computed by calculating the average score across the AHS nine items.

The third scale, developed by Penning et al. [16], is the Alcohol Hangover Severity Scale (AHSS). It consists of 12 items, including "fatigue (being tired)", "clumsiness", "dizziness", "apathy", "sweating", "shivering", "confusion", "stomach pain", "nausea", "concentration problems", "heart pounding", and "thirst". Symptom severity for each item can be rated on a scale ranging from 0 (absent) to 10 (extreme). Overall hangover severity is the average score across the 12 items.

The three hangover symptom scales each present with some shortcomings, and their limitations are discussed in detail elsewhere [16]. For example, scales do not always include true hangover symptoms (e.g., the HSS includes the item "trouble sleeping", which is experienced before the start of the hangover state). The AHS includes the item "hangover", but it is usually not advised to include an item that is identical to the overall concept that one aims to measure with a multi-item scale. Finally, the AHSS does not include the item headache, even though this is a frequently reported hangover symptom. Notwithstanding these limitations, the three hangover symptom scales are currently the most frequently used scales to assess hangover severity. Other researchers such as Hogewoning et al. [17] have been using an extended symptom listing, including all symptoms of the three hangover symptom scales.

Alternatively, overall hangover severity may be assessed with a single item rated on a scale ranging from absent (0) to extreme (10). This 1-item score is hypothesized to encompass all symptoms experienced by the drinker, including their relative impact on daily activities and mood.

3. Comparing Overall Hangover Severity Outcomes of the Different Assessment Methods

When using multi-item instruments, it is important that all items are relevant to the concept under investigation (i.e., alcohol hangover). If irrelevant items were included (e.g., symptoms that are seldomly reported), this would result in an overall hangover severity score that would be biased towards zero, and would therefore tend to underestimate, or even not show, the effects of treatment in intervention studies. In extreme cases, this might even lead to the wrong conclusion that a treatment is ineffective [13].

Penning et al. [3] examined the scientific literature and identified 47 hangover symptoms. However, the AHSS, HSS and AHS each comprise a different selection of these symptoms. This selection of symptoms may have a significant, and potentially biasing effect on the aggregate rating of overall hangover severity. The discrepancy between aggregate symptom scores and 1-item overall hangover severity assessments has been demonstrated previously by Penning et al. [16]. In that study, 947 subjects (Mean (SD) age of 21.1 (2.3) years old, 46 % men) rated the presence and severity of 23 hangover symptoms on a scale ranging from 0 (absent) to 10 (extreme). In addition, overall hangover severity was assessed. Further evaluation of the data showed that mean (SD) scores on a modified HSS and the AHSS were 3.6 (1.4) and 3.7 (1.7), respectively. Of note, the mean (SD) severity score on the 1-item hangover scale, i.e., 5.7 (2.2), was significantly higher ($p < 0.0001$) than both the modified HSS and the AHSS hangover score. These observations suggest that hangover severity scores based on aggregate symptom scores significantly underestimate the true hangover severity assessed with a single overall severity item.

There are three important reasons why aggregate symptom scores may deviate from the true hangover effect. These are related to (1) the relative presence and severity of hangover symptoms, (2) the impact of the experienced symptoms on cognitive functioning, physical activities, and mood, and (3) the fact that several assessed symptoms are also experienced without having a hangover, or even without consuming alcohol at all. These issues are discussed further in the next sections.

4. Presence and Severity of Hangover Symptoms

The occurrence and severity may differ significantly between symptoms experienced in the hangover state. This is illustrated by evaluating the data by Van Schrojenstein Lantman et al. [4], which is depicted in Figure 1. This study surveyed $n = 1837$ social drinkers who reported overall hangover severity and the presence and severity of individual hangover symptoms experienced in their last hangover in the past month, and rated this on a scale ranging from 0 (absent) to 10 (extreme). On this occasion, they reported consuming a mean (SD) of 12.6 (5.5) alcoholic drinks, corresponding to an estimated peak BAC of 0.19 (0.1) %. Figure 1 shows that both the frequency of occurrence and severity differed considerably between individual hangover symptoms. Most individual symptom scores are lower than the 1-item overall hangover severity score, suggesting that a symptom average score will underestimate overall hangover severity. As the three hangover scales comprise different hangover symptoms, it is understandable that the aggregate symptom scores of these scales differ from each other. Several symptoms, such as depression and anxiety (both low frequency/low severity), have a limited contribution to the aggregate scale score, whereas other symptoms, such as concentration problems and being tired (both high frequency/high severity symptoms), have a large contribution to the aggregate score. Including low frequency/low severity symptoms or excluding high frequency/high severity symptoms results in an underestimation of the "true" overall hangover severity. It is evident from Figure 1 that the HSS, especially, contains several low frequency/low presence items. Therefore, HSS scores likely underestimate the true hangover severity to a greater extent than the AHS and AHSS.

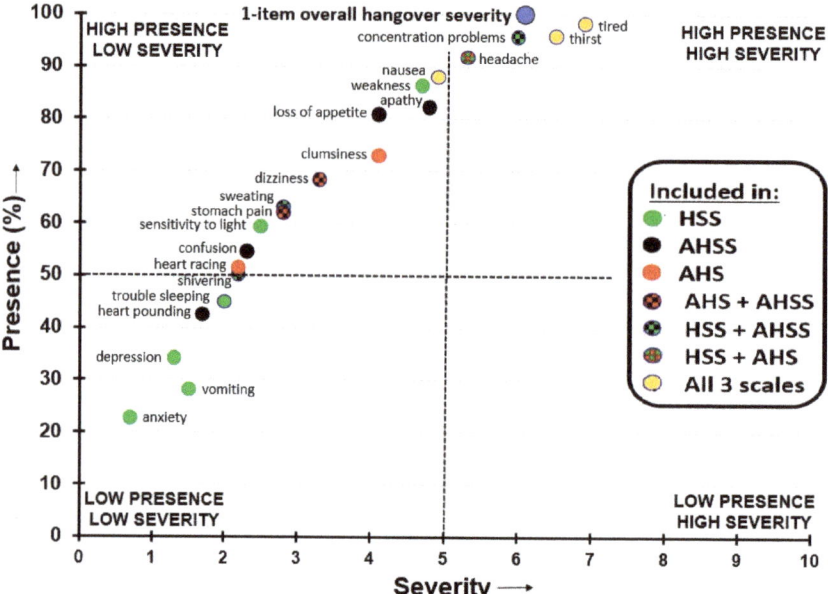

Figure 1. Presence and severity of symptoms included in hangover symptom scales. Data from $n = 1837$ social drinkers who reported on their latest past month hangover [4]. Note: "sensitivity to sound" was not assessed. Abbreviations: HSS = Hangover Symptoms Scale, AHSS = Alcohol Hangover Severity Scale, AHS = Acute Hangover Scale.

A similar variability in the presence and severity of hangover symptoms was recently reported by Van Lawick van Pabst et al. [18]. Omitting relevant items from a scale can have a significant impact on the overall rating of hangover severity. An example from the dataset of Van Schrojenstein Lantman is the item "sleepiness", which is not included in any of the three hangover scales. Sleepiness was reported by 97.1% of participants and its severity was rated as 6.5 out of 10 (extreme). It can be assumed that when completing a single item overall hangover severity item, the subject's rating is influenced by all symptoms and feelings the subject experiences during the hangover state. Therefore, aggregate scale scores of a limited number of symptoms are very likely to underestimate the true overall hangover severity.

5. Negative Impact of Hangover Symptoms

When judging overall hangover severity, it is likely that drinkers will take into account to what extent all experienced individual hangover symptoms negatively affect their cognitive functioning, physical activities, and mood. Symptoms with the largest negative impact on these domains are not necessarily those symptoms that have the highest severity scores. There is also no relationship between the impact symptoms may have and the relative frequency of occurrence in the overall drinking population. For example, heart racing can be a very disturbing effect and have a significant impact on mood. However, the symptom is not frequently reported. Alternatively, severity scores and the presence of ratings of being thirsty are usually high, while effects on cognitive functioning, physical activities, and mood are virtually absent.

Van Schrojenstein Lantman et al. [4] also examined the impact of experiencing hangover symptoms on cognitive and physical functioning, and mood in n = 1837 social drinkers who reported on their last hangover experience in the past month. Negative impact of hangover symptoms on cognitive and physical functioning, and mood was rated on scales ranging from 0 (absent) to 5 (extreme). The results are summarized in Figure 2.

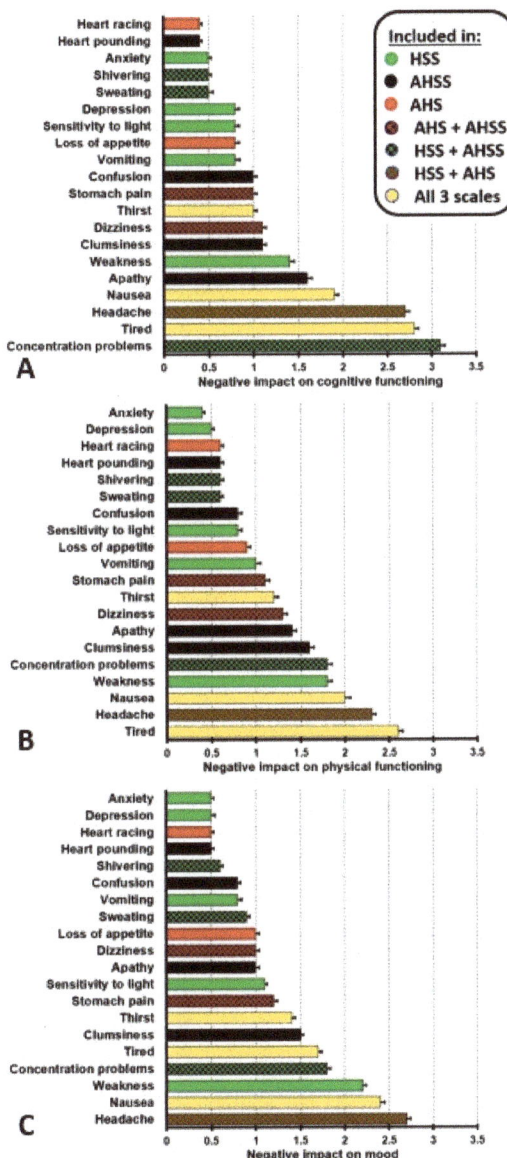

Figure 2. Impact of hangover symptoms on cognitive and physical functioning and mood. Negative impact of hangover symptoms cognitive functioning (**A**), physical functioning (**B**), and mood (**C**) was rated on scales ranging from 0 (absent) to 5 (extreme). Note: "sensitivity to sound" was not assessed and "trouble sleeping" was excluded as not being a true hangover symptom. Data from reference [4].

It is evident from Figure 2, that there are clear differences regarding the extent that hangover symptoms have an impact on cognition, physical activities, and mood. Therefore, the specific items that are included in a hangover symptom scale determine to what degree the true overall hangover effect is accurately reflected in an aggregate score (especially if no item weights are used during the formation of the composite score). As the hangover symptom scales do not include all imaginable hangover symptoms, while at the same time providing items that may not apply to a given participant, they will likely underestimate the "true" overall impact of hangover symptoms on cognition, physical activities, and mood. This can again be illustrated by considering the hangover symptom "sleepiness". Although not included in any of the three hangover symptom scales, Van Schrojenstein Lantman et al. [4] found that sleepiness was reported by 97.1% of drinkers. The mean (SD) impact scores for sleepiness were 2.7 (1.7) for cognitive functioning, 2.5 (1.7) for physical functioning, and 2.4 (1.6) for mood. If this symptom was incorporated in a scale, it would very likely have influenced the aggregate impact score.

6. Symptoms May also be Present without a Hangover or Alcohol Consumption

Hangover symptoms are also experienced when no alcohol is consumed. As a result, aggregate symptom scores may be greater than zero, even when no alcohol has been consumed. A recent study [19] compared hangover symptoms between subjects with and without a hangover and demonstrated that several symptoms are not unique to the hangover state but are also present without having a hangover or consuming alcohol. In this study, $n = 299$ subjects who were on holiday in Greece (mean (SD) age of 38.9 (11.0) years old) completed the AHS in the morning before walking the Samaria Gorge. $n = 47$ subjects consumed alcohol the evening before but reported having no hangover, $n = 176$ consumed alcohol and reported a hangover, and $n = 76$ consumed no alcohol and reported no hangover. Reported hangover symptoms from the three groups are depicted in Figure 3.

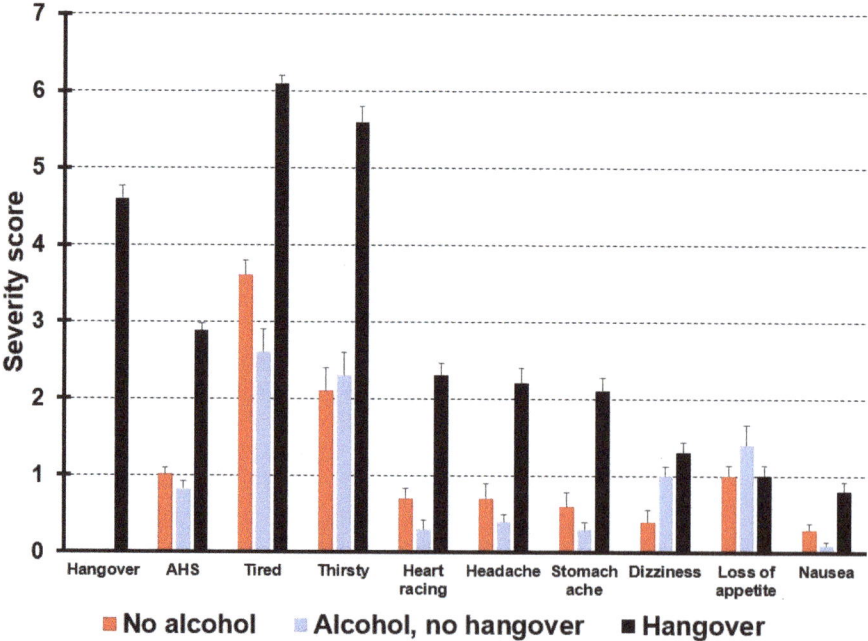

Figure 3. Presence and severity of symptoms related to alcohol hangover. Note: In contrast to the original AHS, scores range from 0 (absent) to 10 (extreme). Data from reference [19].

First of all, these data again demonstrate that the mean (SD) AHS score of 2.9 (1.3) significantly underestimated the true hangover severity of 4.6 (2.1) among subjects with a hangover ($p < 0.0001$). Statistical comparisons of the individual hangover symptom scores between subjects who consumed no alcohol, those who consumed alcohol but reported no hangover, and drinkers with a hangover revealed that no significant differences between the groups were found for nausea and loss of appetite. Severity scores for headache did not significantly differ between drinkers with and without a hangover. The data in Figure 3 suggest that most symptoms that are attributed to the hangover state are always present, irrespective of alcohol consumption or having a hangover. It can be assumed that, when rating overall hangover severity via a 1-item scale, drinkers take into account that some symptoms may already be present on non-drinking days as well. For example, they may usually feel somewhat tired (although perhaps to a lesser extent than during the hangover state). This knowledge is then incorporated in their rating of hangover severity, which is more likely to reflect the difference/changes in symptom severity relative to a normal non-drinking day. Although the latter cannot be proven with the data at hand, we deem it to be a plausible hypothesis. Yet, hangover scales aggregate symptom scores without taking baseline symptom scores into account. As a result, a positive AHS hangover severity score can be obtained in subjects who reported to have no hangover after drinking alcohol (0.9) as well as in subjects who did not consume alcohol at all (1.0). In fact, their 1-item overall hangover severity score was zero. When relying solely on AHS scores, it would incorrectly be assumed that these subjects had a hangover. In this study, 95% of subjects who report no hangover via the 1-item overall hangover severity rating had an AHS score greater than zero and would be incorrectly labelled as having a hangover. When including these subjects in the dataset for statistical analysis, their AHS scores are, however, higher than those assessed with a 1-item severity score (i.e., zero), meaning that the AHS score overestimates the true overall hangover severity, producing false positives. When taken together, the findings that the severity of a true hangover tends to be underestimated (due the fact that not all of the items usually apply), while severity tends to be overestimated in the absence of a hangover, it seems that composite scores might be worse than 1-item overall ratings in differentiating between individuals with severe versus light hangover symptoms, by producing a tendency towards the middle. While it could theoretically be possible to try to identify false positives by the ratio of single-scale scores to overall ratings (even though this currently still remains to be tested), it would likely be impractical in most cases to have participants fill in an entire questionnaire, when a single-item overall hangover rating already provides the required information to a good, if not even better, degree.

7. Day to Day Variability in the Presence and Severity of Hangover Symptoms

Van Wijk et al. [20] examined hangover severity of $n = 22$ students who were on a skiing holiday in Italy. The students experienced multiple hangovers during this period. Each morning at breakfast, subjects completed a modified AHS. The AHS included all nine symptoms, including a 1-item overall hangover severity score, but the severity scoring of items was modified to a range from 0 (absent) to 10 (extreme) In addition, past evening alcohol consumption was recorded and the level of subjective intoxication (i.e., drunkenness) was rated on a scale from 0 (absent) to 10 (extreme). For $n = 13$ subjects, it was possible to match two test days with identical hangover scores, as assessed with a 1-item overall hangover severity score. Several important observations were made when evaluating the data: First of all, the 1-item overall hangover severity score was different between subjects, but identical on the two test days (see Figure 4A). However, the AHS scores of the two test days showed considerable variability for some of the subjects (see Figure 4B), as the severity scores of individual hangover symptoms contributing to the aggregate AHS score differed between the two test days (see Figure 4C–J). In other words, despite having identical 1-item overall hangover severity scores, subjects reported considerable variability in individual symptom scores and overall AHS scores on the two test days. Finally, the data showed that having an identical 1-item overall hangover severity score does not necessarily imply that the same amount of alcohol was consumed, or that the corresponding level of reported intoxication was similar on the evening preceding the test day. Instead, subjects consumed different amounts of alcohol on both test days (see Figure 4K) and reported different levels of subjective intoxication (see Figure 4L). Notwithstanding this, their 1-item overall hangover severity scores on each test day were identical.

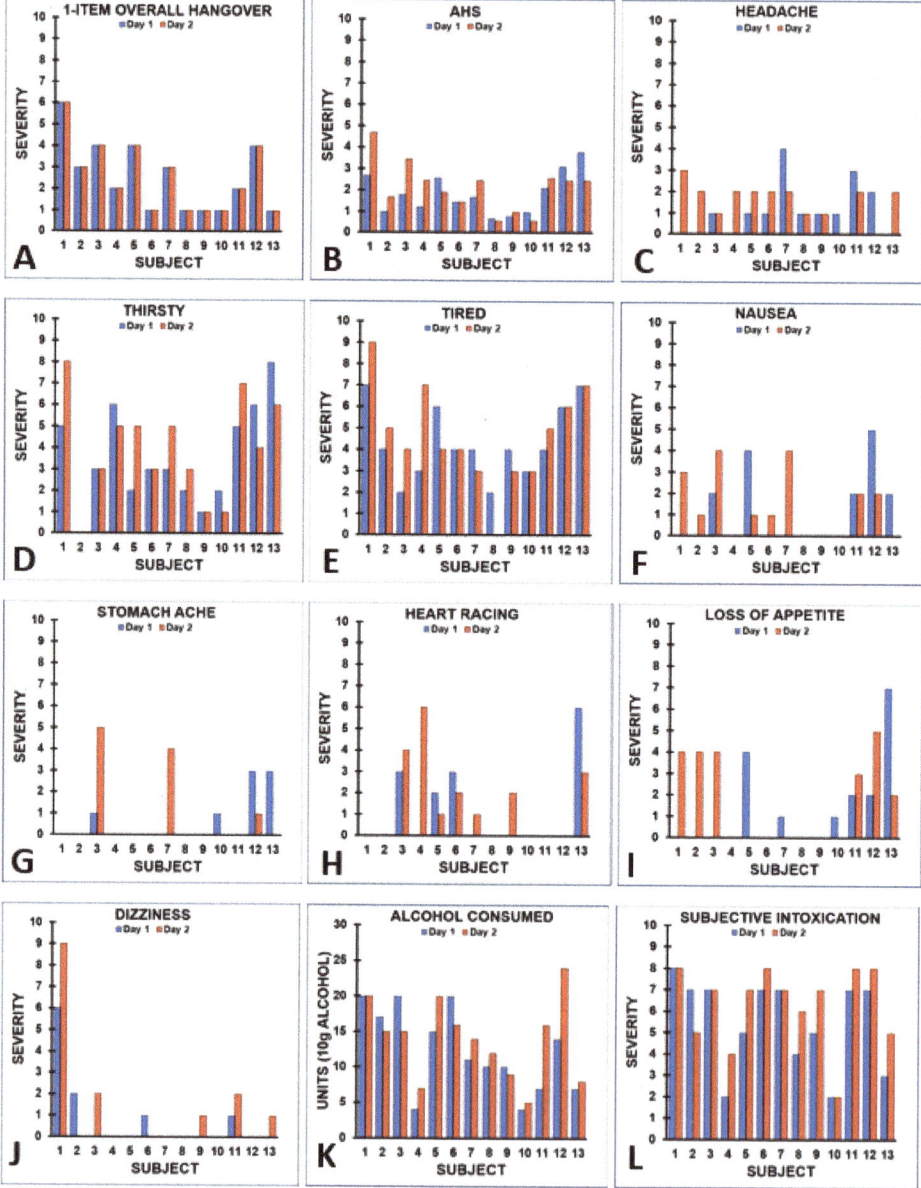

Figure 4. Level of subjective intoxication and alcohol consumption and corresponding next day hangover symptom severity reported on two different test days by the same subjects. Individual subject ratings for 1-item overall hangover severity (**A**), the AHS score (**B**), individual symptom scores (**C–J**), the amount of alcohol consumed the evening before having the hangover (**K**), and the corresponding level of subjective intoxication (**L**) are shown.

Figure 5 shows the test–retest reliability of the AHS and its items. While the test–retest reliability of the 1-item overall hangover severity score was 1.0 (maximal, as test days had been selected to fulfil this criterion), the AHS test–retest reliability (0.69) was below the generally acceptable level of test–retest reliability of 0.7 [21]. The variability in individual hangover severity scores was greatest for headache, thirst and nausea, and none of the symptoms reached the acceptable limit of 0.7 for test–retest reliability, except for dizziness. Applying the more stringent Bland-Altman 95% limits of agreement method [22]—in which 95% of difference scores of day 1 and day 2 item or scale ratings should lie within the range of two standard deviations of the mean difference score to demonstrate agreement between the two assessments—revealed that no agreement was found for the symptoms of headache, heart racing, and loss of appetite.

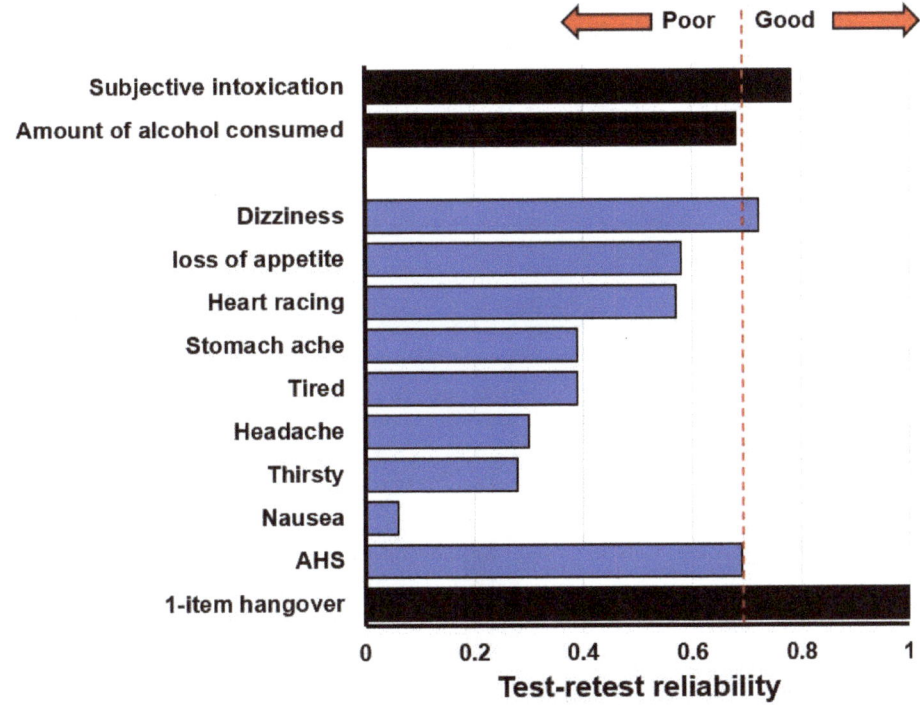

Figure 5. Test–retest reliability. Spearman's correlations are shown. Higher scores suggest a better test–retest reliability. Bootstrapping ($n = 10.000$ samples, bias corrected 95% confidence interval) was applied to adjust correlations for the small sample size. An acceptable test–retest reliability is demonstrated if Spearman's correlation > 0.7 [21].

An alternative way to look at the data is to select test days on which subjects consumed an identical amount of alcohol and then compare the presence and severity of AHS symptom scores. For $n = 18$ subjects, it was possible to match two test days with an identical amount of alcohol consumption. They consumed a mean (SD) of 11.6 (5.7) alcoholic drinks on these test days (range: 2 to 20 alcoholic drinks). Their AHS scores and individual symptom scores are summarized in Table 1 and Figure 6. Despite the fact that subjects consumed the same amount of alcohol on both test days, the data show considerable variability within subjects in both the presence and severity ratings on individual hangover symptoms, including the 1-item hangover severity score.

Table 1. AHS and symptom severity scores for two days on which an equal amount of alcohol was consumed by subjects.

Subject Number	Alcoholic Drinks	AHS Day 1	AHS Day 2	1-item HS Day1	1-item HS Day2
1	20	3.78	2.67	7	6
2	17	1.00	1.33	3	2
3	20	1.89	1.78	3	4
4	12	2.33	1.56	4	0
5	10	2.33	4.11	3	5
6	20	1.44	1.89	2	4
7	2	0.78	0.78	0	0
8	5	0.89	0.33	0	0
9	6	1.00	1.11	0	0
10	10	0.56	0.56	0	0
11	20	1.44	1.00	1	1
12	10	1.00	0.67	2	1
13	10	0.78	0.56	1	0
14	10	1.89	0.56	3	0
15	8	0.56	0.22	0	0
16	14	3.11	0.78	4	0
17	6	1.11	1.00	1	0
18	8	0.56	0.33	0	0

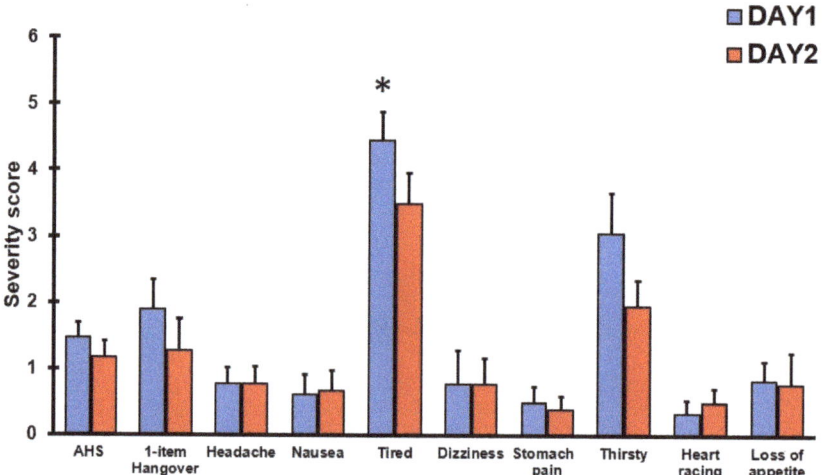

Figure 6. AHS and symptom severity scores for two days on which an equal amount of alcohol was consumed within subjects. Significant differences ($p < 0.05$) between the two days are indicated by an asterisk (*). Abbreviation: AHS = acute hangover scale.

Test-retest reliability for the AHS ($r = 0.731$, $p = 0.001$) was acceptable. With regard to individual symptoms, an acceptable test–retest reliability was, however, only found for the symptom of being tired ($r = 0.775$, $p < 0.0001$). No acceptable test–retest reliability was found for the ratings of overall hangover ($r = 0.537$, $p = 0.022$), and stomach pain ($r = 0.569$, $p = 0.014$). A poor test–retest reliability was found for being thirsty ($r = 0.365$, $p = 0.136$), dizziness ($r = 0.351$, $p = 0.153$), heart racing ($r = 0.240$, $p = 0.337$), headache ($r = 0.186$, $p = 0.460$), and nausea ($r = 0.090$, $p = 0.723$). The low test–retest reliabilities again confirm the fact that the presence and severity of hangover symptoms considerably varies between drinking occasions. In line with this, a recent study showed that there is great intraindividual variability in hangover severity scores between drinking occasions, even when the same amount of alcohol is consumed [23], and regression analyses demonstrated that the amount of consumed alcohol is usually not a strong predictor of hangover severity [24].

Table 1 further shows that the AHS scores on day 1 and day 2 are greater than zero for each subject. This could suggest that all of them experienced a hangover on both days. However, the 1-item overall hangover severity score demonstrated this to be incorrect, as six out of 18 subjects (33.3%) did not report having a hangover on day 1, and eleven subjects (61.1%) reported having no hangover on day 2. Taken together, relying solely on hangover symptom scales to assess the presence and severity of alcohol hangover will likely result in inaccurate results.

8. Should We Abandon the Use of Hangover Symptom Scales?

The fact that the outcomes of composite hangover scales do not appear to accurately reflect overall hangover severity does not imply that we should abandon their use altogether. In many cases, it is very relevant to assess the presence and severity of individual hangover symptoms. For example, if a company claims that treatment X is effective in reducing hangover headaches, it is highly relevant to assess "headache" severity, in addition to an assessment of overall hangover severity. As is evident from Section 2, there is a great variability in the presence and severity of individual hangover symptoms. It is also important to identify those symptoms that are most bothersome and impairing to subjects. For future research, it is therefore recommended to also assess individual hangover symptoms. This can be done by using one of the existing hangover scales, or simply by assessing individual symptoms of interest via symptom-specific 1-item severity scores. However, the judgement of the overall efficacy of a hangover treatment should preferably be based on a 1-item overall hangover severity rating, as this most likely incorporates all experienced symptoms and circumstances of the hangover state. As discussed in Section 5, symptoms experienced during hangover may also be experienced on non-hangover days. Therefore, is advisable to use difference scores for these individual symptoms when comparing their severity on a hangover day versus a no-hangover day to capture the "true" hangover effect. The latter does not apply for intervention studies, where a direct comparison of symptom scores between treatment and placebo should be made to evaluate a possible difference between the two hangover conditions.

Finally, it should be acknowledged that subjective ratings have sometimes been argued to be unreliable per se, thus mandating the assessment of biomarkers in order to have an objective assessment of hangover severity. Indeed, it would be very useful if such a biomarker would be discovered. Biomarkers related to alcohol metabolism and immune function may be promising candidates, but several lines of research have unfortunately not yet been able identified a suitable biomarker. Biomarkers can be assessed in various samples, including blood, saliva, hair, sweat, and stool, but, for practical use, volatiles in expired air would be ideal. However, the traditional breathalyzer readings reflecting ethanol concentrations are not useful, as BAC readings are often zero in the hangover state [25]. Alas, chemical compounds other than ethanol, which should ideally relate to hangover severity and/or functional impairments, would need to be detected by a breathalyzer in order to reliably indicate alcohol hangover. Notwithstanding this, the alcohol hangover is a complex state with many symptoms that may be experienced alone or in combination and may have differential severity and impact on daytime functioning. Given the currently available knowledge on the potential underlying mechanisms, it is still unclear whether a complex concept such as alcohol hangover can be accurately represented by a single biomarker. Further research investigating the suitability of potential volatile biomarkers in order to develop a breathalyzer for the hangover state is currently in progress.

9. Conclusions

Hangover symptoms can vary in their presence and severity between different drinking events in the same individual, and not all symptoms have equal impact in terms of impairment or being bothersome, regardless of their severity. Therefore, researchers should not rely solely on hangover symptom scales to assess overall hangover severity. Based on our reasoning, it is evident that composite, multi-item hangover symptom scales will likely underestimate severe hangovers and, at the same time, overestimate light hangovers, thus partly masking the true hangover effect. The resulting reduction in variance could hamper the ability to assess changes across study appointments or interventions and may thus have serious implications for the interpretation of study outcomes. Furthermore, the use of hangover symptom scales may also contribute to false positives, i.e., misidentifying subjects without a hangover as supposedly suffering from one. We propose that the most important, clinically meaningful endpoint for rating hangovers and the prevention or mitigation of hangovers by an effective product is the 1-item overall hangover severity score. This measurement allows the subject to assess the effect the condition is having on him or her, taking into account the symptoms being experienced, the severity of the symptoms being experienced and, most importantly, how the condition is impacting them in their activities of daily living and interactions with others, regardless of what the individual symptoms comprising his or her condition may be at the moment. The single greatest strength of the 1-item global assessment as a primary outcome measure is that it incorporates the subjects' evaluation of the impact the specific subset of symptoms being experienced at that time, in place of and with greater subject-focused information value than the specific symptom-based sum score can provide. Thus, this single-score approach evaluates the entire constellation of the hangover state, regardless of the individual components contributing to it, in terms of presence, severity, and impact.

Thus, the 1-item overall hangover severity rating represents a self-reported outcome instrument capable of measuring the severity of a condition (i.e., hangover) or the effect of a treatment in concordance with and incorporating all three concepts of an effective Patient-Reported Outcome Measure, namely, assessing the presence of symptoms, their effects on function, and their severity [13]. In addition, as secondary outcome measures (of efficacy), individual symptom presence and severity (or their impact) can be assessed using either hangover symptom scales, or individually. This will identify which individual symptoms are described by subjects as being the most bothersome during the alcohol hangover state.

Author Contributions: Conceptualization, J.C.V., A.J.A.E.v.d.L., S.B., A.S., A.-K.S., formal analysis, J.C.V.; writing—original draft preparation, J.C.V.; writing—review and editing, J.C.V., A.J.A.E.v.d.L., S.B., A.S., A.-K.S. All authors have read and agreed to the published version of the manuscript.

Funding: Open Access Funding by the Publication Fund of the TU Dresden.

Conflicts of Interest: S.B. has received funding from Red Bull GmbH, Kemin Foods, Sanofi Aventis, Phoenix Pharmaceutical and GlaxoSmithKline. Over the past 36 months, A.S. has held research grants from Abbott Nutrition, Arla Foods, Bayer, BioRevive, DuPont, Fonterra, Kemin Foods, Nestlé, Nutricia-Danone, Verdure Sciences. He has acted as a consultant/expert advisor to Bayer, Danone, Naturex, Nestlé, Pfizer, Sanofi, Sen-Jam Pharmaceutical, and has received travel/hospitality/speaker fees Bayer, Sanofi and Verdure Sciences. Over the past 36 months, J.C.V. has held grants from the Dutch Ministry of Infrastructure and the Environment, Janssen, Nutricia, and Sequential, and acted as a consultant/expert advisor to Clinilabs, Morelabs, Red Bull, Sen-Jam Pharmaceutical, Toast!, and ZBiotics. A.K.S. has received funding from the Daimler and Benz Foundation. A.J.A.E.V.D.L. has no conflicts of interest to declare.

References

1. Van Schrojenstein Lantman, M.; van de Loo, A.J.; Mackus, M.; Verster, J.C. Development of a definition for the alcohol hangover: Consumer descriptions and expert consensus. *Curr. Drug Abus. Rev.* **2016**, *9*, 148–154. [CrossRef] [PubMed]
2. Verster, J.C.; Scholey, A.; Stock, A.K. Updating the definition of the alcohol hangover.
3. Penning, R.; McKinney, A.; Verster, J.C. Alcohol hangover symptoms and their contribution to overall hangover severity. *Alcohol Alcohol.* **2012**, *47*, 248–252. [CrossRef] [PubMed]
4. Van Schrojenstein Lantman, M.; Mackus, M.; van de Loo, A.J.A.E.; Verster, J.C. The impact of alcohol hangover symptoms on cognitive and physical functioning, and mood. *Hum. Psychopharmacol.* **2017**, *32*, 2623. [CrossRef] [PubMed]
5. Gunn, C.; Mackus, M.; Griffin, C.; Munafò, M.R.; Adams, S. A systematic review of the next-day effects of heavy alcohol consumption on cognitive performance. *Addiction* **2018**, *113*, 2182–2193. [CrossRef]

6. McKinney, A. A review of the next day effects of alcohol on subjective mood ratings. *Curr. Drug Abus. Rev.* **2010**, *3*, 88–91. [CrossRef]
7. Frone, M.R. Employee psychoactive substance involvement: Historical context, key findings, and future directions. *Annu. Rev. Organ. Psychol. Organ. Behav.* **2019**, *6*, 273–297. [CrossRef]
8. Verster, J.C. Alcohol hangover effects on driving and flying. *Int. J. Disabil. Hum. Dev.* **2007**, *6*, 361–367. [CrossRef]
9. Verster, J.C.; Bervoets, A.C.; de Klerk, S.; Vreman, R.A.; Olivier, B.; Roth, T.; Brookhuis, K.A. Effects of alcohol hangover on simulated highway driving performance. *Psychopharmacology* **2014**, *231*, 2999–3008. [CrossRef]
10. Verster, J.C.; van der Maarel, M.; McKinney, A.; Olivier, B.; de Haan, L. Driving during alcohol hangover among Dutch professional truck drivers. *Traffic Inj. Prev.* **2014**, *15*, 434–438. [CrossRef]
11. Sacks, J.J.; Gonzales, K.R.; Bouchery, E.E.; Tomedi, L.E.; Brewer, R.D. 2010 National and State Costs of Excessive Alcohol Consumption. *Am. J. Prev. Med.* **2015**, *49*, 73–79. [CrossRef]
12. Bhattacharya, A. *Financial Headache. The Cost of Workplace Hangovers and Intoxication to the UK Economy*; Institute of Alcohol Studies: London, UK, 2019.
13. US FDA. Guidance for Industry. Patient-Reported Outcome Measures: Use in Medical Product Development to Support Labeling Claims. December 2009. Available online: https://www.fda.gov/media/77832/download (accessed on 16 February 2020).
14. Slutske, W.S.; Piasecki, T.M.; Hunt-Carter, E.E. Development and initial validation of the Hangover Symptoms Scale: Prevalence and correlates of Hangover Symptoms in college students. *Alcohol. Clin. Exp. Res.* **2003**, *27*, 1442–1450. [CrossRef] [PubMed]
15. Rohsenow, D.J.; Howland, J.; Minsky, S.J.; Greece, J.; Almeida, A.; Roehrs, T.A. The acute hangover scale: A new measure of immediate hangover symptoms. *Addict. Behav.* **2007**, *32*, 1314–1320. [CrossRef] [PubMed]
16. Penning, R.; McKinney, A.; Bus, L.D.; Olivier, B.; Slot, K.; Verster, J.C. Measurement of alcohol hangover severity: Development of the Alcohol Hangover Severity Scale (AHSS). *Psychopharmacology* **2013**, *225*, 803–810. [CrossRef] [PubMed]
17. Hogewoning, A.; Van de Loo, A.J.A.E.; Mackus, M.; Raasveld, S.J.; De Zeeuw, R.; Bosma, E.R.; Bouwmeester, N.H.; Brookhuis, K.A.; Garssen, J.; Verster, J.C. Characteristics of social drinkers with and without a hangover after heavy alcohol consumption. *Subst. Abus. Rehabil.* **2016**, *7*, 161–167. [CrossRef]
18. Van Lawick van Pabst, A.E.; Devenney, L.E.; Verster, J.C. Sex differences in the presence and severity of alcohol hangover symptoms. *J. Clin. Med.* **2019**, *8*, 867. [CrossRef]
19. Verster, J.C.; Anogeianaki, A.; Kruisselbrink, L.D.; Alford, C.; Stock, A.K. Relationship of alcohol hangover and physical endurance performance: Walking the Samaria Gorge. *J. Clin. Med.* **2020**, *9*, 114. [CrossRef]
20. Palmer, E.; Arnoldy, L.; Ayre, E.; Benson, S.; Balikji, S.; Bruce, G.; Chen, F.; van Lawick van Pabst, A.E.; van de Loo, A.J.A.E.; O'Neill, S.; et al. Proceedings of the 11th Alcohol Hangover Research Group meeting in Nadi, Fiji. *Proceedings* **2020**, *43*, 1. [CrossRef]
21. Terwee, C.B.; Bot, S.D.; de Boer, M.R.; van der Windt, D.A.; Knol, D.L.; Dekker, J.; Bouter, L.M.; de Vet, H.C. Quality criteria were proposed for measurement properties of health status questionnaires. *J. Clin. Epidemiol.* **2007**, *60*, 34–42. [CrossRef]
22. Bland, J.M.; Altman, D.G. Measuring agreement in method comparison studies. *Stat. Methods Med. Res.* **1999**, *8*, 135–160. [CrossRef]
23. Hensel, K.O.; Longmire, M.R.; Köchling, J. Should population-based research steer individual health decisions? *Aging* **2019**, *11*, 9231–9233. [CrossRef]
24. Verster, J.C.; Kruisselbrink, L.D.; Slot, K.A.; Anogeianaki, A.; Adams, S.; Alford, C.; Arnoldy, L.; Ayre, E.; Balikji, S.; Benson, S.; et al. Sensitivity to experiencing alcohol hangovers: Reconsideration of the 0.11% blood alcohol concentration (BAC) threshold for having a hangover. *J. Clin. Med.* **2020**, *9*, 179. [CrossRef] [PubMed]
25. Verster, J.C.; Mackus, M.; van de Loo, A.J.A.E.; Garssen, J.; Scholey, A. The breathtaking truth about breath alcohol readings of zero. *Addict. Behav.* **2017**, *70*, 23–26. [CrossRef] [PubMed]

© 2020 by the authors. Licensee MDPI, Basel, Switzerland. This article is an open access article distributed under the terms and conditions of the Creative Commons Attribution (CC BY) license (http://creativecommons.org/licenses/by/4.0/).

Article

Prevalence of Hangover Resistance According to Two Methods for Calculating Estimated Blood Alcohol Concentration (eBAC)

Chantal Terpstra [1,2], Andrew Scholey [1], Joris C. Verster [1,2] and Sarah Benson [1,*]

1   Centre for Human Psychopharmacology, Swinburne University, Melbourne, VIC 3122, Australia; cterpstra@swin.edu.au (C.T.); andrew@scholeylab.com (A.S.); J.C.Verster@uu.nl (J.C.V.)
2   Division of Pharmacology, Utrecht Institute for Pharmaceutical Sciences (UIPS), Utrecht University, 3584 CG Utrecht, The Netherlands
*   Correspondence: sarahbenson@swin.edu.au; Tel.: +61-9214-5212

Received: 31 May 2020; Accepted: 27 August 2020; Published: 31 August 2020

Abstract: Hangover resistance may be linked to an increased risk of continuing harmful drinking behaviours as well as involvement in potentially dangerous daily activities such as driving while hungover, mainly due to the absence of negative consequences (i.e., hangover symptoms) the day after alcohol consumption. The aim of this study was to examine the occurrence of claimed alcohol hangover resistance relative to estimated blood alcohol concentration (eBAC). A total of 1198 participants completed an online survey by answering questions regarding their demographics, alcohol consumption and occurrence of hangover. Two methods were used to calculate eBAC, one based on the modified Widmark Equation (N = 955) and the other from an equation averaging the total body water (TBW) estimates of Forrest, Watson, Seidl, Widmark and Ulrich (males only) (N = 942). The percentage of participants who claimed to be hangover resistant decreased rapidly with increasing eBAC and only a small number of hangover resistant drinkers remained at higher eBACs. Comparisons of the eBACs calculated by the two methods revealed significantly higher BACs when using the modified Widmark equation. These findings suggest that additional research for eBAC calculations is needed to improve accuracy and comprehensiveness of these equations for future alcohol hangover research.

Keywords: hangover; alcohol; BAC; hangover resistance

1. Introduction

Alcohol hangovers are often the result of a night of heavy drinking and are a familiar phenomenon worldwide. The alcohol hangover refers to the combination of negative mental and physical symptoms, which can be experienced after a single episode of alcohol consumption, starting when blood alcohol concentration (BAC) approaches zero [1]. Historical definitions suggested that a blood alcohol concentration level equivalent to 0.11% was required to experience a next day hangover, with a greater percentage of drinkers reporting hangovers at higher BAC levels [2]. However, recent research shows that a considerable number of participants also reported hangover occurrence at moderately lower BAC levels [3]. The most reported hangover symptoms typically include fatigue [4,5], thirst [5], drowsiness [5], headache [5,6], dry mouth [5], nausea [7], reduced alertness [5], weakness [5], and concentration problems [5,8,9]. Other symptoms include vomiting, regret, heart racing, and apathy [5,10].

A hangover can result in impairment in both cognitive and psychomotor functioning [11]. This potential impairment can have repercussions when it comes to daily activities, such as driving a car, studying, or working [12]. A hangover can result in reduced productivity as individuals may be more likely to call in sick for work or study, or work at a reduced level of productivity. Employees with an alcohol hangover reported having significantly more conflicts with supervisors and co-workers, feeling miserable, or falling asleep on the job [13,14]. According to a large survey in The Netherlands, University students are less productive, and half of the students reported not being able to study while hungover. On average, students experience a hangover 2.7 days per month, which adds up to a total study loss of one month per year [15].

Because of the substantial impact of alcohol hangovers on health, economy, and society, it is important to determine the causes of alcohol hangover as well as factors that contribute to symptom severity. Recent research showed that hangover-resistant drinkers may be at increased risk of continuing harmful drinking behaviours as well as involvement in potentially dangerous daily activities, such as driving, mainly because of absence of negative consequences (i.e., severe hangover symptoms) the day after drinking [16,17]. Lower perceptions of hangover severity (thus experiencing less severe hangover symptoms) are associated with stronger beliefs that it is safe to drive the morning after drinking [18]. Despite the belief that one is safe to drive, research showed significant impairment in driving abilities while experiencing a hangover [19].

Identifying how hangover-resistant drinkers differ from non-hangover resistant drinkers may help to unfold the pathology of alcohol hangovers. Thus far, several theories exist as to why hangovers occur and what factors might influence severity, but this phenomenon remains largely unclear [20]. Previous research showed that the presence and severity of alcohol hangovers can differ between and within drinkers since the alcohol hangover is influenced by several other factors other than the amount of alcohol consumed. These factors include, but are not limited to, gender, age, personality, genetics, and health-related behaviours such as smoking, illicit drug use, sleep quality and duration as well as the type of drink consumed [21].

Quantity of alcohol consumed directly impacts the severity of hangovers and associated impairments [22]. As such, it is essential that studies assessing the effects of hangover collect and report data describing alcohol intake. Objective measures of alcohol concentration—such as those collected in blood, breath, saliva, and urine samples—can be obtained relatively easily in laboratory studies. However, when using a naturalistic design this is seldom accessible, as typically the researchers are not present during the drinking session. The naturalistic design in alcohol hangover research describes a methodology where participants consume alcohol on a typical night out and attend the laboratory for testing the following morning. The results are then compared to a non-hangover visit. Participants are questioned regarding their alcohol consumption from the night before, after which estimated blood alcohol concentration (eBAC) is calculated. While forgoing some of the experimental control inherent in laboratory testing, the naturalistic design allows an environment of less interference and free alcohol consumption compared to the typically smaller doses of alcohol approved for controlled studies [23]. Naturalistic studies are likely the most optimum to study hangovers, but it is also necessary to control several intervening factors that complicate the comparison of results among hangover studies. For example, in naturalistic designs, it is difficult to determine when BAC returns to 0. As alcohol hangover commonly utilizes naturalistic designs to ensure ecological validity, it is imperative that methodology to assess (naturalistic) hangovers adapts and develops according to existing information about hangovers [24]. Of relevance to the current study is the fact that the data needed to reconstruct BAC typically involves the number of alcoholic drinks consumed, duration of alcohol consumption, along with morphometric and demographic information such as gender, age, height, and weight [25].

The most widely used eBAC calculation was developed by Widmark [26,27]. This equation is dependent on the amount of alcohol consumed, the relative body water (set value for males (0.68) and females (0.55)), body mass (weight in kg), degradation rate in g/l (0.15), duration of alcohol consumption, and a set value for absorption time of 0.5 h. Please see Appendix A for additional information.

This equation has been modified as research has shown that the original Widmark formula had a tendency to overestimate BAC [25]. Specifically, the modified formula optimised the calculation of total amount of body water and included height and age [28]. The updated equations are found to be reasonably accurate for eBACs under controlled laboratory environments [28–30] and are frequently used in present alcohol hangover research to estimate BACs [3,31–33]. Please refer to Appendix B for additional information.

However, the use of eBAC has limitations when applied to real world drinking experiences. It is likely to introduce more variability of BAC and the accuracy of eBAC equations in naturalistic settings is relatively unknown [25]. Additionally, when reliant on self-report data from intoxicated participants, or retrospectively when remembering intoxication, accuracy of eBAC could be compromised due to poor recall [34]. Furthermore, eBAC calculations based on self-report data are less accurate when estimating higher BACs [35]. Therefore, it is important to compare eBAC calculations to examine whether different calculations are consistent or whether they may differentially affect results.

Approximately 20–25% of alcohol consumers are classed as 'hangover resistant', in that they report no hangover symptoms after a night of heavy drinking [36]. Previous research shows that hangover resistance is tightly coupled to eBAC levels [37,38]. Clearly, the number of hangover resistance claims is highly dependent on the amount of alcohol consumed; as the more drinks are consumed (and the higher the eBAC), the less likely drinkers are to claim hangover resistance. For example, with an eBAC level above 0.20%, only 8.1% of alcohol consumers claim hangover resistance [37].

Another factor that needs to be considered when investigating hangover resistance is the definition that is being used. A Canadian study used two definitions to describe hangover resistance. The first definition stated 'Never having had a hangover during one's lifetime' and the second definition stated 'Never having had a hangover during a certain time period (e.g., 1, 3, 5 years)' [39]. A recent study showed that lifetime hangover resistance amongst heavier drinkers (eBAC 80 mg/dL, heaviest drinking occasion of the past month) occurs in 5–6% of participants [38]. Among people who claim hangover resistance, the symptoms reported by this group of heavy drinkers were limited to sleepiness and tiredness which were rated as less severe than the same items in the hangover sensitive group [40]. Even though hangover resistance seems to be rare and decreases with eBAC, a small number of people persist in claiming hangover resistance.

Due to the frequent use of eBAC in alcohol hangover research, it is vital to explore accuracy, comprehensiveness, and possible improvements for currently used formulas. This study aimed to assess whether the previously discussed calculations differ and explores the use of an additional and more comprehensive eBAC evaluation when assessing hangover resistance to explore similarities and differences in eBAC outcomes within an Australian population when compared with a previously used calculation in international research. This additional and more comprehensive method calculates eBAC by averaging the total body water (TBW) estimates of Forrest [41], Watson [28], Seidl [42], Widmark [26], and Ulrich [43] (males only) calculations. The mean TBW was then used in the following eBAC formula

$$BAC = (G/(TBW)) - \beta \times t \quad (1)$$

where G is the amount of alcohol consumed in grams; β is the metabolic rate in grams per hour; and t is time in hours [44].

In line with previous research, it was predicted that around 50% of the participants would have relatively low eBACs (e.g., <0.08%) [37]. It was further predicted that higher eBACs will lead to a decrease of hangover resistance claims. Despite the decrease of hangover resistance prevalence with higher eBACs, we hypothesised that a small number of hangover resistance would persist. Additionally, the primary aim of the study was to evaluate, for the first time, the use of a more comprehensive eBAC methodology in the context of hangover resistance, and in particular to examine whether it produced different values to previous research.

2. Experimental Section

2.1. Method

The study was approved by the Swinburne University Human Research Ethics Committee (Reference 2012/045) and was conducted in accordance with the Declaration of Helsinki.

2.2. Design

This was an online survey assessing alcohol consumption behaviours and hangover resistance occurrence amongst an Australian population with the use of a commonly used method to calculate eBAC as well as an additional and more comprehensive calculation for eBAC.

2.3. Participants

Overall, 1748 respondents opened the online anonymous questionnaire. Data provided by non-drinkers were removed from the dataset. In total, 1198 participants completed the online survey (N = 444 male and N = 754 female), mostly consisting of students (65.3%). The mean age of the participants was 23.10 years old (SD = 4.68, range = 18–40 years old).

2.4. Measures

2.4.1. Demographics

Participants answered questions regarding demographics (gender, age, weight, height), use of medications, tobacco, and illicit drugs.

2.4.2. Alcohol Consumption Questions

Participants were questioned on their alcohol intake (e.g., in the past 30 days and the past 12 months). Alcohol consumption was defined using standardized Australian alcohol units (one standard drink = 10 g of pure alcohol). The consumption questions assessed frequency and quantity of alcohol consumed across various timescales (i.e., one occasion, 30 days, and 12 months) for each type of drink (beer, wine, spirits, and alcohol mixed with energy drink). Specifically, participants were asked the number of standard drinks and the number of hours that they had spent drinking.

2.4.3. Estimated BAC (eBAC)

Responses to alcohol consumption questions regarding the heaviest drinking session within the previous 30 days were used to calculate eBAC with two different methods. Both methods computed the eBAC separately for males and females. Method 1 is the most frequently used method in previous research and is a modified Widmark equation [26]. The modified Widmark equation considers the number of alcoholic drinks, relative body water volume, weight, gender, and time needed to clear alcohol through metabolism [27], please see Appendix B for additional information [26]. Method 2 calculates eBAC by averaging the total body water (TBW) estimates of Forrest [41], Watson [28] Seidl [42], Widmark [26], and Ulrich [43] (males only) calculations. The mean TBW was then used in the following eBAC formula (1) in Section 1.

2.4.4. Single Item Hangover Sensitivity/Resistance

Two groups were created, depending on the answer to the question: "I have had a hangover (headache, sick stomach) the morning after I had been drinking" over the past year. If the answer was 'yes', the participant was assigned to the hangover sensitive group. If the answer was 'no', the participant was assigned to the hangover resistant group.

2.5. Procedure

Participants were recruited via word of mouth, flyers, and advertisements on social media. Participation was voluntary and anonymous. Participants provided informed consent by agreeing to the survey terms online, after which they completed the online questionnaire. On completion of the survey, participants could choose to go into the draw to win one of two iPads by entering their email address.

2.6. Statistical Analysis

Data were collected online using SurveyMonkey and analysed using the Statistical Package for the Social Sciences version 25 (SPSS Inc., Chicago, IL, USA). Initially, the data were screened for any participants who did not meet the criteria, i.e., surveys completed between 2:00 a.m. and 7:00 a.m., and surveys submitted from the same IP address. Finally, any participant who answered 'no' to the final question, whether they had honestly and correctly answered all questions, were excluded. The mean, standard deviation and frequency distributions were calculated for the hangover sensitive group and the hangover resistant group. Comparisons between the two methods to calculate eBAC were made using nonparametric tests, since the eBAC data were skewed. Pearson's Chi square test was used to measure associations between the different eBAC levels of the two methods and hangover resistance. Effects were regarded as statistically significant if $p < 0.05$.

3. Results

Two methods were used to calculate eBAC. First, the most commonly used method in previous alcohol hangover research was used and described, Method 1 [26,27]. Second, a more comprehensive method based on TBW was calculated (Method 2). Since results differed slightly depending on the eBAC method used, data are described separately.

Participants aged 36 years and older were excluded, since this question was assessed in a multiple-choice format and the exact age was needed in the eBAC calculation. This resulted in the removal of 100 participants, leaving 1098 participants. Participants with an estimated eBAC of 0.00% were excluded in both methods used because this suggests either no or extremely low alcohol consumption. Participants with an estimated eBAC of 0.40% and higher were also excluded in both methods, as this is an unreliably high estimation which suggests overestimation of the alcohol consumption and thus eBAC. This resulted in 20.3% of participant exclusion in Method 1 (leaving 955/1198) and 21.4% in Method 2 (leaving 942/1198). A general description of the typical drinking pattern is provided in Table 1.

Table 1. Participant morphometrics, demographics, and drinking characteristics. Numbers are means with standard deviations in parentheses. * $p < 0.05$ between hangover resistant and hangover sensitive group of participants.

	Full Sample	Hangover Resistant	Hangover Sensitive
Age	23.10(4.7)	22.95(5.1)	23.14(4.6)
Height (m)	1.71(0.1)	1.69(0.1) *	1.71(0.1) *
Weight (kg)	70.57(15.6)	69.93(17.7)	70.77(15.0)
Standard drinks per occasion	5.25(3.9)	3.40(2.6) *	5.83(4.1) *
Days used alcohol (30 days)	6.97(6.7)	4.93(6.1) *	7.60(6.8) *
Days drunk (30 days)	2.43(3.5)	0.95(2.4) *	2.90(3.6) *
Days binge (30 days)	3.22(4.3)	1.50(3.5) *	3.76(4.4) *
Greatest number of drinks (30 days)	7.87(6.1)	4.72(4.8) *	8.85(6.1) *
Consumption duration (hours)	5.16(3.5)	3.68(3.0) *	5.62(3.5) *
Alcohol consumed (grams)	78.67(61.1)	47.23(48.1) *	88.54(61.4) *
eBAC (Method 1) (N = 955)	0.12(0.1)	0.08(0.1) *	0.13(0.1) *
eBAC (Method 2) (N= 942)	0.10(0.1)	0.07(0.1) *	0.11(0.1) *

3.1. Method 1

Of N = 955 remaining participants, 196 (20.5%) reported no hangover the day after drinking while 759 (79.5%) reported experiencing a hangover.

Consistent with previous research [37], participants were divided into seven groups, based on eBAC, to examine the occurrence of reported hangover resistance the day after drinking (Table 2). These eBAC values include those corresponding to the two most common internationally recognized legal driving limits (i.e., 0.05% and 0.08%) and the highest eBAC value (i.e., 0.20% and higher) as described in previous research [37].

Table 2. Percentage of participants who report no hangover symptoms following alcohol consumption across various eBAC ranges for Method 1 and Method 2.

	eBAC Range	% No Hangover Method 1	% No Hangover Method 2	χ^2	p
Group 1	$0 \le$ BAC $< 0.02\%$	48.0 (36/75)	36.1 (53/147)	2.95	0.09
Group 2	$0.02\% \le$ BAC $< 0.05\%$	37.1 (62/167)	29.1 (53/182)	2.53	0.11
Group 3	$0.05\% \le$ BAC $< 0.08\%$	22.6 (33/146)	21.9 (30/137)	0.02	0.89
Group 4	$0.08\% \le$ BAC $< 0.11\%$	17.9 (22/123)	10.8 (13/120)	2.45	0.12
Group 5	$0.11 \le$ BAC $< 0.20\%$	8.0 (22/275)	8.9 (22/246)	0.15	0.70
Group 6	$0.20 \le$ BAC $< 0.30\%$	11.5 (15/131)	15.7 (14/89)	0.85	0.36
Group 7	$0.30 \le$ BAC $< 0.40\%$	16.2 (6/37)	9.5 (2/21)	0.50	0.48

Note. This table also shows results of a chi-square test of independence to investigate associations between the two methods and hangover resistance.

The data show that 39.9 of drinkers fell within an eBAC below 0.08%. When considering only the subset of drinkers with eBACs above 0.08% the proportion of this subset who claimed hangover resistance was 11.5% (65/567). When the smaller subset of drinkers with eBACs above 0.20% was considered, the prevalence of hangover resistance was 12.5% (21/168), respectively.

3.2. Method 2

Of N = 942 remaining participants, 19.9% reported no hangover symptoms the day after drinking (N = 187) and 755 (80.1%) reported experiencing hangover symptoms. The participants were divided into seven groups to examine the occurrence of hangover resistance the day after drinking within different eBAC ranges, which are shown in Table 2. This is consistent with previous research [37] to be able to compare our findings. These eBAC values correspond to the two most common internationally recognized legal driving limits (i.e., 0.05% and 0.08%) and the highest eBAC value (i.e., 0.20% and higher) [37]. The frequency distributions for hangover resistant drinkers across the eBAC continuum of Method 2 are also shown in Figure 1.

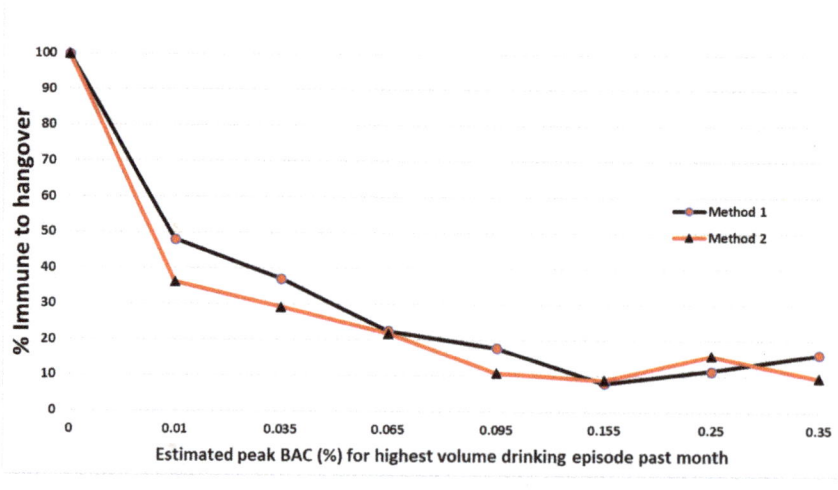

Figure 1. Prevalence of hangover resistance after alcohol consumption at different eBAC levels for Method 1 [26] and Method 2 [44]. $p < 0.05$ with Chi Square test of independence.

The data shows that 49.4% of drinkers fell within an eBAC below 0.08%. When considering only the subset of drinkers with eBACs above 0.08% the proportion of this subset who claimed no next day adverse effects after drinking was 10.7% (51/476). When the smaller subset of drinkers with eBACs above 0.20% was considered, the prevalence of no adverse effects the day after drinking was 14.5% (16/110), respectively.

3.3. Comparison of Two Methods

eBAC data was skewed for both eBAC methods, since most people are distributed along the lower eBAC ranges. Therefore, a Wilcoxon signed-rank test, which evaluated the difference between medians for eBAC Method 1 and eBAC Method 2 was conducted. The results indicated that eBAC levels of Method 1 were statistically significantly higher than eBAC levels of Method 2, $z = -11.07$, $p < 0.001$.

A chi-square test of independence showed that there were no significant differences in reported hangover resistance in the two eBAC methods and across the 7 BAC groups [45]. Results of the chi-square test can be found in Table 2.

4. Discussion

The current study assessed the occurrence of hangover resistance amongst an Australian population and compared two eBAC equations. Several of our hypotheses were supported. Both eBAC methods showed that almost half of the drinkers were distributed at eBAC levels below 0.08% (Method 1: 39.9% and Method 2: 49.4%), which supports the findings of previous research [37,38]. Previous research found that only a small percentage of 8.1% of participants who did not have a hangover in the past year were distributed at an eBAC level of 0.20% and higher [37]. A Canadian study found an even lower level of 4–5% of hangover resistance at an eBAC level of 0.20% and higher [38].

The present study confirms that the percentage of people who claim to be hangover resistant decreases with increasing eBACs, independent of eBAC method. However, contrary to previous research, a slight increase of hangover resistance claims occurred above 0.20% in the present study for both eBAC methods. This is likely a chance fluctuation due to smaller numbers of participants claiming hangover resistance at higher eBACs [36,37,40].

When comparing the two methods to calculate eBAC, several similarities were found. Both methods were able to show a decrease in hangover resistance prevalence with higher eBACs. Both methods showed that between 22–25% of the drinkers reported no hangover, however almost half of these drinkers were distributed at eBAC levels below 0.08%, which is a threshold to determine impaired driving in several countries [37].

The differences between the two methods are found in the mean eBACs, with a significantly higher mean eBAC for Method 1 than for Method 2. Interestingly, hangover resistance prevalence in several eBAC increments was observed to be slightly, but not significantly, lower with the use of Method 2. An explanation for this could possibly be found in the notion that previous research found that Method 1 (based on a modified Widmark formula) [26,27] might lead to an overestimation of eBAC when used in naturalistic settings with the help of self-report measures [29].

Differences found between the two eBAC methods used in this study suggest that additional research exploring the accuracy and comprehensiveness of eBAC equations is needed for future (naturalistic) hangover research. As illustrated earlier, previous alcohol hangover research most commonly uses the modified Widmark calculations for BAC [3,31–33]. Previous research suggests that higher levels of BAC (0.11%) were needed to develop a next day hangover. However, a recent study suggests that hangover prevalence and severity occur at lower BACs [3]. This may mean that previous alcohol hangover research overestimated BACs and that hangover occurrence and severity occur at lower BACs. Interestingly, a significant difference was found between the hangover resistant and hangover sensitive group on all questions related to alcohol consumption with numerically higher means in the hangover sensitive group. This suggests that even though hangover occurs at lower

BACs [3], BAC is a relevant indicator for hangover occurrence. Based on this study only, there appear to be no clinically relevant differences between the two methods of eBAC calculation.

Limitations of this study include relying on self-reported alcohol intake and eBAC calculations to estimate BAC obtained during the drinking episode. Heavy drinking often results in memory impairment, which may lead to less accuracy and recall-bias [46]. This limitation is often observed in (semi) naturalistic hangover research. However, it can be argued that the naturalistic design of this study relates to a representative alcohol consumption in real life situations.

Objective measures of ethanol assessments are absent in this study, which could pose the question whether the calculations for both BAC methods are accurate. It is unclear whether similar results would be found in controlled settings based on these study findings. However, the main purpose of this study was to assess whether the two methods to calculate eBAC differ rather than assess the accuracy of the two methods. An objective measure of ethanol assessment, such as blood and saliva, should be utilized for future research purposes. A controlled setting would allow equal quantity of alcohol consumption in both groups of drinkers. This ensures valid comparison of hangover-resistant and non-hangover-resistant drinkers in a controlled setting, which could then be translated to naturalistic designs. Including measures on hangover frequency and severity in future research will enhance our understanding of differences between hangover-resistant and non-hangover resistant drinkers.

5. Conclusions

The findings presented here show that almost half of drinkers had an eBAC below 0.08%, irrespective of applying Method 1 or Method 2 to calculate estimated BAC. The present results support previous findings and show that the percentage of people who claim to be hangover resistant decreases rapidly with increasing eBACs. However, a small number of hangover resistant drinkers persists, independently of higher eBACs. Differences between the two eBAC methods are found in the mean eBACs, with a significantly higher mean eBAC for Method 1. Future research should assess influences of eBAC on hangover resistance prevalence with the use of an objective and more extensive assessment of hangover (severity) and should further explore research on eBAC calculations to improve their accuracy and comprehensiveness for future research. Notwithstanding the absence of significant findings, this study highlights the importance of continuing assessment of the methodology used to assess hangovers to increase knowledge of alcohol hangover resistance and pathology.

Author Contributions: Conceptualization, A.S., S.B., and C.T.; Methodology, S.B., A.S., J.C.V., and C.T.; Software, formal analysis, C.T., S.B., and A.S.; Data curation, S.B. and C.T.; Writing—original draft preparation, C.T.; Writing—review and editing, C.T., S.B., A.S., and J.C.V.; Visualization, A.S.; Supervision, S.B. and A.S. All authors have read and agreed to the published version of the manuscript.

Funding: This research received no external funding.

Acknowledgments: C.T. is a recipient of a Swinburne University Postgraduate Research Award.

Conflicts of Interest: S.B. has received funding from Red Bull GmbH, Kemin Foods, Sanofi Aventis, Phoenix Pharmaceutical, and GlaxoSmithKline. Over the past 36 months, A.S. has held research grants from Abbott Nutrition, Arla Foods, Bayer, BioRevive, DuPont, Kemin Foods, Nestlé, Nutricia-Danone, and Verdure Sciences. He has acted as a consultant/expert advisor to Bayer, Danone, Naturex, Nestlé, Pfizer, Sanofi, Sen-Jam Pharmaceutical, and has received travel/hospitality/speaker fees Bayer, Sanofi, and Verdure Sciences. Over the past 36 months, J.C.V. has held grants from Janssen, Nutricia, and Sequential, and acted as a consultant/expert advisor to Clinilabs, More Labs, Red Bull, Sen-Jam Pharmaceutical, Toast!, and ZBiotics. C.T. has no conflicts of interest to declare.

Appendix A

$$BAC = A/(r*G) - B*(t - 0.5),$$

For males rw = Total Body Water = $2.447 - 0.09515(y) + 0.1074(h) + 0.3362(W)$

For females rw = Total Body Water = $-2.097 + 0.1069(h) + 0.04666(W)$

BAC = blood alcohol concentration in g/l, A = number of standard drinks consumed in grams, r = relative amount of body water (l/kg), G = weight of subject in kg, B = degradation rate in g/l of 0.15, t = elapsed time during alcohol consumption in hours, r = a set value for males (0.68) and females (0.55), the absorption time is 0.5 h.

Appendix B

$$BAC = A/(W*rw) - (\beta 60*t),$$

For males rw = Total Body Water = 2.447 − 0.09515(y) + 0.1074(h) + 0.3362(W)

For females rw = Total Body Water = −2.097 + 0.1069(h) + 0.04666(W)

BAC = blood alcohol concentration in g/dL, c = number of standard drinks consumed GC = gender constant (9.0 for females and 7.5 for males), w = weight in pounds, β60 = the metabolism rate of alcohol per hour (e.g., 0.017 g/dL), t = time in hours since the first sip of alcohol to the time of assessment, A = total volume (in mL) of drinks consumed multiplied by the percent of alcohol of the drink multiplied by the density of alcohol (0.79 g/mL) divided by 10, W = weight of participant in kg, H = height in meters, y = the age of the participant in years, h = height measured in cm.

References

1. Verster, J.C.; Scholey, A.; van de Loo, A.; Benson, S.; Stock, A.K. Updating the definition of the alcohol hangover. *J. Clin. Med.* **2020**, *9*, 823. [CrossRef] [PubMed]
2. Verster, J.C.; Stephens, R.; Penning, R.; Rohsenow, D.; McGeary, J.; Levy, D.; McKinney, A.; Finnigan, F.; Piasecki, T.M.; Adan, A. The alcohol hangover research group consensus statement on best practice in alcohol hangover research. *Curr. Drug Abuse Rev.* **2010**, *3*, 116–126. [CrossRef] [PubMed]
3. Verster, J.C.; Kruisselbrink, L.D.; Slot, K.A.; Anogeianaki, A.; Adams, S.; Alford, C.; Arnoldy, L.; Ayre, E.; Balikji, S.; Benson, S. Sensitivity to experiencing alcohol hangovers: Reconsideration of the 0.11% Blood Alcohol Concentration (BAC) threshold for having a hangover. *J. Clin. Med.* **2020**, *9*, 179. [CrossRef] [PubMed]
4. Wiese, J.G.; Shlipak, M.G.; Browner, W.S. The alcohol hangover. *Ann. Intern. Med.* **2000**, *132*, 897–902. [CrossRef] [PubMed]
5. Penning, R.; McKinney, A.; Verster, J.C. Alcohol hangover symptoms and their contribution to the overall hangover severity. *Alcohol Alcohol.* **2012**, *47*, 248–252. [CrossRef] [PubMed]
6. Rohsenow, D.J.; Howland, J.; Minsky, S.J.; Greece, J.; Almeida, A.; Roehrs, T.A. The Acute Hangover Scale: A new measure of immediate hangover symptoms. *Addict. Behav.* **2007**, *32*, 1314–1320. [CrossRef]
7. Verster, J.C. The alcohol hangover–a puzzling phenomenon. *Alcohol Alcohol.* **2008**, *43*, 124–126. [CrossRef]
8. van Schrojenstein Lantman, M.; van de Loo, A.J.; Mackus, M.; Verster, J.C. Development of a Definition for the Alcohol Hangover: Consumer Descriptions and Expert Consensus. *Curr. Drug Abuse Rev.* **2016**, *9*, 148–154. [CrossRef]
9. Slutske, W.S.; Piasecki, T.M.; Hunt-Carter, E.E. Development and initial validation of the Hangover Symptoms Scale: Prevalence and correlates of Hangover Symptoms in college students. *Alcohol Clin. Exp. Res.* **2003**, *27*, 1442–1450. [CrossRef]
10. van Schrojenstein Lantman, M.; Mackus, M.; van de Loo, A.; Verster, J.C. The impact of alcohol hangover symptoms on cognitive and physical functioning, and mood. *Hum. Psychopharmacol.* **2017**, *32*, e2623. [CrossRef]
11. Stephens, R.; Grange, J.A.; Jones, K.; Owen, L. A critical analysis of alcohol hangover research methodology for surveys or studies of effects on cognition. *Psychopharmacology* **2014**, *231*, 2223–2236. [CrossRef] [PubMed]
12. Verster, J.C.; Van Der Maarel, M.A.; McKinney, A.; Olivier, B.; De Haan, L. Driving during alcohol hangover among Dutch professional truck drivers. *Traffic Inj. Prev.* **2014**, *15*, 434–438. [CrossRef] [PubMed]
13. Frone, M.R.; Verster, J.C. Alcohol hangover and the workplace: A need for research. *Curr. Drug Abuse Rev.* **2013**, *6*, 177–179. [CrossRef] [PubMed]
14. Ames, G.M.; Grube, J.W.; Moore, R.S. The relationship of drinking and hangovers to workplace problems: An empirical study. *J. Stud. Alcohol* **1997**, *58*, 37–47. [CrossRef] [PubMed]
15. Verster, J. Alcohol hangover frequency, severity and interventions among Dutch college students. *Alcohol Clin. Exp. Res.* **2006**, *30*, 157A.

16. Piasecki, T.M.; Robertson, B.M.; Epler, A.J. Hangover and risk for alcohol use disorders: Existing evidence and potential mechanisms. *Curr. Drug Abuse Rev.* **2010**, *3*, 92–102. [CrossRef] [PubMed]
17. Rohsenow, D.J.; Howland, J.; Winter, M.; Bliss, C.A.; Littlefield, C.A.; Heeren, T.C.; Calise, T.V. Hangover sensitivity after controlled alcohol administration as predictor of post-college drinking. *J. Abnorm. Psychol.* **2012**, *121*, 270–275. [CrossRef]
18. Cameron, E.; French, D.P. Predicting perceived safety to drive the morning after drinking: The importance of hangover symptoms. *Drug Alcohol Rev.* **2016**, *35*, 442–446. [CrossRef]
19. Verster, J.C.; Bervoets, A.C.; de Klerk, S.; Vreman, R.A.; Olivier, B.; Roth, T.; Brookhuis, K.A. Effects of alcohol hangover on simulated highway driving performance. *Psychopharmacology* **2014**, *231*, 2999–3008. [CrossRef]
20. Prat, G.; Adan, A.; Sanchez-Turet, M. Alcohol hangover: A critical review of explanatory factors. *Hum. Psychopharmacol.* **2009**, *24*, 259–267. [CrossRef]
21. Penning, R.; van Nuland, M.; Fliervoet, L.A.; Olivier, B.; Verster, J.C. The pathology of alcohol hangover. *Curr. Drug Abuse Rev.* **2010**, *3*, 68–75. [CrossRef] [PubMed]
22. Scholey, A.; Benson, S.; Kaufman, J.; Terpstra, C.; Ayre, E.; Verster, J.C.; Allen, C.; Devilly, G.J. Effects of Alcohol Hangover on Cognitive Performance: Findings from a Field/Internet Mixed Methodology Study. *J. Clin. Med.* **2019**, *8*, 440. [CrossRef] [PubMed]
23. Verster, J.C.; van de Loo, A.J.; Adams, S.; Stock, A.-K.; Benson, S.; Scholey, A.; Alford, C.; Bruce, G. Advantages and Limitations of Naturalistic Study Designs and Their Implementation in Alcohol Hangover Research. *J. Clin. Med.* **2019**, *8*, 2160. [CrossRef] [PubMed]
24. Prat, G.; Adan, A.; Perez-Pamies, M.; Sanchez-Turet, M. Neurocognitive effects of alcohol hangover. *Addict. Behav.* **2008**, *33*, 15–23. [CrossRef] [PubMed]
25. Hustad, J.T.; Carey, K.B. Using calculations to estimate blood alcohol concentrations for naturally occurring drinking episodes: A validity study. *J. Stud. Alcohol* **2005**, *66*, 130–138. [CrossRef]
26. Widmark, E.M.P. *Principles and Applications of Medicolegal Alcohol Determination*; Biomedical Publications: Davis, CA, USA, 1981.
27. Verster, J.C.; Slot, K.A.; Arnoldy, L.; van Lawick van Pabst, A.E.; van de Loo, A.J.; Benson, S.; Scholey, A. The Association between Alcohol Hangover Frequency and Severity: Evidence for Reverse Tolerance? *J. Clin. Med.* **2019**, *8*, 1520. [CrossRef]
28. Watson, P.E.; Watson, I.D.; Batt, R.D. Prediction of blood alcohol concentrations in human subjects. *Updat. Widmark Equ. J. Stud. Alcohol.* **1981**, *42*, 547–556. [CrossRef]
29. Gullberg, R.G.; Jones, A.W. Guidelines for estimating the amount of alcohol consumed from a single measurement of blood alcohol concentration: Re-evaluation of Widmark's equation. *Forensic Sci. Int.* **1994**, *69*, 119–130. [CrossRef]
30. Stowell, A.R.; Stowell, L.I. Estimation of blood alcohol concentrations after social drinking. *J. Forensic Sci.* **1998**, *43*, 14–21. [CrossRef]
31. van Schrojenstein Lantman, M.; Mackus, M.; Roth, T.; Verster, J.C. Total sleep time, alcohol consumption, and the duration and severity of alcohol hangover. *Nat. Sci. Sleep* **2017**, *9*, 181–186. [CrossRef]
32. van Schrojenstein Lantman, M.; van de Loo, A.J.; Mackus, M.; Kraneveld, A.D.; Brookhuis, K.A.; Garssen, J.; Verster, J.C. Susceptibility to alcohol hangovers: Not just a matter of being resilient. *Alcohol Alcohol.* **2018**, *53*, 241–244. [CrossRef] [PubMed]
33. van de Loo, A.; Mackus, M.; van Schrojenstein Lantman, M.; Kraneveld, A.D.; Brookhuis, K.A.; Garssen, J.; Scholey, A.; Verster, J.C. Susceptibility to Alcohol Hangovers: The Association with Self-Reported Immune Status. *Int. J. Environ. Res. Public Health* **2018**, *15*, 1286. [CrossRef] [PubMed]
34. White, A.; Hingson, R. The burden of alcohol use: Excessive alcohol consumption and related consequences among college students. *Alcohol Res. Health* **2013**, *35*, 201–218.
35. Carey, K.B.; Hustad, J.T. Are retrospectively reconstructed blood alcohol concentrations accurate? Preliminary results from a field study. *J. Stud. Alcohol* **2002**, *63*, 762–766. [CrossRef]
36. Howland, J.; Rohsenow, D.J.; Allensworth-Davies, D.; Greece, J.; Almeida, A.; Minsky, S.J.; Arnedt, J.T.; Hermos, J. The incidence and severity of hangover the morning after moderate alcohol intoxication. *Addiction* **2008**, *103*, 758–765. [CrossRef]
37. Verster, J.C.; de Klerk, S.; Bervoets, A.C.; Darren Kruisselbrink, L. Can hangover immunity be really claimed? *Curr. Drug Abuse Rev.* **2013**, *6*, 253–254. [CrossRef]

38. Kruisselbrink, L.D.; Bervoets, A.C.; de Klerk, S.; van de Loo, A.; Verster, J.C. Hangover resistance in a Canadian University student population. *Addict. Behav. Rep.* **2017**, *5*, 14–18. [CrossRef]
39. Huntley, G.; Treloar, H.; Blanchard, A.; Monti, P.M.; Carey, K.B.; Rohsenow, D.J.; Miranda, R. An event-level investigation of hangovers' relationship to age and drinking. *Exp. Clin. Psychopharmacol.* **2015**, *23*, 314–323. [CrossRef]
40. Hogewoning, A.; Van de Loo, A.; Mackus, M.; Raasveld, S.; De Zeeuw, R.; Bosma, E.; Bouwmeester, N.; Brookhuis, K.; Garssen, J.; Verster, J. Characteristics of social drinkers with and without a hangover after heavy alcohol consumption. *Subst. Abuse Rehabil.* **2016**, *7*, 161–167. [CrossRef]
41. Forrest, A. The estimation of Widmark's factor. *J. Forensic Sci. Soc.* **1986**, *26*, 249–252. [CrossRef]
42. Seidl, S.; Jensen, U.; Alt, A. The calculation of blood ethanol concentrations in males and females. *Int. J. Legal. Med.* **2000**, *114*, 71–77. [CrossRef] [PubMed]
43. Ulrich, L.; Cramer, Y.; Zink, P. Relevance of individual parameters in the calculation of blood alcohol levels in relation to the volume of intake. *Blutalkohol* **1987**, *24*, 192–198. [PubMed]
44. Benson, S.; Ayre, E.; Garrisson, H.; Wetherell, M.A.; Verster, J.C.; Scholey, A. Alcohol Hangover and Multitasking: Effects on Mood, Cognitive Performance, Stress Reactivity, and Perceived Effort. *J. Clin. Med.* **2020**, *9*, 1154. [CrossRef] [PubMed]
45. Statistics, S.S. Available online: https://www.socscistatistics.com/tests/chisquare/default2.aspx (accessed on 22 August 2020).
46. Davis, C.G.; Thake, J.; Vilhena, N. Social desirability biases in self-reported alcohol consumption and harms. *Addict. Behav.* **2010**, *35*, 302–311. [CrossRef] [PubMed]

© 2020 by the authors. Licensee MDPI, Basel, Switzerland. This article is an open access article distributed under the terms and conditions of the Creative Commons Attribution (CC BY) license (http://creativecommons.org/licenses/by/4.0/).

Review

The Role of Alcohol Metabolism in the Pathology of Alcohol Hangover

Marlou Mackus [1,†], Aurora JAE van de Loo [1,2,†], Johan Garssen [1,3], Aletta D. Kraneveld [1], Andrew Scholey [4] and Joris C. Verster [1,2,4,*]

1. Division of Pharmacology, Utrecht Institute for Pharmaceutical Sciences (UIPS), Utrecht University, 3584CG Utrecht, The Netherlands; marloumackus@gmail.com (M.M.); a.j.a.e.vandeloo@uu.nl (A.J.v.d.L.); j.garssen@uu.nl (J.G.); a.d.kraneveld@uu.nl (A.D.K.)
2. Institute for Risk Assessment Sciences (IRAS), Utrecht University, 3584CM Utrecht, The Netherlands
3. Global Centre of Excellence Immunology, Nutricia Danone Research, 3584CT Utrecht, The Netherlands
4. Centre for Human Psychopharmacology, Swinburne University, Melbourne, VIC 3122, Australia; andrew@scholeylab.com
* Correspondence: j.c.verster@uu.nl; Tel.: +31-30-2536-909
† Both authors contributed equally to the manuscript.

Received: 31 August 2020; Accepted: 22 October 2020; Published: 25 October 2020

Abstract: The limited number of available studies that examined the pathology of alcohol hangover focused on biomarkers of alcohol metabolism, oxidative stress and the inflammatory response to alcohol as potentially important determinants of hangover severity. The available literature on alcohol metabolism and oxidative stress is reviewed in this article. The current body of evidence suggests a direct relationship between blood ethanol concentration and hangover severity, whereas this association is not significant for acetaldehyde. The rate of alcohol metabolism seems to be an important determinant of hangover severity. That is, fast elimination of ethanol is associated with experiencing less severe hangovers. An explanation for this observation may be the fact that ethanol—in contrast to acetaldehyde—is capable of crossing the blood–brain barrier. With slower ethanol metabolism, more ethanol is able to reach the brain and elicit hangover symptoms. Hangover severity was also significantly associated with biomarkers of oxidative stress. More oxidative stress in the first hours after alcohol consumption was associated with less severe next-day hangovers (i.e., a significant negative correlation was found between hangover severity and malondialdehyde). On the contrary, more oxidative stress at a later stage after alcohol consumption was associated with having more severe next-day hangovers (i.e., a significant positive correlation was found between hangover severity and 8-isoprostane). In conclusion, assessment of biomarkers of alcohol metabolism suggests that fast elimination of ethanol is associated with experiencing less severe hangovers. More research is needed to further examine the complex interrelationship between alcohol metabolism, the role of acetaldehyde and oxidative stress and antioxidants, and the pathology of the alcohol hangover.

Keywords: alcohol; hangover; ethanol; acetaldehyde; acetate; oxidative stress; malondialdehyde; 8-isoprostane

1. Introduction

The alcohol hangover refers to the combination of negative mental and physical symptoms which can be experienced after a single episode of alcohol consumption, starting when blood alcohol concentration (BAC) approaches zero [1,2]. An increasing body of the scientific literature is addressing the negative consequences of having hangovers for cognitive functioning including memory [3–6], and potentially endangering daily activities such as driving a car [7,8]. In contrast, research on the pathology of the alcohol hangover remains limited. Reviews on this topic suggest a variety of possible causes of the alcohol hangover [9–11], such as the role of the presence of ethanol and its metabolites in the blood, oxidative stress, and the immune response to alcohol consumption. However, the data presented in these reviews to support these hypotheses are limited. In this article, the current scientific evidence is reviewed and hypotheses are formulated on how alcohol metabolism and oxidative stress are related to the development of the alcohol hangover.

1.1. Alcohol Metabolism and Hangover Severity

The majority of ethanol is eliminated via oxidative processes in the liver [12–14]. Metabolism of ethanol is a two-step process, driven by the action of two enzymes, alcohol dehydrogenase (ADH), which oxidizes ethanol to acetaldehyde, and aldehyde dehydrogenase (ALDH), which oxidizes acetaldehyde to acetate. In both cases, the speed of both conversions is determined by the presence of the co-factor nicotinamide adenine dinucleotide (NAD^+). A second pathway for alcohol breakdown, which is especially active in subjects who consume alcohol chronically, or in others after consuming large amounts of alcohol, is the microsomal ethanol oxidizing system (MEOS) [15]. This reaction is catalyzed by CYP2E1 and requires the co-factor nicotinamide adenine dinucleotide phosphate ($NADP^+$), rather than NAD^+, to convert ethanol into acetaldehyde. A third, relatively minor pathway involves the activity of catalase in liver peroxisomes in which ethanol functions as an electron donor for the reduction of hydrogen peroxide to water. Together, these oxidative pathways account for over 90% of alcohol elimination. The other 10% of ethanol is metabolized via non-oxidative pathways [16].

Irrespective of the pathway of alcohol metabolism, ethanol and acetaldehyde are the key compounds researched in relation to alcohol hangover. There is, however, ongoing debate about their role in the pathology of the alcohol hangover [9–11], and given the paucity of empirical data, theoretically both ethanol and acetaldehyde concentrations could have a direct influence on hangover severity. Many core hangover symptoms (e.g., headache, nausea, apathy, and concentration problems) likely involve central processes. While systemic processes clearly play a role in aspects of alcohol hangover, exposure of the brain to ethanol or its metabolites may ultimately determine the pathogenesis of alcohol hangover (symptoms). Given this, it is important to investigate the capability of peripheral ethanol and acetaldehyde to enter the brain and exert central effects, including a hangover.

Ethanol's molecular structure allows it to freely cross the blood–brain barrier (BBB), and peripheral blood ethanol concentrations correlate significantly with brain ethanol concentrations [17]. This is not the case for acetaldehyde. Although original studies claimed that acetaldehyde can cross the BBB [18–21], these studies have been criticized for methodological shortcomings [22,23], and the current consensus is that acetaldehyde does not readily cross the BBB [23,24]. This is primarily due to the abundance of ALDH in the BBB which rapidly converts acetaldehyde into acetate and water before it can pass the membrane [23,25]. Theoretically small amounts of acetaldehyde may enter the brain via the circumventricular organs (CVOs), where no BBB exists [26]. However, under normal drinking circumstances, most acetaldehyde is metabolized systemically (i.e., before it reaches these areas), meaning negligible amounts of acetaldehyde will enter the brain via this pathway [27,28]. As a result, unlike ethanol and acetate, acetaldehyde is physiologically compartmentalized, with independent and different concentrations in the periphery and centrally [29–31]. As such, in theory, peripheral acetaldehyde should have no direct influence on hangover severity. However, it is important to note that acetaldehyde may also be produced centrally, via catalase from ethanol that entered the brain, where it can exert direct effects or be further metabolized and be involved in producing oxidative stress [32].

Only a few studies have reported correlations between hangover severity and acetaldehyde concentrations. This may be because of the fact that acetaldehyde is quickly converted into acetate and water, or because its blood concentration during the hangover state often falls below the detection limit [33,34]. This makes it difficult to accurately detect acetaldehyde.

Ylikahri et al. [35] examined the physiological correlates of a hangover in 23 healthy men. The morning following alcohol consumption (1.5 g/kg), no significant correlations were found between peak blood ethanol, acetaldehyde concentrations, and hangover severity. In a subsequent study, Ylikhari et al. [36] confirmed that blood acetaldehyde concentration was not significantly correlated with hangover severity. In a naturalistic study, Van de Loo et al. [37] found that urine ethanol concentrations assessed the morning following drinking were significantly lower in drinkers who reported no hangover compared to hangover-sensitive drinkers ($p = 0.027$). Overall hangover severity was positively and significantly associated with the amount of urine ethanol in those who reported having a hangover ($r = 0.571$, $p = 0.013$). Finally, using a mixed field and Internet methodology, Scholey et al. [38] found that hangover severity was significantly associated with the previous night's breath alcohol concentration ($r = 0.228$, $p = 0.019$). Together, the majority of these studies suggested a relationship between ethanol concentration and hangover severity.

The end products of ethanol metabolism, acetate (acetic acid) and water, both readily cross the BBB [24,39]. Water is an 'inactive' metabolite of ethanol, while acetate has received very little research attention in the context of alcohol hangovers. One study assessed urine acetate and acetone levels during alcohol hangover and reported significantly increased concentrations [40]. Unfortunately, in the report of this study, no data were presented on the correlations of the concentrations of acetate and acetone with hangover severity. The findings from an animal study do suggest a possible relationship between blood acetate levels and hangover headache [41]. Using a headache model in rats, it was demonstrated that administering acetate contributed to the development of trigeminal pain in rats, whereas a similar correlation was not found for acetaldehyde. On the other hand, acetate is a common additive in food preservation [42] and is, for example, the main acid in vinegar. As such, acetate consumption as food constituent is considered safe and has not been associated with significant adverse effects when consumed at dietary levels. Taken together, the possible role of acetate in the pathology of the alcohol hangover deserves further investigation. Figure 1 summarizes the current hypothesis on how alcohol metabolism may be related to hangover severity.

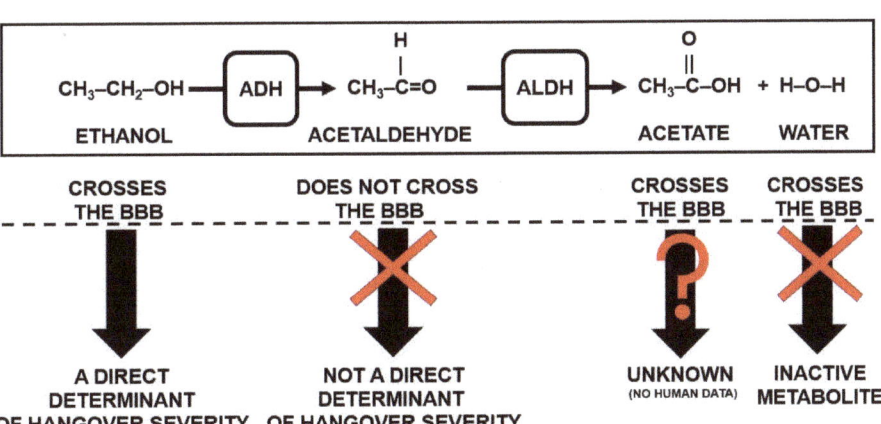

Figure 1. Alcohol metabolites and their hypothesized impact on the presence and severity of alcohol hangover. Abbreviations: BBB = blood brain barrier, ADH = alcohol dehydrogenase, ALDH = aldehyde dehydrogenase.

As summarized in Figure 1, it is hypothesized that the concentration of peripheral ethanol may be the important correlate of hangover severity, as it can freely pass the BBB. In contrast, it is hypothesized that peripheral acetaldehyde does not play a direct role in the development of the alcohol hangover in the brain. However, as discussed elsewhere [43] and in this review, blood acetaldehyde concentrations are suggested to have indirect effects on hangover severity via oxidative stress or via the inflammatory response elicited by alcohol consumption.

1.2. Accelerating Ethanol or Acetaldehyde Breakdown in Reducing Hangover Severity

It has been hypothesized that accelerating alcohol metabolism will reduce hangover severity as ethanol and acetaldehyde are then quickly removed from the body. To further evaluate the literature on alcohol metabolism rate, it is important to take into account the fact that the alcohol breakdown rate is largely determined by the presence of two enzymes vital in catalyzing the alcohol metabolism, ADH and ALDH (see Figure 1), together with their co-factor NAD^+.

There is evidence supporting the hypothesis that accelerating alcohol metabolism may reduce hangover symptom severity. For example, in Korea, red ginseng has been used for medical purposes for thousands of years, and research in rats and dogs revealed that red ginseng extract had short-term effects on ethanol metabolism in that it helped to reduce blood ethanol concentration [44,45]. Findings by Lee et al. [46] in human volunteers suggest that increasing the rate of alcohol metabolism may reduce hangover severity. Lee et al. [46] examined the effects of red ginseng on alcohol hangover severity. Higher blood ethanol concentrations (observed in the placebo condition vs. red ginseng condition at peak concentration, and first 60 min) were associated with significantly more severe hangovers (See Figure 2A). They further observed a rise in blood acetaldehyde levels 120 min after alcohol consumption, which was significantly greater after red ginseng than placebo (See Figure 2B). The reduction in ethanol and increase in acetaldehyde observed after red ginseng administration were followed by a significant reduction in next day hangover severity.

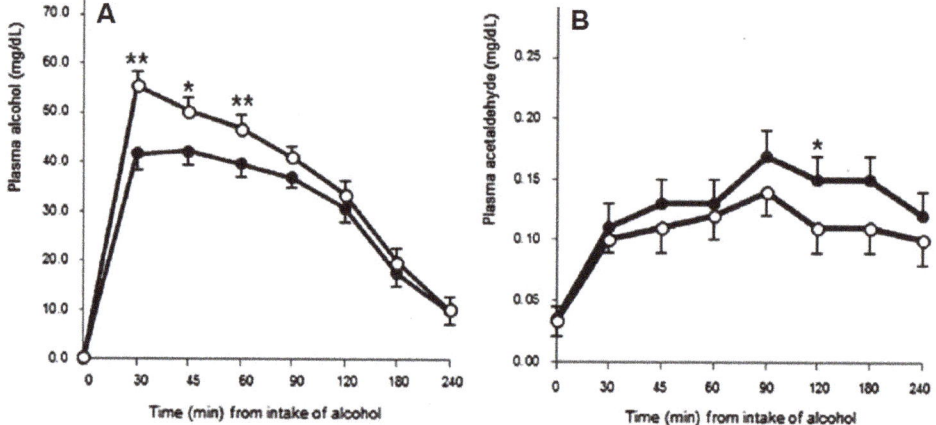

Figure 2. Effects of red ginseng on blood ethanol and acetaldehyde concentration. Alcohol and acetaldehyde levels over 240 min following alcohol intake in placebo (○) and test (●) with red ginseng groups. (**A**) Alcohol levels in plasma; (**B**) acetaldehyde levels in plasma. Values are expressed as means ± SEM. *p*-values are from nonparametric paired *t*-tests. * $p < 0.05$, ** $p < 0.01$. Reproduced from reference [46], with permission from the Royal Society of Chemistry.

Dihydromyricetin (DHM), the active constituent of *Hovenia dulcis*, has been found to induce the expression of ethanol-metabolizing enzymes and reduce ethanol and acetaldehyde concentrations in rats [47]. Kim et al. [48] examined the effects of the fruit of *Hovenia dulcis* versus placebo on hangover severity in N = 26 healthy Asian men with heterozygous ALDH2. Blood samples were taken 1, 4 and 12 h after alcohol consumption and ethanol and acetaldehyde concentrations were determined. Hangover severity was assessed the next morning, 12 h after alcohol consumption. The authors reported a positive, significant relationship (r = 0.410, $p = 0.003$) between blood acetaldehyde concentration assessed 4 h after alcohol consumption and hangover severity. This finding suggests that higher acetaldehyde concentrations in the first hours after drinking are associated with having a more severe hangover the next day. However, critical further analysis of the same data by Van de Loo et al. [43]

revealed a different outcome. Van de Loo et al. [43] examined data from the placebo condition of the study. It appeared that N = 10 subjects did not report a hangover and these were omitted from the analysis. Data from the remaining N = 16 subjects were analyzed, and it appeared that there were no significant correlations between blood acetaldehyde concentration assessed 1, 2 or 4 h after alcohol consumption and next-day hangover severity. In addition, no significant correlations were found between blood ethanol concentration and hangover severity. However, an indirect relationship between the presence of ethanol and hangover severity was found. The presence of ethanol was associated with increased levels of cytokines in the blood (interleukin (IL)-6 and tumor necrosis factor - alpha (TNF-α), of which the concentration was significantly and positively related to next-day hangover severity. This observation provides further evidence that a fast conversion of ethanol into acetaldehyde could be associated with experiencing less severe hangovers.

Finally, disulfiram is an ALDH2 inhibitor used in the treatment of alcoholism to prevent alcoholic patients from consuming alcohol. Its effects are established by inhibiting the breakdown of acetaldehyde into acetate and water, which results in accumulation of acetaldehyde. The effects of disulfiram start approximately 10 min after alcohol consumption and last for 1 h or more. Its effects include experiencing acute intoxication symptoms such as flushing, headache, nausea, vomiting, and sweating. Up to now, no studies have investigated the possible effects of disulfiram in the treatment of alcohol hangover.

Preclinical findings in animal studies and cell lines investigating natural compounds that enhance the activity of ADH and ALDH suggest that these products may possess properties which reduce or prevent alcohol hangovers [49,50]. In this context, recent research in social drinkers revealed that the dietary intake of two essential nutrients for the activation of ADH and ALDH was associated with hangover severity [51]. That is, there was a significant relationship between higher dietary intake of zinc and nicotinic acid and reporting less severe hangovers. Alternatively, acetaldehyde production can be influenced by modifying acetaldehyde producing microbiota [52]. High abundance of several microbiota such as *Rothia* is associated with increased production of acetaldehyde from ethanol [53]. A recent study [54] examined changes in the oral (saliva) microbiome of 15 young drinkers after an evening of heavy alcohol consumption. Compared to an alcohol-free day, the relative abundance of *Rothia*, *Streptococcus*, and *Veillonella* was significantly increased during the hangover state, whereas the relative abundance of *Prevotella*, *Fusobacterium*, *Campylobacter*, and *Leptotrichia* was significantly decreased. The largest change in saliva microbiome after heavy alcohol consumption was an increase in *Rothia*, which correlated significantly and negatively with reported hangover severity (r = −0.564, p = 0.036). Changes in other microbiota did not correlate significantly with hangover severity.

Genetic variety in ADH and ALDH also plays a role in alcohol metabolism [55], and this may affect the presence and severity of alcohol hangover. The ADH variants, of which ADH1B (subvariants *1, *2, and *3) and ADH1C (subvariants *1 and *2) are the most important in this context, differentially influence the breakdown rate of ethanol into acetaldehyde [56]. A relatively quick conversion from ethanol into acetaldehyde is observed in people possessing ADH1B*2, ADH1B*3 and ADH1C*1 alleles, whereas ethanol metabolism is relative slow in in people possessing the ADH1B*1 and ADH1C*2 allele [57]. The ADH1B*2 (common in people of Asian descent) and ADH1B*3 alleles (prevalent in people of African American decent) result in relative high blood concentrations of acetaldehyde (and thus lower blood ethanol concentrations) [58,59]. Additionally, subjects with an ALDH2*2 allele have a slower breakdown of acetaldehyde into acetate and water. Whereas acetaldehyde is usually quickly broken down, possession of the ALDH2*2 allele makes alcohol consumption unpleasant, because due to persistent elevated acetaldehyde concentrations in the blood, adverse effects such as flushing are frequently experienced [58]. Unfortunately, research into the genetics of alcohol hangover is limited. Two studies did report that subjects with Asian descent, possessing the ALDH2*2 allele, typically report significantly worse hangovers, and are more likely to experience hangovers at relatively lower alcohol consumption levels. [58,59]. Twin study by Slutske et al. [60] and Wu et al. [61] revealed that 45% [60] to 55% [61] of the frequency of experiencing hangovers. Forty-three percent of being hangover resistant could be explained by genetic variability [60]. These findings warrant further investigation into the impact of possessing different ADH and ALDH variants on the presence and severity of alcohol hangover.

In conclusion, these studies support the hypothesis that a quick conversion of ethanol into acetaldehyde is associated with having less severe hangovers (see Figure 3). Contrary to popular belief, there is no published scientific evidence showing that any hangover product that claims to enhance the elimination of acetaldehyde is actually effective in reducing hangover severity. This, however, does not rule out the possibility that acetaldehyde may have indirect effects on the presence and severity of alcohol hangover. The fact that possessing the ALDH2*2 allele is associated with experiencing worse hangovers supports this possibility.

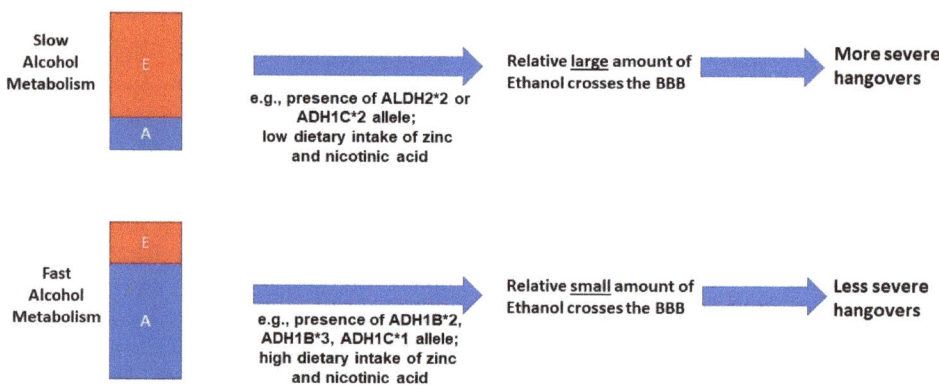

Figure 3. The impact of alcohol metabolism rate on hangover severity. Abbreviations: E = ethanol, A = acetaldehyde, BBB = blood brain barrier, ALDH = aldehyde dehydrogenase.

Up to now, one study directly associated alcohol elimination rate with hangover severity [62]. Data from N = 8 healthy volunteers who participated in both an acute study to investigate alcohol metabolism after alcohol consumption to achieve a BAC of 0.05% [63] and a naturalistic hangover study [64] were combined. In the acute alcohol study, breath alcohol content was assessed with a breathalyzer. Assessments were made every 5 min until subjects reached a BAC of zero. Using these data, the ethanol elimination rate was computed. These data were related to hangover severity reported in the hangover study, applying partial correlations correcting for estimated BAC. Hangover severity was assessed hourly from 09.30 to 15.30, using a one-item hangover severity scale ranging from 0 (absent) to 10 (severe) [65]. The analysis revealed significant negative correlations between hangover severity and ethanol elimination rate. In other words, those with a higher ethanol elimination rate (i.e., a faster conversion from ethanol into acetaldehyde) reported significantly lower hangover severity scores.

1.3. Oxidative Stress

Whereas in the liver the primary pathway of ethanol elimination is via ADH and ALDH activity, there exist alternative pathways for ethanol breakdown via the liver of which the microsomal ethanol oxidizing system (MEOS) is suggested to be the most important (See Figure 4).

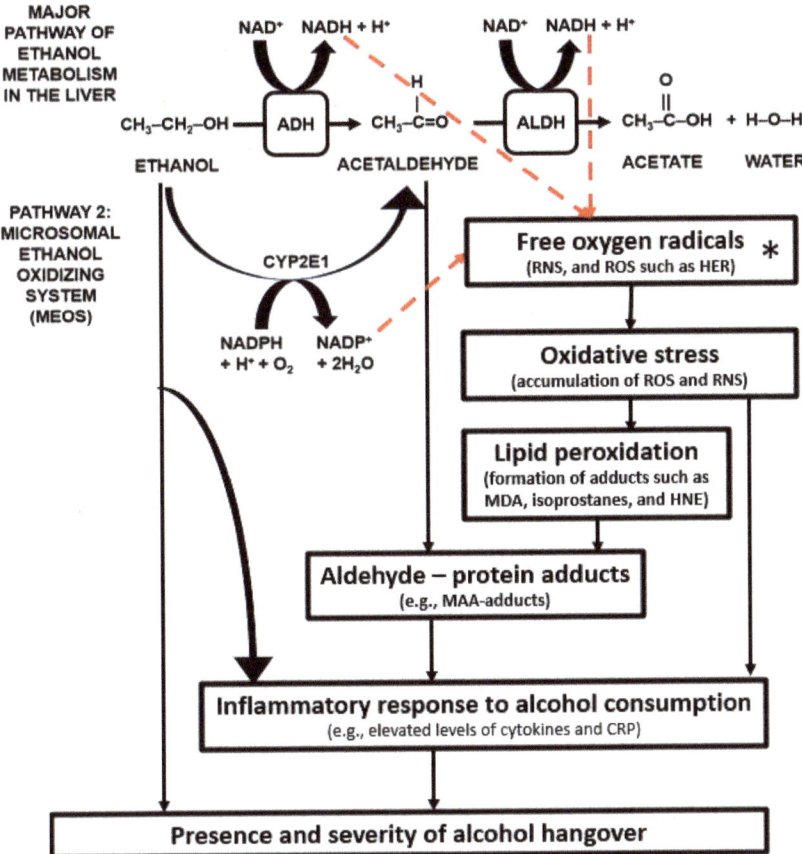

Figure 4. Oxidative stress and the inflammatory response to alcohol consumption. Arrows in red represent the relationship between alcohol metabolism and oxidative stress. Other arrows illustrate the pathways from oxidative stress to an inflammatory response and eliciting the alcohol hangover. Abbreviations: ADH = alcohol dehydrogenase, ALDH = aldehyde dehydrogenase, CRP = C-reactive protein, HER = hydroxyethyl radical, HNE = 4-hydroxy-2-nonenal, MAA = malondialdehyde – acetaldehyde, MDA = malondialdehyde (MDA), NAD^+ = nicotinamide adenine dinucleotide, NADH = nicotinamide adenine dinucleotide (NAD) + hydrogen (H), $NADP^+$ = nicotinamide adenine dinucleotide phosphate, NADPH = nicotinamide adenine dinucleotide phosphate + hydrogen (H), RNS = reactive nitrogen species, ROS = reactive oxygen species. Pathways contributing to ROS production are indicated by red dashed lines. * = free radicals are scavenged by antioxidants such as superoxide dismutase (SOD), catalase, and glutathione peroxidase, which may counteract lipid peroxidation and oxidative stress, and mitigate a subsequent inflammatory response.

The MEOS is more active after chronic alcohol use, as well as after drinking large quantities of alcohol. In addition, the MEOS is known to play a role in the link between alcohol metabolism and the inflammatory response consequent to alcohol consumption [43].

Ethanol is catalyzed by CYP2E1 into acetaldehyde, but in addition the MEOS also produces free (oxygen) radicals [66,67]. Free radicals are oxygen-containing molecules with unpaired electrons which allow them to easily react with other molecules. Free radicals, i.e., reactive oxygen species (ROS), are usually scavenged by antioxidants such as glutathione or superoxide dismutase (SOD). However, consuming large quantities of alcohol causes an imbalance between the amount of free radicals and antioxidants, i.e., excess levels of ROS such as hydroxyethyl radical (HER), called oxidative stress [66,67]. The abundance of free radicals elicits a process called lipid peroxidation, in which various byproducts of alcohol breakdown, called adducts, are formed [68,69]. Adducts are a combination of acetaldehyde or other aldehydes with a protein. Most notable in this regard are the aldehydes malondialdehyde (MDA) and 4-hydroxy-2-nonenal (HNE). Similar to acetaldehyde, these highly reactive aldehydes cannot pass through the blood brain barrier. In the presence of acetaldehyde, MDA can react with various proteins to form malondialdehyde–acetaldehyde adducts (MAA adducts). MAA adducts are known to have proinflammatory properties [70]. The MAA adducts are recognized by the body as foreign substances, and as a result an immune response is elicited [71–73], including increased secretion of cytokines and chemokines, as discussed elsewhere [43].

To counteract oxidative stress, various antioxidants (e.g., glutathione) and enzymes (e.g., catalase and superoxide dismutase) are produced endogenously. Cysteine is one of the precursors of glutathione, functioning as a semi-essential amino acid that binds acetaldehyde [74]. It has therefore been hypothesized that products increasing levels of antioxidants (e.g., glutathione, vitamin C, or vitamin E) may be effective as products to prevent hangovers [53]. However, most data to support these claims are indirect and comes from preclinical studies including animal research. For example, it has been shown that treatments that slowly release L-cysteine in the oral cavity significantly reduce saliva acetaldehyde concentration during alcohol consumption [75]. In rats, it was shown that a preparation of combined glutathione-enriched yeast (GEY) and rice embryo/soybean (RES) extracts significantly reduced both blood alcohol and acetaldehyde concentrations after an alcohol challenge [76]. Other research in rats revealed that administering electrolyzed-reduced water (ERW) was accompanied by activation of glutathione, and a significant increase in ADH and ALDH in the liver [77]. In the context of such preclinical findings, L-cysteine, its precursor N-acetyl-L-cysteine (NAC), and glutathione have been proposed as hangover treatments and are included as ingredients in various marketed hangover treatments. However, supportive peer-reviewed scientific data from clinical trials in humans on the efficacy of antioxidants in reducing or preventing hangovers are currently lacking [78–81]. Indeed, two recent double blind placebo controlled studies found no significant effect of a hangover treatment consisting of L-cysteine. Scholey et al. found no significant improvement of hangover severity after administering L-cysteine (650 mg) and B- and C-vitamins [82], and Eriksson et al. also did not demonstrate a significant reduction in hangover severity after administering L-cysteine (600 mg and 1200 mg) [83,84]. The reported data from a clinical trial on www.clinicaltrials.gov showed that NAC had no significant effects on reducing hangover severity but did produce more adverse effects than placebo (clinicaltrials.gov identifier NCT02541422). Another study examining NAC and hangover was terminated before completion (clinicaltrials.gov identifier NCT03104959), and the results from a third hangover study with glutathione are not reported (clinicaltrials.gov identifier NCT00127309).

To date, only one study has investigated both the biomarkers of oxidative stress and antioxidants in relationship to hangover severity [85]. Mammen et al. [85] examined the effect of a polyphenolic extract of clove buds (Clovinol) versus placebo on hangover severity in N = 16 healthy men, aged 25 to 55 years old. Blood ethanol and acetaldehyde concentration were determined before and at 0.5, 2, 4, and 12 h after alcohol consumption (240 mL of 42.8% McDowell's V.S.O.P. Brandy; United Spirits Limited, Bangalore, India), and a hangover scale was completed 14 h after drinking. To assess oxidative stress, two biomarkers (8-isoprostane and malondialdehyde) were assessed, as well as two antioxidants, glutathione and superoxide dismutase (SOD). No significant difference in ethanol concentrations was found between placebo and Clovinol at any time point. After intake of alcohol alone (i.e., the placebo condition), biomarkers of oxidative stress increased over time, whereas antioxidant concentrations decreased. Mammen et al. [85] observed a significant reduction in hangover severity after Clovinol, which was associated with a significant reduction in acetaldehyde, 8-isoprostane and malondialdehyde concentration and a significant increase in glutathione and SOD. Van de Loo et al. [43] further evaluated the placebo data of the study by Mammen et al. This evaluation revealed no significant correlations between hangover severity and blood ethanol or acetaldehyde concentrations at any time point after alcohol consumption. In addition, concentrations of the antioxidants glutathione and SOD did not significantly correlate with hangover severity at any timepoint after alcohol consumption.

A significant negative correlation was found between hangover severity and blood malondialdehyde concentration at 0.5 h after alcohol consumption ($r_B = -0.707$, BCa 95%CI$_B = -0.96$, -0.16). Correlations between hangover severity and 8-isoprostane in the first hours after drinking were also negative, but did not reach statistical significance. At a later stage after drinking, a significant positive correlation was found between hangover severity and 8-isoprostane at 12 h ($r_B = 0.475$, BCa 95%CI$_B = 0.06, 0.78$), and the association between malondialdehyde at 12 h and hangover severity was also positive, but did not reach significance.

Together, these findings suggest that higher levels of oxidative stress in the first hours after alcohol consumption (i.e., quick metabolism of ethanol) are associated with less severe hangovers (a negative correlation was found), whereas high levels of oxidative stress and inflammation during hangover are associated with more severe hangovers (a positive correlation was found).

2. Conclusions

The data summarized in this review suggest that the ethanol elimination rate is a critical determinant of hangover severity for a number of reasons. First, significant correlations have been found between ethanol concentration (but not acetaldehyde) and hangover severity. Second, nutrients, microbiota, and hangover treatments that speed up the conversion of ethanol into acetaldehyde are associated with less severe hangovers. Taken together, the data suggest that a more rapid conversion of ethanol to acetaldehyde and other aldehydes is associated with less severe hangovers. Based on the currently available data, the hypothesized relationship between alcohol metabolism and the presence and severity of alcohol hangover is schematized in Figure 5.

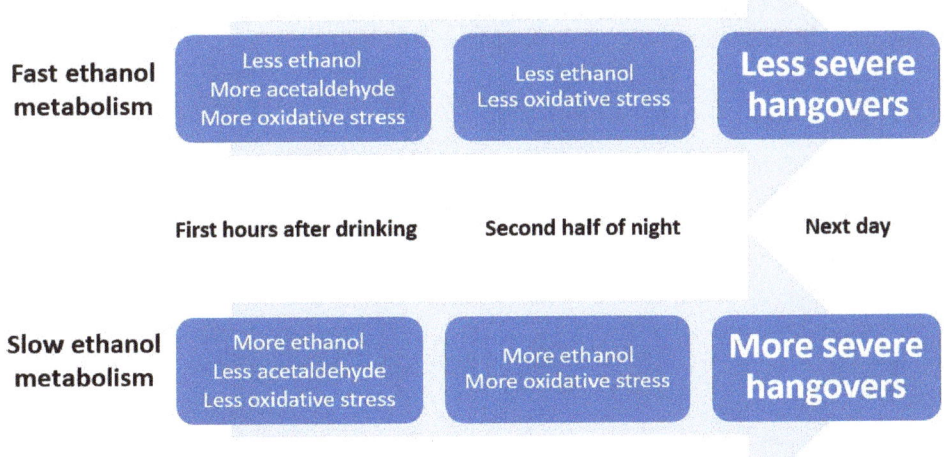

Figure 5. Schematic representation of the hypothesized relationship between alcohol metabolism and hangover severity.

The association between blood ethanol content and hangover severity is physiologically plausible, because unlike acetaldehyde, ethanol can freely pass the BBB. Compared to fast metabolizers of alcohol, slow metabolizers will have relatively large amounts of circulating ethanol that can cross the BBB over a longer time period, increasing the relative probability of a more severe hangover.

Figure 5 also includes oxidative stress, suggesting that high levels of oxidative stress in the first hours after drinking are associated with less severe hangovers, whereas high levels of oxidative stress during hangover are associated with more severe hangovers. It should be taken into account, however, that at this moment limited data are available on the role of oxidative stress and antioxidants in the pathogenesis of alcohol hangover, and this observation was based on only one study [43,85]. Therefore, much more research attention is needed to confirm these findings and elucidate the exact relationship between oxidative stress and alcohol hangover. Additionally, more research is needed to investigate to what extent antioxidants can prevent or reduce alcohol hangover, oxidative stress and the inflammatory response to alcohol consumption are interrelated and may impact hangover severity. Finally, although no direct effects of acetaldehyde on hangover severity have been demonstrated, acetaldehyde does play a role in oxidative stress resulting in hangovers. It has also been shown that subjects with an ALDH2*2 allele report more severe hangovers than other drinkers. Therefore, more research is also needed in the role of acetaldehyde in the development of alcohol hangover. Notwithstanding the limited amount of research, it is evident that alcohol metabolism and ethanol elimination rate play a critical role in the presence and severity of the alcohol hangover.

Author Contributions: Conceptualization, J.C.V., A.J.v.d.L., M.M., J.G., A.D.K. and A.S.; writing—original draft preparation, J.C.V.; writing—review and editing, J.C.V., A.S., A.J.v.d.L., M.M., J.G. and A.D.K. All authors have read and agreed to the published version of the manuscript.

Funding: This research received no external funding.

Conflicts of Interest: Over the past 36 months, A.D.K. has held research grants from H2020, Nutricia-Danone, Netherlands Center of Translational Research, Lungfund, SGF/Health Holland and NWO. J.G. is part-time employee of Nutricia Research. Over the past 36 months, J.G. has held research grants from several profit and non-profit organisations, including EU, NWO, Health Holland, TIFN, and acted as a consultant/expert advisor to Nutricia research Foundation, Friesland Campina, Carbohydrate Competitive Center, International Olympic Committee, Bill Gates Foundation, ID-DLO (WUR). Over the past 36 months, A.S. has held research grants from Abbott Nutrition, Arla Foods, Bayer, BioRevive, DuPont, Kemin Foods, Nestlé, Nutricia-Danone, Verdure Sciences. He has acted as a consultant/expert advisor to Bayer, Danone, Naturex, Nestlé, Pfizer, Sanofi, Sen-Jam Pharmaceutical, and has received travel/hospitality/speaker fees Bayer, Sanofi and Verdure Sciences. Over the past 36 months, J.C.V. has held grants from Janssen, Nutricia, and Sequential Medicine, and acted as a consultant/expert advisor to More Labs, Red Bull, Sen-Jam Pharmaceutical, Toast!, Tomo, and ZBiotics. A.J.v.d.L. and M.M. have no conflict of interest to declare.

References

1. Van Schrojenstein Lantman, M.; van de Loo, A.J.; Mackus, M.; Verster, J.C. Development of a definition for the alcohol hangover: Consumer descriptions and expert consensus. *Curr. Drug Abuse Rev.* **2016**, *9*, 148–154. [CrossRef]
2. Verster, J.C.; Scholey, A.; van de Loo, A.J.A.E.; Benson, S.; Stock, A.K. Updating the definition of the alcohol hangover. *J. Clin. Med.* **2020**, *9*, 823. [CrossRef]
3. Prat, G.; Adan, A.; Pérez-Pàmies, M.; Sànchez-Turet, M. Neurocognitive effects of alcohol hangover. *Addict. Behav.* **2008**, *33*, 15–23. [CrossRef]
4. Prat, G.; Adan, A.; Sánchez-Turet, M. Alcohol hangover: A critical review of explanatory factors. *Hum. Psychopharmacol.* **2009**, *24*, 259–267. [CrossRef]
5. Gunn, C.; Mackus, M.; Griffin, C.; Munafò, M.R.; Adams, S. A systematic review of the next-day effects of heavy alcohol consumption on cognitive performance. *Addiction* **2018**, *113*, 2182–2193. [CrossRef]
6. Kruisselbrink, L.D. The neurocognitive effects of alcohol hangover: Patterns of impairment/nonimpairment within the neurocognitive domain of the Diagnostic and Statistical Manual of Mental Disorders. In *Neuroscience of Alcohol: Mechanisms and Treatment*, 5th ed.; Preedy, V.R., Ed.; Academic Press: Cambridge, MA, USA, 2019; pp. 391–402.
7. Verster, J.C.; Bervoets, A.C.; de Klerk, S.; Vreman, R.A.; Olivier, B.; Roth, T.; Brookhuis, K.A. Effects of alcohol hangover on simulated highway driving performance. *Psychopharmacology* **2014**, *231*, 2999–3008. [CrossRef]
8. Alford, C.; Broom, C.; Carver, H.; Johnson, S.J.; Reece, R.; Lands, S.; Verster, J.C. The impact of alcohol hangover on simulated driving performance during a 'commute to work'—Zero and residual alcohol effects compared. *J. Clin. Med.* **2020**, *9*, 1435. [CrossRef]

9. Penning, R.; van Nuland, M.; Fliervoet, L.A.L.; Olivier, B.; Verster, J.C. The pathology of alcohol hangover. *Curr. Drug Abuse Rev.* **2010**, *3*, 68–75. [CrossRef]
10. Tipple, C.T.; Benson, S.; Scholey, A. A Review of the Physiological Factors Associated with Alcohol Hangover. *Curr. Drug Abuse Rev.* **2016**, *9*, 93–98. [CrossRef]
11. Palmer, E.; Tyacke, R.; Sastre, M.; Lingford-Hughes, A.; Nutt, D.; Ward, R.J. Alcohol Hangover: Underlying Biochemical, Inflammatory and Neurochemical Mechanisms. *Alcohol Alcohol.* **2019**, *54*, 196–203. [CrossRef]
12. Bullock, C. The biochemistry of alcohol metabolism—A brief review. *Biochemical Educ.* **1990**, *18*, 62–66. [CrossRef]
13. Kawai, S.; Murata, K. Structure and function of NAD kinase and NADP phosphatase: Key enzymes that regulate the intracellular balance of NAD(H) and NADP(H). *Biosci. Biotechnol. Biochem.* **2008**, *72*, 919–930. [CrossRef]
14. Cederbaum, A.I. Alcohol metabolism. *Clin. Liver Dis.* **2012**, *16*, 667–685. [CrossRef]
15. Jones, A.W. Evidence-based survey of the elimination rates of ethanol from blood with applications in forensic casework. *Forensic Sci. Int.* **2010**, *200*, 1–20. [CrossRef]
16. Heier, C.; Xie, H.; Zimmermann, R. Nonoxidative ethanol metabolism in humans-from biomarkers to bioactive lipids. *IUBMB Life* **2016**, *68*, 916–923. [CrossRef]
17. Fein, G.; Meyerhoff, D.J. Ethanol in human brain by magnetic resonance spectroscopy: Correlation with blood and breath levels, relaxation, and magnetization transfer. *Alcohol Clin. Exp. Res.* **2000**, *24*, 1227–1235. [CrossRef]
18. Hillbom, M.E.; Lindros, K.O.; Larsen, A. The calcium carbimide-ethanol interaction: Lack of relation between electroencephalographic response and cerebrospinal fluid acetaldehyde. *Toxicol. Lett.* **1981**, *9*, 113–119. [CrossRef]
19. Pösö, A.R.; Hillbom, M.E.; Eriksson, L. Acetaldehyde penetrates the blood-liquor barrier of goats. *Toxicol. Lett.* **1981**, *8*, 57–62. [CrossRef]
20. Heap, L.; Ward, R.J.; Abiaka, C.; Dexter, D.; Lawlor, M.; Pratt, O.; Thomson, A.; Shaw, K.; Peters, T.J. The influence of brain acetaldehyde on oxidative status, dopamine metabolism and visual discrimination task. *Biochem Pharmacol.* **1995**, *50*, 263–270. [CrossRef]
21. Jones, A.W. Measuring and reporting the concentration of acetaldehyde in human breath. *Alcohol Alcohol.* **1995**, *30*, 271–285.
22. Eriksson, C.J.P. Human acetaldehyde levels: Aspects of current interest. ICPEMP Working paper No. 15/3. *Mutation Res.* **1987**, *186*, 235–240. [CrossRef]
23. Zimatkin, S.M. Histochemical study of aldehyde dehydrogenase in the rat CNS. *J. Neurochem.* **1991**, *56*, 1–11. [CrossRef]
24. Deitrich, R.A.; Dunwiddie, T.V.; Harris, R.A.; Erwin, V.G. Mechanism of Action of Ethanol—Initial Central-Nervous-System Actions. *Pharmacol. Rev.* **1989**, *41*, 489–537.
25. Isse, T.; Matsuno, K.; Oyama, T.; Kitagawa, K.; Kawamoto, T. Aldehyde dehydrogenase 2 gene targeting mouse lacking enzyme activity shows high acetaldehyde level in blood, brain, and liver after ethanol gavages. *Alcohol Clin. Exp. Res.* **2005**, *29*, 1959–1964. [CrossRef]
26. Ujihara, I.; Hitomi, S.; One, K.; Kakinoki, Y.; Hashimoto, H.; Ueta, Y.; Inegana, K. The ethanol metabolite acetaldehyde induces water and salt intake via two distinct pathways in the central nervous system of rats. *Neuropharmacology* **2015**, *99*, 589–599. [CrossRef]
27. Eriksson, C.J.P.; Fukunaga, T. Human blood acetaldehyde (update 1992). *Alcohol Alcohol.* **1993**, *S2*, 9–25.
28. Hunt, W.A. Role of acetaldehyde in the actions of ethanol on the brain—A review. *Alcohol* **1996**, *13*, 147–151. [CrossRef]
29. Tabakoff, B.; Anderson, R.A.; Ritzmann, R.F. Brain acetaldehyde after ethanol administration. *Biochem. Pharmacol.* **1976**, *25*, 1305–1309. [CrossRef]
30. Westcott, J.Y.; Weiner, H.; Shultz, J.; Myers, R.D. In vivo acetaldehyde in the brain of the rat treated with ethanol. *Biochem. Pharmacol.* **1980**, *29*, 411–417. [CrossRef]
31. Deitrich, R. Ethanol as a prodrug: Brain metabolism of ethanol mediates its reinforcing effects–a commentary. *Alcohol Clin. Exp. Res.* **2011**, *35*, 581–583. [CrossRef]
32. Hernández, J.A.; López-Sánchez, R.C.; Rendón-Ramírez, A. Lipids and oxidative stress associated with ethanol-induced neurological damage. *Oxidative Med. Cell. Longev.* **2016**, *2016*, 1543809. [CrossRef] [PubMed]

33. Eriksson, C.J. The role of acetaldehyde in the actions of alcohol (update 2000). *Alcohol Clin. Exp. Res.* **2001**, *25*, 15S–32S. [CrossRef]
34. Eriksson, C.J.; Saarenmaa, T.P.; Bykov, I.L.; Heino, P.U. Acceleration of ethanol and acetaldehyde oxidation by D-glycerate in rats. *Metabolism* **2007**, *56*, 895–898. [CrossRef]
35. Ylikahri, R.H.; Huttunen, M.O.; Eriksson, C.J.; Nikkilä, E.A. Metabolic studies on the pathogenesis of hangover. *Eur. J. Clin. Investig.* **1974**, *4*, 93–100. [CrossRef]
36. Ylikahri, R.H.; Leino, T.; Huttunen, M.O.; Pösö, A.R.; Eriksson, C.J.P.; Nikkilä, E.A. Effects of fructose and glucose on ethanol-induced metabolic changes and on the intensity of alcohol intoxication and hangover. *Eur. J. Clin. Investig.* **1976**, *6*, 93–102. [CrossRef] [PubMed]
37. Van de Loo, A.J.A.E.; Mackus, M.; Korte-Bouws, G.A.H.; Brookhuis, K.A.; Garssen, J.; Verster, J.C. Urine ethanol concentration and alcohol hangover severity. *Psychopharmacology* **2017**, *234*, 73–77. [CrossRef] [PubMed]
38. Scholey, A.; Benson, S.; Kaufman, J.; Terpstra, C.; Ayre, E.; Verster, J.C.; Allen, C.; Devilly, G. Effects of alcohol hangover on cognitive performance: A field/internet mixed methodology approach. *J. Clin. Med.* **2019**, *8*, 440. [CrossRef]
39. Frost, G.; Sleeth, M.L.; Sahuri-Arisoylu, M.; Lizarbe, B.; Cerdan, S.; Brody, L.; Anastasovska, J.; Ghourab, S.; Hankir, M.; Zhang, S.; et al. The short-chain fatty acid acetate reduces appetite via a central homeostatic mechanism. *Nat. Commun.* **2014**, *5*, 3611. [CrossRef]
40. Tsukamoto, S.; Kanegae, T.; Saito, M.; Nagoya, T.; Shimamura, M.; Tainaka, H.; Kawagughi, M. Concentrations of blood and urine ethanol, acetaldehyde, acetate and acetone during experimental hangover in volunteers. *Jpn. J. Alcohol Drug Depend.* **1991**, *26*, 500–510.
41. Maxwell, C.R.; Spangenberg, R.J.; Hoek, J.B.; Silberstein, S.D.; Oshinsky, M.L. Acetate causes alcohol hangover headache in rats. *PLoS ONE* **2010**, *5*, e15963. [CrossRef]
42. Pelaez, A.M.L.; Catano, C.; Quintero Yepers, E.A.; Gamba Villaroei, R.R.; de Antoni, G.L.; Giannuzzi, L. Inhibitory activity of lactic and acetic acid on Aspergillus flavus growth for food preservation. *Food Control* **2012**, *24*, 177–183. [CrossRef]
43. Van de Loo, A.J.E.A.; Mackus, M.; Kwon, O.; Krishnakumar, I.M.; Garssen, J.; Kraneveld, A.D.; Scholey, A.; Verster, J.C. The inflammatory response to alcohol consumption and its role in the development of alcohol hangover. *J. Clin. Med.* **2020**, *9*, 2081. [CrossRef] [PubMed]
44. Shin, M.R. Studies on the antidotal effect of red ginseng. *Korean J. Ginseng Sci.* **1976**, *1*, 59–78.
45. Koo, M.W. Effects of ginseng on ethanol induced sedation in mice. *Life Sci.* **1999**, *64*, 153–160. [CrossRef]
46. Lee, M.H.; Kwak, J.H.; Jeon, G.; Lee, J.W.; Seo, J.H.; Lee, H.S.; Lee, J.H. Red ginseng relieves the effects of alcohol consumption and hangover symptoms in healthy men: A randomized crossover study. *Food Funct.* **2014**, *5*, 528–534. [CrossRef]
47. Silva, J.; Yu, X.; Moradian, R.; Folk, C.; Spatz, M.H.; Kim, P.; Bhatti, A.A.; Davies, D.L.; Liang, J. Dihydromyricetin protects the liver via changes in lipid metabolism and enhanced ethanol metabolism. *Alcohol Clin. Exp. Res.* **2020**, *44*, 1046–1060. [CrossRef]
48. Kim, H.; Kim, Y.J.; Jeong, H.Y.; Kim, J.Y.; Choi, E.K.; Chae, S.W.; Kwon, O. A standardized extract of the fruit of Hovenia dulcis alleviated alcohol-induced hangover in healthy subjects with heterozygous ALDH2: A randomized, controlled, crossover trial. *J. Ethnopharmacol.* **2017**, *209*, 167–174. [CrossRef]
49. Cho, M.H.; Shim, S.M.; Lee, S.R.; Mar, W.; Kim, G.H. Effect of Evodiae fructus extracts on gene expressions related with alcohol metabolism and antioxidation in ethanol-loaded mice. *Food Chem. Toxicol.* **2005**, *43*, 1365–1371. [CrossRef]
50. Choi, E.J.; Kwon, H.C.; Sohn, Y.C.; Nam, C.W.; Park, H.B.; Kim, C.Y.; Yang, H.O. Four flavonoids from Echinosophora koreensis and their effects on alcohol metabolizing enzymes. *Arch. Pharm. Res.* **2009**, *32*, 851–855. [CrossRef]
51. Verster, J.C.; Vermeulen, S.A.; van de Loo, A.J.A.E.; Balikji, S.; Kraneveld, A.D.; Garssen, J.; Scholey, A. Dietary nutrient intake, alcohol metabolism, and hangover severity. *J. Clin. Med.* **2019**, *8*, 1316. [CrossRef]
52. Salaspuro, V. Pharmacological treatments and strategies for reducing oral and intestinal acetaldehyde. *Novartis Found. Symp.* **2007**, *285*, 145–153; discussion 153–157, 198–199. [PubMed]
53. Moritani, K.; Takeshita, T.; Shibata, Y.; Ninomiya, T.; Kiyohara, Y.; Yamashita, Y. Acetaldehyde production by major oral microbes. *Oral Dis.* **2015**, *21*, 748–754. [CrossRef] [PubMed]

54. Palmer, E.; Arnoldy, L.; Ayre, E.; Benson, S.; Balikji, S.; Bruce, G.; Chen, F.; van Lawick van Pabst, A.E.; van de Loo, A.J.A.E.; O'Neill, S.; et al. Proceedings of the 11th Alcohol Hangover Research Group meeting in Nadi, Fiji. *Proceedings* **2020**, *43*, 3001. [CrossRef]
55. Edenberg, H.J.; Xuei, X.; Chen, H.-J.; Tian, H.; Flury Wetherill, L.; Dick, D.M.; Almasy, L.; Bierut, L.; Bucholz, K.K.; Goate, A.; et al. Association of alcohol dehydrogenase genes with alcohol dependence: A comprehensive analysis. *Hum. Mol. Genet.* **2006**, *15*, 1539–1549. [CrossRef]
56. Hurley, T.D.; Edenberg, H.J. Genes Encoding Enzymes Involved in Ethanol Metabolism. *Alcohol Res.* **2012**, *34*, 339–344.
57. Jiang, Y.; Zhang, T.; Kusumanchi, P.; Han, S.; Yang, Z.; Liangpunsakul, S. Alcohol Metabolizing Enzymes, Microsomal Ethanol Oxidizing System, Cytochrome P450 2E1, Catalase, and Aldehyde Dehydrogenase in Alcohol-Associated Liver Disease. *Biomedicines* **2020**, *8*, 50. [CrossRef]
58. Yokoyama, M.; Yokoyama, A.; Yokoyama, T.; Funazu, K.; Hamana, G.; Kondo, S.; Yamashita, T.; Nakamura, H. Hangover susceptibility in relation to aldehyde dehydrogenase-2 genotype, alcohol flushing, and mean corpuscular volume in Japanese workers. *Alcohol Clin. Exp. Res.* **2005**, *29*, 1165–1171. [CrossRef]
59. Wall, T.L.; Horn, S.M.; Johnson, M.L.; Smith, T.L.; Carr, L.G. Hangover symptoms in Asian Americans with variations in the aldehyde dehydrogenase (ALDH2) gene. *J. Stud. Alcohol* **2000**, *61*, 13–17. [CrossRef]
60. Slutske, W.S.; Piasecki, T.M.; Nathanson, L.; Statham, D.J.; Martin, N.G. Genetic influences on alcohol-related hangover. *Addiction* **2014**, *109*, 2027–2034. [CrossRef]
61. Wu, S.H.; Guo, Q.; Viken, R.J.; Reed, T.; Dai, J. Heritability of usual alcohol intoxication and hangover in male twins: The NAS-NRC Twin Registry. *Alcohol Clin. Exp. Res.* **2014**, *38*, 2307–2313. [CrossRef]
62. Mackus, M.; Van de Loo, A.J.A.E.; Garssen, J.; Kraneveld, A.D.; Verster, J.C. The association between ethanol elimination rate and hangover severity. *Int. J. Environ. Res. Publ. Health* **2020**, *17*, 4324. [CrossRef] [PubMed]
63. Mackus, M.; van Schrojenstein Lantman, M.; van de Loo, A.J.A.E.; Brookhuis, K.A.; Kraneveld, A.D.; Garssen, J.; Verster, J.C. Alcohol metabolism in hangover sensitive versus hangover resistant social drinkers. *Drug Alcohol Depend.* **2018**, *185*, 351–355. [CrossRef]
64. Hogewoning, A.; van de Loo, A.J.A.E.; Mackus, M.; Raasveld, S.J.; de Zeeuw, R.; Bosma, E.R.; Bouwmeester, N.H.; Brookhuis, K.A.; Garssen, J.; Verster, J.C. Characteristics of social drinkers with and without a hangover after heavy alcohol consumption. *Subst. Abuse Rehab.* **2016**, *7*, 161–167. [CrossRef] [PubMed]
65. Verster, J.C.; van de Loo, A.J.A.E.; Benson, S.; Scholey, A.; Stock, A.-K. The assessment of overall hangover severity. *J. Clin. Med.* **2020**, *9*, 786. [CrossRef]
66. Wu, D.; Cederbaum, A.I. Alcohol, oxidative stress, and free radical damage. *Alcohol Res. Health* **2003**, *27*, 277–284.
67. Das, S.K.; Vasudevan, D.M. Alcohol-induced oxidative stress. *Life Sci.* **2007**, *81*, 177–187. [CrossRef]
68. Niemelä, O. Acetaldehyde adducts in circulation. *Novartis Found. Symp.* **2007**, *285*, 183–192; discussion 193–197. [PubMed]
69. Rahal, A.; Kumar, A.; Singh, V.; Yadav, B.; Tiwari, R.; Chakraborty, S.; Dhama, K. Oxidative stress, prooxidants, and antioxidants: The interplay. *Biomed. Res. Int.* **2014**, *2014*, 761264. [CrossRef]
70. Tuma, D.J. Role of malondialdehyde-acetaldehyde adducts in liver injury. *Free Radic. Biol. Med.* **2002**, *32*, 302–308.
71. Niemela, O. Aldehyde-protein adducts in the liver as a result of ethanol-induced oxidative stress. *Front. Biosci.* **1999**, *4*, d506–d513. [CrossRef]
72. Thiele, G.M.; Worrall, S.; Tuma, D.J.; Klassen, L.W.; Wyatt, T.A.; Nagata, N. The chemistry and biological effects of malondialdehyde-acetaldehyde adducts. *Alcohol Clin. Exp. Res.* **2001**, *25*, 218S–224S. [CrossRef] [PubMed]
73. Tuma, D.J.; Casey, C.A. Dangerous byproducts of alcohol breakdown—Focus on adducts. *Alcohol Res. Health* **2003**, *27*, 285–290. [PubMed]
74. Salaspuro, M. Acetaldehyde and gastric cancer. *J. Dig. Dis.* **2011**, *12*, 51–59. [CrossRef] [PubMed]
75. Salaspuro, V.; Hietala, J.; Kaihovaara, P.; Pihlajarinne, L.; Marvola, M.; Salaspuro, M. Removal of acetaldehyde from saliva by a slow-release buccal tablet of L-cysteine. *Int. J. Cancer* **2002**, *97*, 361–364. [CrossRef] [PubMed]
76. Lee, H.S.; Song, J.; Kim, T.M.; Joo, S.S.; Park, D.; Jeon, J.H.; Shin, S.; Park, H.K.; Lee, W.K.; Ly, S.Y.; et al. Effects of a preparation of combined glutathione-enriched yeast and rice embryo/soybean extracts on ethanol hangover. *J. Med. Food* **2009**, *12*, 1359–1367. [CrossRef]

77. Park, S.K.; Qi, X.F.; Song, S.B.; Kim, D.H.; Teng, Y.C.; Yoon, Y.S.; Kim, K.Y.; Li, J.H.; Jin, D.; Lee, K.J. Electrolyzed-reduced water inhibits acute ethanol-induced hangovers in Sprague-Dawley rats. *Biomed. Res.* **2009**, *30*, 263–269. [CrossRef]
78. Pittler, M.H.; Verster, J.C.; Ernst, E. Interventions for preventing or treating alcohol hangover: Systematic review of randomized trials. *Br. Med. J.* **2005**, *331*, 1515–1518. [CrossRef]
79. Verster, J.C.; Penning, R. Treatment and prevention of alcohol hangover. *Curr. Drug Abuse Rev.* **2010**, *3*, 103–109. [CrossRef]
80. Jayawardena, R.; Thejani, T.; Ranasinghe, P.; Fernando, D.; Verster, J.C. Interventions for treatment and/or prevention of alcohol hangover: Systematic review. *Hum. Psychopharmacol.* **2017**, *32*, e2600. [CrossRef]
81. Mackus, M.; van Schrojenstein Lantman, M.; van de Loo, A.J.A.E.; Nutt, D.J.; Verster, J.C. An effective hangover treatment: Friend or foe? *Drug Sci. Policy Law* **2017**. [CrossRef]
82. Scholey, A.; Ayre, E.; Stock, A.-K.; Verster, J.C.; Benson, S. The effects of Rapid Recovery on alcohol hangover severity: A double-blind, placebo-controlled, randomized and crossover trial. *J. Clin. Med.* **2020**, *9*, 2175. [CrossRef] [PubMed]
83. Eriksson, C.J.P.; Metsälä, M.; Möykkynen, T.; Mäkisalo, H.; Kärkkäinen, O.; Palmén, M.; Salminen, J.E.; Kauhanen, J. L-Cysteine containing vitamin supplement which prevents or alleviates alcohol-related hangover symptoms: Nausea, headache, stress and anxiety. *Alcohol Alcohol.* **2020**. [CrossRef]
84. Benson, S.; Scholey, A.; Verster, J.C. L-cysteine and the treatment of alcohol hangover: A commentary on Eriksson et al. Submitted for publication.
85. Mammen, R.R.; Natinga Mulakal, J.; Mohanan, R.; Maliakel, B.; Krishnakumar, I.M. Clove bud polyphenols alleviate alterations in inflammation and oxidative stress markers associated with binge drinking: A randomized double-blinded placebo-controlled crossover study. *J. Med. Food* **2018**, *21*, 1188–1196. [CrossRef]

Publisher's Note: MDPI stays neutral with regard to jurisdictional claims in published maps and institutional affiliations.

© 2020 by the authors. Licensee MDPI, Basel, Switzerland. This article is an open access article distributed under the terms and conditions of the Creative Commons Attribution (CC BY) license (http://creativecommons.org/licenses/by/4.0/).

Article

The Inflammatory Response to Alcohol Consumption and Its Role in the Pathology of Alcohol Hangover

Aurora J.A.E. van de Loo [1,2,†], Marlou Mackus [1,†], Oran Kwon [3], Illathu Madhavamenon Krishnakumar [4], Johan Garssen [1,5], Aletta D. Kraneveld [1], Andrew Scholey [6] and Joris C. Verster [1,2,3,*]

1. Division of Pharmacology, Utrecht Institute for Pharmaceutical Sciences (UIPS), Utrecht University, 3584CG Utrecht, The Netherlands; a.j.a.e.vandeloo@uu.nl (A.J.A.E.v.d.L.); marloumackus@gmail.com (M.M.); j.garssen@uu.nl (J.G.); a.d.kraneveld@uu.nl (A.D.K.)
2. Institute for Risk Assessment Sciences (IRAS), Utrecht University, 3584CM Utrecht, The Netherlands
3. BioFood Laboratory/BioFood Network, Department of Nutritional Science and Food Management, Ewha Womans University, Seoul 120-750, Korea; orank@ewha.ac.kr
4. Akay Natural Ingredients Private Limited, Ambunad Malaidumthuruthu, Cochin, Aluva, Kerala 683561, India; krishnakumar.im@Akay-group.com
5. Global Centre of Excellence Immunology, Nutricia Danone Research, 3584CT Utrecht, The Netherlands
6. Centre for Human Psychopharmacology, Swinburne University, VIC 3122 Melbourne, Australia; andrew@scholeylab.com
* Correspondence: j.c.verster@uu.nl; Tel.: +313-0253-6909
† Both authors contributed equally to the manuscript.

Received: 30 May 2020; Accepted: 30 June 2020; Published: 2 July 2020

Abstract: An increasing number of studies are focusing on the inflammatory response to alcohol as a potentially important determinant of hangover severity. In this article, data from two studies were re-evaluated to investigate the relationship between hangover severity and relevant biomarkers of alcohol metabolism, oxidative stress and the inflammatory response to alcohol. Hangover severity was significantly and positively correlated with blood concentrations of biomarkers of the inflammatory response to alcohol, in particular, Interleukin-6 (IL-6), tumor necrosis factor-alpha (TNF-α) and C-reactive protein (CRP). At 4 h after alcohol consumption, blood ethanol concentration (but not acetaldehyde) was significantly and positively associated with elevated levels of IL-6, suggesting a direct inflammatory effect of ethanol. In addition, biomarkers of oxidative stress, i.e., malondialdehyde and 8-isoprostane, were significantly correlated with hangover severity, suggesting that oxidative stress also contributes to the inflammatory response. The timing of the assessments suggests initial slow elimination of ethanol in the first hours after alcohol consumption. As a consequence, more ethanol is present in the second half of the night and the next morning, which will elicit more oxidative stress and a more profound inflammatory response. Together, these processes result in more severe hangovers.

Keywords: alcohol; hangover; ethanol; acetaldehyde; acetate; oxidative stress; malondialdehyde; 8-isoprostane; cytokines; C-reactive protein

1. Introduction

The alcohol hangover refers to the combination of negative mental and physical symptoms, which can be experienced after a single episode of alcohol consumption, starting when blood alcohol concentration (BAC) approaches zero [1,2]. Reviews on the pathology of alcohol hangover [3–5] suggest that, although research is limited, both alcohol metabolism and the immune system play an important role in the development of alcohol hangover. The role of alcohol metabolism and oxidative stress are discussed elsewhere in detail [6]. These analyses revealed that higher blood concentrations of ethanol (but not acetaldehyde) and oxidative stress (a combination between depletion of antioxidants and an increase in free radicals, originating in the liver) were associated with having more severe hangovers. These processes are determined by the ethanol elimination rate. In other words, fast elimination of ethanol was associated with experiencing less severe hangovers [7]. The current analysis aimed to further elaborate the role of immune responses elicited by alcohol consumption in the development of the alcohol hangover.

A possible role of the immune system in the pathology of alcohol hangover was first hypothesized in the 1980s [8,9]. Kaivola et al. [8] examined the effects of tolfenamic acid on alcohol hangover in $n = 30$ social drinkers. Tolfenamic acid is a non-steroidal anti-inflammatory drug (NSAID), and prostaglandin (PG) is an inhibitor used for the treatment of pain, and as such, it could be hypothesized that the drug would be helpful in mitigating related hangover symptoms. From $n = 10$ subjects, blood samples were also taken, and prostaglandin PGE_2, $PGF_{2\alpha}$, thromboxane B_2 (TXB_2) and 6-keto $PGF_{1\alpha}$ were determined at 0, 5, 10 and 20 h after drinking. Subjects rated the efficacy of the drug on an 11-point scale ranging from "very ineffective" to "very effective", with subjects rating the treatment as significantly more effective than placebo. The same study further showed a significant reduction in the severity of several hangover symptoms after drug administration, with headache, dry mouth and thirst being the most drug-sensitive items. Alcohol consumption significantly increased PGE_2 and TXB_2 concentrations (the placebo condition). Administering tolfenamic acid partly counteracted the observed increments. No significant changes in $PGF_{2\alpha}$ and 6-keto $PGF_{1\alpha}$ were observed over time, and tolfenamic acid had no significant effect on these assessments.

Paratainen [9] reviewed the evidence for the role of prostaglandins in hangover, which at that time was limited to the study by Kaiviola et al. [8]. The review concluded that alcohol has an effect on prostaglandin synthesis, which can be counteracted by a prostaglandin inhibitor such as tolfenamic acid. It took 20 years for other studies to confirm that alcohol consumption is accompanied by an immune response, and that elevated levels of cytokines in blood and saliva may be related to the presence and severity of the alcohol hangover.

Kim et al. [10] demonstrated an inflammatory response accompanying alcohol consumption that was significantly associated with next-day hangover severity. In their study, twenty male Asian subjects had an evening of alcohol consumption (soju, a distilled beverage of Korean origin), which was compared to an alcohol-free control day. The following morning, blood samples were collected for determination of cytokine concentrations. The hangover scale consisted of two subscales, computing sum scores of ratings of subjective hangover symptoms (i.e., thirst, tension, depression and general discomfort, rated by the participant) or somatic hangover symptoms (i.e., paleness, tremor, perspiration, nystagmus, vomiting and general appearance, rated by two psychiatrists). On the hangover day, compared to the alcohol-free control day, significantly increased blood concentrations were found for interleukin (IL)-10, IL-12 and interferon-gamma (IFN-γ), whereas no significant changes were observed for IL-1β, IL-4, IL-6 or tumor necrosis factor-alpha (TNF-α). Furthermore, the increase in IL-12 and IFN-γ correlated significantly with overall hangover severity. Significant correlations with subjective hangover scale scores were found for difference scores (Δ, alcohol-control day) Δ IFN-γ, and significant correlations with somatic hangover scale scores were found for Δ TNF-α and Δ IL-12.

Wiese et al. [11] examined the effects of *Opuntia ficus indica* (OFI) (prickly pear) on alcohol hangover, administered 5 h before alcohol consumption. OFI is used as a dietary supplement for its antioxidant and anti-inflammatory properties [12]. The authors concluded that the alcohol hangover is mediated by an inflammatory response, as they observed significant correlations between blood C-reactive protein (CRP) concentration and hangover severity. After placebo treatment, CRP levels rose by 50%, accompanied by experiencing an alcohol hangover. After administering OFI, no inflammatory response was seen (CRP levels did not differ from placebo) and several hangover symptoms were alleviated.

George et al. [13] examined the effects of a new hangover treatment (*Phyllanthus amarus*, PHYLLPRO™, 750 mg/day, administered for 10 days) on hangover severity. *Phyllanthus amarus* is traditionally used in India for the treatment of hepatitis and jaundice, and for general liver health, and its reported effectiveness is most likely due to its antioxidant properties [14]. George et al. assessed a variety of blood cytokines the day following alcohol consumption, and hangover severity reported 10 h after drinking. In this study, IL-8, IL-10 and IL-12p70 levels were increased significantly in the active group, denoting an anti-inflammatory effect of the treatment compared to the placebo treatment alcohol challenge. Unfortunately, the authors did not report correlations between inflammatory markers and the presence or severity of alcohol hangover.

Van de Loo et al. [15] examined cytokine concentration in saliva. $n = 36$ healthy social drinkers (18 to 30 years old) participated in a naturalistic study. Based on lifetime self-report, at screening, subjects were allocated to a hangover-sensitive group (those who reported having hangovers), or the hangover-resistant group (those reporting never having hangovers despite consuming large quantities of alcohol). The morning following an evening of alcohol consumption and an alcohol-free control day, saliva samples were collected, and cytokine concentrations were assessed. Significant increases in IL-6 and IL-10 concentrations were observed the morning after heavy drinking. No significant changes were found for IL-1β, IL-8 and TNF-α. Interestingly, no significant differences were observed between the hangover-sensitive group and the drinkers that claimed to be hangover-resistant. The elevation in saliva IL-6 and IL-10 concentration did not significantly correlate with overall hangover severity. However, the correlations between IL-6 and headache ($r = 0.572$, $p = 0.017$) and between IL-6 and concentration problems ($r = 0.536$, $p = 0.027$) showed a trend towards significance (taking into account multiple comparisons, a conservative p-value cut-off was applied for statistical significance, which was set at $p < 0.002$).

Ethanol may elicit an inflammatory response directly, or indirectly via its breakdown products and oxidative stress [6]. The conversion of ethanol into acetaldehyde involves the production of reactive oxygen species (ROS) and reactive nitrogen species (RNS), which are both harmful for the body and thus elicit an immune response [16]. The free radicals are usually neutralized by antioxidants such as superoxide dismutase or glutathione. Oxidative stress refers to the situation when the amount of ROS and RNS is much larger than the present antioxidants. The free radicals are highly reactive. For example, malondialdehyde–acetaldehyde adducts (MAA adducts) are formed, which have proinflammatory properties [17]. The MAA adducts are recognized by the body as foreign substances, and as a result, an immune response is elicited [18–20], including increased secretion of cytokines and chemokines.

Other important biomarkers of oxidative stress are isoprostanes, i.e., prostaglandin-like compounds formed via non-enzymatic free radical-initiated lipid peroxidation of arachidonic acid, and other polyunsaturated fatty acids (PUFAs), including α-linolenic acid, eicosapentaenoic acid (EPA), adrenic acid and docosahexaenoic acid (DHA) [21]. However, studies that have examined oxidative stress in human subjects in the context of alcohol hangover are scarce [6].

Together, previous studies demonstrated that alcohol consumption is followed by an inflammatory response. However, the precise mechanism of how the inflammatory response is elicited is unclear, and there are mixed results on which specific cytokines play a key role in this process. Further, it is uncertain to what extent changes in cytokine concentrations correlate significantly to self-reported hangover severity. Finally, the previous studies also do not provide insight into how cytokine changes are related to biomarkers of alcohol metabolism and oxidative stress. This knowledge is, however, vital to better understand the pathology of the alcohol hangover. Therefore, the aim of the current article was to further investigate and integrate data on alcohol metabolism, oxidative stress and the inflammatory response to alcohol consumption in relation to alcohol hangover. To this extent, we re-evaluated data from the two available studies that collected this combined data in the same subjects, enabling a direct investigation into their interrelationships [22,23].

2. Methods

Data from two studies were re-evaluated (see original articles for details on methodology) [22,23].

In the first double-blind crossover study, Kim et al. [22] examined the effects of the fruit of *Hovenia dulcis* versus placebo on hangover severity in $n = 26$ healthy men (mean ± standard deviation (SD) age of 23.7 ± 0.3 years old), with heterozygous ALDH2. Subjects consumed 360 mL of Korean Soju (50 g alcohol) together with Hovenia dulcis extract (2460 mg) or matched placebo. Blood samples were taken 1, 4 and 12 h after alcohol consumption and ethanol and acetaldehyde concentrations were determined. Blood cytokine concentrations of IL-6, IL-10, IL-12 and TNF-α were assessed at 4 and 12 h after drinking alcohol. Hangover severity was assessed the next morning, 12 h after alcohol consumption, using a 14-item scale on which symptoms could be scored on a 5-point likers scale, which was developed for this study. The sum score of the 14 items was used as the overall hangover severity score. The study was conducted in accordance with the Declaration of Helsinki and was approved by the Institutional Review Board of Chonbuk National University Hospital (World Health Organization International Clinical Trials Registry Platform identification number: KCT0001626).

In a second double-blind crossover study, Mammen et al. [23] examined the effect of a polyphenolic extract of clove buds (Clovinol) versus placebo on hangover severity in $n = 16$ healthy men (mean ± SD age of 29.5 ± 6.2 years old). The study was conducted in India, and genetic polymorphisms in alcohol metabolic enzymes were not considered. Blood ethanol and acetaldehyde concentration were determined before and at 0.5, 2, 4 and 12 h after alcohol consumption (240 mL of 42.8% McDowell's Very Superior Old Pale (V.S.O.P.) Brandy; United Spirits Limited, Bangalore, India). To investigate the inflammatory response, IL-6 and CRP were also assessed at the same time points. Finally, two antioxidants (glutathione and superoxide dismutase (SOD)), and two biomarkers of oxidative stress (8-isoprostane and malondialdehyde) were measured. A hangover scale was completed 14 h after drinking. Hangover severity was measured using a 15-item scale on which symptoms could be scored on scales ranging from 1 (absent) to 10 (extreme). The scale was composed by the authors. The average score of the 15 items was computed and used as overall hangover severity score.

The study was carried out in accordance with the clinical research guidelines established by the Government of India, and protocol was approved by an independent ethical committee (Clinical trial Reg. No. ECR/64/Indt/KA/2013).

Statistical Analysis

Statistical analyses were conducted with SPSS, version 25 (Armonk, IBM Corp, New York, NY, USA). Subjects were only included in the current analysis if they had hangover severity scores greater than zero. Regarding individual data, outliers (mean ± 3 interquartile ranges) were omitted from the analysis. Only data from the placebo condition (i.e., alcohol-only, without treatment) were used. Means and standard deviation (SD) were computed for each variable. Non-parametric Spearman's correlations were used to account for a non-normal distribution.

Due to the relatively small sample size, a bootstrapping technique was applied [24,25] to simulate the population distributions of the partial correlations (r_P). To obtain an adequate resampling size [26], B = 10,000 bootstrapped samples (of $n = 8$ subjects each) were created by randomly drawing cases (resampling), with replacement, from the original sample. For each of the bootstrap samples, the bootstrapped partial correlation, denoted as r_{PB}, was calculated. The standard error (SE) represents how much the r_{PB}s vary across the bootstrap samples, and the reported 'bias' measure represents the deviation of the overall r_{PB} from the r_P that was obtained from the original sample [27]. The bias-corrected and accelerated bootstrapped 95% confidence interval (BCa 95% CI$_B$) was computed for each correlation [28]. The lower and upper limit of the BCa 95% CI$_B$ can range from −1 to +1, with narrower BCa 95% CI$_B$'s implying greater precision. Bootstrapping does not provide *p*-values, but significance is determined by the BCa 95% CI$_B$. If the BCa 95% CI$_B$ does not include zero, the r_{PB} is considered statistically significant (corresponding to a significance level of $\alpha = 5\%$).

3. Results

3.1. Study 1

Only data from the placebo condition (i.e., alcohol-only, without treatment) were considered. Analysis of the raw data revealed that $n = 10$ subjects did not report any hangover symptoms. Data from these $n = 10$ subjects were omitted. The results from the remaining $n = 16$ subjects are summarized in Table 1 and Figure 1.

Table 1. Biomarkers of alcohol metabolism, the inflammatory response to alcohol consumption and their relation to hangover severity.

Biomarker	Time After Alcohol Consumption			
	0 h	1 h	4 h	12 h
Ethanol (mg/dL)	0.0 (0.0)	56.8 (10.0)	38.9 (7.2)	1.1 (0.4)
Acetaldehyde (mg/dL)	0.0 (0.0)	0.13 (0.07)	0.04 (0.02)	0.01 (0.01)
Interleukin (IL)-6 (pg/mL)	0.50 (1.2)	–	0.97 (1.7)	1.03 (1.7) *
Interleukin (IL)-10 (pg/mL)	0.35 (0.5)	–	0.29 (0.4)	0.28 (0.4)
Interleukin (IL)-12 (pg/mL)	0.04 (0.07)	–	0.01 (0.006)	0.01 (0.006)
Interferon-gamma (IFN-γ) (pg/mL)	0.09 (0.2)	–	0.02 (0.01)	0.04 (0.07)
Tumor necrosis factor-alpha (TNF-α) (pg/mL)	3.56 (1.9)	–	3.31 (1.4) *	4.00 (2.0) *

Biomarker assessments after consumption of alcohol to reach a desired breath alcohol concentration (BrAC) of 0.06%. Mean and standard deviation (SD) (between brackets) are shown. Significant correlations with hangover severity ($p < 0.05$), after bootstrapping (B = 10,000 bootstrapped samples), are indicated by *. – = not assessed. Data from Kim et al. [22].

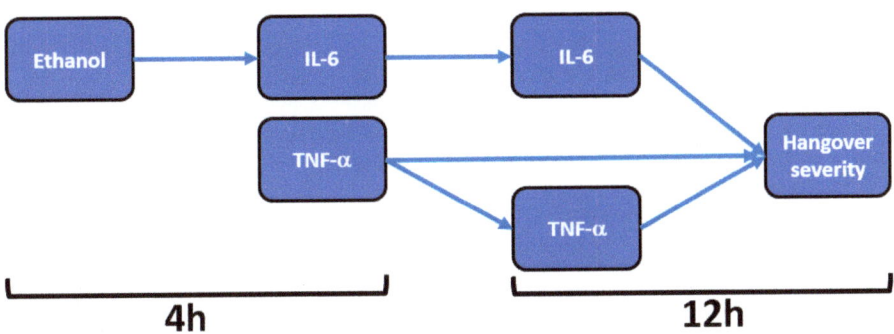

Figure 1. Relationship of hangover severity with blood ethanol concentration and the inflammatory response to alcohol. Note: Arrows represent significant correlations after bootstrapping. Abbreviations: Il = interleukin, TNF = tumor necrosis factor, h = hours after alcohol consumption. Data from Kim et al. [22].

Mean (SD) hangover severity, assessed 12 h after alcohol consumption, was 16.2 (1.8). The analysis revealed no significant correlations between blood ethanol and overall hangover severity at any timepoint. Also, no significant correlations were found between blood acetaldehyde and overall hangover severity at any timepoint. Significant correlations were found between hangover severity and TNF-α at 4 h ($r_B = 0.679$, BCa 95%CI$_B$ = 0.440, 0.835) and TNF-α at 12 h ($r_B = 0.625$, BCa 95%CI$_B$ = 0.252, 0.889). A significant correlation was also found between IL-6 at 12 h and hangover severity ($r_B = 0.459$, BCa 95%CI = 0.036, 0.796).

On the contrary, the inflammatory response seems to be directly related to blood ethanol concentrations. That is, significant positive correlations were found between blood ethanol at 4 h and IL-6 at 4 h ($r_B = 0.482$, BCa 95%CI$_B$ = 0.038, 0.803) and between blood ethanol at 4 h and IL-6 at 12 h ($r_B = 0.573$, BCa 95%CI$_B$ = 0.216, 0.835). Correlations between IL-6 and TNF-α did not reach statistical significance for any timepoint combination. In contrast to ethanol, the inflammatory response was not significantly associated with acetaldehyde concentrations.

3.2. Study 2

Data on hangover severity of three subjects were not collected. Raw data from the remaining $n = 13$ subjects were used for the statistical analysis. Their mean (SD) hangover severity, assessed 14 h after alcohol consumption, was 4.4 (0.5). The results of the assessments in the alcohol-only condition are summarized in Table 2.

Table 2. Biomarkers of alcohol metabolism, oxidative stress and the inflammatory response to alcohol consumption and their relation to hangover severity.

Biomarker	Time After Alcohol Consumption				
	0 h	0.5 h	2 h	4 h	12 h
Ethanol (mg/dL)	–	26.2 (4.5)	14.8 (3.5)	6.8 (1.0)	2.7 (0.4)
Acetaldehyde (mg/dL)	–	0.42 (0.03)	0.28 (0.05)	0.17 (0.02)	–
Interleukin (IL)-6 (pg/mL)	4.2 (0.0)	4.3 (0.8)	5.1 (1.0)	5.8 (1.0)	6.1 (1.0)
C-reactive protein (mg/L)	2.2 (0.6)	2.6 (0.7)	3.3 (0.8)	3.8 (0.7) *	3.8 (1.2)
8-isoprostane (pg/mL)	22.6 (4.5)	25.3 (3.5)	26.2 (4.8)	29.8 (7.1)	30.6 (6.5) *
Malondialdehyde (nmol/dL)	72.0 (7.3)	83.4 (7.2) *	94.4 (7.0)	95.5 (7.0)	92.6 (13.3)
Glutathione (nmol/mg Hb)	23.9 (3.4)	21.7 (3.1)	21.6 (3.9)	23.0 (2.4)	21.7 (3.5)
Superoxide dismutase (U/mL)	0.42 (0.1)	0.35 (0.1)	0.30 (0.1)	0.28 (0.1)	0.27 (0.1)

Biomarker assessments after consumption of alcohol to reach a desired BrAC of 0.03%. Mean and standard deviation (between brackets) are shown. Significant correlations with hangover severity ($p < 0.05$), assessed 14 h after alcohol consumption, are indicated by *. – = ethanol and acetaldehyde were not assessed at 0 h, and acetaldehyde could not be detected 12 h after alcohol consumption. Data from Mammen et al. [23].

Acetaldehyde levels significantly increased directly after alcohol consumption ($p < 0.0001$), but declined steadily thereafter, and at 12 h after drinking, acetaldehyde could no longer be detected, whereas ethanol concentrations were still detectable after this time period. No significant correlations were found between hangover severity and blood ethanol or acetaldehyde concentrations at any time point after alcohol consumption.

CRP concentration at 4 h was significantly associated with hangover severity at 14 h ($r_B = 0.534$, BCa 95%CI$_B$ = 0.120, 0.825). No significant correlations between IL-6 and hangover severity were found at any time point, and the associations between blood ethanol and acetaldehyde concentrations and IL-6 were also not significant at any timepoint. A significant correlation was found between IL-6 at 2 h and CRP at 4 h ($r_B = 0.587$, BCa 95%CI$_B$ = −0.874, −0.130). CRP concentrations did not correlate significantly with blood ethanol or acetaldehyde concentrations at any timepoint.

With regard to antioxidants, the assessments revealed that both glutathione and SOD concentrations showed a decrease after alcohol consumption, but at no timepoint did their concentrations correlate significantly with hangover severity. The concentration of 8-isoprostane increased over time and reached its highest point 12 h after alcohol consumption ($p < 0.0001$).

A significant correlation was found between hangover severity and 8-isoprostane at 12 h ($r_B = 0.475$, BCa 95%CI$_B$ = 0.06, 0.78). Also, the blood malondialdehyde concentration at 0.5 h correlated significantly with hangover severity at 14 h ($r_B = −0.707$, BCa 95%CI$_B$ = −0.96, −0.16). The negative correlation suggests that higher malondialdehyde concentrations directly after drinking are associated with experiencing less severe hangovers. 8-isoprostane at 2 h after alcohol consumption correlated significantly, and negatively, with acetaldehyde assessed 0.5 h ($r_B = −0.612$, BCa 95%CI$_B$ = −0.884, −0.192) and 2 h ($r_B = −0.613$, BCa 95%CI$_B$ = −0.881, −0.191) after alcohol consumption, while ethanol concentrations revealed non-significant positive associations at these timepoints. Assessments at 2 h after alcohol consumption revealed a significant negative association ($r_B = −0.541$, BCa 95%CI$_B$ = 0.113, 0.833) between oxidative stress (8-isoprostane) and antioxidant depletion (glutathione). The results are summarized in Figures 2 and 3.

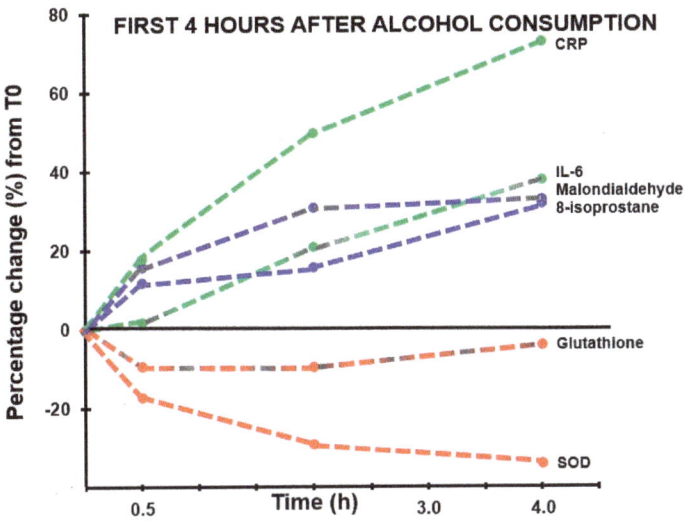

Figure 2. Markers of oxidative stress and inflammatory responses in the first four hours after alcohol consumption. Percentage changes between assessments made in the first 4 h after alcohol consumption relative to T0 (the assessment made before alcohol consumption) are shown. Abbreviations: CRP = C-reactive protein, IL-6 = interleukin-6, SOD = superoxide dismutase.

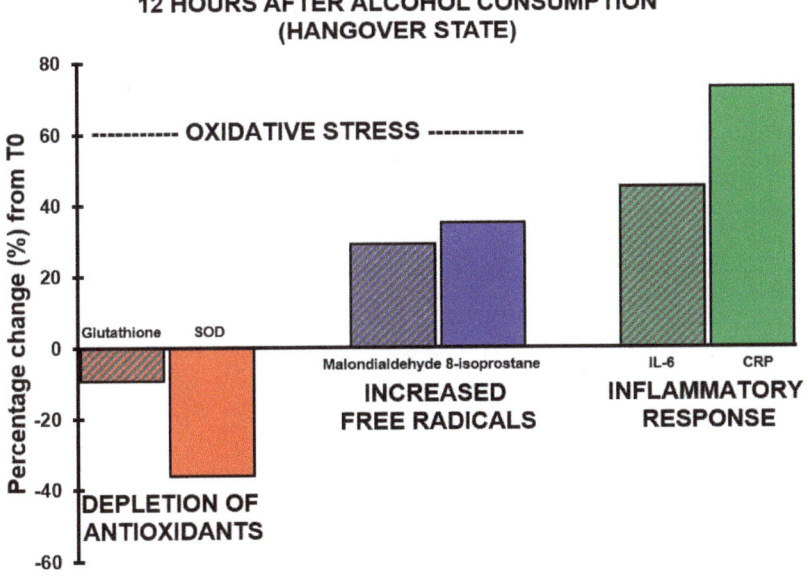

Figure 3. Oxidative stress and the inflammatory response during alcohol hangover. Percentage changes between assessments made 12 h after alcohol consumption relative to T0 (the assessment made before alcohol consumption) are shown. Abbreviations: CRP = C-reactive protein, IL-6 = interleukin-6, SOD = superoxide dismutase.

4. Discussion

Multiple lines of evidence, as summarized elsewhere [6], suggest that the amount of ethanol present in the blood is an important determinant of hangover severity. Specifically, faster conversion of ethanol into acetaldehyde and other aldehydes is associated with having less severe hangovers [7]. Data from the two studies evaluated here confirm and extend previous findings. For example, the data support previously reported relationships between oxidative stress and inflammatory responses to alcohol consumption (i.e., increases in IL-6 and CRP). Blood malondialdehyde concentration at 0.5 h correlated negatively with hangover severity at 14 h. Thus, higher oxidative stress in the first hours after drinking was associated with less severe next-day hangovers. The significant positive correlation between hangover severity and 8-isoprostane at 12 h after alcohol consumption suggests that more oxidative stress, experienced at a later stage after alcohol consumption, is associated with having more severe hangovers. In other words, a quick ethanol metabolism results in more oxidative stress in the first hours of drinking. This is associated with having less severe next-day hangovers. On the other hand, if ethanol metabolism is relatively slow, there will be more oxidative stress the following morning, which is associated with having more severe hangovers. Increased ethanol concentration, but not acetaldehyde, was associated with elevated IL-6 levels. The magnitude of the inflammatory response correlated significantly with hangover severity.

Low-grade inflammation, i.e., relative minor elevations in proinflammatory cytokine concentrations, is a common underlying condition in many diseases [29]. Alcohol consumption is one of the many factors that can provoke such an elevation of proinflammatory cytokines. The studies reviewed in the introduction, and the data from the two studies presented here, demonstrate significant and robust elevations in cytokine concentrations, including IL-6, IL-10, IL-12, IFN-γ and TNF-α, after alcohol consumption and during the hangover state. The relationship between CRP and cytokine production is complex and bi-directional, and CRP production is usually followed in time by cytokine production, particularly IL-6 and TNF-α are induced by the presence of CRP [30].

Results of the two studies were in agreement that an inflammatory response follows alcohol consumption. However, there were also differences. For example, whereas Kim et al. [22] found a significant correlation between Il-6 and hangover severity, this correlation did not reach significance in the study by Mammen et al. [23]. Other studies also reported no significant correlation between IL-6 concentration and hangover severity [10,14]. The finding by Mammen et al. [23] that CRP concentration did correlate significantly with hangover severity is in line with previous findings [11]. In contrast to other studies [10,15], we did not find significant correlations between hangover severity and blood ethanol concentration. The observed differences between the outcomes of the two studies and available literature can be explained by differences in the study designs. For example, with regards to subject recruitment, one study took genetic polymorphism into account [22] whereas the other study did not [23]. The administered type and amount of alcohol differed between the two studies, which resulted in different blood alcohol concentrations. At 4h after alcohol consumption, the difference in blood alcohol levels was roughly 32.1 mg/dL (~0.03%) between the two studies, which might explain why some associations with overall hangover severity were only significant in one study. Further, different hangover assessment scales were used, and the assessments of hangover severity were made at different points in time after drinking (12 versus 14 h after drinking). Similarly, the biomarkers of alcohol metabolism and oxidative stress were assessed at various points in time after alcohol intake, and the timing of the assessments differed between the two studies. Of note, the expression of various cytokines is interrelated and time-dependent. Therefore, different timing of assessments may explain why in some study designs, a relationship of a biomarker of alcohol metabolism or specific cytokines significantly correlated with hangover severity, while in another study, the same correlation does not reach statistical significance. Notwithstanding this, it is clear from all studies that alcohol consumption is followed by oxidative stress and an inflammatory response.

There are several issues that deserve attention when interpreting the results of the two studies we re-analyzed here. First, participants of the studies were young males only. Therefore, it is unclear to what extent the results can be generalized to females and other age groups. Research has shown that age and sex differentially impact both alcohol elimination rate [31] and inflammatory responsiveness [32]. Other studies that did include both male and female subjects (discussed in the introduction section of this article) did not report any sex or age differences in aspects of the pathology of the alcohol hangover. However, these aspects may not have been reported as the investigated samples were not powered to conduct such comparisons. Future research should therefore be conducted in samples comprising both men and women, and also investigate other age groups. These studies should have sufficient sample size to allow comparisons among groups.

Second, the subjects in the re-evaluated studies were of Asian descent. There are differences reported in ethanol metabolism between different ethnic groups. For example, it has been found that in populations of Asian descent, subjects with ALDH2*2 alleles, i.e., those who breakdown acetaldehyde more slowly, typically report significantly worse hangovers, and are more likely to experience hangovers at relatively lower alcohol consumption levels. [33,34]. Therefore, future studies should confirm the current observations in subjects of non-Asian descent.

Third, the administered amount of alcohol consumed was relatively low in the re-evaluated studies. It would be interesting to further investigate the observed effects when higher amounts of alcohol are consumed that may better reflect real-life alcohol consumption. On the other hand, also with the currently applied consumption levels, inflammatory effects were evident. Moreover, recent studies have demonstrated that consuming large amounts of alcohol is not a necessity for having hangovers [35]. In fact, a recent large sample study revealed an average estimated BAC of 0.03% of subjects that reported a hangover [36], which is lower that the achieved BACs in the re-evaluated studies here.

Fourth, both studies assessed hangover severity using a composite scale of hangover symptoms. Recent examinations revealed that the magnitude of hangover severity of composite scales ultimately depends on which symptoms are included in such scales [37]. This may have had an impact on the correlational analyses. Future research should also use single-item hangover severity assessments as these better reflect overall hangover severity, and its impact. It is also important to thoroughly investigate whether potential participants of hangover studies will likely have a hangover when a certain amount of alcohol is consumed. In the study by Kim et al. [22], $n = 10$ subjects were excluded after participation because they reported no hangover during the study. Instead of excluding them after participation, it is more ideal to identify them during screening, and screening tools for this purpose are currently under development.

To conclude, it should be noted that the current body of data examining the pathology of the alcohol hangover is limited, and much more research is needed to further elucidate the exact nature of how ethanol metabolism, oxidative stress and the immune response are interrelated and may impact hangover severity. In this context, it is important to note that the two studies in this article comprised only male participants. As it is known that ethanol elimination rate is influenced by sex, future studies should be conducted to further investigate the role of the immune system on hangover in samples comprising both men and women, of different ethnicity and age.

Taken together, the timing of the observed assessments (see Figures 2 and 3) suggests that initial slow elimination of ethanol in the first hours after drinking results in more ethanol present in the second half of the night and the next morning, which will elicit more oxidative stress and a stronger inflammatory response. Together, these processes result in having more severe next-morning hangovers (See Figure 4).

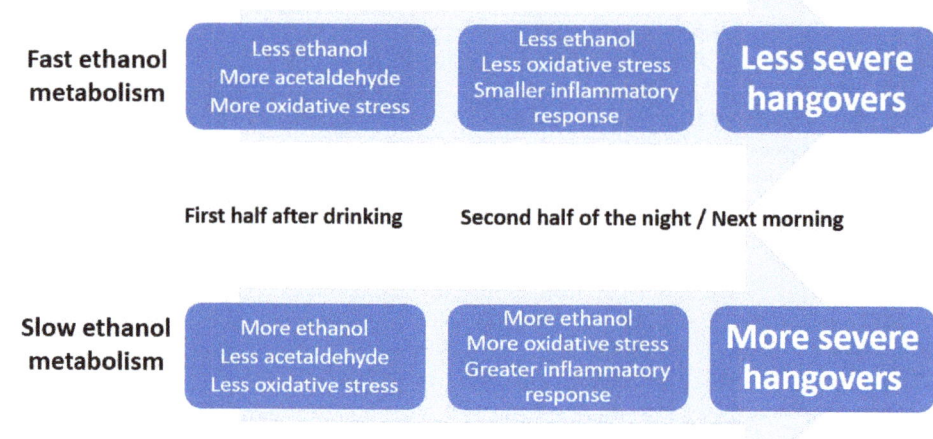

Figure 4. The relationship of alcohol metabolism, oxidative stress and the inflammatory response with hangover severity. Note: The rate of alcohol metabolism, i.e., the conversion of ethanol into acetaldehyde and other aldehydes, seems to be the primary determinant of hangover severity.

Author Contributions: Conceptualization, J.C.V., A.J.A.E.v.d.L., M.M., J.G., A.D.K., and A.S.; investigation, O.K. and I.M.K.; formal analysis, J.C.V.; writing—original draft preparation, J.C.V.; writing—review and editing, J.C.V., A.S., A.J.A.E.v.d.L., M.M., J.G., A.D.K., O.K., and I.M.K. All authors have read and agreed to the published version of the manuscript.

Funding: This research received no external funding.

Conflicts of Interest: Over the past 36 months, A.D.K. has held research grants from H2020, Nutricia-Danone, The Netherlands Center of Translational Research, Lungfund, SGF/Health Holland and NWO. J.G. is part-time employee of Nutricia Research. Over the past 36 months, J.G. has held research grants from several profit and non-profit organizations, including EU, NWO, Health Holland, TIFN, and acted as a consultant/expert advisor to Nutricia research Foundation, Friesland Campina, Carbohydrate Competitive Center, International Olympic Committee, Bill Gates Foundation, ID-DLO (WUR). I.M.K. is a member of Akay Flavours and Aromatics Pvt. Ltd., that market Clovinol (Clove bud extract). Over the past 36 months, A.S. has held research grants from Abbott Nutrition, Arla Foods, Bayer, BioRevive, DuPont, Kemin Foods, Nestlé, Nutricia-Danone, Verdure Sciences. He has acted as a consultant/expert advisor to Bayer, Danone, Naturex, Nestlé, Pfizer, Sanofi, Sen-Jam Pharmaceutical, and has received travel/hospitality/speaker fees from Bayer, Sanofi and Verdure Sciences. Over the past 36 months, J.C.V. has held grants from Janssen, Nutricia, and Sequential Medicine, and acted as a consultant/expert advisor to Clinilabs, More Labs, Sen-Jam Pharmaceutical, Toast!, and ZBiotics. A.E.A.E.v.d.L., M.M., and O.K. have no conflict of interest to declare.

References

1. Van Schrojenstein Lantman, M.; van de Loo, A.J.; Mackus, M.; Verster, J.C. Development of a definition for the alcohol hangover: Consumer descriptions and expert consensus. *Curr. Drug Abuse Rev.* **2016**, *9*, 148–154. [CrossRef] [PubMed]
2. Verster, J.C.; Scholey, A.; van de Loo, A.J.A.E.; Benson, S.; Stock, A.K. Updating the definition of the alcohol hangover. *J. Clin. Med.* **2020**, *9*, 823. [CrossRef] [PubMed]
3. Penning, R.; van Nuland, M.; Fliervoet, L.A.L.; Olivier, B.; Verster, J.C. The pathology of alcohol hangover. *Curr. Drug Abuse Rev.* **2010**, *3*, 68–75. [CrossRef] [PubMed]
4. Tipple, C.T.; Benson, S.; Scholey, A. A Review of the Physiological Factors associated with alcohol hangover. *Curr. Drug Abuse Rev.* **2016**, *9*, 93–98. [CrossRef]

5. Palmer, E.; Tyacke, R.; Sastre, M.; Lingford-Hughes, A.; Nutt, D.; Ward, R.J. Alcohol hangover: Underlying biochemical, inflammatory and neurochemical mechanisms. *Alcohol Alcohol.* **2019**, *54*, 196–203. [CrossRef]
6. Mackus, M.; van de Loo, A.J.A.E.; Garssen, J.; Kraneveld, A.D.; Scholey, A.; Verster, J.C. The role of alcohol metabolism in the pathology of alcohol hangover. Submitted.
7. Mackus, M.; van de Loo, A.J.E.A.; Garssen, J.; Kraneveld, A.D.; Scholey, A.; Verster, J.C. The association between ethanol elimination rate and hangover severity. *Int. J. Environ. Res. Publ. Health* **2020**, *17*, 4324. [CrossRef]
8. Kaivola, S.; Parantainen, J.; Osterman, T.; Timonen, H. Hangover headache and prostaglandins: Prophylactic treatment with tolfenamic acid. *Cephalalgia* **1983**, *3*, 31–36. [CrossRef]
9. Parantainen, J. Prostaglandins in alcohol intolerance and hangover. *Drug Alcohol Depend.* **1983**, *11*, 239–248. [CrossRef]
10. Kim, D.J.; Kim, W.; Yoon, S.J.; Choi, B.M.; Kim, J.S.; Go, H.J.; Kim, Y.K.; Jeong, J. Effects of alcohol hangover on cytokine production in healthy subjects. *Alcohol* **2003**, *31*, 167–170. [CrossRef]
11. Wiese, J.; McPherson, S.; Odden, M.C.; Shlipak, M.G. Effect of *Opuntia ficus indica* on symptoms of the alcohol hangover. *Arch. Intern. Med.* **2004**, *164*, 1334–1340. [CrossRef] [PubMed]
12. Attanzio, A.; Tesoriere, L.; Vasto, S.; Pintaudi, A.M.; Livrea, M.A.; Allegra, M. Short-term cactus pear [*Opuntia ficus-indica* (L.) Mill] fruit supplementation ameliorates the inflammatory profile and is associated with improved antioxidant status among healthy humans. *Food Nutr. Res.* **2018**, *62*. [CrossRef]
13. George, A.; Udani, J.K.; Yusof, A. Effects of Phyllanthus amarus PHYLLPROTM leaves on hangover symptoms: A randomized, double-blind, placebo-controlled crossover study. *Pharm. Biol.* **2019**, *57*, 145–153. [CrossRef]
14. Karuna, R.; Reddy, S.S.; Baskar, R.; Saralakumari, D. Antioxidant potential of aqueous extract of Phyllanthus amarus in rats. *Indian J. Pharmacol.* **2009**, *41*, 64–67. [PubMed]
15. Van de Loo, A.J.A.E.; Raasveld, S.J.; Hogewoning, A.; de Zeeuw, R.; Bosma, E.R.; Bouwmeester, N.H.; Lukkes, M.; Knipping, K.; Brookhuis, K.A.; Mackus, M.; et al. Immune responses after heavy alcohol consumption: Cytokine changes in drinkers with and without a hangover. Submitted.
16. Hernández, J.A.; López-Sánchez, R.C.; Rendón-Ramírez, A. Lipids and oxidative stress associated with ethanol-induced neurological damage. *Oxid. Med. Cell. Longev.* **2016**, *2016*. [CrossRef] [PubMed]
17. Tuma, D.J. Role of malondialdehyde-acetaldehyde adducts in liver injury. *Free Radic. Biol. Med.* **2002**, *32*, 302–308.
18. Niemela, O. Aldehyde-protein adducts in the liver as a result of ethanol-induced oxidative stress. *Front. Biosci.* **1999**, *4*, d506–d513. [CrossRef]
19. Thiele, G.M.; Worrall, S.; Tuma, D.J.; Klassen, L.W.; Wyatt, T.A.; Nagata, N. The chemistry and biological effects of malondialdehyde-acetaldehyde adducts. *Alcohol Clin. Exp. Res.* **2001**, *25*, 218S–224S. [CrossRef]
20. Tuma, D.J.; Casey, C.A. Dangerous byproducts of alcohol breakdown—Focus on adducts. *Alcohol Res. Health* **2003**, *27*, 285–290.
21. Milne, G.L.; Yin, H.; Hardy, K.D.; Davies, S.S.; Roberts, L.J., 2nd. Isoprostane generation and function. *Chem. Rev.* **2011**, *111*, 5973–5996. [CrossRef]
22. Kim, H.; Kim, Y.J.; Jeong, H.Y.; Kim, J.Y.; Choi, E.K.; Chae, S.W.; Kwon, O. A standardized extract of the fruit of Hovenia dulcis alleviated alcohol-induced hangover in healthy subjects with heterozygous ALDH2: A randomized, controlled, crossover trial. *J. Ethnopharmacol.* **2017**, *209*, 167–174. [CrossRef]
23. Mammen, R.R.; Natinga Mulakal, J.; Mohanan, R.; Maliakel, B.; Krishnakumar, I.M. Clove bud polyphenols alleviate alterations in inflammation and oxidative stress markers associated with binge drinking: A randomized double-blinded placebo-controlled crossover study. *J. Med. Food* **2018**, *21*, 1188–1196. [CrossRef] [PubMed]
24. Efron, B. Bootstrap methods: Another look at the jackknife. *Ann. Stat.* **1979**, *7*, 1–26. [CrossRef]
25. Efron, B.; Tibshirani, R. *An Introduction to the Bootstrap*; Chapman & Hall/CRC: Boca Raton, FL, USA, 1993.
26. Rahman, S.; Majumber, A.K. *Use of Bootstrapping in Hypothesis Testing: Bootstrapping for Estimation and Hypothesis Testing*; LAP LAMBERT Academic Publishing: Riga, Latvia, 2013.
27. Sideridis, G.D.; Simos, P. What is the actual correlation between expressive and receptive measures of vocabulary? Approximating the sampling distribution of the correlation coefficient using the bootstrapping method. *Int. J. Educ. Psychol. Assess.* **2010**, *5*, 117–133.
28. Efron, B. Better bootstrap confidence intervals. *J. Am. Stat. Assoc.* **1987**, *82*, 171–185. [CrossRef]

29. Eklund, C.M. Proinflammatory cytokines in CRP baseline regulation. *Adv. Clin. Chem.* **2009**, *48*, 111–136. [PubMed]
30. Sproston, N.R.; Ashworth, J.J. Role of C-Reactive Protein at Sites of Inflammation and Infection. *Front. Immunol.* **2018**, *9*, 754. [CrossRef]
31. Fiorentino, D.D.; Moskowitz, H. Breath alcohol elimination rate as a function of age, gender, and drinking practice. *Forensic Sci. Int.* **2013**, *233*, 278–282. [CrossRef]
32. Murtaj, V.; Belloli, S.; Di Grigoli, G.; Pannese, M.; Ballarini, E.; Rodriguez-Menendez, V.; Marmiroli, P.; Cappelli, A.; Masiello, V.; Monterisi, C.; et al. Age and Sex Influence the Neuro-inflammatory Response to a Peripheral Acute LPS Challenge. *Front. Aging Neurosci.* **2019**, *11*, 299. [CrossRef]
33. Wall, T.L.; Horn, S.M.; Johnson, M.L.; Smith, T.L.; Carr, L.G. Hangover symptoms in Asian Americans with variations in the aldehyde dehydrogenase (ALDH2) gene. *J. Stud. Alcohol* **2000**, *61*, 13–17. [CrossRef]
34. Yokoyama, M.; Yokoyama, A.; Yokoyama, T.; Funazu, K.; Hamana, G.; Kondo, S.; Yamashita, T.; Nakamura, H. Hangover susceptibility in relation to aldehyde dehydrogenase-2 genotype, alcohol flushing, and mean corpuscular volume in Japanese workers. *Alcohol Clin. Exp. Res.* **2005**, *29*, 1165–1171. [CrossRef]
35. Verster, J.C.; Kruisselbrink, L.D.; Slot, K.A.; Anogeianaki, A.; Adams, S.; Alford, C.; Arnoldy, L.; Ayre, E.; Balikji, S.; Benson, S.; et al. Sensitivity to experiencing alcohol hangovers: Reconsideration of the 0.11% blood alcohol concentration (BAC) threshold for having a hangover. *J. Clin. Med.* **2020**, *9*, 179. [CrossRef] [PubMed]
36. Verster, J.C.; Kruisselbrink, L.D.; Anogeianaki, A.; Alford, C.; Stock, A.K. Relationship of alcohol hangover and physical endurance performance: Walking the Samaria Gorge. *J. Clin. Med.* **2020**, *9*, 114. [CrossRef] [PubMed]
37. Verster, J.C.; van de Loo, A.J.A.E.; Benson, S.; Scholey, A.; Stock, A.-K. The assessment of overall hangover severity. *J. Clin. Med.* **2020**, *9*, 786. [CrossRef] [PubMed]

© 2020 by the authors. Licensee MDPI, Basel, Switzerland. This article is an open access article distributed under the terms and conditions of the Creative Commons Attribution (CC BY) license (http://creativecommons.org/licenses/by/4.0/).

Article

Dietary Nutrient Intake, Alcohol Metabolism, and Hangover Severity

Joris C. Verster [1,2,3,*], **Sterre A. Vermeulen** [1], **Aurora J. A. E. van de Loo** [1,2], **Stephanie Balikji** [1], **Aletta D. Kraneveld** [1,2], **Johan Garssen** [1,4] **and Andrew Scholey** [3]

1. Division of Pharmacology, Utrecht Institute for Pharmaceutical Sciences, Utrecht University, 3584CG Utrecht, The Netherlands
2. Institute for Risk Assessment Sciences (IRAS), Utrecht University, 3584CM Utrecht, The Netherlands
3. Centre for Human Psychopharmacology, Swinburne University, Melbourne, VIC 3122, Australia
4. Nutricia Research, 3584CT Utrecht, The Netherlands
* Correspondence: j.c.verster@uu.nl; Tel.: +31-030-253-6909

Received: 14 July 2019; Accepted: 24 August 2019; Published: 27 August 2019

Abstract: Several dietary components have been shown to influence alcohol metabolism and thereby potentially affect the development of a hangover. From the literature, it is evident that dietary nicotinic acid and zinc play a pivotal role in the oxidation of ethanol into acetaldehyde. The aim of the current study was to associate dietary intake of nicotinic acid and zinc with hangover severity. To this end, data from $n = 23$ healthy social drinkers who participated in a naturalistic hangover study were analyzed. $n = 10$ of them reported to be hangover-resistant (the control group), whereas $n = 13$ reported to have regular hangovers (the hangover-sensitive group). Two 24 h dietary recall records were completed, one for the day of alcohol consumption and another one for an alcohol-free control day. Dietary nutrient intake was averaged and did not significantly differ between hangover-sensitive and hangover-resistant drinkers. For the hangover-sensitive drinkers, partial correlations with overall hangover severity were computed, controlling for estimated blood alcohol concentration. A bootstrapping technique was applied to account for the relatively small sample size. The results showed that dietary intake of nicotinic acid ($r_{PB} = -0.521$) and zinc ($r_{PB} = -0.341$) were significantly and negatively associated ($p < 0.002$) with overall hangover severity. Dietary zinc intake was also significantly and negatively associated with severity of vomiting ($r_{PB} = -0.577$, $p < 0.002$). No significant associations with hangover severity were found for other nutrients, such as fat and fibers. In conclusion, this study suggests that social drinkers who have a higher dietary intake of nicotinic acid and zinc report significantly less severe hangovers. As hangover-resistant and hangover-sensitive drinkers had a similar dietary nutrient intake, the claim of being hangover-resistant must be based on other unknown biopsychosocial factors. These findings should be replicated in a larger sample and include more elaborate food frequency questionnaires or nutrient-specific dietary intake records for zinc and nicotinic acid, and preferably accompanied by nutrient assessments in urine and/or blood.

Keywords: alcohol; hangover; nutrients; zinc; nicotinic acid; bootstrapping

1. Introduction

The alcohol hangover refers to the combination of mental and physical symptoms, experienced the day after a single episode of heavy drinking, starting when blood alcohol concentration (BAC) approaches zero [1]. The pathology underlying alcohol hangover is not well understood [2,3], and increasingly the subject of scientific investigation. In parallel, research has also been directed at the development of alcohol hangover treatments. This has led to the study of compounds that can influence the immune response to heavy alcohol consumption (which is assumed to contribute to alcohol hangover). Several hangover treatments have been reported to attenuate the rise in blood cytokines concentrations seen after heavy drinking, as well as reducing selective next day hangover symptoms. For example, Kim et al. [4] showed that *Hovenia dulcis Thunb* fruit extract (containing dihydromyricetin and heteropolysaccharides) significantly reduced blood cytokine concentrations that were increased due to heavy drinking. This was accompanied by a significant reduction in overall hangover severity. Interestingly, *Hovenia dulcis Thunb* fruit extract had no effect on alcohol metabolism (i.e., blood ethanol and acetaldehyde concentrations were not different from the alcohol only condition).

A different approach is to develop compounds that accelerate alcohol metabolism. The rationale for this approach is that more rapid elimination of ethanol and acetaldehyde could reduce the presence and severity of alcohol hangover symptoms. This hypothesis is supported by recent research showing that urine ethanol concentration was significantly lower in drinkers claiming to have no hangover after heavy alcohol consumption compared to drinkers who reported a hangover [5]. Although overall hangover severity was positively associated with the amount of ethanol found in urine of those who reported having a hangover, the partial correlation controlling for eBAC was not statistically significant. Nevertheless, this finding suggests that drinkers with slower alcohol metabolism, i.e., those with more ethanol in their urine, report significantly more frequent, and more severe hangovers. In other words, speeding up alcohol metabolism may have a beneficial effect on reducing hangover severity.

Another approach to the development of hangover treatments is to examine whether dietary nutrient intake has an effect on hangover severity. Two review papers have addressed this [6,7]. Firstly, Min et al. [6] argued that various minerals, including selenium, zinc, copper, vanadium, iron, and magnesium, may have a direct effect on either alcohol metabolism, on glutamatergic activity, or may influence the presence and severity of alcohol hangover via their antioxidant and/or anti-inflammatory properties. Secondly, Wang et al. [7] described the proposed mechanism of action of a number of natural products that might alleviate alcohol hangover symptoms. Both authors stressed that their hypotheses were based on limited animal research, and that research in humans is necessary to investigate the actual efficacy of minerals and herbal supplements in reducing or preventing alcohol hangover symptoms.

The scientific literature indicates that food intake can indeed have a significant effect on alcohol metabolism, both quantitatively and qualitatively. For example, relative to fasting, the consumption of foods before or together with alcohol reduces peak blood alcohol concentration (BAC), decreases absorption and slows metabolism [8–10]. In particular, 'high-energy' meals may slow down alcohol metabolism and reduce subjective intoxication [11–13]. Specific food products or nutrients have also been investigated. However, mixed results have been reported in relation to alcohol metabolism. For example, Kim et al. [14] found that consuming a mixed fruit and vegetable juice (Angelica keiskei/green grape/pear juice) significantly reduced peak BAC. In another study, Hong [15] examined the effect of a purported hangover treatment (DTS20, a mixture that consists of *Viscum album L., Lycium chinense L., Inonotus obliquus*, and *Acanthopanax senticosus H.*). The proposed active ingredients of DTS20 are sugar, uronic acid, and polyphenols. Relative to placebo, DTS20 significantly reduced BAC at 2 h after drinking alcohol in the form of Soju. The reduction in blood acetaldehyde levels, however, did not reach statistical significance. Taken together, there is limited evidence to date to support the notion that acute intake of specific nutrients can alter alcohol metabolism. Thus, further research into nutrients that can accelerate alcohol metabolism is warranted. Research on the possible impact of regular nutrient intake on hangover susceptibility to hangovers or relating nutrient intake to hangover symptom severity is currently lacking.

Alcohol is metabolized primarily in the liver via a two-step reaction (see Figure 1) [16,17]. First, ethanol is oxidized into acetaldehyde, which is highly toxic. Although the first step in alcohol metabolism is reversible, acetaldehyde is usually metabolized rapidly. In this second step, acetaldehyde enters the mitochondria where it is oxidized into acetate and water. This process is facilitated by mitochondrial aldehyde dehydrogenase (ALDH). For both steps, nicotinamide adenine dinucleotide

(NAD$^+$) is essential to provide the necessary energy for the conversion, which becomes available when NAD$^+$ is converted into NADH + H$^+$.

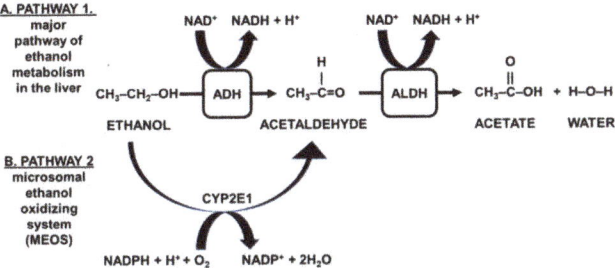

Figure 1. Pathways involved in alcohol metabolism. In the major metabolic pathway (**A**) ethanol is oxidized into acetaldehyde. This oxidative process is facilitated by alcohol dehydrogenase (ADH), which is present in high concentration in the cytosol of hepatocytes. In this second step, acetaldehyde enters the mitochondria where it is oxidized into acetate and water. This process is facilitated by mitochondrial aldehyde dehydrogenase (ALDH). For both steps, nicotinamide adenine dinucleotide (NAD$^+$) is essential to provide the necessary energy for the conversion, which becomes available when NAD$^+$ is converted into NADH + H$^+$. A second major pathway for alcohol breakdown, especially active in subjects who chronically drink alcohol, is the microsomal ethanol oxidizing system (MEOS, see (**B**)). The reaction is catalyzed by CYP2E1 and requires nicotinamide adenine dinucleotide phosphate (NADP$^+$) instead of NAD$^+$ to convert ethanol into acetaldehyde.

NADP$^+$ can be formed from NAD$^+$, and differs from NAD$^+$ in the presence of an additional phosphate group [18]. The conversion of acetaldehyde into acetate and water is similar to that overserved in the major alcohol metabolism pathway and requires NAD$^+$. A third minor pathway oxidizes ethanol into acetaldehyde via catalase (not shown in Figure 1) [17]. Together, the oxidative pathways account for over 90% of alcohol elimination [19]. Thus, ADH thus plays a vital role in alcohol metabolism.

Two nutrients are known to play an important role in alcohol metabolism, namely nicotinic acid and zinc [20,21]. Dietary intake of these micronutrients is necessary, as the body is unable to synthesize them itself [22,23]. Other nutrients do not seem to have an important direct influence on alcohol metabolism.

Zinc (Zn^{2+}) is absorbed from the small intestine. Most zinc can be found in tissue, with only 0.1% of total bodily zinc present in blood, where most (~70%) is bound to serum albumin [24]. From here zinc is transported as needed to body tissues. Since zinc is essential in the conversion of ethanol into acetaldehyde [20,21], we hypothesize that drinkers who consume abundant amounts of dietary zinc metabolize alcohol faster than those who consume relatively lower levels.

Niacin and its equivalents are the main dietary source of NAD$^+$. Tryptophan is also a source of NAD$^+$, and its relative contribution is estimated as 60 mg of tryptophan equaling 1 mg of nicotinic acid and other niacin equivalents, although a 30% individual variability in the conversion from tryptophan into nicotinic acid has been observed [25]. For the MEOS alcohol metabolism pathway (see Figure 1), NADP$^+$ is required. Nicotinic acid and its equivalents are the dietary sources of both NAD$^+$ and NADP$^+$, which together catalyze alcohol metabolism.

We hypothesize that when abundant amounts of nicotinic acid are present in the daily diet of a drinker, alcohol is metabolized faster than in drinkers who have lower levels of dietary nicotinic acid intake. To investigate the hypothesis that higher levels of dietary nicotinic acid and zinc may be protective against alcohol hangover of healthy social drinkers, dietary food intake was recorded during an experimental hangover study. Dietary nicotinic acid and zinc intake were computed and related to hangover severity.

2. Materials and Methods

A naturalistic study was conducted, including a 'control day' (no alcohol consumed) and a 'hangover day' (alcohol consumed the evening before), separated by approximately one week. Written informed consent was obtained from each participant, and The University of Groningen Psychology Ethics Committee approved the study.

2.1. Subjects

n = 23 subjects participated in the study. Of them, n = 13 subjects (n = 7 men and n = 6 women) regularly have hangovers when they consume alcohol and n = 10 (n = 4 men and n = 6 women) drinkers claimed to be hangover-resistant. They were recruited by local advertisement at Utrecht University. After indicating interest, participants were contacted over the phone and underwent an initial screening. Subjects were included if they were between 18–30 years old, mentally and physically healthy, social drinkers. Recreational drug users and smokers were excluded from participation. Further exclusion criteria comprised a positive urine drug or pregnancy screen, the use of medicinal drugs (including over-the-counter pain killers), and alcohol consumption within 24 h before the start of the control test day. To maximize the likelihood that subjects would actually have a hangover during the study, subjects were included only if it was demonstrated that they consume sufficient amounts of alcohol that can produce a hangover per se. To check this, estimated blood alcohol concentration (eBAC) was calculated for the alcohol consumption they reported for 'a regular night out'. This was done by applying a modified Widmark equation [26], taking into account gender and body weight. To be enrolled in the study, an eBAC of 0.08% or higher was required, based on hangover susceptibility likelihood calculations by Verster et al. [27] and Kruisselbrink et al. [28].

2.2. Procedures

The naturalistic design was chosen to closely mimic a realistic real-life hangover experience which has relatively higher ecological validity compared with other methodologies [29]. Researchers were not present during alcohol consumption, and thus had no influence on the participants' (drinking) behavior. Participants, therefore, dictated their own time period of drinking, types of alcoholic beverages consumed, and their activities during drinking (e.g., staying at home, going to a bar, dancing, etc.). Subjects were asked not to change their lifestyle while participating in the study. They were asked to refrain from consuming any alcohol three days prior to the control day. Both test days started at 9 a.m. Subjects completed a 24 h dietary recall diary [30] and the presence and severity of hangover symptoms were assessed. A urine drug screen (Instant-View, determining the presence of amphetamines, barbiturates, cannabinoids, benzodiazepines, cocaine, and opiates) was conducted according to the manufacturer's instructions (Alfa Scientific Designs, Inc., Poway, CA, USA).

2.3. Alcohol Consumption and Hangover Severity

At 09:45 a.m., previous night alcohol consumption was recorded (number of units), including the times at which drinking commenced and ceased. The adjusted Widmark formula [26] was used to calculate eBAC. Overall hangover severity was assessed with a single one-item rating on an 11-point scale ranging from 0 (absent) to 10 (extreme) [31]. In addition, using the same scale, severity of 23 hangover symptoms was assessed, including headache, nausea, concentration problems, regret, sleepiness, heart pounding, vomiting, being tired, shaking/shivering, clumsiness, weakness, dizziness, apathy, sweating, stomach pain, confusion, sensitivity to light, sensitivity to sound, thirst, heart racing, anxiety, depression, and reduced appetite.

2.4. Dietary Recall

The participants were asked to complete a 24-h dietary recall diary [31]. This assessed the 24 h dietary intake before both test days. Participants wrote down the amount and type of food as precisely as possible. Subjects were asked to record all food and beverages consumed for breakfast, lunch, dinner, and 'in-between' during the past 24 h. They were instructed to write down the time of consumption and the amount consumed (i.e., an estimate of the portion size). Examples were provided on how to complete the diary. Participants were urged not to forget smaller, 'less significant' food items, such as pieces of candy and to include details such as whether or not they buttered any bread consumed. Nutrient calculations were performed using the 'eetmeter' ('eating meter' in English) [32], developed by the 'Voedingscentrum' of 'Rijksinstituut voor Volksgezondheid en Milieu' (RIVM). The assessments take into account quantities, frequency, and portion sizes of consumed food and beverages. Nutrients obtained from alcoholic and non-alcoholic beverages, consumed during the drinking session, were included in the calculation models. The 'eetmeter' provided data on nutrient intake of nicotinic acid, zinc, total fat (triglycerides, esters derived from glycerol and fatty acids, and fatty constituents such as phosphatides and sterols), saturated fat (total of saturated fatty acids), carbohydrates (total of mono- and disaccharides, starch, dextrin, and glycogen), proteins, fibers, water, sugar, vitamin A (retinol), vitamin B1 (thiamine), vitamin B2 (riboflavin), vitamin B6 (pyridoxine, including pyridoxal and pyridoxamine), vitamin B11 (folic acid), vitamin B12 (cobalamins), vitamin C (ascorbic acid, including L-ascorbic acid and L-dehydro-ascorbic acid), vitamin D (cholecalciferol and 25-hydroxy vitamin D), vitamin E, salt, sodium, potassium, calcium, magnesium, iron, selenium, iodine, and phosphorus [33]. Dietary intake was computed for the alcohol and the control day. The average dietary intake of the two days was computed to better represent daily dietary nutrient intake.

2.5. Statistical Analysis

Statistical analyses were conducted with SPSS, version 25 (Armonk, IBM Corp, New York, NY, USA). Mean and standard deviation (SD) were computed for each variable. Hangover-sensitive drinkers were included in the analysis if they had an overall hangover score of 2 or higher. Hangover-resistant drinkers were included in the analysis is they had an overall hangover score of 0 or 1. Firstly, nutrient intake was computed separately for the alcohol test day and alcohol-free control day respectively. These were combined into a two-day average score to better reflect regular nutrient intake. Secondly, dietary nutrient intake of hangover-sensitive and hangover-resistant drinkers was compared. Third, dietary nutrient intake of hangover-sensitive drinkers was correlated with overall hangover severity, by computing partial correlations (r_P), controlling for eBAC. This approach controlled for the different alcohol consumption levels between participants. Thirdly, nutrient intake levels that correlated significantly with overall hangover severity were further correlated with the 23 individual hangover severity measures, again by computing r_P, controlling for eBAC.

To further examine the data and account for the relatively small sample size, a bootstrapping analysis [34,35] was conducted to simulate the population distributions of the partial correlations (r_P). Bootstrapping and interpretation of its results are summarized in Figure 2. In bootstrapping, data from the original sample (Figure 2A) are used to generate B new datasets (Figure 2B). The new samples have the same sample size as the original sample. The new samples are constructed by randomly drawing cases (resampling), with replacement, from the original sample. In the current analysis, B = 10.000 samples (of n = 13 subjects each) were created (Figure 2C), a recommended resampling size for bootstrapped CI estimation [36]. For each of the bootstrap samples a new r_P is then computed (the bootstrapped partial correlation, denoted as r_{PB}). Composing a histogram of all r_{PB} (Figure 2D) results in a histogram with a normal distribution that usually mimics the population distribution [36]. Subsequently, it can then be calculated how much the r_{PB}'s vary across the bootstrap samples (Standard Error, SE). The reported 'Bias' measure represents the deviation of the overall r_{PB} from the r_P that was obtained from the original sample [37]. To compute the corresponding bootstrapped confidence interval (CIB), a Bias Corrected and accelerated (BCa) correction was applied [38]. This correction was applied to adjust for the observed Bias, and to account for potential skewness of the bootstrap distribution (operationalized as SE). In the case of (partial) correlations, CIB's can range from −1 to +1. Narrow CIB's imply greater precision, and if the CIB does not include zero, the r_{PB} is considered statistically significant (Figure 2E). Usually, 95% CIB's are computed (corresponding to a significance level of α = 5%). However, to correct for the multiple comparisons in the current study, a 99.8% CIB was computed (corresponding to a significance level of $p < 0.002$).

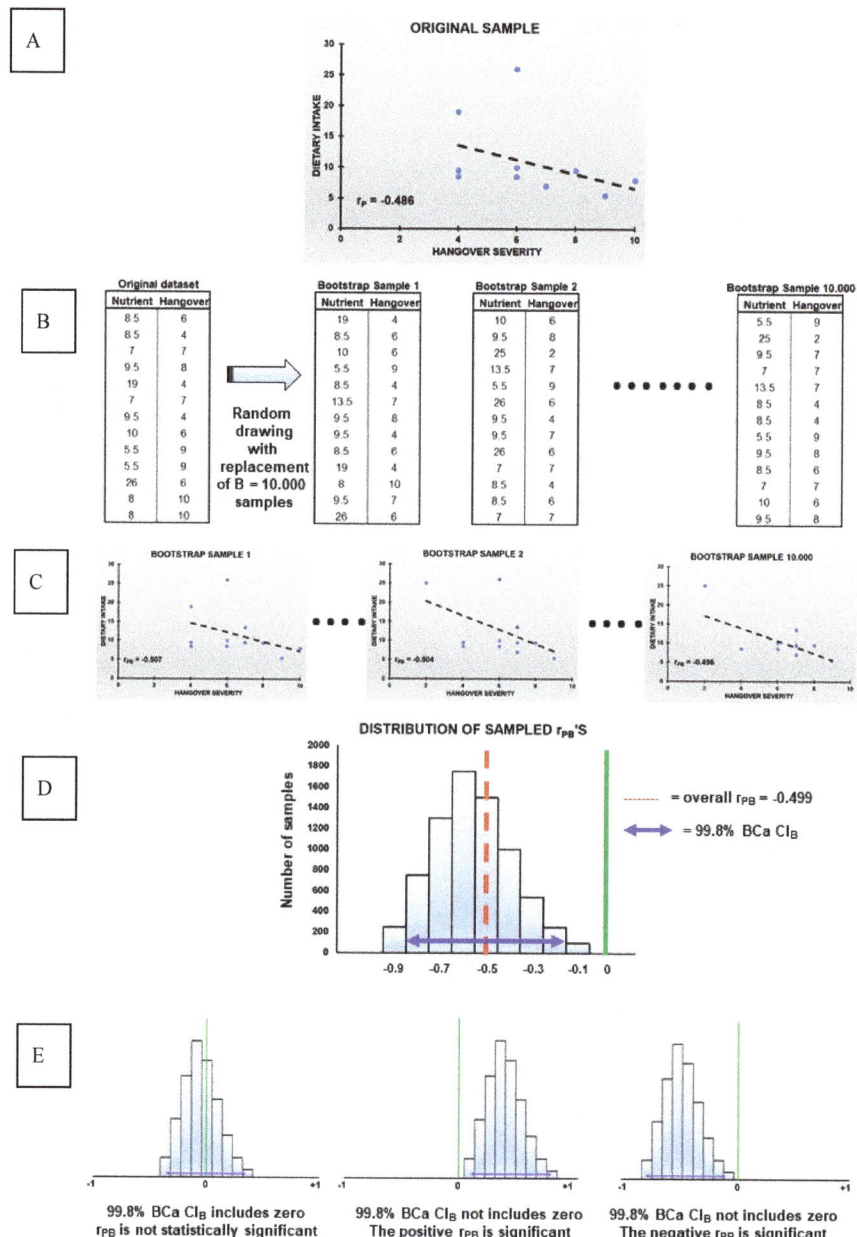

Figure 2. The bootstrapping technique. (**A**) original sample, (**B**) new datasets, (**C**) B = 10.000 samples, (**D**) Composing a histogram of all rPB, (**E**) rPB is considered statistically significant. Abbreviations: r_P = partial correlation, r_{PB} = bootstrapped partial correlation, BCa = bias-corrected and accelerated, C_{IB} = bootstrapped confidence interval.

3. Results

Data from $n = 13$ hangover-sensitive subjects (seven men and six women) were included in the analysis. Their mean (±SD) age was 20.8 (±1.4) years old. Their mean height and weight were 1.76 (±0.07) m and 70.8 (±7.9) kg, respectively. They consumed a mean of 11.3 (±3.8) alcoholic drinks on the alcohol test day, resulting in an eBAC of 0.20 (±0.07)%, and a next-day overall hangover score of 6.2 (±2.2). $n = 10$ hangover-resistant drinkers (four men and six women) served as control group in this study. Demographics and drinking variable outcomes were comparable to those of the hangover-sensitive group. Their mean (±SD) age was 21.2 (±2.4) years old. Their mean height and weight were 1.76 (±0.11) m and 73.0 (±15.9) kg, respectively. They consumed a mean of 13.0 (±6.2) alcoholic drinks on the alcohol test day, resulting in an eBAC of 0.22 (±0.12)%, and a next-day overall hangover score of 0.2 (±0.4). Dietary nutrient intake of hangover-sensitive drinkers did not significantly differ from that of hangover-resistant drinkers. Moreover, alcohol consumption and eBAC did not significantly differ between the two groups. The next statistical analysis was, therefore, conducted only for the hangover-sensitive group. Dietary nutrient intake on the alcohol and control day are summarized in Table 1.

Table 1. Dietary nutrient intake.

Nutrient	Control Day	Alcohol Day	p-Value	2-Day Average
Nicotinic acid (mg)	22.0 (17.1)	34.8 (19.4)	0.075	28.4 (14.9)
Zinc (mg)	9.7 (3.4)	12.6 (8.3)	0.134	11.2 (5.3)
Total fat (g)	83.6 (19.2)	103.1 (47.6)	0.221	93.4 (26.7)
Saturated fat (g)	27.3 (10.2)	35.5 (20.2)	0.249	31.4 (11.6)
Carbohydrates (g)	208.0 (57.8)	330.3 (86.7)	0.002	269.2 (62.2)
Proteins (g)	91.6 (38.3)	108.2 (69.0)	0.650	99.9 (45.9)
Fibers (g)	22.7 (7.5)	26.2 (5.5)	0.158	24.5 (4.8)
Salt (g)	7.5 (3.0)	10.1 (5.6)	0.055	8.8 (3.8)
Alcohol (g)	0 (0)	128.2 (45.3)	0.001 *	64.1 (22.6)
Water (ml)	942.7 (347.6)	3165.2 (933.9)	0.001 *	2053.9 (533.8)
Sodium (mg)	2989.3 (1121.5)	3843.4 (1946.8)	0.016	3416.3 (1396.2)
Potassium (mg)	2987.1 (695.6)	4228.5 (1287.1)	0.003	3607.8 (901.2)
Calcium (mg)	749.4 (356.7)	863.2 (472.4)	0.552	806.3 (285.3)
Magnesium (mg)	301.6 (86.9)	454.4 (168.3)	0.003	378.0 (111.4)
Iron (mg)	10.2 (2.2)	11.7 (4.7)	0.311	11.0 (2.9)
Selenium (mg)	52.8 (30.4)	60.8 (34.5)	0.916	56.8 (26.4)
Iodine (mg)	188.8 (51.4)	187.1 (76.6)	0.600	187.3 (52.6)
Phosphorus (mg)	1464.3 (504.9)	1622.2 (1134.5)	0.701	1543.3 (693.4)
Vitamin A (mg)	665.6 (640.6)	769.3 (521.8)	0.345	717.5 (451.8)
Vitamin B1 (mg)	0.9 (0.2)	1.2 (0.9)	0.421	1.1 (0.5)
Vitamin B2 (mg)	1.3 (0.5)	1.8 (0.9)	0.133	1.5 (0.5)
Vitamin B6 (mg)	1.7 (0.9)	2.7 (1.3)	0.011	2.2 (0.9)
Vitamin B11 (mg)	238.4 (100.8)	349.4 (129.0)	0.033	293.9 (86.3)
Vitamin B12 (mg)	7.4 (12.4)	4.3 (2.8)	0.861	5.8 (6.0)
Vitamin C (mg)	89.2 (52.6)	97.1 (95.4)	0.969	93.2 (65.4)
Vitamin D (mg)	2.0 (1.9)	3.1 (1.8)	0.084	2.5 (1.4)
Vitamin E (mg)	13.3 (5.7)	14.4 (7.0)	0.807	13.8 (4.7)
Energy (Kcal)	2003.7 (406.8)	3655.8 (1030.0)	0.001 *	2829.8 (621.1)

Related-samples Wilcoxon singed rank test. Results are significant if $p < 0.0017$, after Bonferroni's correction for multiple comparisons, indicated by *.

On the alcohol day, alcohol was consumed, and this was accompanied by significantly increased water intake (included in the beverages). As a result, the energy intake on the alcohol day was also significantly greater compared to the control day. However, dietary nutrient intake data show no significant differences between the alcohol day and the control day. Therefore, the statistical analysis that follows the two-day average nutrient intake is used. The association of dietary nutrient intake and overall hangover severity is summarized in Table 2.

Table 2. Association between dietary nutrient intake (two-day average) and overall hangover severity.

Nutrients	Original Sample				Bootstrapping Results		
	r_P	p-Value	Bias	SE	r_{PB}	Lower CI$_B$ Limit	Upper CI$_B$ Limit
Nicotinic acid *	−0.512	0.089	0.009	0.185	−0.521	−0.893	−0.032
Zinc *	−0.393	0.206	−0.052	0.219	−0.341	−0.829	−0.109
Total fat	−0.014	0.967	0.018	0.338	−0.032	−0.993	+0.998
Saturated fat	0.021	0.948	−0.011	0.331	−0.010	−0.917	+0.910
Carbohydrates	−0.223	0.485	−0.029	0.274	−0.204	−0.991	+0.940
Proteins	−0.285	0.370	0.000	0.269	−0.285	−0.997	+0.939
Fibers	−0.157	0.627	0.004	0.319	−0.161	−1.000	+0.999
Salt	−0.059	0.855	−0.073	0.424	−0.014	−0.981	+0.973
Alcohol	0.138	0.669	−0.026	0.294	0.112	−0.858	+0.910
Water	−0.103	0.749	−0.037	0.277	−0.066	−0.880	+0.764
Sodium	−0.157	0.626	−0.041	0.365	−0.116	−0.973	+0.856
Potassium	−0.409	0.187	0.004	0.266	−0.413	−0.985	+0.846
Calcium	−0.045	0.890	0.006	0.282	−0.051	−0.933	+0.946
Magnesium	−0.499	0.098	0.007	0.233	−0.506	−0.954	+0.328
Iron	−0.250	0.433	0.008	0.282	−0.258	−1.000	+1.000
Selenium	−0.356	0.256	0.017	0.292	−0.373	−0.971	+0.912
Iodine	−0.347	0.270	0.024	0.255	−0.371	−0.967	+0.885
Phosphorus	−0.355	0.258	0.034	0.247	−0.389	−1.000	+0.872
Vitamin A	0.125	0.698	0.012	0.243	0.137	−0.848	+0.962
Vitamin B1	0.019	0.952	0.017	0.272	0.036	−0.883	+0.874
Vitamin B2	−0.193	0.549	0.000	0.288	−0.193	−0.986	+0.723
Vitamin B6	−0.407	0.189	0.000	0.293	−0.407	−0.994	+0.950
Vitamin B11	−0.175	0.586	−0.014	0.354	−0.161	−0.996	+0.897
Vitamin B12	0.096	0.768	−0.046	0.279	0.050	−0.875	+0.697
Vitamin C	0.489	0.107	−0.125	0.417	0.364	−0.885	+0.947
Vitamin D	−0.146	0.651	−0.023	0.341	−0.123	−0.976	+0.956
Vitamin E	0.083	0.797	−0.003	0.273	0.080	−0.770	+0.917
Energy (Kcal)	−0.145	0.652	−0.009	0.259	−0.136	−0.961	+0.925

Bootstrapping was conducted with B = 10.000 samples. A BCa 99.8% CI$_B$ (corresponding to $p < 0.002$) was used to correct for multiple comparisons. Partial correlations control for eBAC and are significant if the BCa 99.8% CI$_B$ does not contain zero, indicated by *. Abbreviations: r_P = partial correlation, r_{PB} = bootstrapped partial correlation, eBAC = estimated blood alcohol concentration, SE = standard error, BCa = bias-corrected and accelerated, CI$_B$ = bootstrapped confidence interval.

The association between both the two-day average dietary nicotinic acid and zinc intake and hangover severity is shown in Figure 3. It is evident from Figure 3 that higher levels of dietary nicotinic acid and zinc are associated with less severe alcohol hangovers. After bootstrapping with 10.000 samples the r_{PB}'s were statistically significant ($p < 0.002$). Increasing the bootstrap sample size to B = 100.000 samples did not alter the results. Bootstrapping analysis of other nutrients revealed no significant r_{PB}s between overall hangover severity and dietary nutrient intake (see Table 2).

Figure 4 summarizes the mean (SD) severity scores on the individual hangover symptoms. A bootstrapping analysis was conducted to investigate the r_{PB}'s, controlling for eBAC, between individual hangover symptom severity scores and dietary intake of nicotinic acid and zinc. The bootstrapping results are summarized in Tables 3 and 4.

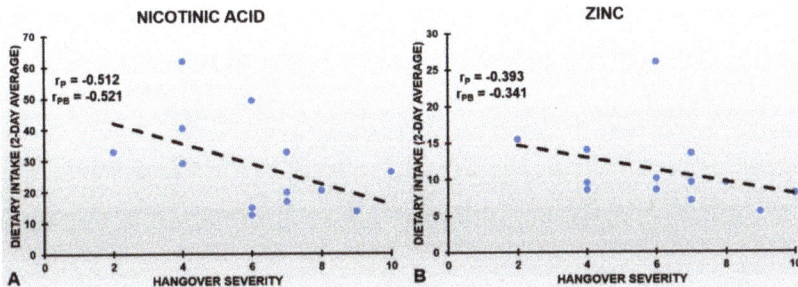

Figure 3. The negative association of dietary nicotinic acid intake (two-day average) and dietary zinc intake with overall hangover severity. Figure 3 shows the partial correlation (r_P), controlled for eBAC, between (**A**) dietary nicotinic acid and (**B**) zinc intake (two-day average) and overall hangover severity.

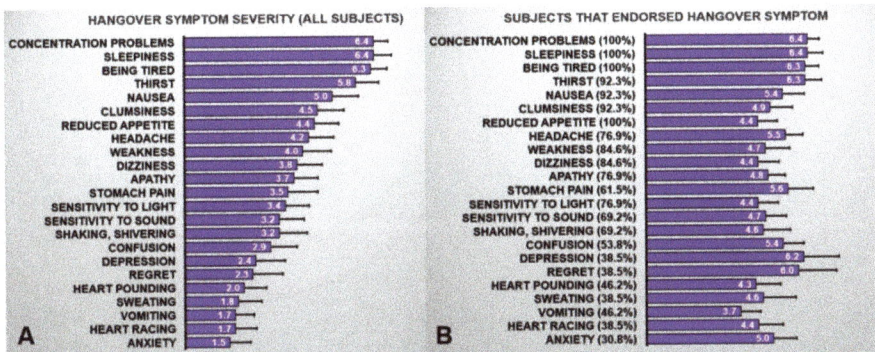

Figure 4. Hangover symptom severity. The average score across all subjects is shown for each hangover symptom in (**A**). (**B**) shows the hangover severity reported by only those subjects that endorsed the hangover symptom. Error bars represent the standard error.

Table 3. Association between dietary nicotinic acid intake (two-day average) and hangover symptom severity.

	Original Sample				Bootstrapping Results		
Hangover Symptoms	r_P	p-Value	Bias	SE	r_{PB}	Lower CI_B Limit	Upper CI_B Limit
Concentration problems	−0.163	0.612	0.049	0.329	−0.212	−0.940	+0.948
Sleepiness	−0.248	0.437	−0.022	0.386	−0.226	−0.985	+0.885
Being tired	−0.448	0.145	0.043	0.348	−0.531	−1.000	+0.998
Thirst	−0.157	0.625	−0.013	0.255	−0.144	−0.897	+0.634
Nausea	−0.447	0.145	0.103	0.347	−0.550	−1.000	+1.000
Clumsiness	−0.272	0.392	0.029	0.302	−0.301	−0.975	+0.942
Reduced appetite	−0.181	0.573	0.080	0.360	−0.261	−1.000	+0.998
Headache	−0.561	0.058	0.049	0.234	−0.610	−0.987	+0.911
Weakness	−0.141	0.662	0.028	0.348	−0.169	−0.998	+1.000
Dizziness	−0.281	0.376	0.062	0.334	−0.343	−0.940	+0.792
Apathy	−0.303	0.339	0.120	0.419	−0.423	−0.987	+0.994
Stomach pain	−0.363	0.246	0.027	0.299	−0.390	−0.973	+0.813
Sensitivity to light	−0.352	0.262	0.082	0.345	−0.434	−0.985	+0.919
Sensitivity to sound	−0.406	0.190	0.064	0.349	−0.470	−1.000	+0.865
Shaking, shivering	−0.430	0.163	0.018	0.248	−0.448	−0.999	+0.940
Confusion	−0.204	0.524	0.047	0.334	−0.251	−0.999	+0.987
Depression	−0.335	0.287	0.061	0.354	−0.396	−0.982	+0.961
Regret	−0.324	0.304	0.078	0.365	−0.402	−0.994	+0.998
Heart pounding	−0.514	0.087	0.054	0.282	−0.568	−1.000	+0.997

Table 3. Cont.

	Original Sample				Bootstrapping Results		
Hangover Symptoms	r_P	p-Value	Bias	SE	r_{PB}	Lower CI_B Limit	Upper CI_B Limit
Sweating	−0.347	0.268	0.071	0.366	−0.418	−0.994	+0.973
Vomiting	−0.506	0.093	0.016	0.235	−0.522	−0.999	+0.985
Heart racing	−0.439	0.154	0.066	0.325	−0.505	−1.000	+0.980
Anxiety	−0.379	0.224	0.059	0.346	−0.438	−0.977	+0.842

Reported r and p-value are from the original r_P, controlling for eBAC. Bootstrapping was conducted with B = 10.000 samples. A 99.8% CI_B (corresponding to $p < 0.002$) was used to correct for multiple comparisons. None of the correlations are significant, as their BCa 99.8% CI_B contains zero. Abbreviations: r_P = partial correlation, eBAC = estimated blood alcohol concentration, SE = standard error, BCa = bias-corrected and accelerated, CI_B = bootstrapped confidence interval.

Table 4. Association between dietary zinc intake (two-day average) and hangover symptom severity.

	Original Sample				Bootstrapping Results		
Hangover Symptoms	r_P	p-Value	Bias	SE	r_{PB}	Lower CI_B Limit	Upper CI_B Limit
Concentration problems	−0.128	0.691	−0.007	0.314	−0.121	−0.854	+0.887
Sleepiness	−0.055	0.865	−0.072	0.326	0.023	−0.855	+0.741
Being tired	−0.195	0.544	−0.020	0.305	−0.175	−0.979	+0.933
Thirst	−0.045	0.888	0.000	0.224	−0.045	−0.816	+0.659
Nausea	−0.077	0.813	−0.050	0.370	−0.027	−0.904	+0.806
Clumsiness	−0.209	0.514	−0.054	0.275	−0.155	−0.910	+0.703
Reduced appetite	0.105	0.745	−0.048	0.404	0.057	−0.956	+0.939
Headache	−0.260	0.414	−0.022	0.241	−0.238	−0.882	+0.611
Weakness	−0.002	0.995	−0.128	0.473	0.126	−0.963	+0.932
Dizziness	−0.005	0.988	−0.052	0.403	0.047	−0.952	+0.933
Apathy	0.060	0.853	−0.019	0.373	0.041	−0.972	+0.965
Stomach pain	−0.117	0.718	−0.103	0.389	−0.014	−0.978	+0.765
Sensitivity to light	−0.138	0.669	−0.031	0.292	−0.107	−0.845	+0.674
Sensitivity to sound	−0.157	0.627	−0.047	0.298	−0.110	−0.851	+0.699
Shaking, shivering	−0.370	0.236	−0.076	0.290	−0.294	−0.958	+0.492
Confusion	−0.113	0.726	−0.071	0.390	−0.042	−0.961	+0.982
Depression	−0.160	0.618	−0.085	0.425	−0.075	−0.960	+0.984
Regret	−0.150	0.641	−0.061	0.431	−0.089	−0.980	+0.987
Heart pounding	−0.298	0.347	−0.067	0.384	−0.231	−0.995	+0.952
Sweating	−0.136	0.674	−0.061	0.479	−0.075	−0.964	+0.972
Vomiting *	−0.609	0.035	−0.032	0.179	−0.577	−0.944	−0.059
Heart racing	−0.211	0.511	−0.076	0.411	−0.135	−0.976	+0.962
Anxiety	−0.214	0.505	−0.065	0.466	−0.149	−0.990	+0.993

Reported r_P and p-value are from the original partial correlation, controlling for eBAC. Bootstrapping was conducted with B = 10.000 samples. A 99.8% CI_B (corresponding to $p < 0.002$) was used to correct for multiple comparisons. The r_{PB}'s are significant if the BCa 99.8% CI_B does not contain zero, indicated by *. Abbreviations: r_P = partial correlation, eBAC = estimated blood alcohol concentration, SE = standard error, BCa = bias-corrected and accelerated, CI_B = bootstrapped confidence interval.

The r_{PB}'s between hangover symptom severity and dietary intake of nicotinic acid were negative, suggesting that increased dietary nicotinic acid intake is beneficial for reducing hangover symptom severity. However, in contrast to overall hangover severity, the r_{PB}'s with individual hangover symptoms were not statistically significant (see Table 3).

The r_{PB}'s between hangover symptom severity and dietary intake of zinc were also negative, suggesting that also increased dietary nicotinic acid intake might be beneficial for reducing hangover symptom severity. Dietary zinc intake (two-day average) was significantly associated with the severity of vomiting (see Table 4 and Figure 5A). The negative r_{PB} implies that higher levels of dietary zinc are associated with less severe vomiting. Increasing the bootstrap sample size to B = 100.000 samples did not alter the results. Figure 5B shows that the severity of vomiting is an important determinant of overall hangover severity ($r_P = 0.661$, $p = 0.019$, $r_{PB} = 0.635$, significant at the $p < 0.002$ level). No significant r_{PB}'s were found between two-day average dietary zinc intake and the severity of the other 21 hangover symptoms (see Table 4).

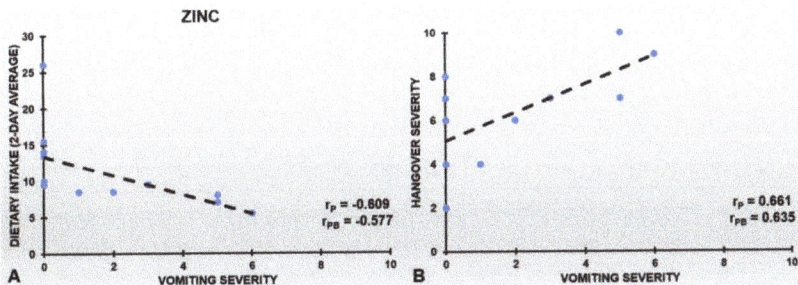

Figure 5. Severity of vomiting, dietary zinc intake, and overall hangover severity. (**A**) shows the partial correlation (r_P), controlled for eBAC, between dietary zinc intake (two-day average) and vomiting severity during alcohol hangover. (**B**) shows the r_P, controlled for eBAC, between overall hangover severity and vomiting severity. After bootstrapping, both the corresponding r_{PB}'s are significant at the $p < 0.002$ level.

Finally, although the number of men ($n = 7$) and women ($n = 6$) did not allow for reliable sex comparisons, we conducted some exploratory analysis. The analysis revealed that men had a higher dietary nicotinic acid intake than women (38.0 mg and 17.2 mg, respectively, $p = 0.002$) as well as a higher intake of zinc (13.8 mg and 8.1 mg, respectively, $p = 0.022$). The differences did not reach significance at the $p < 0.0017$ level. Men reported less severe hangovers than women (scores of 5.0 and 7.5, respectively, $p = 0.073$), while their eBAC did not differ significantly (eBAC 0.20% and 0.19%, respectively, $p = 0.628$). The partial correlations between dietary nicotinic acid and zinc intake and hangover severity calculated for men and women separately were not statistically significant.

4. Discussion

The current study demonstrated that dietary nicotinic acid and zinc intake were significantly and negatively associated with overall hangover severity. Both nutrients are essential in effective alcohol metabolism and, therefore, the current findings suggest that more rapid and efficient oxidation of ethanol into acetaldehyde, and acetaldehyde into acetate, may be associated with less severe hangovers. There was no similar association found for the other nutrients that were examined.

Sufficient dietary intake of zinc and nicotinic acid are important to maintain health. Examples of food rich in zinc are meat, shellfish (e.g., oysters), and legumes, such as lentils and beans. The recommended dietary allowance for zinc is 11 mg per day for men and 8 mg per day for women [39]. Deficiency of zinc can have serious health consequences and negatively impact immune defense [23]. Zinc deficiency is relatively more common among the elderly. For example, a US study found that only 42.5% of the elderly had an adequate level of dietary zinc intake [40].

Examples of food rich in nicotinic acid include those containing high levels of niacin or tryptophan such as meat, fish and poultry, avocado, peanuts, whole grains, and mushrooms. The recommended dietary allowance (RDA) for niacin and its equivalents is 16 mg per day for men and 14 mg per day for women [41,42]. Pellagra (pigmented skin rash) is a common consequence of severe niacin deficiency [43], and, although uncommon in the general population of Western countries, it is seen among chronic alcoholics [41]. In this context, niacin supplementation has been suggested as a treatment for alcoholism [44]. Our data suggest that this should be explored in more detail in different types of social drinkers, given the obvious relationship between frequent heavy drinking and hangover frequency.

The current finding may have implications for the development of effective hangover treatments. Although there is a buoyant market for so-called hangover treatment among social drinkers [45], currently they lack robust evidence of efficacy [46–48]. Several newly developed putative hangover treatments are comprised of natural ingredients, such as plant extracts, herbs, minerals and vitamins [48]. For example, studies were conducted to investigate the effects of, Korean pear juice [49] and red ginseng [50], which both showed some positive effects in reducing hangover severity. However, other products, such as artichoke extract [51], showed no beneficial effects on hangover. Kelly et al. [52] found that intravenous vitamin B complex and Vitamin C had no significant effect on alcohol metabolism. Kahn et al. [53] reported that pyritinol (1200 mg oral vitamin B6) significantly reduced the number of reported hangover symptoms, but unfortunately, no assessments were conducted with regard to the severity of hangover symptoms. Laas [54] conducted a double-blind placebo-controlled study to examine the efficacy of 'Morning Fit' (dried yeast, thiamine nitrate (Vitamin B1), pyridozine hydrochloride (Vitamin B6) and riboflavin (Vitamin B2) and found no significant differences in either blood alcohol or acetaldehyde concentrations between the Morning Fit and placebo condition. Although significant improvements were reported for certain individual symptoms, namely 'uncomfortable feelings', 'restlessness', and 'impatience', no significant improvement was found on general wellbeing. A study conducted by Ylikahri et al. found no significant effect of sugars such as fructose and glucose on alcohol metabolism or hangover severity [55]. A more recent study by Bang et al. examined the effects on hangover of a polysaccharide-rich extract of *Acanthopanax senticosus* (PEA) [56]. While blood sample analysis revealed no significant effect on alcohol metabolism, PEA did, however, significantly reduce alcohol-induced next-day changes in glucose and c-reactive protein levels (i.e., it was effective in reducing alcohol-induced hypoglycemia and inhibiting the inflammatory response, respectively). Overall hangover severity, and the individual hangover symptoms, such as tiredness, headache, dizziness, stomachache and nausea, significantly improved after administering PEA.

Taken together, there is mixed evidence on acute effects of dietary nutrients on the presence and severity of hangover symptoms. The current findings are also in contrast to anecdotal evidence that suggests that taking fiber-rich food, consuming water, or eating fat-rich meals may reduce the severity of alcohol hangovers.

In the current study, dietary zinc and nicotinic acid intake (or any other nutrient that was assessed) did not significantly differ between hangover-sensitive drinkers and hangover-resistant drinkers. Thus, it is unlikely that supplementing diet with high levels of nicotinic acid and zinc makes a hangover sensitive drinker immune to hangovers. However, the data of hangover-sensitive drinkers clearly show that higher dietary intake of both nutrients is associated with experiencing less severe hangovers. The issue of drinkers claiming hangover resistance is a complex one. Data show that this claim heavily depends on how much alcohol drinkers consume, but even at higher eBAC levels a small proportion of drinkers claim not to have hangovers [27,28]. At the same time, other hangover research showed no significant differences between the two groups of drinkers on several biomarkers such as urine ethyl glucuronide (EtG) and ethyl sulfate (EtS) [57] or methanol [58], saliva cytokine levels [59], sensitivity to acute alcohol effects [60], demographics [31], or psychological characteristics such as mental resilience [61]. Additionally, the current study could not differentiate hangover-sensitive and hangover-resistant drinkers based on their dietary nutrient intake. Thus, there must be different unknown biopsychosocial factors (e.g., alexithymia) that may explain why some drinkers claim to be hangover-resistant. Research did show that experiencing alcohol hangovers (compared to claiming to be resistant) was associated with significantly poorer self-reported immune function [62] and having higher urine ethanol concentrations during the hangover state [5]. Future research should further investigate the puzzling phenomenon of hangover-resistance.

The study has several limitations. Firstly, it had a small sample size. Future research should, therefore, aim to replicate the current findings in larger samples. The use of bootstrapping techniques in hypothesis testing is increasingly popular [36] and was used in the current analysis to mitigate the small sample size. Secondly, the sample size was too small to reliably assess possible gender differences. Explorative analysis revealed that men had a higher intake of dietary nicotinic acid and zinc. Moreover, women reported non-significantly higher hangover severity than men, and eBAC levels did not differ significantly. These findings are in line with a recent analysis showing that the presence and severity of hangover symptoms experienced at the same eBAC levels show no relevant sex differences [63]. However, studies have shown sex differences in cognitive functioning and driving performance the morning following bedtime administration of other psychoactive drugs, such as hypnotics [64]. Therefore, future replication research with larger sample size is required to further investigate possible sex differences during the hangover state in these domains. Thirdly, eBAC assessments were based on subjective retrospective recall of the number of alcoholic drinks consumed. These reports may to some extent be inaccurate. In naturalistic study designs, researchers are not present during the drinking occasion, to ensure the real-life 'natural' drinking setting, and retrospective assessments of alcohol consumption are common practice. However, very recently real-time objective BAC assessment devices have been developed that are capable of continuously recording transdermal BAC. It would be useful to use such devices in future naturalistic studies. Fourthly, there were many statistical comparisons made in the analysis. Although the primary aim was to investigate the association between dietary nicotinic acid and zinc intake with overall hangover severity, we also collected data on 25 other nutrients and 23 individual hangover symptoms. A strict Bonferroni's correction ($p < 0.002$) was applied to account for this, and therefore we are confident with the reported statistical significance thresholds.

Dietary nutrient intake was collected via 24-h dietary recall records. These were completed by the subjects the day after drinking, and thus recall bias may have resulted in underestimation or overestimation of food portions, omitting food items or erroneously adding others. On the other hand, there was good correspondence between the diary measures across the two collection periods. Furthermore, the group average nutrient intake in the current study corresponds well to large scale studies that assessed nutrient intake via elaborate food frequency questionnaires [41,42]. This provides some confidence that recall bias played a minor role in the current study. Clearly, future research should utilize more elaborate food frequency questionnaires, or nutrient-specific dietary records for nicotinic acid or zinc. Additionally, assessments of nutrient status in blood or urine would provide an objective measure of nutrient status.

It is relevant to note that dietary nutrients can impact alcohol metabolism via the gut and oral microbiome. Dietary nutrient intake, as well as alcohol consumption, have an influence on the composition of the microbiome. Several studies have reported the effects of alcohol consumption and dietary intake on microbiota composition [65,66]. The effect of these on hangover is not well understood, but a high abundance of several microbiota, including *Rothia*, *Neisseria*, and *Streptococcus*, is associated with accelerated alcohol metabolism by producing relatively higher amounts of acetaldehyde [67]. Future research should investigate the relationship between the microbiome, and the presence and severity of alcohol hangover. Moreover, there are several other factors that may influence alcohol metabolism that were not assessed in the current study. These include, for example, various genetic and environmental factors, sex, age, race, biological rhythms (time of day), and medicinal and recreational drug use (e.g., compounds which inhibit ADH such as pyrazoles or isobutyramine), Antabuse (disulfiram, which inhibits the elimination of acetaldehyde), or other alcohols that compete with ethanol for ADH (e.g., methanol) [17]. These are also important topics for future research.

Finally, the oxidative pathways account for over 90% of alcohol elimination [19]. In addition, there are also nonoxidative pathways for alcohol metabolism, producing metabolites such as ethyl-glucuronide (EtG), ethyl-sulfate (EtS), phosphatidyl-ethanol (PEth) and fatty acid ethyl ester (FAEE) [19]. As these pathways usually only process a very limited amount of alcohol, and thus the overall impact of nutrients on alcohol metabolism via these pathways can be considered as marginal, they were not taken into account in the current paper.

5. Conclusions

In conclusion, this study suggests that social drinkers who have a higher dietary intake of nicotinic acid and zinc report significantly less severe hangovers.

Author Contributions: Conceptualization, J.C.V., S.A.V., A.J.A.E.v.d.L., S.B., A.D.K., J.G. and A.S.; formal analysis, J.C.V.; investigation, S.A.V., A.J.A.E.v.d.L. and S.B.; methodology, J.C.V., S.A.V., A.D.K., J.G. and A.S.; writing—original draft, J.C.V. and A.S.; writing—review & editing, S.A.V., A.J.A.E.v.d.L., S.B., A.D.K., J.G. and A.S.

Conflicts of Interest: J.G. is part-time employee of Nutricia Research and received research grants from Nutricia research foundation, Top Institute Pharma, Top Institute Food and Nutrition, GSK, STW, NWO, Friesland Campina, CCC, Raak-Pro, and EU. A.D.K. has received grants/research support from Top Institute Pharma, NWO, Janssen, GSK, Nutricia Research, and Friesland Campina. J.C.V. has received grants/research support from the Dutch Ministry of Infrastructure and the Environment, Janssen, Nutricia, Red Bull, Sequential, and Takeda, and has acted as a consultant for Canadian Beverage Association, Centraal Bureau Drogisterijbedrijven, Clinilabs, Coleman Frost, Danone, Deenox, Eisai, Janssen, Jazz, More Labs, Purdue, Red Bull, Sanofi-Aventis, Sen-Jam Pharmaceutical, Sepracor, Takeda, Toast!, Transcept, Trimbos Institute, Vital Beverages, and ZBiotics. A.S. has held research grants from Abbott Nutrition, Arla Foods, Bayer Healthcare, Cognis, Cyvex, GlaxoSmithKline, Naturex, Nestle, Martek, Masterfoods, Wrigley, and has acted as a consultant/expert advisor to Abbott Nutrition, Barilla, Bayer Healthcare, Danone, Flordis, GlaxoSmithKline Healthcare, Masterfoods, Martek, Novartis, Unilever, and Wrigley.The other authors have not potential conflicts of interest to disclose.

References

1. Van Schrojenstein Lantman, M.; van de Loo, A.J.; Mackus, M.; Verster, J.C. Development of a definition for the alcohol hangover: Consumer descriptions and expert consensus. *Curr. Drug Abuse Rev.* **2016**, *9*, 148–154. [CrossRef] [PubMed]
2. Palmer, E.; Tyacke, R.; Sastre, M.; Lingford-Hughes, A.; Nutt, D.; Ward, R.J. Alcohol Hangover: Underlying Biochemical, Inflammatory and Neurochemical Mechanisms. *Alcohol Alcohol.* **2019**, *54*, 196–203. [CrossRef] [PubMed]
3. Penning, R.; van Nuland, M.; Fliervoet, L.A.L.; Olivier, B.; Verster, J.C. The pathology of alcohol hangover. *Curr. Drug Abuse Rev.* **2010**, *3*, 68–75. [CrossRef] [PubMed]
4. Kim, H.; Kim, Y.J.; Jeong, H.Y.; Kim, J.Y.; Choi, E.K.; Chae, S.W.; Kwon, O. A standardized extract of the fruit of Hovenia dulcis alleviated alcohol-induced hangover in healthy subjects with heterozygous ALDH2: A randomized, controlled, crossover trial. *J. Ethnopharmacol.* **2017**, *209*, 167–174. [CrossRef] [PubMed]
5. Van de Loo, A.J.A.E.; Mackus, M.; Korte-Bouws, G.A.H.; Brookhuis, K.A.; Garssen, J.; Verster, J.C. Urine ethanol concentration and alcohol hangover severity. *Psychopharmacology* **2017**, *234*, 73–77. [CrossRef] [PubMed]
6. Min, J.A.; Lee, K.; Ki, D.J. The application of minerals in managing alcohol hangover: A preliminary review. *Curr. Drug Abuse Rev.* **2010**, *3*, 110–115. [CrossRef] [PubMed]
7. Wang, F.; Li, Y.; Zhang, Y.J.; Zhou, Y.; Li, S.; Li, H.B. Natural Products for the Prevention and Treatment of Hangover and Alcohol Use Disorder. *Molecules* **2016**, *21*, 64. [CrossRef] [PubMed]
8. Sadler, D.W.; Fox, J. Intra-individual and inter-individual variation in breath alcohol pharmacokinetics: The effect of food on absorption. *Sci. Justice* **2011**, *51*, 3–9. [CrossRef] [PubMed]
9. Watkins, R.L.; Adler, E.V. The effect of food on alcohol absorption and elimination patterns. *J. Forensic Sci.* **1993**, *38*, 285–291. [CrossRef]
10. Lin, Y.; Weidler, D.J.; Garg, D.C.; Wagner, J.G. Effects of solid food on blood levels of alcohol in man. *Res. Commun. Chem. Pathol. Pharmacol.* **1976**, *13*, 713–722. [PubMed]
11. Millar, K.; Hammersley, R.H.; Finnigan, F. Reduction of alcohol-induced performance impairment by prior ingestion of food. *Br. J. Psychol.* **1992**, *83 Pt 2*, 261–278. [CrossRef]
12. Pikaar, N.A.; Wedel, M.; Hermus, R.J. Influence of several factors on blood alcohol concentrations after drinking alcohol. *Alcohol Alcohol.* **1988**, *23*, 289–297. [PubMed]
13. Finnigan, F.; Hammersley, R.; Millar, K. Effects of meal composition on blood alcohol level, psychomotor performance and subjective state after ingestion of alcohol. *Appetite* **1998**, *31*, 361–375. [CrossRef] [PubMed]
14. Kim, M.J.; Lim, S.W.; Kim, J.H.; Choe, D.J.; Kim, J.I.; Kang, M.J. Effect of Mixed Fruit and Vegetable Juice on Alcohol Hangovers in Healthy Adults. *Prev. Nutr. Food Sci.* **2018**, *23*, 1–7. [CrossRef] [PubMed]
15. Hong, Y.H. Effects of the herb mixture, DTS20, on oxidative stress and plasma alcoholic metabolites after alcohol consumption in healthy young men. *Integr. Med. Res.* **2016**, *5*, 309–316. [CrossRef] [PubMed]
16. Bullock, C. The biochemistry of alcohol metabolism—A brief review. *Biochem. Educ.* **1990**, *18*, 62–66. [CrossRef]
17. Cederbaum, A.I. Alcohol metabolism. *Clin. Liver Dis.* **2012**, *16*, 667–685. [CrossRef]

18. Kawai, S.; Murata, K. Structure and function of NAD kinase and NADP phosphatase: Key enzymes that regulate the intracellular balance of NAD(H) and NADP(H). *Biosci. Biotechnol. Biochem.* **2008**, *72*, 919–930. [CrossRef]
19. Heier, C.; Xie, H.; Zimmermann, R. Nonoxidative ethanol metabolism in humans-from biomarkers to bioactive lipids. *IUBMB Life* **2016**, *68*, 916–923. [CrossRef]
20. Goodsell, D.S. Molecule of the Month: Alcohol Dehydrogenase. 2001. Available online: https://pdb101.rcsb.org/motm/13 (accessed on 7 July 2019).
21. Kägi, J.H.; Vallee, B.L. The role of zinc in alcohol dehydrogenase. V. The effect of metal-binding agents on the structure of the yeast alcohol dehydrogenase molecule. *J. Biol. Chem.* **1960**, *235*, 3188–3192.
22. Kirkland, J.B.; Meyer-Ficca, M.L. Niacin. *Adv. Food Nutr. Res.* **2018**, *83*, 83–149. [PubMed]
23. Plum, L.M.; Rink, L.; Haase, H. The essential toxin: The impact of zinc on human health. *Int. J. Environ. Res. Public Health* **2010**, *7*, 1342–1365. [CrossRef] [PubMed]
24. Roohani, N.; Hurrell, R.; Kelishadi, R.; Schulin, R. Zinc and its importance for human health: An integrative review. *J. Res. Med. Sci.* **2013**, *18*, 144–157. [PubMed]
25. Horwitt, M.K.; Harper, A.E.; Henderson, L.M. Niacin-tryptophan relationships for evaluating niacin equivalents. *Am. J. Clin. Nutr.* **1981**, *34*, 423–427. [CrossRef] [PubMed]
26. Watson, P.E.; Watson, I.D.; Batt, R.D. Prediction of blood alcohol concentrations in human subjects. Updating the Widmark Equation. *J. Stud. Alcohol Drugs* **1981**, *42*, 547–556. [CrossRef]
27. Verster, J.C.; de Klerk, S.; Bervoets, A.C.; Kruisselbrink, L.D. Can hangover immunity really be claimed? *Curr. Drug Abuse Rev.* **2013**, *6*, 253–254. [CrossRef]
28. Kruisselbrink, L.D.; Bervoets, A.C.; de Klerk, S.; van de Loo, A.J.A.E.; Verster, J.C. Hangover resistance in a Canadian university student population. *Addict. Behav. Rep.* **2017**, *5*, 14–18. [CrossRef]
29. Scholey, A.; Benson, S.; Kaufman, J.; Terpstra, C.; Ayre, E.; Verster, J.C.; Allen, C.; Devilly, G. Effects of alcohol hangover on cognitive performance: A field/internet mixed methodology approach. *J. Clin. Med.* **2019**, *8*, 440. [CrossRef]
30. Fernstrand, A.M.; Bury, D.; Garssen, J.; Verster, J.C. Dietary intake of fibers: Differential effects in men and women on general health and perceived immune functioning. *Food Nutr. Res.* **2017**, *61*, 1297053. [CrossRef]
31. Hogewoning, A.; van de Loo, A.J.A.E.; Mackus, M.; Raasveld, S.J.; de Zeeuw, R.; Bosma, E.R.; Bouwmeester, N.H.; Brookhuis, K.A.; Garssen, J.; Verster, J.C. Characteristics of social drinkers with and without a hangover after heavy alcohol consumption. *Subst. Abuse Rehabil.* **2016**, *7*, 161–167. [CrossRef]
32. Eetmeter. Available online: https://mijn.voedingscentrum.nl/nl/eetmeter/ (accessed on 7 July 2019).
33. List of Components in NEVO Online 2016. Available online: https://www.voedingswaardetabel.nl (accessed on 7 July 2019).
34. Efron, B. Bootstrap methods: Another look at the jackknife. *Ann. Stat.* **1979**, *7*, 1–26. [CrossRef]
35. Efron, B.; Tibshirani, R. *An Introduction to the Bootstrap*; Chapman & Hall/CRC: Boca Raton, FL, USA, 1993.
36. Rahman, S.; Majumber, A.K. *Use of Bootstrapping in Hypothesis Testing: Bootstrapping for Estimation and Hypothesis Testing*; LAP LAMBERT Academic Publishing: Riga, Latvia, 2013.
37. Sideridis, G.D.; Simos, P. What is the actual correlation between expressive and receptive measures of vocabulary? Approximating the sampling distribution of the correlation coefficient using the bootstrapping method. *Int. J. Educ. Psychol. Assess.* **2010**, *5*, 117–133.
38. Efron, B. Better bootstrap confidence intervals. *J. Am. Stat. Assoc.* **1987**, *82*, 171–185. [CrossRef]
39. Trumbo, P.; Yates, A.A.; Schlicker, S.; Poos, M. Dietary reference intakes: Vitamin, A. vitamin K, arsenic, boron, chromium, copper, iodine, iron, manganese, molybdenum, nickel, silicon, vanadium, and zinc. *J. Am. Diet. Assoc.* **2001**, *101*, 294–301. [CrossRef]
40. Briefel, R.R.; Bialostosky, K.; Kennedy-Stephenson, J.; McDowell, M.A.; Ervin, R.B.; Wright, J.D. Zinc intake of the U.S. population: Findings from the third National Health and Nutrition Examination Survey, 1988–1994. *J. Nutr.* **2000**, *130*, 1367S–1373S. [CrossRef]
41. Institute of Medicine (US) Standing Committee on the Scientific Evaluation of Dietary Reference Intakes and its Panel on Folate, Other B Vitamins, and Choline. *Dietary Reference Intakes for Thiamin, Riboflavin, Niacin, Vitamin B6, Folate, Vitamin B12, Pantothenic Acid, Biotin, and Choline*; National Academies Press: Washington, DC, USA, 1998.
42. U.S. Food and Drug Administration. Food labelling: Revision of the nutrition and supplement facts labels. *Fed. Regist.* **2016**, *81*, 33741.

43. Carpenter, K.J.; Lewin, W.J. A reexamination of the composition of diets associated with pellagra. *J. Nutr.* **1985**, *115*, 543–552. [CrossRef]
44. Prousky, J.E. The Treatment of Alcoholism with Vitamin B3. *J. Orthomol. Med.* **2014**, *29*, 123–131.
45. Mackus, M.; van Schrojenstein Lantman, M.; van de Loo, A.J.A.E.; Nutt, D.J.; Verster, J.C. An effective hangover treatment: Friend or foe? *Drug Sci. Policy Law* **2017**. [CrossRef]
46. Pittler, M.H.; Verster, J.C.; Ernst, E. Interventions for preventing or treating alcohol hangover: Systematic review of randomized trials. *Br. Med. J.* **2005**, *331*, 1515–1518. [CrossRef]
47. Verster, J.C.; Penning, R. Treatment and prevention of alcohol hangover. *Curr. Drug Abuse Rev.* **2010**, *3*, 103–109. [CrossRef] [PubMed]
48. Jayawardena, R.; Thejani, T.; Ranasinghe, P.; Fernando, D.; Verster, J.C. Interventions for treatment and/or prevention of alcohol hangover: Systematic review. *Hum. Psychopharmacol.* **2017**, *32*, e2600. [CrossRef] [PubMed]
49. Lee, H.S.; Isse, T.; Kawamoto, T.; Baik, H.W.; Park, J.Y.; Yang, M. Effect of Korean pear (Pyruspyrifolia cv. Shingo) juice on hangover severity following alcohol consumption. *Food Chem. Toxicol.* **2013**, *58*, 101–106. [CrossRef] [PubMed]
50. Lee, M.H.; Kwak, J.H.; Jeon, G.; Lee, J.W.; Seo, J.H.; Lee, H.S.; Lee, J.H. Red ginseng relieves the effects of alcohol consumption and hangover symptoms in healthy men: A randomized crossover study. *Food Funct.* **2014**, *5*, 528–534. [CrossRef] [PubMed]
51. Pittler, M.H.; White, A.R.; Stevinson, C.; Ernst, E. Effectiveness of artichoke extract in preventing alcohol-induced hangovers: A randomized controlled trial. *CMAJ* **2003**, *169*, 1269–1273. [PubMed]
52. Kelly, M.; Myrsten, A.-L.; Goldberg, L. Intravenous vitamins in acute intoxication: Effects on physiological and psychological functions. *Br. J. Addict.* **1971**, *66*, 19–30. [CrossRef]
53. Kahn, M.A.; Jensen, K.; Krogh, H.J. Alcohol-induced hangover. A double-blind comparison of pyritinol and placebo in preventing hangover symptoms. *Q. J. Stud. Alcohol* **1973**, *34*, 1195–1201.
54. Laas, I. A double-blind placebo-controlled study on the effects of Morning Fit on hangover symptoms after a high level of alcohol consumption in healthy volunteers. *J. Clin. Res.* **1999**, *2*, 9–15.
55. Ylikahri, R.H.; Leino, T.; Huttunen, M.O.; Pösö, A.R.; Eriksson, C.J.P.; Nikkilä, E.A. Effects of fructose and glucose on ethanol-induced metabolic changes and on the intensity of alcohol intoxication and hangover. *Eur. J. Clin. Investig.* **1976**, *6*, 93–102. [CrossRef]
56. Bang, J.S.; Chung, Y.H.; Chung, S.J.; Lee, H.S.; Song, E.H.; Shin, Y.K.; Lee, Y.J.; Kim, H.C.; Nam, Y.; Jeong, J.H. Clinical effect of a polysaccharide-rich extract of Acanthopanax senticosus on alcohol hangover. *Pharmazie* **2015**, *70*, 269–273.
57. Mackus, M.; Van de Loo, A.J.A.E.; Raasveld, S.J.; Hogewoning, A.; Sastre Toraño, J.; Flesch, F.M.; Korte-Bouws, G.A.H.; Van Neer, R.H.P.; Wang, X.; Nguyen, T.T.; et al. Biomarkers of the alcohol hangover state: Ethyl glucuronide (EtG) and ethyl sulfate (EtS). *Hum. Psychopharmacol.* **2017**, *32*, e2624. [CrossRef] [PubMed]
58. Mackus, M.; van de Loo, A.J.A.E.; Korte-Bouws, G.A.H.; van Neer, R.H.P.; Wang, X.; Nguyen, T.T.; Brookhuis, K.A.; Garssen, J.; Verster, J.C. Urine methanol concentration and alcohol hangover severity. *Alcohol* **2017**, *59*, 37–41. [CrossRef] [PubMed]
59. Van de Loo, A.J.A.E.; Knipping, K.; Mackus, M.; Kraneveld, A.D.; Garssen, J.; Scholey, A.; Bruce, G.; Verster, J.C. Differential effects on acute saliva cytokine response following alcohol consumption and alcohol hangover: Preliminary results from two independent studies. *Alcohol. Clin. Exp. Res.* **2018**, *42* (Suppl. S2), 20A.
60. Mackus, M.; van Schrojenstein Lantman, M.; van de Loo, A.J.A.E.; Brookhuis, K.A.; Kraneveld, A.D.; Garssen, J.; Verster, J.C. Alcohol metabolism in hangover sensitive versus hangover resistant social drinkers. *Drug Alcohol Depend.* **2018**, *185*, 351–355. [CrossRef] [PubMed]
61. Van Schrojenstein Lantman, M.; van de Loo, A.J.A.E.; Mackus, M.; Brookhuis, K.A.; Kraneveld, A.D.; Garssen, J.; Verster, J.C. Susceptibility to alcohol hangovers: Not just a matter of being resilient. *Alcohol Alcohol.* **2018**, *53*, 241–244. [CrossRef] [PubMed]
62. Van de Loo, A.J.A.E.; Mackus, M.; van Schrojenstein Lantman, M.; Kraneveld, A.D.; Garssen, J.; Scholey, A.; Verster, J.C. Susceptibility to alcohol hangovers: The association with self-reported immune status. *Int. J. Environ. Res. Public Health* **2018**, *15*, 1286. [CrossRef]
63. Van Lawick van Pabst, A.E.; Devenney, L.E.; Verster, J.C. Sex differences in the presence and severity of alcohol hangover symptoms. *J. Clin. Med.* **2019**, *8*, 867. [CrossRef] [PubMed]

64. Verster, J.C.; Roth, T. Gender differences in highway driving performance after administration of sleep medication: A review of the literature. *Traffic Inj. Prev.* **2012**, *13*, 286–292. [CrossRef] [PubMed]
65. Leclercq, S.; Matamoros, S.; Cani, P.D.; Neyrinck, A.M.; Jamar, F.; Stärkel, P.; Windey, K.; Tremaroli, V.; Bäckhed, F.; Verbeke, K.; et al. Intestinal permeability, gut-bacterial dysbiosis, and behavioral markers of alcohol-dependence severity. *Proc. Natl. Acad. Sci. USA* **2014**, *111*, E4485–E4493. [CrossRef]
66. Engen, P.A.; Green, S.J.; Voigt, R.M.; Forsyth, C.B.; Keshavarzian, A. The Gastrointestinal Microbiome: Alcohol Effects on the Composition of Intestinal Microbiota. *Alcohol Res.* **2015**, *37*, 223–236.
67. Moritani, K.; Takeshita, T.; Shibata, Y.; Ninomiya, T.; Kiyohara, Y.; Yamashita, Y. Acetaldehyde production by major oral microbes. *Oral Dis.* **2015**, *21*, 748–754. [CrossRef] [PubMed]

© 2019 by the authors. Licensee MDPI, Basel, Switzerland. This article is an open access article distributed under the terms and conditions of the Creative Commons Attribution (CC BY) license (http://creativecommons.org/licenses/by/4.0/).

Article

Exacerbation of Hangover Symptomology Significantly Corresponds with Heavy and Chronic Alcohol Drinking: A Pilot Study

Vatsalya Vatsalya [1,2,3,4,5,*], Hamza Z. Hassan [1], Maiying Kong [6], Bethany L. Stangl [4], Melanie L. Schwandt [4], Veronica Y. Schmidt-Teron [4], Joris C. Verster [7,8,9], Vijay A. Ramchandani [4,†] and Craig J. McClain [1,2,3,5,10,†]

1. Department of Medicine, University of Louisville, Louisville, KY 40202, USA; hamza.hassan@louisville.edu (H.Z.H.); craig.mcclain@louisville.edu (C.J.M.)
2. Alcohol Research Center, University of Louisville, Louisville, KY 40202, USA
3. Hepatobiology & Toxicology Center, University of Louisville, Louisville, KY 40202, USA
4. National Institute on Alcohol Abuse and Alcoholism, NIH, Bethesda, MD 20892, USA; bethany.stangl@nih.gov (B.L.S.); melanies@mail.nih.gov (M.L.S.); vschmidt@uchc.edu (V.Y.S.-T.); vijayr@mail.nih.gov (V.A.R.)
5. Robley Rex Louisville VAMC, Louisville, KY 40206, USA
6. Department of Bioinformatics and Biostatistics, SPHIS, University of Louisville, Louisville, KY 40202, USA; maiying.kong@louisville.edu
7. Division of Pharmacology, Utrecht Institute for Pharmaceutical Sciences (UIPS), Utrecht University, 3584 CG Utrecht, The Netherlands; J.C.Verster@uu.nl
8. Faculty of Veterinary Medicine, Institute for Risk Assessment Sciences (IRAS), Utrecht University, 3511 CM Utrecht, The Netherlands
9. Centre for Human Psychopharmacology, Swinburne University, Melbourne, VIC 3211, Australia
10. Department of Pharmacology & Toxicology, University of Louisville, Louisville, KY 40202, USA
* Correspondence: v0vats01@louisville.edu; Tel.: +1-502-488-0446
† Senior authors.

Received: 28 September 2019; Accepted: 6 November 2019; Published: 12 November 2019

Abstract: Alcohol hangover is a combination of mental, sympathetic, and physical symptoms experienced the day after a single period of heavy drinking, starting when blood alcohol concentration approaches zero. How individual measures/domains of hangover symptomology might differ with moderate to heavy alcohol consumption and how these symptoms correlate with the drinking markers is unclear. We investigated the amount/patterns of drinking and hangover symptomology by the categories of alcohol drinking. We studied males and females in three groups: 12 heavy drinkers (HD; >15 drinks/week, 34–63 years old (y.o.)); 17 moderate drinkers (MD; 5–14 drinks/week, 21–30 y.o.); and 12 healthy controls (social/light drinkers, SD; <5 drinks/week, 25–54 y.o.). Demographics, drinking measures (Timeline followback past 90 days (TLFB90), Alcohol Use Disorders Identification Test (AUDIT)), and alcohol hangover scale (AHS) were analyzed. Average drinks/day was 5.1-times greater in HD compared to MD. Average AHS score showed moderate incapacity, and individual measures and domains of the AHS were significantly elevated in HD compared to MD. Symptoms of three domains of the AHS (mental, gastrointestinal, and sympathetic) showed domain-specific significant increase in HD. A domain-specific relation was present between AUDIT and specific measures of AHS scores in HD, specifically with the dependence symptoms. Exacerbation in hangover symptomology could be a marker of more severe alcohol use disorder.

Keywords: alcohol hangover scale (AHS); alcohol use disorders identification test (AUDIT); dependence symptoms of AUDIT (DS-AUDIT); hangover; heavy drinking

1. Introduction

Excessive alcohol consumption is a leading cause of preventable mortality in the United States [1]. However, alcohol consumption continues to steadily rise in the United States; 2014 year estimates (52.7% of people aged 12 or older drank in the past month of reporting) were higher than the estimates in most years between 2002 and 2008 [2]. The most frequently reported consequence of excessive alcohol consumption is experiencing a hangover. The alcohol hangover refers to the combination of mental and physical symptoms, experienced the day after a single episode of heavy drinking, starting when blood alcohol concentration approaches zero [3]. The number of people experiencing alcohol hangover is high, with studies reporting 78% and higher prevalence of alcohol induced hangover [4,5] including social/light, moderate, and heavy drinkers. Hangover symptoms can be severe enough to impair daily routine, reduce productivity, and cause other associated complications of alcohol consumption [6]. Thus, understanding these symptoms and how they relate to the severity of alcohol abuse is important.

Adverse effects of excess alcohol intake manifest acutely as hangover symptoms. "Hangover" consists of a wide array of mental, physical, and neuropsychological (including sympathetic) symptoms that occur usually within hours of alcohol consumption and are recorded through the following day starting when blood alcohol concentration approaches zero [7–9]. Some of the common symptoms are fatigue, headaches, nausea, sleepiness, shakiness, weakness, excessive thirst and dry mouth, mood disturbances, and apathy [10–12].

Although hangover has been linked to heavy alcohol consumption and binge drinking, it can also be seen in individuals with moderate alcohol drinking [13,14]. This observation illustrates the gaps in our knowledge of how hangover symptoms manifest differently in moderate and heavy drinkers. Further, we do not know how hangover symptomology changes with the increased/altered drinking volume/patterns. Recent drinking history (Timeline follow-back, TLFB), and Alcohol Use Disorders Identification Test (AUDIT) are validated measures for assessing drinking patterns and quantity. In recent investigations, AUDIT and TLFB were used to study hangover symptoms [12,15] and they showed a close association with the hangover spectrum. Such associations have not been investigated in individuals with a high level of heavy drinking (≥15 drinks/per day), nor have there been studies of how hangover symptoms change with increased levels of alcohol drinking and altered patterns of consumption.

The primary objective of this pilot study was to identify the domains and individual measures of hangover symptoms that are different in individuals who consume alcohol in a moderate fashion and those who are heavy drinkers. Another aim of this study was to identify the associations of hangover symptoms and heavy drinking markers (derived from drinking history assessments) in heavy drinkers. We also included some of the potential modifiers (sex, family history, comorbid conditions) contributing to the changes in drinking patterns and hangover symptoms and assessed their interactions.

2. Study Participants and Methods

2.1. Patient Recruitment

The specific investigations reported here were conducted under two different larger protocols, one approved by the Combined Neuroscience Institutional Review Board at the NIH and other approved by the University of Louisville IRB. IRBs of both institutions approved the study once it was concluded that the study objectives met the ethical standards/regulations. The study at the NIH was indexed at ClinicalTrials.gov with identifier #NCT00713492, and the studies carried out at the University of Louisville were indexed at ClinicalTrials.gov, with the identifier numbers: #NCT01922895; #NCT01809132. Study participants who met diagnosis criteria for alcohol use disorder (AUD) and alcoholic liver disease (ALD) were approached at the outpatient and inpatient settings of the University of Louisville. Moderate drinkers and healthy volunteers were approached by advertisement using flyers, newspaper column, and word of mouth at both sites. All participants meeting eligibility criteria were enrolled after providing written informed consent. All participants' data were available securely to the PIs only. De-identified data were used to perform statistical analyses. Heavy drinkers and social/light drinkers were recruited at the University of Louisville; and moderate drinkers were recruited at NIH using cohort specific eligibility criteria as mentioned further.

A total of forty-one male and female individuals aged 21 to 64 years participated in this study. Study participants were classified based on their alcohol consumption pattern as follows: heavy drinkers ((HD), $n = 12$, aged 34–64 years); moderate drinkers ((MD), $n = 17$, aged 21–30 years); and social/light drinkers (healthy volunteers (SD); $n = 12$, aged 25–54 years). In our study, heavy drinkers (NIAAA guideline: ≥15 drinks/week for females; ≥20 drinks/week for males) drank around 15 drinks per day on an average (>90 drinks/week); all participants also met criteria for alcohol use disorder diagnosis according to DSM 5 criteria. Moderate drinkers drank in between 5–14 alcoholic drinks per week. Social/light drinkers (SD) drank four or fewer drinks on an average per week; they were not abstinent to alcohol drinking and were actively drinking at the time of screening. Inclusion criteria for heavy drinkers were that the females drank 15 or more drinks/week, and males drank 20 or more drinks/week meeting the NIAAA guidelines. Inclusion criteria for heavy drinkers also included being age 21 years or older with reported heavy drinking for at least the past six months. Inclusion criteria for the social/light drinkers (SD) in this study were: (a) age 21 years or older, (b) without any reported heavy or moderate drinking for at least the past six months, (c) normal comprehensive metabolic panel (normal liver and kidney panel, specifically serum albumin, total bilirubin, aspartate aminotransferase (AST), and alanine aminotransferase (ALT), and (d) no ongoing inflammation/infection or occurrence in the last three months. Exclusion criteria for heavy drinkers were: (a) unwilling or unable to provide informed consent, (b) significant comorbid conditions (heart, kidney, lung, neurological or psychiatric illnesses, sepsis) or active drug abuse, (c) pregnant or lactating women, (d) other known liver disease (except alcoholic liver disease—ALD), and/or (e) prisoners or other vulnerable subjects. Exclusion criteria for social/light drinkers included clinical diagnosis of any kind of liver disease and alcohol consumption meeting heavy or moderate drinking classification plus the four criteria above for heavy drinkers. Moderate drinkers (MD) had the same inclusion and exclusion as of social/light drinkers apart from the drinking profile (clause b). All study participants had a complete history and physical examination and laboratory evaluation upon study enrollment.

2.2. Study Paradigm

This investigation was a single time point assessment of AUD patients and healthy control participants. All participating individuals were consented for this study prior to collection of data and bodily samples. We collected demographic data, clinical data, medical history, and biochemical measures of liver injury and dysfunction. We also collected drinking history information (using timeline followback (TLFB), and Alcohol Use Disorders Identification Test (AUDIT)) from heavy drinkers and TLFB from moderate and social/light drinkers. Hangover symptomology was assessed using the Alcohol Hangover Scale (AHS). Participants reported on hangover after the last alcohol consumption until 10:00 am of the following day. The following drinking markers were calculated from the TLFB assessment for the past 90 days: total drinks past 90 days (TD90), number of drinking days past 90 days (NDD90), average drinks per drinking day past 90 Days (AvgDPD90), and heavy drinking days past 90 Days (HDD90) [16]. We used the total AUDIT score [17] and scores for its individual domains: hazardous alcohol use (frequency of drinking, typical quantity, and frequency of heavy drinking (HzAU)), dependence symptoms (impaired control over drinking, increased salience of drinking, and morning drinking (DS)) and harmful alcohol use (guilt after drinking, blackouts, alcohol-related injuries (HAU)). AHS is one of the standardized symptom scales used to quantify intensity of hangover symptoms based on the two publications we used as our assessment for hangover symptoms [18,19]. We used the guideline for AHS data based on our previous publication [12]. Symptoms measured by the AHS included mental domain ("hangover", "dizziness", and "craving"), physical domain measures ("thirsty", "tired", and "headache"), gastrointestinal domain symptoms ("nausea", "loss of appetite", and "stomach ache") and sympathetic ("heart-racing") that were analyzed individually and domain-wise. Participants rated the hangover symptoms they experienced until 10 am of the following morning after alcohol consumption on a scale from 0–7; as described in a previous publication from our group [12]. Categories of severity for hangover scale are n = 0 as "none", n = 1–3 as "mild", n = 4–6 as "moderate", and n = 7 as "incapacitated".

Family history of alcoholism (FHA) assessment was performed on all the subjects [20]. Subjects were asked for a family history of alcohol use disorders. A positive history (FHP) meant having at least one or more first degree relatives with a history of alcohol use disorder. A negative family history (FHN) was confirmed as being without any first- or second-degree relatives with alcohol use disorder.

Liver injury markers alanine transaminase (ALT), aspartate aminotransferase (AST), the ratio of AST:ALT, albumin, and total bilirubin were examined. All these measures were analyzed under comprehensive metabolic panel of the standard of care order. Laboratory testing were performed by the hospital laboratory at both the institutions.

2.3. Analysis

Demographics, drinking history and hangover data were analyzed using one-way analysis of variance (ANOVA) along with post-hoc t-tests for group comparisons. Group means and standard deviations were tabulated. Assessment of the association of hangover symptoms (itemized hangover measures, their domains, and average hangover score (avgAHS)) with AUDIT scores, TLFB measures, and family history of alcoholism were evaluated using multiple linear regression model, where sex and age were selected as co-variables; and liver function markers using liver panel tests (serum albumin and serum total bilirubin) were incorporated as modifying factors in multivariate linear regression models. SPSS 25.0 (IBM, Chicago IL, USA), GraphPad Prism 7.0 (GraphPad Software, San Diego CA, USA), and Microsoft Excel 2016 (Microsoft Corp., Redmond WA, USA) software were used for data processing, statistical analyses, and plot/figure development. Descriptive data are presented as mean ± standard deviation (M ± SD). Effect size is shown in Figures 2–4 as a model fit (goodness-of-fit) of the relation (adjusted R^2). Statistical significance was set at $p < 0.05$.

3. Results

3.1. Demographics and Drinking Profile

More females than males (almost double) were enrolled in both the moderate drinkers (MD) group ($n = 11$ and $n = 6$, respectively) and the social/light drinkers (SD) groups ($n = 7$ and $n = 5$, respectively). However, only approximately 1/4 of the heavy drinkers (HD) were female ($n = 3$ of 12; see Table 1). HD individuals had significantly higher ($p = 0.023$) BMI and were significantly ($p \leq 0.001$) older. In our study, most of the HD individuals were family history positive for alcohol use disorder. The likelihood ratio for positive family history of alcohol use disorder in HD group was considerably higher (+30%, 6.971) at $p = 0.008$ (2-sided) compared to MD. By race, SD group had only one subject as Asian, and two subjects as African American. One subject was African American in the HD group. One individual reported multi-race background in the MD group. All study subjects but one in MD group by ethnicity were Non-Hispanic or Latino. As anticipated, all TLFB90 markers were significantly higher in HD group compared to SD or MD groups. The MD group showed numerically higher levels of all the TLFB90 measures compared to SD; however, only AvgDPD90 scores were statistically significant (Table 1). The mean AUDIT score in the HD group was >20, thus meeting criteria for diagnosis of alcohol dependence [17]. Ten of the 12 HD subjects had an AUDIT score >20. We did not evaluate sex differences since there were only three female heavy drinkers (Table 1).

Table 1. Demographics, family history of alcoholism, and drinking measures.

Measures	Heavy Drinkers (n = 12)	Moderate Drinkers (n = 17)	Social Drinkers (n = 12)	Heavy vs. Moderate Drinking Group Significance
Demographics and Family History of Alcoholism				
Sex (M or F)	n(M) = 9, n(F) = 3	n(M) = 6, n(F) = 11	n(M) = 5, n(F) = 7	NA
Age (years.) [a,b]	49.8 ± 9.8	25.1 ± 3.1	31.6 ± 10.6	≤0.001
BMI [a]	30.6 ± 7.6	25.1 ± 4.7	25.0 ± 3.1	0.023
FHA	FHP, n = 11; FHN, n = 1	FHP, n = 8; FHN, n = 9	FHP, n = 5; FHN, n = 7	NA
Heavy Drinking Markers				
TD90 [a,b]	1153.92 ± 765.56	74.41 ± 31.39	23.67 ± 18.14	≤0.001
HDD90 [a]	67.33 ± 34.88	4.82 ± 9.47	0.58 ± 1.38	≤0.001
NDD90 [a]	72.5 ± 24.4	31.47 ± 16.62	18.25 ± 18.70	≤0.001
AvgDPD90 [a,b]	13.57 ± 8.1	2.5 ± 0.99	1.4 ± 0.72	≤0.001
AUDIT Scores and Its Domains				
AUDIT	23.08 ± 8.19	NC	NC	NA
AUDIT>20	10/12	NC	NC	NA
HzAU	9.92 ± 2.15	NC	NC	NA
DS	5.83 ± 3.86	NC	NC	NA
HAU	7.33 ± 4.5	NC	NC	NA

Abbreviations; M: male; F: female; BMI: body mass index; FHA: Family history of alcoholism; FHP: family history positive; FHN: family history negative; TD90: total drinks past 90 days; HDD90: heavy drinking days past 90 days; NDD90: number of drinking days past 90 days; AvDPD90: average drinks per drinking day past 90 days; AUDIT: alcohol use disorder identification test; DS: dependence symptoms; HAU: harmful alcohol use; HzAU: hazardous alcohol use. NA: not applicable, NC: not collected. a: between group comparisons of heavy and social/light drinkers; all comparisons found to have $p \leq 0.001$. b: between group comparison of moderate and social drinkers (age: $p = 0.023$; BMI: not significant; TD90: $p \leq 0.001$; HDD90: not significant; AvgDPD90: $p = 0.002$; NDD90: $p = 0.055$).

3.2. Assessment of Hangover Symptoms

There was a stepwise pattern to the occurrence of hangover symptomology wherein almost all of the hangover measures increased numerically from the social to moderate drinker groups and were the highest in heavy drinkers (Table 2). We found a large variability in some of the measures (as observed by the standard deviations) (Figure 1). This could be due to the individual variability in how the individual measures and domains of hangover manifest.

Table 2. Scores of hangover symptoms in social, moderate, and heavy drinkers. Effects of difference between moderate and heavy drinkers.

Symptoms	Social Drinkers (SD) n = 12	Moderate Drinkers (MD) n = 17	Heavy Drinkers (HD) n = 12	MD vs. HD Effects (Adjusted R^2)
Hangover [a]	0.33 ± 0.89	1.12 ± 1.22	2.67 ± 2.87	*p = 0.056*; R^2 = 0.129
Thirsty [a,b]	0.33 ± 0.65	2.94 ± 1.89	4.50 ± 1.93	$p = 0.039$; R^2 = 0.149
Tired [a,b]	0.83 ± 1.64	3.35 ± 1.73	3.08 ± 2.54	*p = 0.736*; R^2 = 0.004
Headache	0.33 ± 1.16	1.59 ± 1.91	1.92 ± 2.54	*p = 0.693*; R^2 = 0.006
Dizziness [a]	0.00	0.18 ± 0.53	2.25 ± 0.53	$p = 0.005$; R^2 = 0.259
Nausea [a]	0.00	0.59 ± 1.12	3.50 ± 2.61	$p \leq 0.001$; R^2 = 0.214
Stomachache [a]	0.00	0.35 ± 0.86	3.17 ± 2.29	$p \leq 0.001$; R^2 = 0.445
Heart-racing [a]	0.00	0.00	3.08 ± 2.78	$p \leq 0.001$; R^2 = 0.441
Loss of Appetite [a]	0.00	0.18 ± 0.73	4.50 ± 2.15	$p \leq 0.001$; R^2 = 0.689
Craving [a]	0.00	0.41 ± 1.18	4.0 ± 2.83	$p \leq 0.001$; R^2 = 0.451
AvgHS [a,b]	0.18 ± 0.41	1.14 ± 0.64	3.3 ± 1.58	$p \leq 0.001$; R^2 = 0.489

Abbreviations; SD: social drinkers (or light drinkers); MD: moderate drinkers; HD: heavy drinkers; AvgAHS: average hangover score. a significant statistical difference in hangover symptoms between social and heavy drinkers. b significant statistical difference in hangover symptoms between social and moderate drinkers. In italics: not significant.

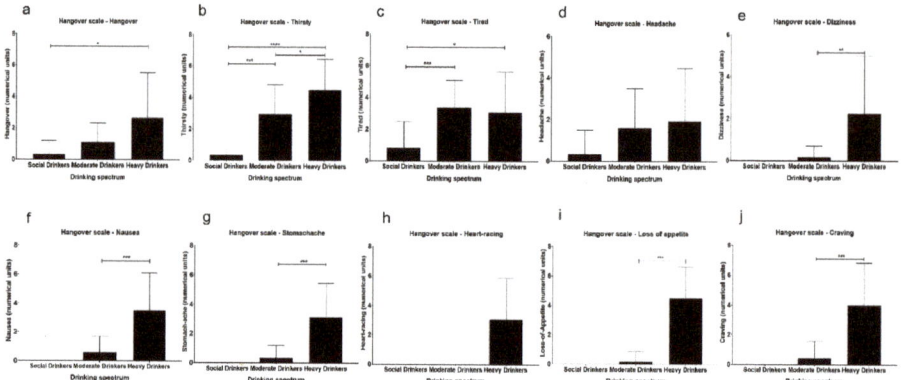

Figure 1. Levels of hangover measures in social/light drinkers (SD), moderate drinkers (MD), and heavy drinkers (HD). (**a**): Hangover measure "Hangover". (**b**): Hangover measure "Thirsty". (**c**): Hangover measure "Tired". (**d**): Hangover measure "Headache". (**e**): Hangover measure "Dizziness". (**f**): Hangover measure "Nausea". (**g**): Hangover measure "Stomachache". (**h**): Hangover measure "Heart-racing" (**i**): Hangover measure "Loss of appetite". (**j**): "Craving", "loss of appetite", and "thirsty" measures showed moderate level of hangover severity in heavy drinkers (Table 2). Data are presented as mean ± standard deviation. Statistical significance was set at $p \leq 0.05$. Statistical significance is not described in comparisons when the SD group individuals reported "zero", as in some measures.

3.3. AUDIT Domains, and Association of Drinking Markers in Heavy Drinkers

The HzAU domain of the AUDIT was significantly higher compared to DS and HAU domains in HD group (Figure 2a). Neither the DS nor the HAU domain exerted as great an effect as the HzAU (Figure 2a). The AUDIT score and the recent heavy drinking marker, HDD90, showed a significant association in the HD group (Figure 2b). Among the domains of AUDIT, only HzAU domain was significantly associated with HDD90 marker at moderate effects in the HD group (Figure 2c), thus suggesting overall AUDIT vs. heavy drinking relation was primarily due to the hazardous domain of AUDIT.

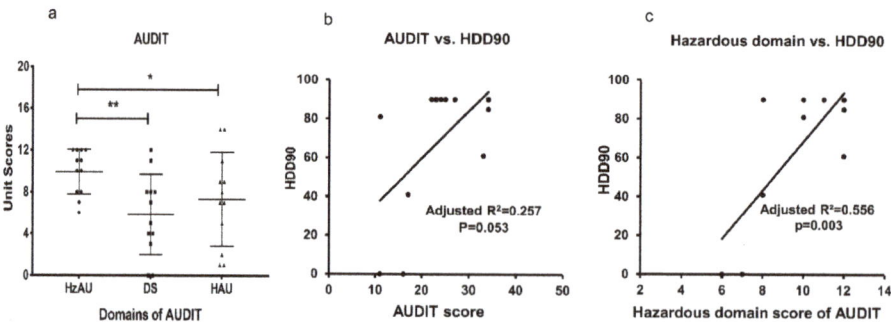

Figure 2. AUDIT and its association with timeline followback measures in heavy drinkers (HD). (**a**) Presentation of hazardous, dependency, and harmful domains that constitute AUDIT. (**b**) AUDIT score and heavy drinking days past 90 days (HDD90) drinking marker. (**c**) Scores of "hazardous-domain" of AUDIT and HDD90 (There are two points each for 11 and 12 [hazardous domain] that have corresponding 90 units of HDD90). Statistical significance was set at $p \leq 0.05$.

3.4. Internal Consistency of Hangover Measures in Heavy Drinkers

Mental domain measures of "hangover", "craving", and "dizziness" together showed very strong main effects, unadjusted $R^2 = 0.707$ at high significance $p = 0.009$. Physical domain measures of "headache", "thirsty", and "tired" together showed the strongest main effects, unadjusted $R^2 = 0.862$ at very high significance $p \leq 0.001$. Gastrointestinal domain measures of "stomachache", "nausea", and "loss of appetite" together also showed strong main effects, unadjusted $R^2 = 0.842$ at high significance $p = 0.001$. "Stomachache" and "nausea" measures were associated at high effects adjusted $R^2 = 0.642$ and high significance $p = 0.001$. No other measures within any of the domains showed any such association.

3.5. Association of AUDIT and Hangover Measures in Heavy Drinkers

In the HD group, the AUDIT showed a significant association with the average hangover scores (Figure 3a) and with three hangover measures: "heart-racing" (Figure 3b), "craving" (Figure 3c), and "thirsty" (Figure 3d). The AUDIT score showed significant high main effects with the symptoms of mental domain ("hangover", "dizziness", and "craving") (unadjusted $R^2 = 0.727$, $p \leq 0.001$), as well as physical domain ("thirsty", "tired", and "headache" together) (unadjusted $R^2 = 0.778$, $p \leq 0.001$) in this group. Gastrointestinal domain symptoms ("nausea", "loss of appetite", and "stomachache") showed significant, though moderate, main effects with AUDIT scores (unadjusted $R^2 = 0.685$, $p \leq 0.001$) in these heavy drinkers.

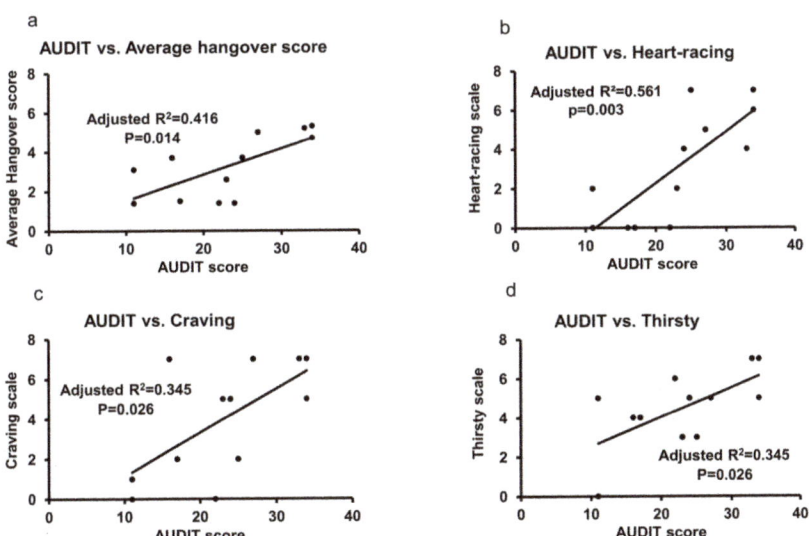

Figure 3. Association of AUDIT and hangover measures in heavy drinkers. (**a**) AUDIT and "Average hangover score" (AvgAHS). (**b**) AUDIT and "Heart-racing" (sympathetic domain) measure. (**c**) AUDIT and "Craving" (mental domain) measure. (**d**) AUDIT and "Thirsty" (physical domain) measure. Statistical significance was set at $p \leq 0.05$.

3.6. Role of AUDIT and Heavy Drinking TLFB Markers on Hangover Symptoms

There was no direct association of hangover measures and timeline followback markers (data not plotted). However, specific TLFB drinking markers and AUDIT scores together showed the augmented association with hangover measures, and that effect was higher than those exhibited by the univariate associations of AUDIT and specific hangover measures (Figure 3a,c). The average hangover score and AUDIT score showed a strong association (adjusted $R^2 = 0.578$) at high significance, $p = 0.021$, when adjusted for NDD90. The "craving" measure and the AUDIT showed an association (adjusted $R^2 = 0.462$) at a high significance ($p = 0.025$), when adjusted for HDD90.

We further evaluated the association of hangover measures and domains of AUDIT to identify domain-specific relations in heavy drinkers. The hangover measure, "heart-racing", was closely associated with the combined effect of hazardous alcohol use (HzAU) and the "dependence symptoms" (DS) domains of the AUDIT (Figure 4a,d). Notably, the association of "heart-racing" values and DS scores was of very high effect (Figure 4d). DS scores also correlated with average hangover scores (Figure 4b), for "craving" (Figure 4c), and for "stomachache" (Figure 4e) in heavy drinkers. The association of DS and craving was likely significant (Figure 4c); when this association was tested with HDD90 as a covariate, both the significance ($p = 0.025$) and effect size (Adjusted $R^2 = 0.462$) showed substantial augmentation. The harmful alcohol use (HAU) domain was significantly associated with only one hangover measure, "thirsty", in heavy drinkers (Figure 4f). Domain specific responses of the AUDIT showed close associations with specific hangover symptoms.

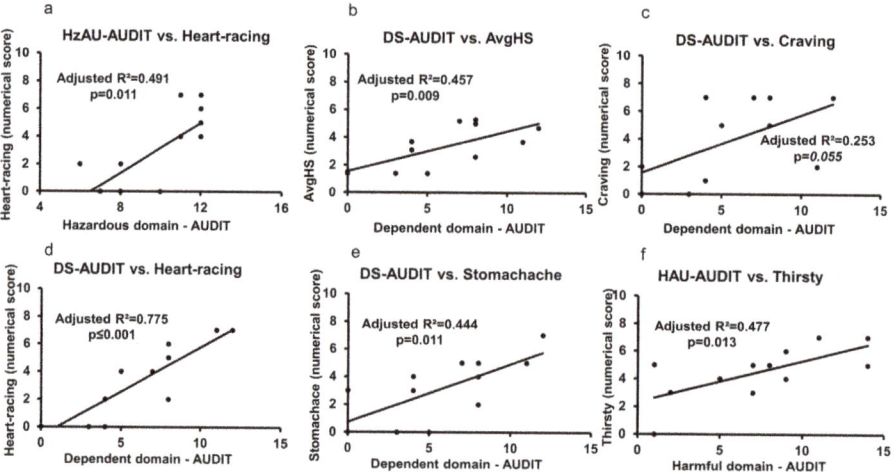

Figure 4. Association of AUDIT domains (Hazardous alcohol use (HzAU), dependence score (DS), and harmful alcohol use (HAU)) and individual hangover measures in heavy drinkers. (**a**) Association of HzAU and "Heart-racing" measure. (**b**) Association of DS and "Average hangover score" (AvgAHS) measure. (**c**) Association of DS and "Craving" measure. (**d**) Association of DS and "Heart-racing" measure. (**e**) Association of DS and "stomachache" measure. (**f**) Association of HAU and "Thirsty" measure. Statistical significance was set at $p \leq 0.05$.

3.7. Role of Liver Dysfunction on Hangover Symptoms

Impaired liver health, characterized by serum total bilirubin and serum albumin (Table 3), along with drinking profile (using AUDIT) was predictive of the average hangover score (Figure 5). However, we did not find major effects of liver injury markers (ALT, AST, and AST:ALT ratio) in combination with AUDIT on hangover severity.

In heavy drinking AUD patients, more severe form of ALD was a comorbid condition (with a diagnosis of alcoholic hepatitis) (Table 3). Heavy drinkers with comorbid ALD showed signs of both clinical range of liver injury and function markers. AUDIT scores showed significant association with average hangover score and other hangover symptoms (Figure 3), and thus, we investigated if there is any mediating role of liver function/injury in the hangover symptomology (Figure 5).

One of the hypotheses for our observation is that impaired liver function contributes to suboptimal metabolism of alcohol and alcohol metabolites. This could lead to both higher blood alcohol concentrations and longer durations in the system. Aldehyde, a metabolite of alcohol metabolism primarily in liver, has toxic effects on the system that lead to pathological (oxidative stress), and this may be related to psychiatric manifestations (withdrawal/hangover). We found that higher hangover scores are associated with greater effects by heavy drinking (AUDIT) mediated by worsening indications of liver function (Figure 5). Further investigation on the roles of alcohol dehydrogenase and aldehyde dehydrogenase with respect to hangover symptomology could help define this relation.

Table 3. Presentation of candidate liver panel markers.

Measures	Heavy Drinkers ($n = 12$)	Moderate Drinkers ($n = 17$)	Social Drinkers ($n = 12$)	Heavy vs. Moderate Drinking Group Significance
ALT (U/L)	73.83 ± 61.00	18.82 ± 7.95	28.5 ± 28.77	0.001
AST (U/L)	174.75 ± 82.11	24.53 ± 6.27	28.7 ± 13.92	≤0.001
AST:ALT ratio	3.53 ± 2.56	1.39 ± 0.48	1.03 ± 0.56	0.002
Total Bilirubin (μmol/L)	9.72 ± 7.77	0.55 ± 0.35	0.610 ± 0.31	≤0.001
Albumin (g/dL)	2.66 ± 0.29	4.17 ± 0.26	4.08 ± 0.24	≤0.001

Abbreviations; ALT: alanine aminotransferase, AST: aspartate aminotransferase. All markers in social and moderate drinkers were not clinically significant.

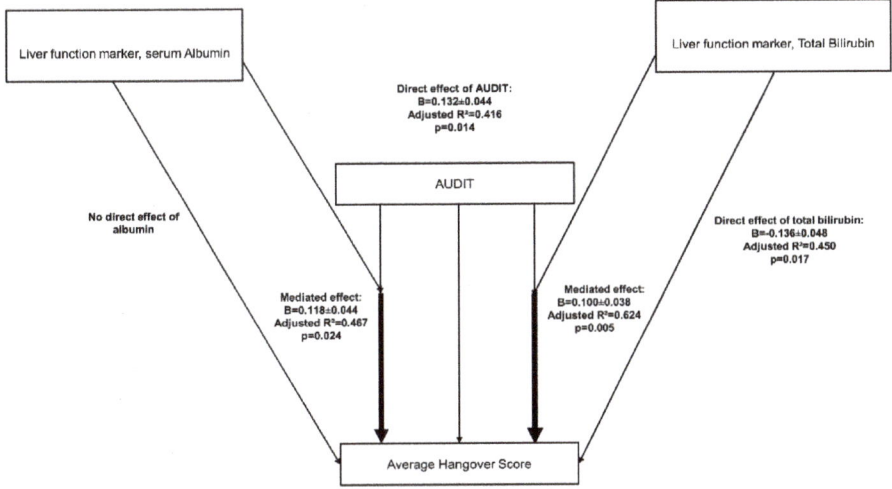

Figure 5. Schema of the effects of liver status (impaired liver function) and drinking severity on hangover symptomology. Liver function markers serum albumin and serum total bilirubin showed potential mediating roles in the effects of AUDIT on hangover symptoms.

4. Discussion

All heavy drinkers in this study reported hangover symptoms, whereas only some moderate drinker subjects reported hangover symptoms. The average hangover score in heavy drinkers was >3 of a possible 7, which was, by definition, a moderate level of incapacity. Three measures of hangover, "thirsty", "craving", and "loss of appetite", were numerically well into the moderate range of hangover intensity. Mental, gastrointestinal and sympathetic domains of the hangover scale were significantly higher in heavy drinkers compared to the moderate drinkers. Almost all of the HD patients reported a positive family history of alcoholism in our study. The AUDIT score has been shown to have a significant association with future alcohol use disorder, with up to 61% of higher AUDIT scores linked to alcohol use disorder in comparison to 10% with lower scores [21]. In our study, we found the majority of the heavy drinkers scored very high (>20) on the AUDIT scale. We found a close association of heavy drinking days past 90 days (HDD90, a component of TLFB90) with the AUDIT score and with its hazardous domain in particular. In our previous findings, we found a close association of recent drinking and hangover in moderate drinkers [12]. In the heavy drinkers, hangover measures showed a close association with AUDIT and its domains. Recent studies have linked higher hangover symptoms to a higher risk of alcohol use disorder and delirium tremens [22]. However, there is a significant gap in the understanding of the intensity of hangover symptoms and their association with patterns of heavy alcohol drinking. Our previous study reported a mild level of hangover symptoms in moderate drinkers [12]. In this study, we found that the heavy drinkers, who also exhibited remarkably heavier alcohol intake, showed a moderate level of incapacity in overall hangover assessment.

AUDIT scores (and specifically hazardous and dependence domains of the AUDIT scale) of heavy drinkers were most closely associated with the "heart-racing" measure of hangover. "Heart-racing" is a measure of the sympathetic domain of hangover, and manifestations of sympathetic hyperactivity could be related to vasoconstriction and tachycardia [23]. Stimulant and sedative use can potentiate sympathetic nervous system hyperactivity that could be related to elevated "heart-racing" and may result in severe withdrawal symptoms in alcohol use disorder.

In our study, the AUDIT score and its harmful domain showed a close association with the physical ("thirsty") domain. This association is consistent with a previous publication stating that hangover could be an indicator of risk for physical dependence [24]. Likewise, the AUDIT score and its dependence domain showed a significant association with the mental domain "craving" measure. Subsequent drinking to relieve hangover could lead to heavier alcohol consumption and higher AUD symptoms [25]. Further, craving and its association with the dependence domain of AUDIT might influence subsequent drinks and their timing, as has been reported previously [26], also suggesting increased predisposition for alcohol use disorder (AUD). Hangover symptomology is a direct and immediate adverse effect of alcohol intake, which we have now assessed in all categories of drinking (social, moderate, and heavy) in this pilot study. Importantly, we have shown a stepwise relation between levels of drinking and symptomology of hangover. However, in the alcohol use disorder spectrum, much emphasis is given on dependence and withdrawal categories of diagnosis. In heavy drinkers, these adverse symptoms of alcohol-associated hangover are worse than in moderate or social/light drinkers. Furthermore, the dependence domain of the AUDIT and specific domains of the hangover scale correlate highly in heavy drinkers. Thus, these findings provide the basis for including alcohol-associated adverse effects scales, such as the AHS, in the diagnosis of alcohol use disorder (detailing item 3 of the "DSM-5" concerning the aftereffects: https://pubs.niaaa.nih.gov/publications/dsmfactsheet/DSMfact.pdf). Validation of these findings and their effects from a large study cohort of heavy and moderate drinkers may elucidate the role of specific adverse effects of alcohol intake in the development and progression of alcohol dependence.

Most studies report that males consumed a higher amount of alcohol in comparison to females; thus, males are likely to show higher incapacity in the hangover symptoms scale [27]. In our study, 3/4 of the participants in HD group were males, and they exhibited numerically higher severity both in alcohol consumption and hangover symptoms. We examined other factors that could be associated with higher alcohol consumption. Positive family history of alcoholism was a significant factor associated with heavy alcohol consumption and higher hangover symptom scores, as reported previously [28–30]. Our study was consistent with those findings.

There were limitations in this study. This study was performed as a proof of principle design to address the primary aims. Given the results showing considerable effect sizes, this study supports the utility of these outcomes in larger populations. We acknowledge that these initial results would benefit from further testing with a larger sample size, and we are pursuing such study presently. We did not collect the length of time for each of the hangover symptoms. This should likely be used as another parameter in the ongoing and future studies. There were few females in the heavy drinking group, therefore we could not determine any sex differences. Variability in ethnicity was limited, most of the participants were Caucasian, and few were Hispanic-Latino; thus, any meaningful race/ethnicity-based comparisons is not within the scope of this study. Gamma-glutamyl transferase (GGT), a marker of liver injury was not tested in this study. Our focus in this study was on the heavy drinkers; 11 of 12 participants were positive for family history of alcoholism. Importantly, the sample size for each of the subgroups is not large, and thus, we focused on the effect sizes for interpreting the results. There are more factors that can modify/potentiate hangover symptoms, and assessing those was not in the scope of this study. For example, dehydration and electrolyte imbalances can worsen significantly under the effects of alcohol, and we did not study either of these factors. Nor did we evaluate acute withdrawal symptoms (Clinical Institute Withdrawal Assessment for Alcohol Scale-revised (CIWA-Ar)) in heavy drinkers for comparisons with the AHS, the AUDIT, or the TLFB measures. This is a future direction.

We found an interesting association between liver function markers (albumin and total bilirubin) and higher hangover scale results. One of the hypotheses for our observation is that impaired liver function contributes to slower metabolism of alcohol and alcohol metabolites. This could lead to both higher and more persistent blood alcohol/metabolite concentrations. We did not find any direct (or modifying) association of liver injury and higher hangover symptoms. This could inform that liver function (not liver injury) might be a better predictor of alcohol associated adverse effects. Further study could identify an altered mechanistic response that causes delayed and slower alcohol metabolism that would be consequential in higher and prolonged levels of alcohol consumption.

The presence and severity of hangover symptoms varied substantially across the three drinking groups. Hangover in heavy drinkers was debilitating, while in moderate drinkers, it was reported only as uncomfortable. Domains of hangover symptoms corresponded with the specific markers/domains of heavy alcohol drinking in heavy drinkers. Specific assessment of drinking measures should be used for assessing hangover symptomology in different drinking groups; AUDIT seems to better correlate with hangover measures in heavy drinkers. Elevated domains of symptoms assessed by the AUDIT in heavy drinkers, and their relevance with domains of the AHS, support the potential use of the AHS in determining severity of alcohol use disorder. Our findings point to the need for further studies on larger groups of drinkers with regard to patterns of alcohol consumption, levels of alcohol use, and adverse symptomology. Alcohol associated hangover, adverse effects, and liver injury are most common immediate manifestations that are observed together very frequently. AUD characterization based on DSM-5 criteria addresses some of the hangover symptomology in the diagnosis; however, it does not specify the integrated role of the aforementioned measures. Available treatment options for AUD are also limited. Shared pathology of AUD and ALD require focused treatment for both AUD and ALD. Disulfiram and Acamprosate can be useful in AUD patients who are abstinent at the time of treatment; and Naltrexone cannot be prescribed in AUD patients with liver injury (black box warning identifies adverse interaction with liver health). Thus, intervention for AUD patients with liver injury who are actively drinking heavily is limited and new drugs are under investigation, for example, Varenicline tartrate, which does not interfere with liver [31]. Specifically, larger groups are needed to evaluate any potential sex differences; we have reported about the differences in liver injury and drinking patterns in AUD females in a larger study population previously [32].

5. Conclusions

Alcohol hangover symptoms increase with corresponding increases in the level of alcohol consumption. Specific domains of hangover symptomology are exacerbated with heavy drinking. Symptoms of hangover are associated with long term assessment of heavy drinking rather than the recent drinking history. AUDIT (Alcohol Use Disorders Identification Test) and its hazardous domain are associated with the heavy drinking marker, HDD90 (heavy drinking days in the past 90 days). Dependence scores of AUDIT scale were closely associated with the presentation of adverse after-effects of heavy drinking as characterized by the hangover scale (chiefly in the mental, sympathetic, and gastrointestinal domains). Large population studies are needed to confirm and more precisely illustrate the findings in this study and would also allow for analysis of differences between sexes.

Author Contributions: V.V. is the study PI and responsible for the study concept and design. V.V., V.Y.S.-T., H.Z.H., M.L.S., and B.L.S. contributed to the acquisition of clinical data and conducted data validation and quality assurance. V.V., M.K., and V.Y.S.-T. performed analysis of the data. V.V., C.J.M., B.L.S., J.C.V., and H.Z.H. drafted the manuscript. C.J.M., V.V., M.L.S., J.C.V., B.L.S., and V.A.R. contributed scientifically. All authors critically reviewed content and approved final version for publication.

Funding: This research was funded by Z99-AA999999 (VV); U01AA021901, U01AA021893-01, U01AA022489-01A1 (CJM); Z1A AA000466 (VAR). Research reported in this publication was supported by an Institutional Development Award (IDeA) from the National Institute of General Medical Sciences of the National Institutes of Health under grant number: P20GM113226 (CJM); and the National Institute on Alcohol Abuse and Alcoholism of the National Institutes of Health under award number: P50AA024337 (CJM). The content is solely the responsibility of the authors and does not necessarily represent the official views of the National Institutes of Health.

Acknowledgments: Authors acknowledge clinical and research staff of U-01 consortia affiliated clinical centers, and NIH clinical for their support in patient care and research support for this project. We also acknowledge ARC P60 (Indiana University) for CAIS support and development of alcohol administration paradigm. We thank Marion McClain for editing this manuscript.

Conflicts of Interest: Over the past three years, Joris Verster has received grants/research support from the Dutch Ministry of Infrastructure and the Environment, Janssen, Nutricia, and Sequential, and acted as a consultant/advisor for Clinilabs, More Labs, Red Bull, Sen-Jam Pharmaceutical, Toast!, and ZBiotics. The other authors declare no conflicts of interest.

References

1. Stahre, M.; Roeber, J.; Kanny, D.; Brewer, R.D.; Zhang, X. Contribution of excessive alcohol consumption to deaths and years of potential life lost in the United States. *Prev. Chronic Dis.* **2014**, *11*, E109. [CrossRef] [PubMed]
2. Abuse, S.; M.H.S. Administration. *National Survey on Drug Use and Health*; M.H.S.: Everett, WA, USA, 2014.
3. Van Schrojenstein Lantman, M.; van de Loo, A.J.; Mackus, M.; Verster, J.C. Development of a definition for the alcohol hangover: Consumer descriptions and expert consensus. *Curr. Drug Abus. Rev.* **2016**, *9*, 148–154. [CrossRef] [PubMed]
4. Howland, J.; Rohsenow, D.J.; Greece, J.; Almeida, A.; Minsky, S.J.; Allensworth-Davies, D.; Arnedt, J.T.; Hermos, J. The Incidence and Severity of Hangover the Morning after Moderate Alcohol Intoxication. *Addiction* **2008**, *103*. [CrossRef] [PubMed]
5. Verster, J.C.; de Klerk, S.; Bervoets, A.C.; Kruisselbrink, L.D. Can hangover immunity really be claimed? *Curr. Drug Abuse Rev.* **2013**, *6*, 253–254. [CrossRef] [PubMed]
6. Sacks, J.J.; Gonzales, K.R.; Bouchery, E.E.; Tomedi, L.E.; Brewer, R.D. 2010 National and State Costs of Excessive Alcohol Consumption. *Am. J. Prev. Med.* **2015**, *49*, e73–e79. [CrossRef] [PubMed]
7. Verster, J.C. The alcohol hangover—A puzzling phenomenon. *Alcohol Alcohol.* **2008**, *43*, 124–126. [CrossRef] [PubMed]
8. Van Schrojenstein Lantman, M.; Mackus, M.; van de Loo, A.J.A.E.; Verster, J.C. The impact of alcohol hangover symptoms on cognitive and physical functioning, and mood. *Hum. Psychopharmacol. Clin. Exp.* **2017**, *32*, e2623. [CrossRef] [PubMed]
9. Penning, R.; McKinney, A.; Verster, J.C. Alcohol hangover symptoms and their contribution to the overall hangover severity. *Alcohol Alcohol.* **2012**, *47*, 248–252. [CrossRef] [PubMed]
10. Verster, J.C.; Alford, C.; Bervoets, A.C.; de Klerk, S.; Grange, J.A.; Hogewoning, A.; Jones, K.; Kruisselbrink, D.L.; Owen, L.; Piasecki, T.M.; et al. Hangover Research Needs: Proceedings of the 5(th) Alcohol Hangover Research Group Meeting. *Curr. Drug Abuse Rev.* **2013**, *6*, 245–251. [CrossRef] [PubMed]
11. Verster, J.C.; Stephens, R.; Penning, R.; Rohsenow, D.; McGeary, J.; Levy, D.; McKinney, A.; Finnigan, F.; Piasecki, T.M.; Adan, A.; et al. The Alcohol Hangover Research Group Consensus Statement on Best Practice in Alcohol Hangover Research. *Curr. Drug Abus. Rev.* **2010**, *3*, 116–126. [CrossRef]
12. Vatsalya, V.; Stangl, B.L.; Schmidt, V.Y.; Ramchandani, V.A. Characterization of hangover following intravenous alcohol exposure in social drinkers: Methodological and clinical implications. *Addict. Biol.* **2018**, *23*, 493–502. [CrossRef] [PubMed]
13. Wiese, G.J.; Shlipak, M.G.; Browner, W.S. The alcohol hangover. *Ann. Intern. Med.* **2000**, *132*, 897–902. [CrossRef] [PubMed]
14. Kruisselbrink, L.D.; Bervoets, A.C.; de Klerk, S.; van de Loo, A.J.A.E.; Verster, J.C. Hangover resistance in a Canadian university student population. *Addict. Behav. Rep.* **2017**, *5*, 14–18. [CrossRef] [PubMed]

15. Farokhnia, M.; Lee, M.R.; Farinelli, L.A.; Ramchandani, V.A.; Akhlaghi, F.; Leggio, L. Pharmacological manipulation of the ghrelin system and alcohol hangover symptoms in heavy drinking individuals: Is there a link? *Pharmacol. Biochem. Behav.* **2018**, *172*, 39–49. [CrossRef] [PubMed]
16. Sobell, L.C.; Sobell, M.B. Timeline Follow-Back. In *Measuring Alcohol Consumption*; Springer: Berlin, Germany, 1992; pp. 41–72.
17. Saunders, J.B.; Aasland, O.G.; Babor, T.F.; de la Fuente, J.R.; Grant, M. Development of the alcohol use disorders identification test (AUDIT): WHO collaborative project on early detection of persons with harmful alcohol consumption-II. *Addiction* **1993**, *88*, 791–804. [CrossRef] [PubMed]
18. Stephens, R.; Grange, J.A.; Jones, K.; Owen, L. A critical analysis of alcohol hangover research methodology for surveys or studies of effects on cognition. *Psychopharmacology* **2014**, *231*, 2223–2236. [CrossRef] [PubMed]
19. Rohsenow, D.J.; Howland, J.; Minsky, S.J.; Greece, J.; Almeida, A.; Roehrs, T.A. The Acute Hangover Scale: A New Measure of Immediate Hangover Symptoms. *Addict. Behav.* **2007**, *32*, 1314–1320. [CrossRef] [PubMed]
20. Mann, R.E.; Sobell, L.C.; Sobell, M.B.; Pavan, D. Reliability of a family tree questionnaire for assessing family history of alcohol problems. *Drug Alcohol Depend.* **1985**, *15*, 61–67. [CrossRef]
21. Conigrave, M.K.; Saunders, J.B.; Reznik, R.B. Predictive capacity of the AUDIT questionnaire for alcohol-related harm. *Addiction* **1995**, *90*, 1479–1485. [CrossRef] [PubMed]
22. Piasecki, M.T.; Robertson, B.M.; Epler, A.J. Hangover and Risk for Alcohol Use Disorders: Existing Evidence and Potential Mechanisms. *Curr. Drug. Abus. Rev.* **2010**, *3*, 92–102. [CrossRef]
23. Tryba, M. Alpha2-adrenoceptor agonists in intensive care medicine: Prevention and treatment of withdrawal. *Best Pract. Res. Clin. Anaesthesiol.* **2000**, *14*, 459–470. [CrossRef]
24. Piasecki, T.M.; Sher, K.J.; Slutske, W.S.; Jackson, K.M. Hangover frequency and risk for alcohol use disorders: Evidence from a longitudinal high-risk study. *J. Abnorm. Psychol.* **2005**, *114*, 223. [CrossRef] [PubMed]
25. Hunt-Carter, E.; Slutske, W.; Piasecki, T. Characteristics and correlates of drinking to relieve hangover in a college sample. In *Alcoholism-Clinical and Experimental Research*; Lippincott Williams & Wilkins: Philadelphia, PA, USA, 2005.
26. Epler, A.J.; Tomko, R.L.; Piasecki, T.M.; Wood, P.K.; Sher, K.J.; Shiffman, S.; Heath, A.C. Does hangover influence the time to next drink? An investigation using ecological momentary assessment. *Alcohol. Clin. Exp. Res.* **2014**, *38*, 1461–1469. [CrossRef] [PubMed]
27. McKinney, A.; Coyle, K. Alcohol hangover effects on measures of affect the morning after a normal night's drinking. *Alcohol. Alcohol.* **2006**, *41*, 54–60. [CrossRef] [PubMed]
28. Slutske, S.W.; Piasecki, T.M.; Hunt-Carter, E.E. Development and initial validation of the Hangover Symptoms Scale: Prevalence and correlates of Hangover Symptoms in college students. *Alcohol. Clin. Exp. Res.* **2003**, *27*, 1442–1450. [CrossRef] [PubMed]
29. Dawson, A.D.; Harford, T.C.; Grant, B.F. Family history as a predictor of alcohol dependence. *Alcohol. Clin. Exp. Res.* **1992**, *16*, 572–575. [CrossRef] [PubMed]
30. Newlin, B.D.; Pretorius, M.B. Sons of alcoholics report greater hangover symptoms than sons of nonalcoholics: A pilot study. *Alcohol. Clin. Exp. Res.* **1990**, *14*, 713–716. [CrossRef] [PubMed]
31. Vatsalya, V.; Gowin, J.L.; Schwandt, M.L.; Momenan, R.; Coe, M.A.; Cooke, M.E.; Hommer, D.W.; Bartlett, S.; Heilig, M.; Ramchandani, V.A. Effects of varenicline on neural correlates of alcohol salience in heavy drinkers. *Int. J. Neuropsychopharmacol.* **2015**, *18*. [CrossRef] [PubMed]
32. Vatsalya, V.; Song, M.; Schwandt, M.L.; Cave, M.C.; Barve, S.S.; George, D.T.; Ramchandani, V.A.; McClain, C.J. Effects of sex, drinking history, and omega-3 and omega-6 fatty acids dysregulation on the onset of liver injury in very heavy drinking alcohol-dependent patients. *Alcohol. Clin. Exp. Res.* **2016**, *40*, 2085–2093. [CrossRef] [PubMed]

© 2019 by the authors. Licensee MDPI, Basel, Switzerland. This article is an open access article distributed under the terms and conditions of the Creative Commons Attribution (CC BY) license (http://creativecommons.org/licenses/by/4.0/).

Article

The Association between Alcohol Hangover Frequency and Severity: Evidence for Reverse Tolerance?

Joris C. Verster [1,2,3,*], Karin A. Slot [1], Lizanne Arnoldy [1], Albertine E. van Lawick van Pabst [1], Aurora J. A. E. van de Loo [1,2], Sarah Benson [3] and Andrew Scholey [3]

1   Division of Pharmacology, Utrecht Institute for Pharmaceutical Sciences (UIPS), Utrecht University, 3584CG Utrecht, The Netherlands
2   Institute for Risk Assessment Sciences (IRAS), Utrecht University, 3584CM Utrecht, The Netherlands
3   Centre for Human Psychopharmacology, Swinburne University, VIC 3122 Melbourne, Australia
*   Correspondence: j.c.verster@uu.nl; Tel.: +31-30-253-6909

Received: 31 August 2019; Accepted: 18 September 2019; Published: 21 September 2019

Abstract: Although hangover is a common consequence of heavy alcohol consumption, the area is heavily under-researched. Hangover frequency is a potential predictor of future alcohol use disorder that may be affected by hangover severity, yet the relationship between hangover frequency and severity has not been investigated. Using different methodologies and assessment instruments, two surveys, and one naturalistic study collected data on hangover frequency, hangover severity, and alcohol consumption. The relationship between hangover frequency and severity was investigated via correlational analysis, considering potentially moderating variables including alcohol intake, estimated blood alcohol concentration, demographics, and personality characteristics. In all the three studies, a positive and significant association between hangover frequency and severity was found, which remained significant after correcting for alcohol intake and other moderating factors. These findings suggest that hangover severity increases when hangovers are experienced more frequently and may be driven by sensitization or reverse tolerance to this aspect of alcohol consumption. Future research should further investigate the relationship between hangover frequency and severity and alcohol use disorder and its implications for prevention.

Keywords: alcohol; hangover; frequency; severity; tolerance

1. Introduction

Alcohol hangover refers to the combination of mental and physical symptoms, experienced the day after a single episode of heavy drinking, starting when the blood alcohol concentration (BAC) approaches zero [1]. Although hangover is a common consequence of heavy alcohol consumption, research investigating hangover frequency is sparse [2], and perhaps surprisingly, the relationship between hangover frequency and hangover severity has not been investigated.

This is potentially important, as experiencing frequent hangovers has been associated with significant health consequences and mortality. For example, there is a significant positive association between hangover frequency and cardiovascular mortality [3]. A negative relationship has also been identified between intelligence quotient (IQ) at age 11 and hangover frequency in middle-age [4]. Whereas childhood socioeconomic status had no relevant influence, the association between childhood IQ and later life hangover frequency was significantly attenuated by middle age socioeconomic status. Other reports have identified a significant relationship between hangover frequency and experiencing depressive symptoms [5,6]. These studies did not explore the relationship between hangover frequency and hangover severity. The latter is important as it has been suggested that having hangovers frequently could be a predictor of future alcohol use disorder [7].

If tolerance develops when hangovers are experienced more frequently, i.e., their severity diminishes, this could feasibly lead to higher alcohol consumption. Alternatively, if hangover severity remains constant or increases with greater frequency, this may have a protective effect in that drinkers would consume less alcohol on future drinking occasions. A prospective study using the sensitivity to the effects of alcohol (SRE) scale, showed that drinkers who have a low sensitivity to alcohol (i.e., more drinks are needed to feel an effect), report hangovers less frequently following a given number of drinks [8]. These findings suggest that tolerance develops to experiencing hangovers in subjects who report frequent heavy drinking episodes. Conversely, Courtney et al. [9] found that having more frequent hangovers was a marker for increased future numbers of alcoholic drinks per drinking day. Importantly, neither study assessed hangover severity, which may play a pivotal role in the relationship between hangover frequency and drinking quantity.

Recently, Köchling et al. [10] examined the impact of the sequence of consuming beer and wine on hangover severity in 90 participants. On two test days, beer, wine, or both were consumed to reach a BAC of 0.12%. Overall hangover severity was assessed as a composite score of eight individual items, including thirst, fatigue, headache, dizziness, nausea, stomach ache, tachycardia, and loss of appetite. The intensity of these symptoms was scored on a seven-point scale, with the sum score representing the overall hangover severity ranging from 0 (absent) to 56 (extreme hangover). Hangover frequency was assessed using five categories, including 'rarely', 'once-monthly', 'more than once-monthly, and less than once-weekly', 'once-weekly', and 'more than once-weekly'. The authors reported that hangover frequency was not a significant predictor of hangover severity. These findings suggest that hangover frequency and severity are unrelated. However, it should be considered that (1) a very crude measure of hangover frequency was used which provides little differentiation between the subjects, and (2) overall hangover severity was computed as a sum score from individual hangover symptoms, and no overall hangover severity rating was obtained. It is, thus important to further investigate the relationship between hangover frequency and severity using more precise measures.

Therefore, the current analysis investigated whether tolerance develops for experiencing alcohol hangovers. We interrogated databases from three independent studies which utilized different methodological approaches to evaluate hangover severity and frequency. It was hypothesized that hangover frequency would negatively correlate with hangover severity. In other words, if hangovers are experienced more frequently, their intensity diminishes, that is "hangover tolerance" develops.

2. Methods

The relationship between hangover frequency and severity was analyzed using data from three independent studies. Study 1 was a survey by Penning et al. [11] among Dutch students that retrospectively assessed hangover frequency and severity. Study 2 was an international survey among young adults who recorded hangover severity on three different days and related these scores to hangover frequency. The retrospective assessment for multiple days is important as it has been suggested that hangover severity not only varies between drinkers but also within the same individual [2,12]. Using the average hangover severity score over the three days accounts for any intra-individual variation. Study 3 was a semi-naturalistic laboratory study with real-time assessments of hangover severity.

Study 1 was approved by the Institutional Review Board of the Utrecht Institute for Pharmaceutical Sciences of Utrecht University. No formal medical ethics approval was required to conduct this survey according to the Central Committee of Research Involving Human Subjects, The Netherlands. Study 2 received ethics approval from the Ethics Committee of the Faculty of Social and Behavioral Sciences of Utrecht University (FETC17-063). Study 3 also received ethics approval from the Ethics Committee of the Faculty of Social and Behavioral Sciences of Utrecht University. Study 3 was registered at www.clinicaltrials.gov (ClinicalTrials.gov Identifier: NCT01400204).

2.1. Study 1: Dutch Students Survey

Study 1 was a survey among Dutch students [11]. The survey comprised questions about drinking behavior, and the nature and severity of symptoms experienced during the past month latest hangover. Demographic data were collected (e.g., age, sex, height, and weight) and questions regarding usual drinking behavior were assessed, including a question on how many hangovers participants had experienced during the past month. The number of alcoholic drinks consumed for the evening before their past month latest hangover was recorded. The start and stop time of drinking alcohol was not measured in this study; therefore, the estimated blood alcohol concentration (eBAC) could not be computed. Subjects were asked to rate the severity of a list of 47 hangover symptoms they possibly experienced on their most recent hangover experience within the past month. From these, three outcome measures of overall hangover severity were computed. First, a modified version of the Hangover Symptoms Scale (HSS) [13] was compiled. This included the following 12 items; being tired, headache, nausea, vomiting, weakness, thirst, concentration problems, sensitivity to light, sweating, anxiety, depression, and shaking/shivering. For this study, the HSS item 'trouble sleeping' was omitted. The rationale for excluding this item was that it is not a true hangover symptom [14]. Trouble sleeping, if anything, is part of the cause, rather than the effect of hangover. As such, it may have an impact on the presence and severity of hangover symptoms [15]. The original HSS assessed the frequency of occurrence of hangover symptoms. To obtain a severity score, in our modified version, the severity of each item was rated on an 11-point scale from 0 (absent) to 10 (extremely severe). The mean score of the 12 items was used as a measure of overall hangover severity. Second, the Acute Hangover Scale (AHS) score was computed [16]. Third, the Alcohol Hangover Severity Scale (AHSS) score was computed [14]. Data from those subjects aged 18–30 who reported a past month alcohol hangover were included in the statistical analyses, which were conducted with SPSS (SPSS, version 25.0, SPSS Inc., Armonk, NY: IBM Corp). The analyses included correlating past month hangover frequency with overall hangover severity scores, using partial correlations, adjusting for the number of alcoholic drinks consumed.

2.2. Study 2: International Survey

An international survey was conducted in Nadi, Fiji among people working or on holiday. Subjects were approached to complete the survey at Wailoaloa Beach, Nadi, Fiji. The survey was anonymous, and participants did not receive an incentive for completing the survey. Potential participants were approached between 8 a.m. and 4 p.m. They were asked if they had consumed alcohol during the past three days. Those who were willing to participate and understood the English language were handed the survey. The investigator was present to address any queries with regards to language comprehension of the participants (who included international holidaymakers).

The survey collected demographic data such as gender, age, and country of citizenship. Usual weekly alcohol consumption (at home, before going to Fiji) was recorded, as well as information on how many day subjects were abroad from their home country, and how long they had been in Fiji.

Data regarding alcohol consumption and hangover severity were recorded for three consecutive days. Both the number of alcoholic drinks and the timeframe of consumption were recorded. The survey contained guidance about standard drinking sizes and how to convert, for example, bottles of wine into standardized alcohol units. Estimated blood alcohol concentration (eBAC) was computed, applying a modified Widmark formula [17]. On each test day, subjects reported their overall hangover severity on an 11-point scale ranging from 0 (absent) to 10 (extreme). Further questions asked about alcohol consumption when at home (before coming to Fiji), how many days per month they experienced a hangover, and how many times they had a hangover during their stay in Fiji.

Statistical analysis was performed using the SPSS statistical program (SPSS, version 25.0, SPSS Inc., Armonk, NY: IBM Corp). The participants who reported they did not have a hangover were excluded. Data from the three days were averaged to provide a more reliable average measure of alcohol intake and hangover severity. The hangover frequency at home and in Fiji was correlated with the average overall hangover severity score, using a partial correlation, adjusting for eBAC.

2.3. Study 3: Naturalistic Study

Students, aged 18 to 30 years old, were recruited at different locations (university canteen and colleges) of the university campus of Utrecht University, The Netherlands. The subjects remained anonymous, and after finalizing the study, they were paid 20 Euros for their participation. This naturalistic, observational study comprised two assessments: one the day after an evening of alcohol consumption, and the second assessment was made after an alcohol-free day. The investigators took steps to not influence the subjects' drinking behavior or activities on the test days. There were no lifestyle or other guidelines imposed on the subjects, except that they were asked to complete the online survey between 2 p.m. and 10 p.m. The test days were scheduled one to three weeks apart, and the exact dates of the test days were chosen by the subjects themselves, depending on whether they choose to consume alcohol or not.

Demographic data were collected (e.g., age, sex, height, and weight) and questions regarding usual drinking behavior were assessed, including a question on how many hangovers participants had experienced during the past year. The Five-Shot questionnaire alcohol screening test was used to analyze general drinking behavior [18]. Personality was assessed with the Brief Symptom Inventory (BSI) [19] and the RT18 risk-taking questionnaire [20].

The BSI is a broad-used multidimensional symptom self-report inventory. It consists of 53 items; each rated on a five-point scale of distress from 1 (not-at-all) to 5 (extremely). The instrument is scored on nine symptom dimensions; somatization, obsession-compulsion, interpersonal sensitivity, depression, anxiety, hostility, phobic anxiety, paranoid ideation, and psychoticism.

The RT-18 consists of 18 items that can be scored as 'yes' or 'no' [19]. Higher scores imply greater levels of risk-taking. Two subscales can be computed, namely 'level of risk-taking behavior' and 'risk assessment'.

The Five-Shot questionnaire alcohol screening test was used to detect heavy drinking [18]. It is a short, self-report inventory, composed of two questions from the Alcohol Use Disorders Identification Test (AUDIT) [21,22] and three questions from the CAGE test [23]. A score of 2.5 or greater indicates possible alcohol misuse.

The number of alcoholic drinks consumed on the night prior to the test day was recorded, including the start and stop time of drinking alcohol. The estimated blood alcohol concentration (eBAC) was computed using a modified Widmark equation [17]. The modified Widmark equation takes into account the number of alcoholic drinks, relative body water volume, weight, gender, time taken to clear alcohol through metabolism. Participants were also asked how many alcoholic drinks they had consumed on this occasion, relative to a 'regular' drinking occasion. For each test day, subjects reported their overall hangover severity on an 11-point scale ranging from 0 (absent) to 10 (extreme). In addition, subjects rated their alcohol hangover via the modified 13-item version of the HSS [13], on which the severity of each item was rated on an 11-point scale from 0 (absent) to 10 (extremely severe).

Statistical analysis was performed using the SPSS statistical program (SPSS, version 25.0, SPSS Inc., Armonk, NY: IBM Corp). Subject reporting not having a hangover were excluded from the analysis. Past month hangover frequency was correlated with overall hangover severity scores, using partial correlations, adjusting for eBAC.

3. Results

3.1. Study 1: Dutch Students Survey

A total of 1410 participants completed the questionnaire, of which 56.1% ($n = 791$) had experienced at least one hangover during the last month. Their data were used for the analysis. The mean (SD) age of the sample was 20.4 (3.4) years, and 68.7% of the sample were women. The results of the study are summarized in Table 1.

Overall hangover severity, measured using the HSS, correlated significantly and positively with the frequency of past month hangovers ($r = 0.188$, $p = 0.000$). When adjusting for the number of alcoholic drinks consumed, the partial correlation remained significant ($r = 0.126$, $p = 0.001$).

Overall AHS hangover severity correlated significantly and positively with the frequency of past month hangovers ($r = 0.145$, $p = 0.000$). When adjusting for the number of alcoholic drinks consumed, the partial correlation remained significant ($r = 0.125$, $p = 0.001$).

Overall AHSS hangover severity correlated significantly and positively with the frequency of past month hangovers ($r = 0.198$, $p = 0.000$). When adjusting for the number of alcoholic drinks consumed, the partial correlation remained significant ($r = 0.153$, $p = 0.001$).

Table 1. The association between drinking variables and the severity and frequency of hangovers.

Hangover	Mean (SD)	Frequency	Severity (AHS Score)	Severity (AHSS Score)	Severity (HSS Score)
Usual consumption					
weekly number of alcoholic drinks	16.8 (14.6)	$r = 0.583, p = 0.000$	$r = 0.170, p = 0.000$		$r = 0.167, p = 0.000$
number of past month hangovers	2.4 (2.2)	---	$r = 0.145, p = 0.000$		$r = 0.188, p = 0.000$
Hangover day					
number of alcoholic drinks	10.6 (5.9)	$r = 0.329, p = 0.000$	$r = 0.164, p = 0.000$	$r = 0.211, p = 0.000$	$r = 0.196, p = 0.000$
HSS hangover severity	3.1 (1.5)	$r = 0.188, p = 0.000$	$r = 0.873, p = 0.000$	$r = 0.913, p = 0.000$	---
AHSS hangover severity	3.0 (1.6)	$r = 0.153, p = 0.000$	$r = 0.874, p = 0.000$	---	$r = 0.913, p = 0.000$
AHS hangover severity	3.7 (1.7)	$r = 0.145, p = 0.000$	---	$r = 0.188, p = 0.000$	$r = 0.873, p = 0.000$

Note: Partial correlations, adjusting for the number of alcoholic drinks consumed before the hangover day, were computed to relate the hangover frequency with severity. For other associations, non-parametric (Spearman's rho) correlations were computed.

3.2. Study 2: International Survey

The international survey was completed by n = 333 subjects (145 men and 188 women). Subjects originated from 20 different countries, with most of them coming from the UK (34.6%), Australia (11%), France (9.4%), USA (8.7%), and New Zealand (7.1%). Their mean (± SD) age was 23.5 (± 4.2) years old, and they reported weekly alcohol consumption of 11.5 (± 10.9) drinks. They reported experiencing 3.9 (± 3.4) hangovers per month at home, and 4.8 (± 13.2) while at Fiji. There was no significant difference in hangover frequency reported at home or in Fiji ($t = -1.26$, $p = 0.208$). On average they were 78 (± 102) days abroad from their home country, of which the past 26 (± 26) days were spent at Fiji. In Fiji, alcohol consumption and hangover data were gathered for three consecutive days, and the results are summarized in Table 2. A partial correlation, adjusting for eBAC, revealed that the associations between hangover severity and hangover frequency at home ($r = 0.197$, $p = 0.024$) and hangover frequency at Fiji ($r = 0.276$, $p = 0.001$) remained significant.

Table 2. The association between drinking variables, hangover severity, and hangover frequency.

Hangover (3 Day Average)	Mean (SD)	Frequency at Home	Frequency at Fiji	Severity (1-Item)
Number of alcoholic drinks consumed	6.0 (5.2)	$r = 0.188, p = 0.001$	$r = 0.228, p = 0.000$	$r = 0.504, p = 0.000$
eBAC (%)	0.11 (0.1)	$r = 0.216, p = 0.012$	$r = 0.250, p = 0.004$	$r = 0.454, p = 0.000$
Drinking duration (h)	5.3 (3.5)	$r = 0.062, p = 0.443$	$r = 0.197, p = 0.014$	$r = 0.294, p = 0.000$
1-item hangover severity score	1.2 (1.5)	$r = 0.309, p = 0.000$	$r = 0.468, p = 0.000$	---

Note: data represent the average scores of three consecutive days of alcohol consumption and experiencing hangovers. Abbreviation: eBAC = estimated blood alcohol concentration.

3.3. Study 3: Naturalistic Study

A total of n = 99 subjects participated in the study. Subjects were excluded if they used drugs of abuse on the test days (n = 3) or alcohol on the control day (n = 3). N = 12 other subjects were excluded because they reported no hangover on the alcohol test day. N = 81 subjects were eligible for the present analysis, of which 36 (44.4%) were men. The subjects had a mean (SD) age of 21.2 (2.9) years old. The associations between drinking variables and hangover frequency and severity are summarized in Table 3.

Table 3. Association between drinking variables and severity and frequency of hangovers.

Hangover	Mean (SD)	Frequency	Severity (HSS Score)	Severity (1-Item)
Usual consumption				
weekly number of alcoholic drinks	8.4 (7.4)	$r = 0.695, p = 0.000$	$r = 0.324, p = 0.002$	$r = 0.695, p = 0.000$
the Five Shot score	2.6 (1.2)	$r = 0.507, p = 0.000$	$r = 0.206, p = 0.048$	$r = 0.507, p = 0.000$
Number of hangovers per month	1.4 (1.7)	---	$r = 0.452, p = 0.000$	$r = 0.529, p = 0.000$
Hangover day				
number of alcoholic drinks consumed	9.2 (4.6)	$r = 0.452, p = 0.000$	$r = 0.413, p = 0.000$	$r = 0.452, p = 0.000$
eBAC (%)	0.16 (0.1)	$r = 0.416, p = 0.000$	$r = 0.421, p = 0.000$	$r = 0.416, p = 0.000$
drinking duration (h)	6.3 (2.2)	$r = 0.174, p = 0.095$	$r = 0.236, p = 0.023$	$r = 0.174, p = 0.095$
HSS hangover severity score	2.3 (1.4)	$r = 0.452, p = 0.000$	---	$r = 0.718, p = 0.000$
1-item hangover severity	3.5 (2.5)	$r = 0.529, p = 0.000$	$r = 0.718, p = 0.000$	---

Note: Non-parametric (Spearman's rho) correlations were computed. These are considered statistically significant if $p < 0.004$, after Bonferroni's correction for multiple comparisons.

As expected, hangover frequency was significantly correlated with a weekly number of alcoholic drinks consumed, the Five-Shot score, and the number of alcoholic drinks consumed, and eBAC on the evening before the hangover. Similarly, both HSS and the one-item hangover severity was significantly associated with the weekly number of alcoholic drinks consumed, the Five-Shot score, and the number of alcoholic drinks consumed, and the eBAC on the evening before the hangover. Figure 1 shows the distribution of hangover frequencies per month (Figure 1A) and its correlation with the one-item hangover severity (Figure 1B).

A positive and significant Spearman's correlation was found between the hangover frequency and HSS hangover severity ($r = 0.452$, $p = 0.000$) and between hangover frequency and the one-item hangover severity ($r = 0.529$, $p = 0.000$). When conducting a partial correlation, adjusting for eBAC, the correlations between the hangover frequency and the severity remained significant and positive (HSS: $r = 0.301$, $p = 0.004$, 1-item: $r = 0.297$, $p = 0.004$).

The partial correlation between the hangover frequency and the severity also remained significant when in addition to eBAC, was also corrected for the usual alcohol intake variables (weekly number of alcoholic drinks consumed and the Five-Shot score) (HSS: $r = 0.297$, $p = 0.005$, 1-item: $r = 0.341$, $p = 0.001$). When, in addition to eBAC, the correlation also adjusted for the hangover test day variables (number of alcoholic drinks consumed on the drinking occasion and the drinking duration), the correlation between the hangover frequency and severity also remained significant (HSS: $r = 0.318$, $p = 0.002$, 1-item: $r = 0.283$, $p = 0.007$).

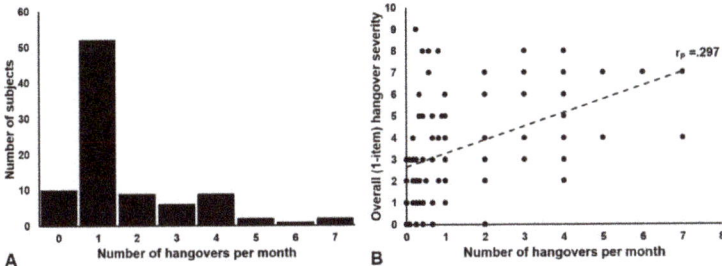

Figure 1. Distribution of the hangover frequency scores (**A**) and their association with the overall hangover severity (**B**). Note: a partial correlation (r_P), adjusting for estimated blood alcohol concentration, was computed between the hangover frequency and severity.

3.4. Individual Hangover Symptoms

The severity of individual hangover symptoms and their frequency of being reported are summarized in Table 4. An exploratory analysis was conducted to identify individual hangover symptoms of which the reported severity was associated with hangover severity. Partial correlations, adjusting for eBAC, were computed, and a Bonferroni's correction was applied to account for the multiple comparisons (p-values are considered statistically significant if $p < 0.004$). The analysis revealed that only the severity score of headache correlated significantly with hangover frequency (See Table 4).

Table 4. The severity of individual hangover symptoms and past year's hangover frequency.

Hangover Symptom	Mean (SD)	Frequency Reported	Correlation with Hangover Frequency
Being tired	5.6 (2.6)	97.8%	$r = 0.059, p = 0.580$
Thirst	4.7 (3.0)	90.3%	$r = 0.176, p = 0.095$
Weakness	4.1 (3.0)	83.9%	$r = 0.238, p = 0.023$
Concentration problems	3.6 (2.7)	81.7%	$r = 0.242, p = 0.021$
Headache	2.8 (3.2)	58.1%	$r = 0.340, p = 0.001$ *
Nausea	2.3 (2.8)	57.0%	$r = 0.212, p = 0.044$
Shaking, shivering	1.3 (2.2)	40.9%	$r = 0.184, p = 0.081$
Sleep problems	1.3 (2.1)	36.6%	$r = 0.025, p = 0.811$
Sensitivity to light	1.3 (2.2)	37.6%	$r = 0.045, p = 0.672$
Sweating	1.2 (2.2)	32.3%	$r = 0.263, p = 0.012$
Depression	0.6 (1.6)	15.1%	$r = 0.182, p = 0.084$
Anxiety	0.4 (1.3)	11.8%	$r = 0.219, p = 0.037$
Vomiting	0.4 (1.7)	6.5%	$r = -0.047, p = 0.660$

Note: Partial correlations, adjusted for eBAC, were computed. These are considered statistically significant if $p < 0.004$, after Bonferroni's correction for multiple comparisons. Significance is indicated by *.

3.5. Personality Characteristics

It is unlikely that the association between the hangover frequency and the severity could be explained by personality characteristics, including somatization. In Study 3, partial correlations, adjusting for eBAC and all BSI subscales, also revealed significant correlations between hangover frequency and severity (HSS: $r = 0.217, p = 0.050$, 1-item: $r = 0.361, p = 0.000$). Also, partial correlations, adjusting for eBAC and RT-18 risk-taking subscales, revealed significant correlations between hangover frequency and severity (HSS: $r = 0.277, p = 0.008$, 1-item: $r = 0.278, p = 0.008$).

3.6. Age

It could be argued that the hangover frequency is a proxy measure of age. As age progresses, the lifetime number of experienced hangovers increases. Also, research has suggested that the severity of hangover declines when age progresses [24], along with a steady reduction in the number of binge drinking days when age progresses [25]. To investigate the possible impact of age on the association between the hangover frequency and severity, the partial correlations in study 1, 2, and 3 were now corrected for both eBAC and age. It should be taken into account that the age range for the three studies, 18-30 years old, was relatively small. In study 1 the partial correlation, adjusting for the number of drinks consumed and age, between hangover frequency and hangover severity remained significant when assessed with either the HSS ($r = 0.125$, $p = 0.001$), AHS ($r = 0.121$, $p = 0.002$), or the AHSS ($r = 0.151$, $p = 0.000$). In study 2, the partial correlation, adjusting for eBAC and age, between hangover severity and hangover frequency at home or at Fiji remained significant ($r = 0.195$, $p = 0.026$, and $r = 0.274$, $p = 0.002$, respectively). In study 3, the partial correlation, adjusting for eBAC and age, between hangover frequency and hangover severity assessed with the HSS or the one-item hangover rating also remained significant ($r = 0.286$, $p = 0.006$, and $r = 0.300$, $p = 0.004$, respectively).

4. Discussion

Across studies, a significant and positive correlation was found between hangover frequency and severity, suggesting that when hangovers are experienced more frequently, their severity increases. These findings run counter to our prediction that tolerance develops to the effects of alcohol hangover. Therefore our hypothesis was not supported.

Our findings, are in contrast to those of the recent study by Köchling et al. [10], who reported that hangover frequency and severity are unrelated. The studies differed, however, in the methodologies to assess hangover frequency. Whereas Köchling et al. [10] used five relative crude frequency categories, in the current studies, the actual number of hangovers per month was calculated, which gives a more precise measure. Also, as opposed to our studies, Köchling et al. [10] used an unvalidated hangover scale to assess hangover severity, and an overall hangover severity rating was not obtained. Alternatively, one could argue that the study by Köchling et al. [10] was a controlled experiment, while our data was gathered via survey research and a study utilizing a less controlled naturalistic design. These methodological differences may account for the different outcomes of the study. Future research is, therefore, necessary to elucidate the exact nature of the relationship between hangover frequency and severity.

Our findings also contrast with the literature pertaining to acute intoxication, which suggests that, to some extent, tolerance develops to the acute effects of alcohol. This can be conceptualized as a rightward shift in the dose-response curve whereby repeated exposure to alcohol is manifested in two ways. The first is by individuals becoming less sensitive to the behavioral effects of the same dose of alcohol; the other is by requiring a greater amount of alcohol to achieve previous effects.

The physiological evidence of this effect includes data showing that more frequent drinkers develop pharmacokinetic tolerance. This is illustrated by reports of lower breath alcohol concentration (BAC) to the same level of alcohol compared with less frequent drinkers [26]. Consistent with this finding, after consuming the same dose of alcohol, frequent drinkers may report fewer adverse sedative effects [27] and feel less intoxicated [28]. In addition, behavioral tolerance to the acute effects of alcohol may develop differentially, depending on the task domain [29]. Behavioral tolerance, i.e., the observation that greater experience with drinking to intoxication leads to less impaired cognitive and psychomotor performance, has been observed in several studies [29–33]. In this context, a recent study showed that drinkers with higher levels of performance maintenance and those who experienced less severe intoxication effects, self-administered higher dosages of intravenous alcohol and reached higher peak BACs than drinkers who were more sensitive to the effects of alcohol [34]. Finally, research has shown that the sensitivity to acute alcohol effects varies with age such that, across several domains, young drinkers are less sensitive than older drinkers to alcohol intoxication effects [35].

Taken together, there is evidence that more frequent consumption of greater alcohol quantities results in tolerance to the acute effects of alcohol. In contrast to the assumption that alcohol hangover would exhibit similar characteristics [7], the reduced sensitivity to acute alcohol effects is not reflected in lower sensitivity to hangover effects. Instead, the opposite was observed: the data suggest that with increased hangover frequency, the severity of hangovers becomes worse. In the context of the current findings, it appears that repeated exposure to alcohol differentially affects intoxication and hangover. That is, repeated exposure to alcohol was related to increases in the magnitude of the hangover effect. This leftward shift in the dose-response is characteristic of sensitization or reverse tolerance.

A strength of the current paper is that the results were consistent across several studies using different methodologies (surveys and naturalistic study). Different instruments were applied to assess overall hangover severity (one-item score, HSS, AHS, and AHSS, and aggregate severity scores over three days). Despite the variation in these measures, in all cases, a significant and positive association was found between hangover frequency and severity. Additionally, the sample sizes were sufficiently large to be confident about this outcome. This finding persists when adjusting for alcohol consumption variables (e.g., the amount of alcohol consumed and eBAC), age, or personality characteristics.

The current findings have several implications. Firstly, as no tolerance to hangover severity develops but rather the opposite, this may have consequences for the functional outcomes of the alcohol hangover. That is, behavioral effects may also further deteriorate in drinkers who exhibit more frequent hangovers. This is an important issue for further research, as reverse tolerance may have a significant impact on the magnitude of impairments seen in the hangover state on common daily activities such as driving a car [36,37].

Secondly, the findings suggest that alcohol hangovers do not act as a deterrent to further alcohol consumption. In fact, even though hangovers become worse with frequency, drinkers persist in consuming alcohol to levels that produce hangovers. Previous studies [38,39] also noted that experiencing hangovers does not have a relevant impact on future drinking behavior. One implication of this is that there may be value in investigating harm reduction strategies which act as 'hangover treatments' [40–42].

The studies also have limitations that should be addressed. The first one is the possibility of recall bias. The survey by Penning et al. [11] gathered data retrospectively. Therefore, hangover severity may not have been accurately recalled. However, the naturalistic study and International survey recorded hangover severity in real-time and showed similar results. Moreover, the observed correlation between hangover frequency and severity in real-time was of greater magnitude in the naturalistic study than the correlations observed when data was collected retrospectively. Secondly, hangover frequency was recorded via a single question. In future research, it could be considered to apply alternative methods such as the Time Line Follow Back approach [43] to reduce the possible impact of recall bias. Thirdly, hangover severity assessment may be biased, as this is a single measurement rather than an average over multiple drinking sessions. A single assessment may not accurately represent the typical hangover severity experienced by drinkers. However, the Fiji survey made hangover severity assessments on three consecutive days, and average scores were used for the analysis to account for intrapersonal differences and day-to-day fluctuations in alcohol consumption and hangover severity. Analysis of the Fiji dataset revealed similar results as the single timepoint survey and the naturalistic study. Thus, the observations cannot be regarded as coincidental, depending on the unique unknown features affecting hangover severity of one hangover occasion.

Fourthly, there is the issue of experimental control. None of the three datasets were drawn from controlled experiments with a set amount of alcohol consumed, nor were drink types standardized or controlled by the investigators. Therefore, hangover severity scores are influenced by several factors such as total amount of alcohol consumed, drinking duration, congener content, and eBAC. This may also explain in part why our findings differ from the controlled study conducted by Köchling et al. [10]. In the statistical analysis, the variability in drinking patterns was controlled by computing partial correlations, adjusting the observed associations between hangover frequency and severity for eBAC or the number of alcoholic drinks consumed.

Fifthly, the assessment of hangover frequency was done by simply asking subjects how many hangovers they had experienced per month, or during the past year. It is likely, however, that hangover frequency varies across a lifetime, along with periods of higher and lower alcohol intake. The current analysis did not take this variability into account as this data was not collected. There are alternative measures that could be applied, such as lifetime number of experienced hangovers that may be more informative in this regard and could be applied in future research.

Sixthly, the age range of study participants, 18–30 years old, was small. This may have an impact on the relationship between hangover frequency and hangover severity. Future studies should, therefore, also include older-aged drinkers. Although longitudinal research showed that reported past year hangover frequency remains relatively stable from when assessed one year apart [6], Piasecki et al. [44] also followed student drinkers for 11 years and showed a steady decline in past year hangover frequency. Tolstrup et al. [24] also reported that during adulthood (18–65 years old) with increasing age, the frequency of hangovers gradually ameliorates. Although the title of the paper suggests otherwise, a closer look at the data revealed that they did not assess hangover severity. Instead, the frequency of experiencing nine hangover symptoms was assessed. The observation that the occurrence of hangovers declined with increasing age persisted after correction for the usual amount of alcohol intake, frequency of binge drinking, and the proportion of alcohol consumed with meals. There are various socioeconomic and cultural factors that may contribute to the reduction in hangover frequency when age progresses. For example, the start of job and family life responsibilities at the transition from student to adult life may reduce the frequency and quantity of alcohol consumption. Research has shown that the number of binge drinking days reduces with progressing age [25]. Hence, when age progresses, given the reduced amount of alcohol consumed per drinking session, there is a reduction of the opportunities to have a hangover per se. As stated above, the association between hangover frequency and age is complex, and there may be many moderating variables, which should be the subject of future research. For example, it has also been found that infrequent drinkers (14 or fewer days per month) consumed more alcohol on drinking days and were more frequently involved in binge drinking (13.4%) compared to frequent drinkers (4.3%) [45]. Taken together, future research on the association between hangover frequency and severity should take age into account as a moderating factor.

Finally, there is no validated, reliable assessment scale to determine the vulnerability and sensitivity of alcohol hangover. It should be the aim of future research to develop such a scale, to be able to, for example, select study subjects that are sensitive to hangover effects per se at a given number of alcoholic drinks, or to create homogenous research samples. The current findings can then be replicated in prospective studies or controlled experimental studies with standardized alcohol intake.

In addition to the limitations described above, future research should also elucidate the possible reasons for variability in the presence and severity of alcohol hangover. This is important because although a consistent positive association was shown between hangover frequency and hangover severity, correlational analysis does not imply causality. There may be other factors than hangover frequency that may be the actual cause of variability in hangover severity. Similarly, the regression analysis by Kochling et al. [10] cannot prove the absence of a causal relationship between hangover frequency and severity. The association between hangover frequency and severity is complex, and there may be many moderating variables. Future research should address biopsychological age-related factors that may impact the association between hangover frequency and severity, such as deterioration of liver function, psychological changes, motives for alcohol consumption, and cognitive decline.

Slutske et al. [13] found that only 13% of 1265 students reported having experienced no hangover symptoms during the past year. Most subjects reported 1–2 past year hangovers (27%) followed by 3–11 past year hangovers (34%). Hangover resistance is reported by around 25% of drinkers who reach BACs around 0.10–0.12% [46], but the percentage depends on the amount of alcohol consumed and at higher dosages significantly fewer drinkers claim to be hangover resistant [47–49].

Studies reported that subjects with a positive family history of alcoholism reported more frequent hangovers than drinkers with a negative family history of alcoholism [13,44,49]. In an experimental study, Span and Earlywine [50] found that subjects with a positive family history of alcoholism reported more severe hangovers than subjects with a negative family history of alcoholism.

Genetic profiling may elucidate why these differences exist. An Australian twin study revealed that 43% of hangover resistance could be explained by genetic influences [51]. Genetic factors accounted for 40% (men) to 45% (women) of variability in hangover frequency. A US twin study reported 55% heritability of the frequency of having hangovers [52]. In both studies, a close relationship was observed with genetic variability in the frequency of being intoxicated. The severity of hangovers is also influenced by genetic factors, for example by variability in alleles decoding for aldehyde dehydrogenase (ALDH2). For example, it has been found that Asian American students with ALDH2*2 alleles may experience more severe hangovers [53], and similar findings were reported for Japanese workers [54]. Unfortunately, in both studies, a direct relationship between hangover frequency and hangover severity was not assessed. Nevertheless, Wall et al. [53] suggested that tolerance develops to the risk of having hangovers, as they found that higher levels of usual alcohol intake were associated with increased hangover frequency and with reduced hangover severity. This suggestion is, however, not supported by the current findings.

5. Conclusions

A positive and significant association between hangover frequency and severity was found, which remained significant after correcting for alcohol intake. This finding suggests that hangovers become worse when they are experienced more often. Future research should further investigate this, and factors mediating the observed association, including its implications for alcohol prevention.

Author Contributions: Conceptualization—J.C.V., A.E.v.L.v.P., S.B. and A.S.; formal analysis—J.C.V.; investigation—L.A., A.E.v.L.v.P. and K.A.S.; methodology—J.C.V., writing–original draft—J.C.V. and A.S.; writing–review & editing—J.C.V. A.S.; A.J.A.E.v.d.L., L.A., A.E.v.L.v.P., K.A.S., S.B. and A.S.

Funding: This research received no external funding.

Conflicts of Interest: S.B. has received funding from Red Bull GmbH, Kemin Foods, Sanofi Aventis, Phoenix Pharmaceutical and GlaxoSmithKline. A.S. has held research grants from Abbott Nutrition, Arla Foods, Bayer Healthcare, Cognis, Cyvex, GlaxoSmithKline, Naturex, Nestle, Martek, Masterfoods, Wrigley, and has acted as a consultant/expert advisor to Abbott Nutrition, Barilla, Bayer Healthcare, Danone, Flordis, GlaxoSmithKline Healthcare, Masterfoods, Martek, Novartis, Unilever, and Wrigley. J.C.V. has received grants/research support from the Dutch Ministry of Infrastructure and the Environment, Janssen, Nutricia, Red Bull, Sequential, and Takeda, and has acted as a consultant for Canadian Beverage Association, Centraal Bureau Drogisterijbedrijven, Clinilabs, Coleman Frost, Danone, Deenox, Eisai, Janssen, Jazz, More Labs, Purdue, Red Bull, Sanofi-Aventis, Sen-Jam Pharmaceutical, Sepracor, Takeda, Toast!, Transcept, Trimbos Institute, Vital Beverages, and ZBiotics. The other authors have not potential conflicts of interest to disclose.

References

1. Van Schrojenstein Lantman, M.; van de Loo, A.J.; Mackus, M.; Verster, J.C. Development of a definition for the alcohol hangover: Consumer descriptions and expert consensus. *Curr. Drug Abuse Rev.* **2016**, *9*, 148–154. [CrossRef] [PubMed]
2. Verster, J.C.; Stephens, R.; Penning, R.; Rohsenow, D.; McGeary, J.; Levy, D.; McKinney, A.; Finnigan, F.; Piasecki, T.M.; Adan, A.; et al. The alcohol hangover research group consensus statement on best practice in alcohol hangover research. *Curr. Drug Abuse Rev.* **2010**, *3*, 116–126. [CrossRef] [PubMed]
3. Kauhanen, J.; Kaplan, G.A.; Goldberg, D.D.; Cohen, R.D.; Lakka, T.A.; Salonen, J.T. Frequent Hangovers and Cardiovascular Mortality in Middle-Aged Men. *Epidemiology* **1997**, *8*, 310–314. [CrossRef] [PubMed]
4. Batty, G.D.; Deary, I.J.; MacIntyre, S. Childhood IQ and life course socioeconomic position in relation to alcohol induced hangovers in adulthood: The Aberdeen children of the 1950s study. *J. Epidemiol. Community Health* **2006**, *60*, 872–874. [CrossRef] [PubMed]

5. Paljärvi, T.; Koskenvuo, M.; Poikolainen, K.; Kauhanen, J.; Sillanmäki, L.; Mäkelä, P. Binge drinking and depressive symptoms: A 5-year population-based cohort study. *Addiction* **2009**, *104*, 1168–1178. [CrossRef] [PubMed]
6. Piasecki, T.M.; Trela, C.J.; Mermelstein, R.J. Hangover Symptoms, Heavy Episodic Drinking, and Depression in Young Adults: A Cross-Lagged Analysis. *J. Stud. Alcohol Drugs* **2017**, *78*, 580–587. [CrossRef] [PubMed]
7. Rohsenow, D.J.; Howland, J.; Winter, M.; Bliss, C.A.; Littlefield, C.A.; Heeren, T.C.; Calise, T.V. Hangover sensitivity after controlled alcohol administration as predictor of post-college drinking. *J. Abnorm. Psychol.* **2012**, *121*, 270–275. [CrossRef]
8. Piasecki, T.M.; Alley, K.J.; Slutske, W.S.; Wood, P.K.; Sher, K.J.; Shiffman, S.; Heath, A.C. Low Sensitivity to Alcohol: Relations with Hangover Occurrence and Susceptibility in an Ecological Momentary Assessment Investigation. *J. Stud. Alcohol Drugs* **2012**, *73*, 925–932. [CrossRef]
9. Courtney, K.E.; Worley, M.; Castro, N.; Tapert, S.F. The effects of alcohol hangover on future drinking behavior and the development of alcohol problems. *Addict. Behav.* **2018**, *78*, 209–215. [CrossRef]
10. Köchling, J.; Geis, B.; Wirth, S.; Hensel, K.O. Grape or grain but never the twain? A randomized controlled multiarm matched-triplet crossover trial of beer and wine. *Am. J. Clin. Nutr.* **2019**, *109*, 345–352. [CrossRef]
11. Penning, R.; McKinney, A.; Verster, J.C. Alcohol Hangover Symptoms and Their Contribution to the Overall Hangover Severity. *Alcohol Alcohol.* **2012**, *47*, 248–252. [CrossRef]
12. Verster, J.C.; Lantman, M.V.S.; Mackus, M.; Van De Loo, A.J.; Garssen, J.; Scholey, A. Differences in the Temporal Typology of Alcohol Hangover. *Alcohol. Clin. Exp. Res.* **2018**, *42*, 691–697. [CrossRef] [PubMed]
13. Slutske, W.S.; Piasecki, T.M.; Hunt-Carter, E.E. Development and Initial Validation of the Hangover Symptoms Scale: Prevalence and Correlates of Hangover Symptoms in College Students. *Alcohol. Clin. Exp. Res.* **2003**, *27*, 1442–1450. [CrossRef] [PubMed]
14. Penning, R.; McKinney, A.; Bus, L.D.; Olivier, B.; Slot, K.; Verster, J.C. Measurement of alcohol hangover severity: Development of the Alcohol Hangover Severity Scale (AHSS). *Psychopharmacology* **2013**, *225*, 803–810. [CrossRef]
15. Devenney, L.E.; Coyle, K.B.; Roth, T.; Verster, J.C. Sleep after Heavy Alcohol Consumption and Physical Activity Levels during Alcohol Hangover. *J. Clin. Med.* **2019**, *8*, 752. [CrossRef]
16. Rohsenow, D.J.; Howland, J.; Minsky, S.J.; Greece, J.; Almeida, A.; Roehrs, T.A. The Acute Hangover Scale: A new measure of immediate hangover symptoms. *Addict. Behav.* **2007**, *32*, 1314–1320. [CrossRef]
17. Watson, P.E.; Watson, I.D.; Batt, R.D. Prediction of blood alcohol concentrations in human subjects. Updating the Widmark Equation. *J. Stud. Alcohol* **1981**, *42*, 547–556. [CrossRef]
18. Seppä, K.; Lepistö, J.; Sillanaukee, P. Five-Shot Questionnaire on Heavy Drinking. *Alcohol. Clin. Exp. Res.* **1998**, *22*, 1788–1791. [CrossRef]
19. Derogatis, L.R. *Brief Symptom Inventory*; Clinical Psychometric Research: Baltimore, Maryland, 1975.
20. De Haan, L.; Kuipers, E.; Kuerten, Y.; Van Laar, M.; Olivier, B.; Verster, J.C. The RT-18: A new screening tool to assess young adult risk-taking behavior. *Int. J. Gen. Med.* **2011**, *4*, 575–584.
21. Babor, T.F.; de la Fuente, J.R.; Saunders, J.; Grant, M. AUDIT. The Alcohol Use Disorders Identification Test. In *Guidelines for Use in Primary Health Care*; World Health Organization: Geneva, Switzerland, 1992.
22. Saunders, J.B.; Aasland, O.G.; Babor, T.F.; de la Puente, J.R.; Grant, M. Development of the Alcohol Use Disorders Screening Test (AUDIT). WHO collaborative project on early detection of persons with harmful alcohol consumption. II. *Addiction* **1993**, *88*, 791–804. [CrossRef]
23. O'Brien, C.P. The CAGE questionnaire for detection of alcoholism. *JAMA* **2008**, *300*, 2054–2056. [CrossRef] [PubMed]
24. Tolstrup, J.S.; Stephens, R.; Grønbaek, M. Does the severity of hangovers decline with age? Survey of the incidence of hangover in different age groups. *Alcohol. Clin. Exp. Res.* **2014**, *38*, 466–470. [CrossRef] [PubMed]
25. Naimi, T.S.; Brewer, R.D.; Mokdad, A.; Denny, C.; Serdula, M.K.; Marks, J.S. Binge drinking among US adults. *JAMA* **2003**, *289*, 70–75. [CrossRef] [PubMed]
26. Whitfield, J.B.; Martin, N.G. Alcohol Consumption and Alcohol Pharmacokinetics: Interactions Within the Normal Population. *Alcohol. Clin. Exp. Res.* **1994**, *18*, 238–243. [CrossRef] [PubMed]
27. Holdstock, L.; King, A.C.; Wit, H. Subjective and Objective Responses to Ethanol in Moderate/Heavy and Light Social Drinkers. *Alcohol. Clin. Exp. Res.* **2000**, *24*, 789–794. [CrossRef] [PubMed]

28. Nathan, P.E.; Lipscomb, T.R. Studies in blood alcohol level discrimination: Etiologic cues to alcoholism. *NIDA Res. Monogr.* **1979**, *25*, 178–190.
29. Brumback, T.; Cao, D.; McNamara, P.; King, A. Alcohol-induced performance impairment: A 5-year re-examination study in heavy and light drinkers. *Psychopharmacology* **2017**, *234*, 1749–1759. [CrossRef] [PubMed]
30. Hollingworth, H.L. The influence of alcohol (part II). *J. Abnorm. Psychol. Soc. Psychol.* **1924**, *18*, 311–333. [CrossRef]
31. Goldberg, L. Quantitative studies on alcohol tolerance in man. *Acta Physiol. Scand.* **1943**, *5*, 1–128.
32. Frankenhaeuser, M.; Goldberg, L.; Hagdahl, R.; Myrsten, A.-L. Subjective and objective effects of alcohol as functions of dosage and time. *Psychopharmacology* **1964**, *6*, 399–409.
33. Comley, R.E.; Dry, M.J. Acute behavioral tolerance to alcohol. *Exp. Clin. Psychopharmacol.* **2019**. [CrossRef] [PubMed]
34. Vaughan, C.L.; Stangl, B.L.; Schwandt, M.L.; Corey, K.M.; Hendershot, C.S.; Ramchandani, V.A. The relationship between impaired control, impulsivity, and alcohol self-administration in nondependent drinkers. *Exp. Clin. Psychopharmacol.* **2019**, *27*, 236–246. [CrossRef] [PubMed]
35. Spear, L.P.; Varlinskaya, E. Adolescence. Alcohol sensitivity, tolerance, and intake. *Recent Dev. Alcohol.* **2005**, *17*, 143–159. [PubMed]
36. Verster, J.C.; Bervoets, A.C.; De Klerk, S.; Vreman, R.A.; Olivier, B.; Roth, T.; Brookhuis, K.A. Effects of alcohol hangover on simulated highway driving performance. *Psychopharmacology* **2014**, *231*, 2999–3008. [CrossRef] [PubMed]
37. Verster, J.C.; Van Der Maarel, M.A.; McKinney, A.; Olivier, B.; De Haan, L. Driving During Alcohol Hangover Among Dutch Professional Truck Drivers. *Traffic Inj. Prev.* **2014**, *15*, 434–438. [CrossRef]
38. Mallett, K.A.; Lee, C.M.; Neighbors, C.; Larimer, M.E.; Turrisi, R. Do We Learn from Our Mistakes? An Examination of the Impact of Negative Alcohol-Related Consequences on College Students' Drinking Patterns and Perceptions. *J. Stud. Alcohol* **2006**, *67*, 269–276. [CrossRef]
39. Mackus, M.; Van De Loo, A.J.; Nutt, D.; Verster, J.C.; Lantman, M.V.S. An effective hangover treatment: Friend or foe? *Drug Sci. Policy Law* **2017**, *3*. [CrossRef]
40. Pittler, M.H.; Verster, J.C.; Ernst, E. Interventions for preventing or treating alcohol hangover: Systematic review of randomized trials. *BMJ* **2005**, *331*, 1515–1518. [CrossRef]
41. Verster, J.C.; Penning, R. Treatment and prevention of alcohol hangover. *Curr. Drug Abus. Rev.* **2010**, *3*, 103–109. [CrossRef]
42. Jayawardena, R.; Thejani, T.; Ranasinghe, P.; Fernando, D.; Verster, J.C. Interventions for treatment and/or prevention of alcohol hangover: Systematic review. *Hum. Psychopharmacol. Clin. Exp.* **2017**, *32*, e2600. [CrossRef]
43. Sobell, M.B.; Sobell, L.C.; Klajner, F.; Pavan, D.; Basian, E. The reliability of a timeline method for assessing normal drinker college students' recent drinking history: Utility for alcohol research. *Addict. Behav.* **1986**, *11*, 149–161. [CrossRef]
44. Piasecki, T.M.; Sher, K.J.; Slutske, W.S.; Jackson, K.M. Hangover Frequency and Risk for Alcohol Use Disorders: Evidence from a Longitudinal High-Risk Study. *J. Abnorm. Psychol.* **2005**, *114*, 223–234. [CrossRef] [PubMed]
45. Naimi, T.S.; Xuan, Z.; Brown, D.W.; Saitz, R. Confounding and studies of 'moderate' alcohol consumption: The case of drinking frequency and implications for low-risk drinking guidelines. *Addiction* **2013**, *108*, 1534–1543. [CrossRef] [PubMed]
46. Howland, J.; Rohsenow, D.J.; Edwards, E.M. Are some drinkers resistant to hangover? A literature review. *Curr. Drug Abus. Rev.* **2008**, *1*, 42–46. [CrossRef]
47. Verster, J.C.; de Klerk, S.; Bervoets, A.C.; Kruisselbrink, L.D. Can hangover immunity be really claimed? *Curr. Drug Abuse Rev.* **2013**, *6*, 253–254. [CrossRef] [PubMed]
48. Kruisselbrink, L.D.; Bervoets, A.C.; De Klerk, S.; Van De Loo, A.J.; Verster, J.C. Hangover resistance in a Canadian University student population. *Addict. Behav. Rep.* **2017**, *5*, 14–18. [CrossRef]
49. Newlin, D.B.; Pretorius, M.B. Sons of alcoholics report greater hangover symptoms than sons of nonalcoholics: A Pilot Study. *Alcohol. Clin. Exp. Res.* **1990**, *14*, 713–716. [CrossRef]
50. Span, S.A.; Earleywine, M. Familial risk for alcoholism and hangover symptoms. *Addict. Behav.* **1999**, *24*, 121–125. [CrossRef]

51. Slutske, W.S.; Piasecki, T.M.; Nathanson, L.; Statham, D.J.; Martin, N.G. Genetic influences on alcohol-related hangover. *Addiction* **2014**, *109*, 2027–2034. [CrossRef]
52. Wu, S.-H.; Guo, Q.; Viken, R.J.; Reed, T.; Dai, J. Heritability of Usual Alcohol Intoxication and Hangover in Male Twins: The NAS-NRC Twin Registry. *Alcohol. Clin. Exp. Res.* **2014**, *38*, 2307–2313. [CrossRef]
53. Wall, T.L.; Horn, S.M.; Johnson, M.L.; Smith, T.L.; Carr, L.G. Hangover symptoms in Asian Americans with variations in the aldehyde dehydrogenase (ALDH2) gene. *J. Stud. Alcohol* **2000**, *61*, 13–17. [CrossRef] [PubMed]
54. Yokoyama, M.; Yokoyama, A.; Yokoyama, T.; Funazu, K.; Hamana, G.; Kondo, S.; Yamashita, T.; Nakamura, H. Hangover susceptibility in relation to aldehyde dehydrogenase-2 genotype, alcohol flushing, and mean corpuscular volume in Japanese workers. *Alcohol. Clin. Exp. Res.* **2005**, *29*, 1165–1171. [CrossRef] [PubMed]

© 2019 by the authors. Licensee MDPI, Basel, Switzerland. This article is an open access article distributed under the terms and conditions of the Creative Commons Attribution (CC BY) license (http://creativecommons.org/licenses/by/4.0/).

Article

Pain Catastrophising Predicts Alcohol Hangover Severity and Symptoms

Sam Royle *, Lauren Owen, David Roberts and Lynne Marrow

Department of Psychology, School of Health and Society, University of Salford, Frederick Road, Salford M6 6PU, UK; L.J.Owen2@salford.ac.uk (L.O.); D.J.Roberts@salford.ac.uk (D.R.); L.Marrow@salford.ac.uk (L.M.)
* Correspondence: W.S.S.Royle@salford.ac.uk; Tel.: +44-161-2950-278

Received: 8 January 2020; Accepted: 15 January 2020; Published: 20 January 2020

Abstract: Alcohol hangover is a cause of considerable social and economic burden. Identification of predictors of alcohol hangover severity have the potential to contribute to reductions in costs associated with both absenteeism/presenteeism and health care. Pain catastrophising (PC) is the tendency to ruminate and describe a pain experience in more exaggerated terms. The current study examines the possibility that this cognitive coping strategy may influence experience of alcohol hangover. The aims of the current study were to (1) examine the relationship between hangover severity and PC, (2) explore and identify discreet factors within the Acute Hangover Scale (AHS) and (3) explore whether independent factors/dimensions of acute hangover are differentially predicted by PC. A retrospective survey ($n = 86$) was conducted in which participants completed the Acute Hangover Scale (AHS); the Pain Catastrophising Scale (PCS); a questionnaire pertaining to the amount of alcohol consumed; and a demographic information questionnaire. Regression analyses showed a significant relationship between PC and hangover severity scores and demonstrated that PC was, in fact, a stronger predictor of perceived hangover severity than estimated peak blood alcohol concentrations (eBACs). Factor analysis of the AHS scale, resulted in the identification of two distinct symptom dimensions; 'Headache and thirst', and 'Gastric and cardiovascular' symptoms. Regression analyses showed that both eBAC and PCS score were significantly associated with 'Headache and thirst'. However, only PCS score was associated with 'Gastric and cardiovascular' symptoms. These novel findings implicate a role for cognitive coping strategies in self-reports of alcohol hangover severity, and may have implications for understanding behavioural response to hangover, as well as suggesting that hangover and PC may be important factors mediating the motivation to drink and/or abuse alcohol, with potential implications in addiction research. Furthermore, these findings suggest that distinct alcohol hangover symptoms may be associated with different mechanisms underlying the experience of alcohol hangover.

Keywords: hangover; catastrophising; alcohol; veisalgia; acute hangover scale

1. Introduction

1.1. Alcohol Hangover, Symptoms and Economic Burden

Alcohol hangover is a phenomenon that occurs the day after the ingestion of alcohol, once the blood alcohol concentration (BAC) is approaching nil [1], and it is associated with a wide variety of symptoms, such as headache, nausea, and concentration problems [2,3]. Hangover is thought to be a considerable cause of economic loss through workplace absenteeism and lost productivity [4]. Researchers have also speculated that the severity of alcohol hangover is linked to the development of alcohol use disorders (AUDs) [5,6], indicating that a better understanding of the individual hangover experience and its mediators may offset the associated financial and social burden of AUD.

A number of explanations for the variance seen in alcohol hangover presentation have been suggested, including gene associations of alcohol metabolism [7,8], gender differences [9] inflammatory response to alcohol consumption [10,11], immunological functioning [12], and congener content of alcoholic drinks [13], as well as individual differences in psychosocial factors such as anxiety and mood [14], or guilt related to the actions carried out whilst drinking [3]. There is, however, little consensus regarding the biological mechanisms that underpin the experience of alcohol hangover [4], and this is particularly true for psycho-social variables. Identification of mediating factors of alcohol hangover severity may thus inform mechanistic investigations of hangover, as well as having the potential to reduce costs associated with absenteeism/presenteeism and improve health care outcomes.

1.2. Hangover and Risk of Alcohol Abuse

Despite the lack of mechanistic explanations for the influence of predictor variables on the experience of hangover, and mixed findings regarding relationships between familial risk for addiction and experience of hangover [15,16], there is some evidence that alcohol hangover experience may be a potential risk factor for alcohol use disorder (AUD) [15]. In this regard, hangover has been conceptualised as affecting cognitive control processes that influence local drinking behaviour. Evidence suggests that people who experience a more severe hangover will drink less, when they engage in drinking the day of a hangover [17], and that hangover can increase the time before the next alcoholic drink is consumed in frequent drinkers [18], with hangover occurrence predicting a 6 hour delay to next drink when used as sole predictor in a survival model. It is notable, however, that hangover occurrence was only associated with a delayed time to next drink in multivariate models when interacting with the onset of financial stressors, or the presence of high levels of craving at the end of the drinking episode (pre-hangover). This may implicate a role for the hangover in the delay of further engagement with drinking, when experienced alongside a continued desire/motivation to drink. The investigation of differences in factors related to motivational and inhibitive processes, such as cognitive coping strategies, during hangover, therefore has the potential to contribute to understanding of possible relationships between the hangover experience and propensity for development of AUDs.

1.3. Alcohol Hangover and Pain Catastrophising

Pain catastrophising (PC) has been broadly defined as an exaggerated negative orientation towards actual or anticipated pain experiences [19] and has been described as the tendency to recall pain experiences in more exaggerated terms, to feel helpless and ruminate over painful events. PC appears to be moderated, to some degree, by gender [20], psychosocial and dispositional factors [19,20]. However, despite these moderating factors, PC contributes unique and significant variance to the prediction of self-reported pain intensity, as well as to neural processing of pain [21]. Evidence has shown that the relationship between PC and pain ratings is partially mediated by diminished diffuse noxious inhibitory controls (a measure of endogenous pain inhibition), indicating a disruption in pain inhibition and suggesting a relationship between PC and pain inhibition [22]. Neurological evidence (utilising functional magnetic resonance imaging) has demonstrated that PC predicts the experience of pain, in that, during exposure to a painful stimulus, pain specific response activation in the dorsolateral prefrontal cortex (dlPFC) and medial prefrontal cortex (mPFC) correlate negatively with PC [23]. The effect of PC on brain activity in the mPFC and dlPFC also seems to be mediated by the severity of pain experienced, with reduced activity during more intense pain [24]. Additionally, the dlPFC shows greater bilateral activation during response inhibition, in comparison to interference monitoring and suppression [25], indicating some anatomical overlap between inhibitive processes and PC. It has been argued that PC may heighten pain experiences by reducing the efficiency of inhibitory pathways, though evidence for this position is indirect [26].

Alcohol hangover is characterised by pain symptoms. Indeed, the medical term for alcohol hangover "veisalgia" comes from the Norwegian kveis, which refers to the uneasiness following debauchery, and algia, the Greek term for pain. Cytokines, proteins produced during immune response that are involved in both the initiation and persistence of pathologic pain [27], are altered during hangover. Interleukin (IL-2; IL-10) and interferon (IFN-γ) cytokines have been shown to be elevated in blood during hangover [10]. In saliva, elevations of IL-2, IL-4, IL-5, IL-6, IL-10, IFN-γ, and TNF-α have been observed during hangover, and in urine, elevations of IL-4 and IL-6, as well as decreases in IL-8 have been observed in comparison to non-hangover-days [28]. These differences in cytokine levels between hangover and non-hangover days do not, however, appear to explain variance in the experience of hangover, with similar changes observed in both those reporting hangover, and those reporting hangover resistance [29]. Headache, a continuous pain in the head which has also been associated with changes in cytokine levels [29,30], represents the 3rd most common symptom of alcohol hangover [31] and symptomatic ratings of headache severity have large statistical effects in measures of hangover severity [32]. PC therefore presents a good candidate for potentially explaining some of the variance in self-reported hangover severity scores. Consequently, the current study hypothesises that greater PC will be associated with elevated hangover severity scores.

It has also been argued that hangover lacks in mechanistic explanations [4], despite the wide variety of symptoms associated with the hangover experience [2]. Certainly, dehydration is thought to represent one potential mechanism, with thirst being one of the most commonly reported hangover symptoms [2,31], due to the diuretic effects of alcohol [33]. Vasopressin levels, a biological marker of dehydration, do not, however, necessarily correlate with overall hangover severity [12]. It is possible that this is due to dehydration representing a mechanism of hangover that explains only a particular symptom cluster. Factor analysis of measures of hangover severity may therefore provide some direction for investigations of the mechanisms that give rise to symptom clusters. Furthermore, certain symptom clusters may be independently moderated by PC and may represent better predictors of alcohol abuse risk.

The current retrospective survey was therefore designed with three main aims: (i) to examine whether increased PC scores are associated with elevated hangover severity scores, (ii) to explore the factor structure of the acute hangover scale (AHS), and (iii) to explore whether different dimensions of the AHS are independently associated with PC.

2. Materials and Methods

2.1. Participants

Participants were recruited through opportunity sampling—both at a university in the north-west United Kingdom, and online via social media. Ninety-one participants completed the survey, with 3 participants excluded from analysis because they reported a non-binary gender, which presents issues for calculating blood alcohol concentrations. A further 2 participants were excluded from analyses based on age, with one reporting an age of 5 years, violating exclusion criteria, and one reporting an age of 80 years, representing an extreme outlier in the current sample (+6.24 SDs from the mean). The remaining sample of 86 had an overall mean age of 25.93 years (SD: 6.03, range: 18–46), with 51 female respondents (59%). No incentive was provided for participation. Exclusion criteria for the study were: below 18 years of age (not of legal drinking age for the area in which they reside); having not experienced hangover in the last 6 months. Exclusion was based on self-report: independent verification of these criteria was not possible with an online survey. Ethnicity was not analysed as a variable as the majority of participants (88%) self-identified as white.

2.2. Materials

Participants were required to complete an online survey that included the Acute Aangover Scale (AHS) [32], to measure hangover severity retrospectively, and the Pain Catastrophising Scale (PCS) [19], to measure PC associated with the experience. The PCS consists of 13 items addressing the experience of catastrophic thoughts related to the experience of pain, rated on a scale from 0 (not at all) to 4 (all the time), with total score calculated as the sum of all items. The PCS has shown a high level of reliability, with Cronbach's α scores typically exceeding 0.8 [19,30,31], and the PCS has been validated in a variety of samples, including those 'seeking treatment' (vs. not seeking treatment) [34], pain outpatients (vs. community participants) [35], those with back pain (Norwegian version) [36], and those with chronic pain (Korean and Brazilian-Portuguese versions) [37,38]. The AHS consists of 9 items addressing the severity of 8 hangover symptoms, plus an overall hangover severity rating, all given on a scale from 0 (None) to 7 (Incapacitating), with AHS total score calculated as the mean of all items. The AHS has been validated for concurrent hangover severity measurement, but also shows very high correlations with scales designed for measurement of the most recent (retrospective) hangover experience [39], as well as having been utilised for recent hangover severity measurement in other research [40]. Participants reported the drinking that led to their most recent hangover using the items from McKinney and Coyle's investigation, which asks about the number of a variety of standardised drinks consumed (e.g., pints of beer, bottles of beer, alcopops, etc.), and allows for the number of units of ethanol consumed to be calculated [41]. Demographics were also collected, including age (which was recorded for use as a covariable since previous evidence has suggested a role in the experience of hangover [42]), height and weight, (for the calculation of estimated blood alcohol concentrations), and ethnicity. Estimated blood alcohol concentrations were calculated using the method from Siedl et al. [9], resulting in the estimated peak blood alcohol concentration assuming no elimination (eBAC), and assuming a 15% elimination rate (eBAC15%). Both the AHS and PCS were adapted to reference how the participants felt during their most recent hangover.

2.3. Procedure

Participants, recruited via social media (Twitter, Reddit) and posters located around a university in the north-west, UK, completed an online survey hosted on Online Surveys by Jisc (https://www.onlinesurveys.ac.uk/). Participants completed the survey during their own time. The total time to complete the study was approximately 15 min. No incentives were offered for participation.

2.4. Ethics

The materials and methods utilised in this procedure were approved by the University of Salford Health Sciences Research ethics board (HSR1617-15), and all participants provided informed consent. The use of Online Surveys for data collection allowed for participant anonymity, and the system adheres to high ethical standards (e.g., no use of 'cookies' which store files to the local PC used to complete the survey).

3. Results

All analyses were carried out in IBM SPSS 25.0.0.1 (IBM Corp., Armonk, NY, USA).

3.1. Factor Analysis

A dimension reduction procedure was carried out on responses to the items of the AHS to establish whether symptom clusters existed. The item 'hangover' was excluded from this analysis, as this is non-symptomatic and thought to capture a broad rating of hangover severity [32]. Descriptive statistics for the remaining items are presented in Table 1. Inter-item correlations are presented in Table 2.

As all items failed to meet at least parametric assumption of normality in univariate analyses, principal axis factor analysis was utilised for dimension reduction [43]. Reduction was carried out using a direct obliminal rotation with a delta of 0, given the likelihood of correlations between dimensions of hangover, and in line with the recommendations of Costello and Osborne [43]. Factors with an eigenvalue above 1 were retained, with results checked visually using the Scree test. Results indicated a solution with 2 dimensions, one factor consisting of symptoms linked to dehydration, and one to stress responses. Kaiser–Meyer–Olkin statistics indicated good sampling adequacy (KMO = 0.722), though KMO values for individual variables indicated a potential issue with the item 'nausea' (KMO = 0.495; removal of this item did not result in changes to the factor structure). Bartlett's test of sphericity indicated acceptable deviance from an identity matrix ($x\ 2(28) = 138.762$, $p < 0.001$), and 35% of residuals between observed and reproduced correlations had absolute values above 0.05, indicating moderate model fit. The factor model is summarized in Table 3.

Table 1. Descriptive statistics for ratings of hangover symptom severity on the acute hangover scale (AHS).

Item.	Mean	SD	Median	Normality	95% CI Lower	95% CI Upper
Tired	5.94	1.24	6	<0.001 ***	5.68	6.21
Thirsty	5.31	1.528	5	0.001 ****	4.99	5.64
Headache	4.71	2.057	5	<0.001 ****	4.27	5.15
Nausea	3.66	2.433	3	<0.001 ****	3.14	4.18
Loss of appetite	3.33	2.078	3	<0.001 ****	2.88	3.77
Dizziness/faintness	3.22	2.008	3	<0.001 ****	2.79	3.65
Stomach ache	3.06	2.054	2	<0.001 ****	2.62	3.5
Heart racing	2.72	2.096	2	<0.001 ****	2.27	3.17

SD—standard deviation; normality—p-Value for Shapiro–Wilk analysis of normality. The n for all items was 86. Significant results indicated by *** $p < 0.01$, **** $p < 0.001$.

Table 2. Correlations (and significance) of items included in factor analysis of AHS.

	Thirsty	Headache	Nausea	Loss of Appetite	Dizziness/ Faintness	Stomach Ache	Heart Racing
Tired	0.239 (0.013) *	0.312 (0.002) ***	0.192 (0.038) *	0.176 (0.052)	0.156 (0.075)	0.223 (0.020) *	0.161 (0.069)
Thirsty		0.276 (0.005) **	0.140 (0.100)	0.112 (0.152)	−0.031 (0.390)	0.084 (0.221)	0.226 (0.018) *
Headache			0.208 (0.027) *	0.344 (0.001) ***	0.241 (0.013) *	0.182 (0.047) *	0.183 (0.046) *
Nausea				0.473 (<0.001) *	0.497 (<0.001) ****	0.529 (<0.001) ****	0.434 (<0.001) ****
Loss of appetite					0.369 (<0.001) *	0.211 (0.026) *	0.305 (0.002) *
Dizziness/faintness						0.251 (0.010) *	0.409 (<0.001) ****
Stomach ache							0.362 (<0.001) ****

Significant correlations indicated by * $p < 0.05$, ** $p < 0.01$, *** $p < 0.005$, **** $p < 0.001$.

Table 3. Factor loadings of AHS items based on principal axis factoring.

Item	Headache and Thirst	Gastric and Cardiovascular Symptoms
Headache	**0.587**	0.082
Thirsty	**0.506**	−0.069
Tired	**0.456**	0.072
Nausea	−0.102	**0.894**
Dizziness/faintness	−0.065	**0.641**
Heart racing	0.088	**0.535**
Stomach ache	0.024	**0.532**
Loss of appetite	0.164	**0.47**
Eigenvalues	1.252	2.906
Factor correlation	0.452	

Composite scores were calculated for each factor based on the mean of the items which had their primary loadings on each factor (Headache and thirst: mean = 5.32, SD = 1.17; Gastric and cardiovascular symptoms: mean = 3.20, SD = 1.53) with higher scores indicating greater severity of symptoms within the cluster. The bold indicates most relevance.

3.2. Regression Models

Three initial regression models of the AHS score were formed, with PCS score and age used as predictor variables in all three models, and the contribution of 'measures of drinking' assessed across separate models. A further two regression models were formed to assess dimensions of the AHS identified in factor analysis. Descriptive statistics for the variables included across these regression models are presented in Table 4. To minimize overfitting of the data, rather than utilize an automated variable selection method, for each model, each variable was entered into the regression concurrently [44], with a model formed for each of the three 'measures of drinking' obtained; units consumed, eBAC, and eBAC15%. Gender was also entered into the model with the number of units consumed, since this is controlled for in the calculations used to derive eBAC and eBAC15% scores and approximates differences in the body fat composition of different genders which influences alcohol distribution through body water during consumption.

Table 4. Descriptive statistics for variables included across the five regression models constructed.

Variable	Mean	SD	Median	Normality	95% CI	
					Lower	Upper
Acute Hangover Scale (AHS)	4.1	1.2	4.11	0.101	3.85	4.36
Pain Catastrophizing scale (PCS)	28.81	11.64	26.5	<0.001 ****	26.32	31.31
Age	25.93	6.03	25	<0.001 ****	24.64	27.22
Total units	15.62	8.64	13.4	<0.001 ****	13.77	17.48
eBAC	0.26	0.14	0.22	<0.001****	0.23	0.29
eBAC15%	0.18	0.13	0.14	<0.001 ****	0.15	0.21
Headache and thirst	5.32	1.17	5.33	0.06	5.07	5.57
Gastric and cardiovascular symptoms	3.2	1.53	3	0.002 ***	2.87	3.53

Variables: AHS—acute hangover scale; PCS—pain catastrophising scale; total units—units of alcohol consumed, calculated from self-report; eBAC—the estimated blood alcohol concentration assuming no elimination; eBAC15%—the estimated blood alcohol concentration assuming 15% elimination rate; Headache and thirst—mean score for items identified within dimension 1 of the factor analysis; Gastric and cardiovascular symptoms—mean score for items identified within dimension 2 of the factor analysis; SD—standard deviation; normality—significance of Shapiro–Wilk analysis. The n for all measures was 86, *** $p < 0.005$, **** $p < 0.001$.

Results indicated that eBAC represented the drinking measure that explained the most variance in AHS scores (power = 0.50, calculated post-hoc), and as such the model containing this variable was carried forward for regression analyses of factor scores calculated during factor axis analysis. A regression model containing eBAC, PCS score, and age, as predictor variables, was therefore formed for each of the two factor scores derived. Summaries of regression models are presented in Table 5, with a summary of individual variable contributions presented in Table 6.

Tolerance, variance inflation factors (VIFs), and collinearity diagnostics indicated no issues with multicollinearity. Manual examination of standardized residuals plotted against standardized predicted values suggested no issues with heteroscedasticity in the data, and multivariate normality was present in all models. Durbin–Watson statistics indicated independence of errors. Some issues were identified in casewise diagnostics, with a small number of cases indicating issues with either problematic covariance ratios or high leverage values in regression models.

Given that a number of cases presented potential issues with covariance and leverage, and in line with recommendations made by Babyak [44], validation of the final models (those containing the eBAC drinking measure) was carried out using bootstrap methods with 2000 random resamples drawn. Bootstrapped models are summarised in Table 7.

Table 5. Summary of model statistics for regression analyses.

Model	DV	IV	R	R^2	Adj R^2	F	F Sig.	Durbin–Watson
1	Acute Hangover Scale (AHS)	Units consumed Gender PCS score Age	0.432	0.187	0.147	4.656	0.002 ***	1.789
2	AHS	eBAC15 PCS score Age	0.397	0.158	0.127	5.118	0.003 ***	1.771
3	AHS	eBAC PCS score Age	0.429	0.184	0.154	6.163	0.001 ***	1.802
4	Headache and thirst	eBAC PCS score Age	0.307	0.094	0.061	2.844	0.043 *	1.612
5	Gastric and cardiovascular symptoms	eBAC PCS score Age	0.398	0.158	0.128	5.141	0.003 ***	1.906

DV (dependent variable): AHS—Acute hangover scale; Headache and thirst—mean score for items identified within dimension 1 of the factor analysis; Gastric and cardiovascular symptoms—mean score for items identified within dimension 2 of the factor analysis. R—value of r for model; R2—value of r squared for model; Adj R2—adjusted r squared for model; F—F value for model; F Sig.—Significance of the F value for the model; Durbin-Watson—Durbin-Watson statistic for the model. Significant results indicated by *, *, $p < 0.05$, *** $p < 0.005$.

Table 6. Summary of statistics determining independent variable contributions to regression effects.

Model	DV	IV	B	SE B	β	t	t Sig.	95% CI Lower Bound	95% CI Upper Bound	Zero-Order	Partial	Part	Tolerance	VIF	Pratt
1	Acute Hangover Scale (AHS) score	Constant	2.955	0.709		4.17	<0.001 ****	1.545	4.365						
		Units consumed	0.033	0.014	0.239	2.342	0.022 *	0.005	0.061	0.175	0.252	0.235	0.967	1.034	0.042
		Gender	0.366	0.25	0.151	1.46	0.148	−0.133	0.864	0.186	0.16	0.146	0.939	1.065	0.028
		PCS score	0.032	0.011	0.314	3.038	0.003 ***	0.011	0.053	0.326	0.32	0.304	0.94	1.063	0.102
		Age	−0.02	0.02	−0.1	−0.991	0.324	−0.06	0.02	−0.148	−0.109	−0.099	0.981	1.019	0.015
2	AHS score	Constant	3.255	0.695		4.682	<0.001 ****	1.872	4.637						
		eBAC15	1.867	0.954	0.2	1.957	0.054	−0.03	3.765	0.209	0.211	0.198	0.982	1.018	0.042
		PCS score	0.033	0.011	0.317	3.106	0.003 ***	0.012	0.054	0.326	0.324	0.315	0.985	1.015	0.103
		Age	−0.017	0.02	−0.083	−0.809	0.421	−0.057	0.024	−0.148	−0.089	−0.082	0.968	1.033	0.012
3	AHS score	Constant	3.005	0.7		4.292	<0.001 ****	1.612	4.398						
		eBAC	2.262	0.881	0.257	2.568	0.012 *	0.51	4.014	0.26	0.273	0.256	0.99	1.01	0.067
		PCS score	0.033	0.01	0.321	3.189	0.002 ***	0.012	0.054	0.326	0.332	0.318	0.984	1.016	0.105
		Age	−0.017	0.02	−0.085	−0.839	0.404	−0.057	0.023	−0.148	−0.092	−0.084	0.976	1.025	0.013
4	Headache and thirst	Constant	4.288	0.721		5.943	<0.001 ****	2.852	5.723						
		eBAC	1.876	0.907	0.218	2.067	0.042 *	0.07	3.681	0.216	0.223	0.217	0.99	1.01	0.047
		PCS score	0.022	0.011	0.216	2.039	0.045 *	0.001	0.043	0.214	0.22	0.214	0.984	1.016	0.046
		Age	−0.003	0.021	−0.015	−0.141	0.888	−0.044	0.038	−0.062	−0.016	−0.015	0.976	1.025	0.001
5	Gastric and cardiovascular symptoms	Constant	2.323	0.908		2.559	0.012 *	0.517	4.129						
		eBAC	2.258	1.142	0.201	1.978	0.051	−0.013	4.529	0.208	0.213	0.2	0.99	1.01	0.042
		PCS score	0.039	0.013	0.299	2.931	0.004 ***	0.013	0.066	0.311	0.308	0.297	0.976	1.025	0.093
		Age	−0.032	0.026	−0.128	−1.245	0.217	−0.084	0.019	−0.183	−0.136	−0.126	0.976	1.025	0.023

DV (dependent variable): AHS—Acute hangover scale; Headache and thirst—mean score for items identified within dimension 1 of the factor analysis; Gastric and cardiovascular symptoms—mean score for items identified within dimension 2 of the factor analysis. B—Beta coefficient; SE B—Standard error of beta coefficient; β—standardized beta coefficient; t—t-statistic value for parameter; t Sig.—significance of t-statistic for parameter; VIF—Variance inflation factor; Pratt—Pratt statistic for parameter. Significant results indicated by *, * $p < 0.05$, *** $p < 0.005$, **** $p < 0.001$.

Table 7. Summary of bootstrapped regression model coefficients.

Model	DV	IV	B	Bias	SE	Sig. (Two-Tailed)	BCa 95% CI Lower Bound	BCa 95% CI Upper Bound
3	Acute Hangover Scale (AHS) score	Constant	3.005	0.009	0.725	<0.001 ****	1.612	4.491
		eBAC	2.262	0.027	0.959	0.022 *	0.483	4.280
		PCS score	0.033	<0.001	0.009	<0.001 ****	0.015	0.051
		Age	−0.017	<0.001	0.020	0.401	−0.058	0.021
4	Headache and thirst	Constant	4.288	0.046	0.658	<0.001 ****	2.975	5.840
		eBAC	1.876	0.032	0.950	0.046 *	0.049	3.812
		PCS score	0.022	<0.001	0.010	0.026 *	0.003	0.040
		Age	−0.003	−0.002	0.022	0.890	−0.047	0.035
5	Gastric and cardiovascular symptoms	Constant	2.323	−0.002	0.922	0.014 *	0.577	4.119
		eBAC	2.258	0.007	1.233	0.066	−0.060	4.654
		PCS score	0.039	<0.001	0.013	0.004 ***	0.013	0.064
		Age	−0.032	<0.001	0.024	0.165	−0.080	0.015

DV—dependent variable; IV—independent variable; B—beta weight; SE—standard error; BCa 95% CI—bias-corrected accelerated 95% confidence interval. Bootstrap results based on 2000 bootstrap samples. Significant results indicated by *, * $p < 0.05$ = *, *** $p < 0.005$, **** $p < 0.001$.

Results of bootstrap analyses have a fairly high level of agreement with original regression models, with significant predictor variables remaining constant. Both eBAC and total PCS score demonstrated significant relationships with total AHS score, and a composite score based on symptoms of the AHS related to 'Headache and thirst'. Only total PCS score demonstrated a relationship with a composite score based on 'Gastric and cardiovascular' symptoms. eBAC approached significance in this model. However, bootstrapping did indicate some bias toward significance of this variable with this sample.

4. Discussion

4.1. Summary of the Main Findings

Hangover, the mental and physical symptoms experienced the day after drinking and once BAC is approaching 0, has previously been associated with a variety of other factors, including genetic influences on alcohol metabolism [7,8], gender [9], inflammatory responses [10,11], immunological function [12] and congeners [13]. However, few psychosocial predictors of hangover have been identified so far.

The three main aims of the current study were: (i) to examine whether increased PC scores are associated with elevated hangover severity scores, (ii) to explore the factor structure of the acute hangover scale (AHS), and (iii) to explore whether different dimensions of the AHS were independently associated with PC. The current study demonstrated that PC was a predictor of perceived hangover severity and was, in fact, a stronger predictor than the estimated peak blood alcohol concentration (eBAC). Exploration of the dimensions of the AHS revealed two distinct symptom dimensions; 'Headache and thirst'; and 'Gastric and cardiovascular' symptoms. While both eBAC and PC were significantly associated with 'Headache and thirst', only PC was associated with 'Gastric and cardiovascular' symptoms.

4.2. Relationships between Pain Catastrophising and Hangover Severity

Relationships between PC and AHS scores were investigated using multiple linear regression. Initial models indicated that, of the drinking measures, a calculated blood alcohol concentration that did not account for elimination (eBAC) was the best predictor of the AHS score. One potential explanation is that the preferential metabolism of ethanol limits downstream action to eliminate ethanol metabolites leading to a build-up of biologically active compounds [45]. This would be consistent with the time course of alcohol hangover, with symptomology extending beyond the period of acute ethanol intoxication.

Regression models (and calculated product measures [46]) indicated PC was a better predictor of perceived hangover severity than eBAC. Given relationships between PC and other psychosocial variables such as depression and anxiety [20,21], PC could provide a mechanism through which other psychosocial variables influence self-report hangover severity scores. Furthermore, given links between PC and inhibitive processes [26], this cognitive strategy may influence motivational responses to hangover, providing a potential link between hangover experience and local behaviour, such as engagement with further drinking. Such effects could have implications in addiction research [15].

The results of the current investigation may also have implications for the measurement of hangover, given its reliance on self-report measures, such as the AHS [32]. This issue has been largely ignored in hangover research for purposes of practicality, with a lack of other approaches available. Results from the regression models developed as part of this investigation indicate a moderate effect of PC on AHS score, comparable to the effect observed for measures of alcohol consumption, and support the view that self-report hangover questionnaires contain a significant subjective element. This may reinforce the need for an objective measure of hangover. However, research into biomarkers of hangover severity has yet to find a reliable indicator [1,28]. An alternative approach to measuring hangover severity in a more objective manner may be to examine the cognitive effects of hangover. A meta-analytic examination of the next-day cognitive effects of hangover published in 2018 suggested that effects can be seen during hangover on short- and long-term memory, sustained attention, and psychomotor speed [47]. Differences in performance on tasks examining these functions between hangover and non-hangover days could therefore present a measure of functional hangover severity.

Questions can be raised regarding the value of any of these measurement approaches. Arguably, the subjective experience of hangover is likely to influence the behavioural response to the experience, and may provide value over 'objective' measurements of hangover, such as cognitive performance measures or biomarkers, in particular contexts (e.g., the investigation of absenteeism/presenteeism and other acute behaviours). In comparison, objective measures may be more useful in investigations examining the biological correlates of alcohol hangover. Further research will need to examine the comparative value of different measurement approaches in relation to different outcomes. However, controlling for PC in future analyses may also aid in understanding the hangover experience, particularly with regard to the investigation of biomarkers.

Dimensions of the AHS

A recent review of the physiology of hangover identified alcohol metabolites, neurotransmitter alterations, inflammatory factors, and mitochondrial (metabolic) dysfunction as the most likely factors involved in hangover symptomology [48]. PC has also been associated with alterations to immune responses, with heightened reactivity of cytokine IL-6 related to increased levels of PC as measured immediately after painful stimulation [49]. This relationship between PC and IL-6 also appeared to be independent of pain ratings given during stimulation. Likewise, immune responses during hangover have been shown to include increases in IL-6 levels [28], with IL-6 thought to have particular importance as a messenger molecule that connects peripheral regulatory processes with the central nervous system during responses to both physiological and psychological stress [49].

In this study, factor analysis of AHS responses resulted in two symptom dimensions; (1) Headache and thirst ('headache', 'tired', 'thirsty'), and (2) Gastric and cardiovascular symptoms ('nausea', 'dizziness/faintness', 'heart racing', 'loss of appetite', and 'stomach ache'). The 'Headache and thirst' symptom cluster could be related to the diuretic properties of alcohol [29], which can lead to dehydration. Dehydration has been linked to headache [50], with tiredness and thirst being considered common symptoms. Headache may also be the result of cytokine release prompted by physiological stress associated with alcohol consumption [29,30], or indeed physiological stress may be caused by dehydration. However, there is potential for overlap in the causes of symptom clusters.

Speculation regarding the biological mechanisms underlying symptom cluster experience is, however, not possible based on the current investigation. Future work will be needed to identify specific biological associations with the experience of hangover symptom clusters. Penning et al.'s factor analysis also identified dehydration ('disturbed water balance') as a dimension of the hangover experience [2]. However, in their investigation, the item 'headache' was not loaded on this dimension. Dehydration causes physiological changes, e.g., to electrolyte balance, which have proposed associations with hangover. However, evidence for relationships between physiological changes and hangover severity is lacking [12], though they have not been investigated in relation to specific symptom clusters. One potential explanation for the 'Gastric and cardiovascular' symptom cluster emerging is that they can all be linked to physiological stress responses. Effects of stress response on the autonomic nervous system are well established [51], and acute physiological stress can also induce various responses in the gastrointestinal system [52].

An alternative explanation of the factor structure of the AHS identified in this investigation relates to the prevalence of symptoms. Tiredness, thirst and headache, the items loaded within the 'Headache and thirst' dimension, represent the three most commonly reported hangover symptoms [31]. It is possible that these symptom clusters are thus representative of different groups that either experience one or both of the symptom clusters. An extension of this reasoning, given the prevalence of headache and thirst symptoms in hangover, could be that less severe hangovers consist of symptoms included within the 'Headache and thirst' symptom cluster, with more severe hangovers including 'Gastric and cardiovascular' symptoms.

4.3. Dimensions of the AHS Independently Associated with PC

Composite scores based on 'Headache and thirst' symptoms, and 'Gastric and cardiovascular' symptoms, identified during factor analysis of AHS responses, were also assessed using regression. Both eBAC and PCS score significantly predicted 'Headache and thirst' symptom scores with approximately equal contributions. The observation of PC score as a significant predictor in this model is possibly due to the inclusion of headache severity ratings in the construction of this score, with PC having previously been linked with both the presence of weekly headache [53], and the severity of migraine symptoms, a phenomenon associated with headache [54]. Given the diuretic effects of alcohol [33], it follows that measures of alcohol consumption would be related to symptoms associated with dehydration.

Finally, only PCS score significantly predicted composite scores based on 'Gastric and cardiovascular' symptoms, though eBAC was only marginally non-significant. Product measures supported the interpretation that PC was more strongly related with 'Gastric and cardiovascular' symptoms than eBAC, and robust regression provided some validation of this model. PC has been related to activity in the mPFC [24], an anatomical area that has also been shown to mediate stress response [55]. This may provide a link through which this cognitive strategy can influence stress responses occurring as a result of hangover. The exclusion of eBAC from this model may indicate that these symptoms are not direct products of alcohol consumption, or that this symptom set is not associated linearly with the volume of alcohol consumed (e.g., threshold effects). It has, however, been previously suggested that increased levels of fatty acids seen during hangover are products of a stress response concurrent with hangover [12], which could indicate that 'Gastric and cardiovascular' symptoms in hangover are somewhat independent of the amount of alcohol consumed.

4.4. Conclusions and Directions for Future Research

Hangover represents a considerable economic toll due to its influence on local behaviour, such as lost productivity and workplace absenteeism [4]. Furthermore, the experience of hangover may be related to downstream health consequences by promoting deviant drinking practices [12]. The current investigation revealed, for the first time, that PC predicts alcohol hangover severity and that this effect occurs in a symptom specific manner. PC may also provide a cognitive strategy through which other psychosocial variables can influence hangover.

Exploratory factor analysis provided evidence of two distinct sub-structures of the AHS, 'Headache and thirst', and 'Gastric and cardiovascular' symptoms. Results of this investigation could be interpreted as suggesting that dehydration and physiological stress responses represent areas that warrant further examination, with differences in regression models based on composite hangover scores for symptom clusters providing some evidence that symptom clusters are somewhat independent. This may provide an explanation for why markers of dehydration have not always correlated with overall hangover severity [12], as well as why thirst had the lowest item-total correlation during development of the AHS [32]. Further research will be required to establish whether particular covariables correlate with symptom clusters either derived from dimension reduction procedures or theoretical mechanistic relationships. The AHS also measures a somewhat limited sample of hangover symptoms, and recent research has adopted the approach of combining the symptoms identified in a number of validated hangover measures, in order to capture the diversity of the hangover experience [56]. These measures consist largely of different symptoms, but show high correlations, and further research will be needed to examine whether the dimensions of the hangover experience suggested here are evident within this broader context, as well as their relationships with PC.

As noted previously, hangover has also been associated with effects on local drinking behaviour [17,18], with an ecological momentary assessment conducted by Epler et al. in 2014 indicating that the presence of hangover delayed the onset of the next drinking episode when interacting with either the onset of financial stressors, or the presence of craving at the end of the drinking episode [18]. Epler et al.'s (2014) sample consisted of participants with a reasonably low risk of alcohol problems (average AUDIT score = 12.21), but no research has addressed this relationship in high-risk or clinical groups. Evidence has suggested relationships between craving and AUD symptomology in a sample containing a high proportion of participants meeting criteria for diagnosis of AUD. However, no relationship was found between craving and drinking habits in this sample [57]. This may suggest that interactions between craving, hangover, and local drinking behaviour do not exist in those at a high risk for AUD. Greater craving in the high-risk sample also showed a relationship with increased impulsive discounting (a devaluation of future reward) [57], which may provide a mechanism for observed losses of inhibitory response control in alcohol disorders, as well as other disorders, such as depression [58]. Weaker inhibition processes have also been noted in those with a family history of AUD [59], and in young-adult binge drinkers [60], a form of drinking associated with an increased incidence of hangover. Inhibition is also inherently linked with impulsivity [61], which has itself been strongly associated with AUD [58]. Given the links between PC and motivational/inhibitive processes [26,62], future research should consider PC and hangover alongside factors related to motivation/inhibition, such as performance on inhibition dependent tasks, and craving. Vatsalya et al.'s (2018) investigation found no relationship between hangover severity (as measured by the AHS) and a single item measure of craving [40]. However, this craving measurement is unlikely to capture the theoretical complexity of the phenomenon and future research would benefit from the use of context appropriate, validated craving measures [63].

Future research should therefore seek to elucidate the potential interaction between PC and cognitive processing systems mediating inhibitory control and the craving response during alcohol hangover.

Author Contributions: Conceptualisation, S.R., D.R. and L.M.; methodology, S.R., D.R. and L.M.; formal analysis, S.R.; investigation, S.R.; data curation, S.R.; writing—original draft preparation, S.R.; writing—review and editing, S.R., D.R., L.O. and L.M.; visualisation, S.R., L.O. and L.M.; supervision, D.R., L.O. and L.M.; project administration, S.R.; funding acquisition, D.R., L.O. and L.M. All authors have read and agreed to the published version of the manuscript.

Funding: This research received no external funding.

Conflicts of Interest: The authors declare no conflict of interest.

References

1. Merlo, A.; Adams, S.; Benson, S.; Devenney, L.; Gunn, C.; Iversen, J.; Van De Loo, A.J. Proceeding of the 9th Alcohol Hangover Research Group Meeting. *Curr. Drug Abus. Rev.* **2017**, *10*, 68–75. [CrossRef]
2. Penning, R.; McKinney, A.; Verster, J.C. Alcohol hangover symptoms and their contribution to the overall hangover severity. *Alcohol Alcohol.* **2012**, *47*, 248–252. [CrossRef]
3. Verster, J.C.; Alford, C.; Bervoets, A.C.; de Klerk, S.; Grange, J.A.; Hogewoning, A.; Raasveld, S.J. Hangover research needs: Proceedings of the 5th alcohol hangover research group meeting. *Curr. Drug Abus. Rev.* **2013**, *6*, 245–251. [CrossRef]
4. Prat, G.; Adan, A.; Sánchez-Turet, M. Alcohol hangover: A critical review of explanatory factors. *Hum. Psychopharm.* **2009**, *24*, 259–267. [CrossRef]
5. Dudley, R. Fermenting fruit and the historical ecology of ethanol ingestion: Is alcoholism in modern humans an evolutionary hangover? *Addiction* **2002**, *97*, 381–388. [CrossRef]
6. Piasecki, T.M.; Sher, K.J.; Slutske, W.S.; Jackson, K.M. Hangover frequency and risk for alcohol use disorders: Evidence from a longitudinal high-risk study. *J. Abnorm. Psychol.* **2005**, *114*, 223–234. [CrossRef] [PubMed]
7. Wall, T.L.; Shea, S.H.; Luczak, S.E.; Cook, T.A.; Carr, L.G. Genetic associations of alcohol dehydrogenase with alcohol use disorders and endophenotypes in white college students. *J. Abnorm. Psychol.* **2005**, *114*, 456–465. [CrossRef] [PubMed]
8. Edenberg, H.J. The genetics of alcohol metabolism: Role of alcohol dehydrogenase and aldehyde dehydrogenase variants. *Alcohol. Res. Health* **2007**, *30*, 5–13. [PubMed]
9. Seidl, S.; Jensen, U.; Alt, A. The calculation of blood ethanol concentrations in males and females. *Int. J. Leg. Med.* **2000**, *114*, 71–77. [CrossRef]
10. Kim, D.J.; Kim, W.; Yoon, S.J.; Choi, B.M.; Kim, J.S.; Go, H.J.; Jeong, J. Effects of alcohol hangover on cytokine production in healthy subjects. *Alcohol* **2003**, *31*, 167–170. [CrossRef]
11. Verster, J.C.; Van Doornen, L.J.P.; Kleinjan, M.; Garssen, J.; de Haan, L.; Penning, R.; Olivier, B.; Slot, K.A. Biological, psychological, and behavioral correlates of the alcohol hangover. *Alcohol Alcohol.* **2013**, *48* (Suppl. 1), i15–i16.
12. Penning, R.; van Nuland, M.; Fliervoet, L.A.L.; Olivier, B.; Verster, J.C. The pathology of alcohol hangover. *Curr. Drug Abus. Rev.* **2010**, *3*, 68–75. [CrossRef] [PubMed]
13. Verster, J.C. Congeners and alcohol hangover: Differences in hangover severity among Dutch college students after consuming beer, wine, or liquor. *Alcohol. Clin. Exp. Res.* **2006**, *30*, 53A.
14. McKinney, A. A review of the next day effects of alcohol on subjective mood ratings. *Curr. Drug Abus. Rev.* **2010**, *3*, 88–91. [CrossRef] [PubMed]
15. Piasecki, T.; Robertson, B.; Epler, A. Hangover and risk for alcohol use disorders: Existing evidence and potential mechanisms. *Curr. Drug Abus. Rev.* **2010**, *3*, 92–102. [CrossRef] [PubMed]
16. Stephens, R.; Holloway, K.; Grange, J.A.; Owen, L.; Jones, K.; Kruisselbrink, D. Does familial risk for alcohol use disorder predict alcohol hangover? *Psychopharmacology* **2018**, *234*, 1795–1802. [CrossRef]
17. Huntley, G.; Treloar, H.; Blanchard, A.; Monti, P.M.; Carey, K.B.; Rohsenow, D.J.; Miranda, R., Jr. An event-level investigation of hangovers' relationship to age and drinking. *Exp. Clin. Psychopharm.* **2015**, *23*, 314. [CrossRef]
18. Epler, A.J.; Tomko, R.L.; Piasecki, T.M.; Wood, P.K.; Sher, K.J.; Shiffman, S.; Heath, A.C. Does Hangover Influence the Time to Next Drink? An Investigation Using Ecological Momentary Assessment. *Alcoholism* **2014**, *38*, 1461–1469. [CrossRef]
19. Sullivan, M.J.; Bishop, S.R.; Pivik, J. The pain catastrophizing scale: Development and validation. *Psychol. Assess.* **1995**, *7*, 524–532. [CrossRef]
20. Thorn, B.E.; Clements, K.L.; Ward, L.C.; Dixon, K.E.; Kersh, B.C.; Boothby, J.L.; Chaplin, W.F. Personality factors in the explanation of sex differences in pain catastrophizing and response to experimental pain. *Clin. J. Pain* **2004**, *20*, 275–282. [CrossRef]
21. Quartana, P.J.; Campbell, C.M.; Edwards, R.R. Pain catastrophizing: A critical review. *Expert Rev. Neurother.* **2009**, *9*, 745–758. [CrossRef] [PubMed]
22. Jensen, M.P.; Ehde, D.M.; Day, M.A. The behavioral activation and inhibition systems: Implications for understanding and treating chronic pain. *J. Pain* **2016**, *17*, 529-e1. [CrossRef] [PubMed]

23. Henderson, L.A.; Akhter, R.; Youssef, A.M.; Reeves, J.M.; Peck, C.C.; Murray, G.M.; Svensson, P. The effects of catastrophizing on central motor activity. *Eur. J. Pain* **2016**, *20*, 639–651. [CrossRef]
24. Seminowicz, D.A.; Davis, K.D. Cortical responses to pain in healthy individuals depends on pain catastrophizing. *Pain* **2006**, *120*, 297–306. [CrossRef] [PubMed]
25. Blasi, G.; Goldberg, T.E.; Weickert, T.; Das, S.; Kohn, P.; Zoltick, B.; Mattay, V.S. Brain regions underlying response inhibition and interference monitoring and suppression. *Eur. J. Neurosci.* **2006**, *23*, 1658–1664. [CrossRef] [PubMed]
26. Goodin, B.R.; McGuire, L.; Allshouse, M.; Stapleton, L.; Haythornthwaite, J.A.; Burns, N.; Edwards, R.R. Associations between catastrophizing and endogenous pain-inhibitory processes: Sex differences. *J. Pain* **2009**, *10*, 180–190. [CrossRef]
27. Zhang, J.M.; An, J. Cytokines, inflammation and pain. *Int. Anesth. Clin.* **2007**, *45*, 27. [CrossRef]
28. Raasveld, S.J.; Hogewoning, A.; Van de Loo, A.J.A.E.; De Zeeuw, R.; Bosma, E.R.; Bouwmeester, N.H.; Verster, J.C. Cytokine concentrations after heavy alcohol consumption in people with and without a hangover. *Eur. Neuropsychopharm.* **2015**, *25*, 228. [CrossRef]
29. Bø, S.H.; Davidsen, E.M.; Gulbrandsen, P.; Dietrichs, E.; Bovim, G.; Stovner, L.J.; White, L.R. Cerebrospinal fluid cytokine levels in migraine, tension-type headache and cervicogenic headache. *Cephalalgia* **2009**, *29*, 365–372. [CrossRef]
30. Zhu, S.; Wang, X.; Wang, X.; Shiwen, W.U. Analysis of Inflammatory cytokine expression in cluster headache. *Chin. J. Nerv. Ment. Dis.* **2017**, *43*, 274–278.
31. Slutske, W.S.; Piasecki, T.M.; Hunt-Carter, E.E. Development and initial validation of the Hangover Symptoms Scale: Prevalence and correlates of hangover symptoms in college students. *Alcoholism* **2003**, *27*, 1442–1450. [CrossRef] [PubMed]
32. Rohsenow, D.J.; Howland, J.; Minsky, S.J.; Greece, J.; Almeida, A.; Roehrs, T.A. The Acute Hangover Scale: A new measure of immediate hangover symptoms. *Addict. Behav.* **2007**, *32*, 1314–1320. [CrossRef] [PubMed]
33. Hobson, R.M.; Maughan, R.J. Hydration status and the diuretic action of a small dose of alcohol. *Alcohol Alcohol.* **2010**, *45*, 366–373. [CrossRef]
34. Osman, A.; Barrios, F.X.; Kopper, B.A.; Hauptmann, W.; Jones, J.; O'Neill, E. Factor structure, reliability, and validity of the Pain Catastrophizing Scale. *J. Behav. Med.* **1997**, *20*, 589–605. [CrossRef]
35. Osman, A.; Barrios, F.X.; Gutierrez, P.M.; Kopper, B.A.; Merrifield, T.; Grittmann, L. The Pain Catastrophizing Scale: Further psychometric evaluation with adult samples. *J. Behav. Med.* **2000**, *23*, 351–365. [CrossRef] [PubMed]
36. Fernandes, L.; Storheim, K.; Lochting, I.; Grotle, M. Cross-cultural adaptation and validation of the Norwegian pain catastrophizing scale in patients with low back pain. *BMC Musculoskelet. Dis.* **2012**, *13*, 111. [CrossRef]
37. Cho, S.; Kim, H.Y.; Lee, J.H. Validation of the Korean version of the Pain Catastrophizing Scale in patients with chronic non-cancer pain. *Qual. LIFE Res.* **2013**, *22*, 1767–1772. [CrossRef]
38. Sehn, F.; Chachamovich, E.; Vidor, L.P.; Dall-Agnol, L.; Custódio de Souza, I.C.; Torres, I.L.; Caumo, W. Cross-cultural adaptation and validation of the Brazilian Portuguese version of the pain catastrophizing scale. *Pain Med.* **2012**, *13*, 1425–1435. [CrossRef]
39. Stephens, R.; Grange, J.A.; Jones, K.; Owen, L.A. critical analysis of alcohol hangover research methodology for surveys or studies of effects on cognition. *Psychopharmacology* **2014**, *231*, 2223–2236. [CrossRef]
40. Vatsalya, V.; Stangl, B.L.; Schmidt, V.Y.; Ramchandani, V.A. Characterization of hangover following intravenous alcohol exposure in social drinkers: Methodological and clinical implications. *Addict. Biol.* **2018**, *23*, 493–502. [CrossRef]
41. McKinney, A.; Coyle, K. Alcohol hangover effects on measures of affect the morning after a normal nights drinking. *Alcohol Alcohol.* **2006**, *41*, 54–60. [CrossRef] [PubMed]
42. Tolstrup, J.S.; Stephens, R.; Grønbæk, M. Does the severity of hangovers decline with age? Survey of the incidence of hangover in different age groups. *Alcoholism* **2014**, *38*, 466–470. [CrossRef] [PubMed]
43. Costello, A.B.; Osborne, J.W. Best practices in exploratory factor analysis: Four recommendations for getting the most from your analysis. *Pract. Assess. Res. Eval.* **2005**, *10*, 7.
44. Babyak, M.A. What you see may not be what you get: A brief, nontechnical introduction to overfitting in regression-type models. *Psychosom. Med.* **2004**, *66*, 411–421. [PubMed]
45. Cederbaum, A.I. Alcohol metabolism. *Clin. Liver Dis.* **2012**, *16*, 667–685. [CrossRef]

46. Nathans, L.L.; Oswald, F.L.; Nimon, K. Interpreting multiple linear regression: A guidebook of variable importance. *Pract. Assess. Res. Eval.* **2012**, *17*, 1–19.
47. Gunn, C.; Mackus, M.; Griffin, C.; Munafò, M.R.; Adams, S. A systematic review of the next-day effects of heavy alcohol consumption on cognitive performance. *Addiction* **2018**, *113*, 2182–2193. [CrossRef]
48. Palmer, E.; Tyacke, R.; Sastre, M.; Lingford-Hughes, A.; Nutt, D.; Ward, R.J. Alcohol Hangover: Underlying Biochemical, Inflammatory and Neurochemical Mechanisms. *Alcohol Alcohol.* **2019**, *54*, 196–203. [CrossRef]
49. Edwards, R.R.; Kronfli, T.; Haythornthwaite, J.A.; Smith, M.T.; McGuire, L.; Page, G.G. Association of catastrophizing with interleukin-6 responses to acute pain. *Pain* **2008**, *140*, 135–144. [CrossRef] [PubMed]
50. Blau, J.N.; Kell, C.A.; Sperling, J.M. Water-deprivation headache: A new headache with two variants. *Headache* **2004**, *44*, 79–83. [CrossRef]
51. Chrousos, G.P. Stress and disorders of the stress system. *Nat. Rev. Endocrinol.* **2009**, *5*, 374. [CrossRef] [PubMed]
52. Soderholm, J.D.; Perdue, M.H., II. Stress and intestinal barrier function. *Am. J. Physiol-Gastrointest. Liver Physiol.* **1997**, *280*, G7–G13. [CrossRef] [PubMed]
53. Drahovzal, D.N.; Stewart, S.H.; Sullivan, M.J. Tendency to catastrophize somatic sensations: Pain catastrophizing and anxiety sensitivity in predicting headache. *Cogn. Behav. Ther.* **2006**, *35*, 226–235. [CrossRef] [PubMed]
54. Hubbard, C.S.; Khan, S.A.; Keaser, M.L.; Mathur, V.A.; Goyal, M.; Seminowicz, D.A. Altered brain structure and function correlate with disease severity and pain catastrophizing in migraine patients. *Eneuro* **2014**, *1*, e2014. [CrossRef] [PubMed]
55. Yang, X.; Garcia, K.M.; Jung, Y.; Whitlow, C.T.; McRae, K.; Waugh, C.E. vmPFC activation during a stressor predicts positive emotions during stress recovery. *Soc. Cogn. Affect Neur.* **2018**, *13*, 256–268. [CrossRef]
56. Van Schrojenstein Lantman, M.; Mackus, M.; van de Loo, A.J.; Verster, J.C. The impact of alcohol hangover symptoms on cognitive and physical functioning, and mood. *Hum. Psychopharm. Clin.* **2017**, *32*, e2623. [CrossRef]
57. MacKillop, J.; Miranda, R., Jr.; Monti, P.M.; Ray, L.A.; Murphy, J.G.; Rohsenow, D.J.; Gwaltney, C.J. Alcohol demand, delayed reward discounting, and craving in relation to drinking and alcohol use disorders. *J. Abnorm. Psychol.* **2010**, *119*, 106–114. [CrossRef]
58. Dick, D.M.; Smith, G.; Olausson, P.; Mitchell, S.H.; Leeman, R.F.; O'malley, S.S.; Sher, K. Understanding the construct of impulsivity and its relationship to alcohol use disorders. *Addict. Boil.* **2010**, *15*, 217–226. [CrossRef]
59. Nigg, J.T.; Wong, M.M.; Martel, M.M.; Jester, J.M.; Puttler, L.I.; Glass, J.M.; Zucker, R.A. Poor response inhibition as a predictor of problem drinking and illicit drug use in adolescents at risk for alcoholism and other substance use disorders. *J. Am. Acad Child Psychiatry* **2006**, *45*, 468–475. [CrossRef]
60. Czapla, M.; Simon, J.J.; Friederich, H.C.; Herpertz, S.C.; Zimmermann, P.; Loeber, S. Is binge drinking in young adults associated with an alcohol-specific impairment of response inhibition? *Eur. Addict. Res.* **2015**, *21*, 105–113. [CrossRef]
61. Bari, A.; Robbins, T.W. Inhibition and impulsivity: Behavioural and neural basis of response control. *Prog. Neurobiol.* **2013**, *108*, 44–79. [CrossRef] [PubMed]
62. Verhoeven, K.; Crombez, G.; Eccleston, C.; Van Ryckeghem, D.M.; Morley, S.; Van Damme, S. The role of motivation in distracting attention away from pain: An experimental study. *Pain* **2010**, *149*, 229–234. [CrossRef] [PubMed]
63. Sayette, M.A.; Shiffman, S.; Tiffany, S.T.; Niaura, R.S.; Martin, C.S.; Schadel, W.G. The measurement of drug craving. *Addiction* **2000**, *95*, 189–210. [CrossRef]

© 2020 by the authors. Licensee MDPI, Basel, Switzerland. This article is an open access article distributed under the terms and conditions of the Creative Commons Attribution (CC BY) license (http://creativecommons.org/licenses/by/4.0/).

Article

Sex Differences in the Presence and Severity of Alcohol Hangover Symptoms

Albertine E. van Lawick van Pabst [1], Lydia E. Devenney [2] and Joris C. Verster [1,3,*]

1. Division of Pharmacology, Utrecht University, 3584 CG Utrecht, The Netherlands; albertinevanlawick@live.nl
2. School of Psychology, Life and Health Sciences Ulster University, Londonderry BT52 1SA, Northern Ireland, UK; Devenney-l2@ulster.ac.uk
3. Centre for Human Psychopharmacology, Swinburne University, Melbourne VIC 3122, Australia
* Correspondence: J.C.Verster@uu.nl; Tel.: +31-30-253-6909

Received: 20 May 2019; Accepted: 13 June 2019; Published: 17 June 2019

Abstract: Studies have demonstrated significant sex differences in alcohol intoxication effects. In contrast, the majority of studies on the alcohol hangover phase did not investigate sex differences. Therefore, the current study examined possible sex differences in the presence and severity of alcohol hangover symptoms. Data from n = 2446 Dutch students (male = 50.7%, female = 49.3%) were analyzed. They reported the presence and severity of 22 hangover symptoms experienced after their past month heaviest drinking occasion. Subjects were categorized according to their estimated peak blood alcohol concentration (eBAC) and presence and severity of the hangover symptoms were compared between men and women. In the lowest eBAC group (0% ≤ eBAC < 0.08%), no significant sex differences were found. In the subsequent eBAC group (0.08% ≤ eBAC < 0.11%), severity of nausea was significantly higher in women than in men. In the third eBAC group (0.11% ≤ eBAC < 0.2%), women reported higher severity scores on nausea, tiredness, weakness, and dizziness than men. Men reported the presence of confusion significantly more often than women, and women reported the presence of shivering significantly more often than men. In the fourth eBAC group (0.2% ≤ eBAC < 0.3%), women reported higher severity scores on nausea and tiredness than men. In the highest eBAC group (0.3% ≤ eBAC < 0.4%), no significant sex differences were found. In conclusion, across the eBAC groups, severity scores of nausea and tiredness were higher in women than in men. However, albeit statistically significant, the observed sex differences in presence and severity of hangover symptoms were of small magnitude, and therefore, have little clinical relevance.

Keywords: alcohol; hangover; symptoms; sex differences; presence; severity

1. Introduction

The alcohol hangover is defined as the combination of mental and physical symptoms that are experienced the day after an episode of heavy alcohol consumption, starting when blood alcohol concentration (BAC) approaches zero [1]. Alcohol hangovers may negatively impact people's psychological and physical well-being by increasing accidents and injury [2,3] and impairing daily activities such as driving a car [4,5] or riding a bicycle [6].

Sex differences in acute alcohol effects have been well documented. Females have more body fat and less water than men of the same body weight [7]. Since alcohol is dispensed in body water, women reach higher BAC levels than men despite consuming an identical number of alcohol units [7]. Moreover, women usually have increased bioavailability and faster disappearance rates than men [8]. Also, alcohol appears to impair cognitive and psychomotor functioning in women more than in men [9,10].

Research usually shows that men consume more alcohol than women on a single drinking session (e.g., [11]). Therefore, in most observational and naturalistic studies men have higher BACs and experience more negative alcohol related consequences and performance impairment then women. However, studies employing the self-rating of the effects of alcohol (SRE) form show that at the same BAC levels, women usually are more sensitive to alcohol effects than men [12]. Lower scores on the SRE imply that fewer alcoholic drinks are needed to achieve a certain effect, meaning that subjects are more sensitive to alcohol. Usually, men have significantly higher SRE scores than women [13], suggesting that at the same BAC level men are less sensitive to (the adverse effects of) alcohol compared to women. As sex differences in pharmacokinetics and pharmacodynamics sex have been observed after acute alcohol intake, it is important to further investigate possible sex differences in the presence and severity of alcohol hangover effects.

A search of the literature revealed that the majority of studies on alcohol hangovers did not investigate possible sex differences. In fact, many older studies included men only. For example, until 2004, studies on the pathology and physiological correlates of alcohol hangover were solely conducted in male subjects [14–28].

Also, many studies investigating cognitive impairments that may accompany the alcohol hangover used samples that consisted of men only [29–44]. In addition, some studies have focused on women only [45], while others did not mention the sex of their participants [46,47]. Moreover, several studies on cognitive performance included both men and women but omitted any statistical analysis on possible sex effects [6,48–55], leaving it unknown whether sex differences were either absent or not investigated. Up to now, six controlled studies examining performance on a variety of cognitive, psychomotor, and memory tests found no significant sex differences on any of the administered tests [56–61], and two other studies reported no significant sex differences in driving performance [4,62]. In experimental studies where sex-adjusted dosing was used, with few exceptions [58], usually no sex differences in the presence and severity of hangover symptoms was found [60,63]. Most commonly, however, possible sex differences were not investigated. Also, with few exceptions [64,65], survey studies usually did not report on sex differences.

Insight into possible sex differences in the symptomatology of the alcohol hangover may help to increase our understanding of the pathology of the alcohol hangover and be of guidance in the development of sex-tailored alcohol hangover treatments, targeting those core symptoms that are characteristic for the hangover state in relation to different peak eBAC levels. The aim of the current study was, therefore, to systematically investigate possible sex differences in the presence and severity of alcohol hangover symptoms at various estimated blood alcohol concentration (eBAC) ranges. The latter is a novel approach compared to previous research but differentiating BAC levels is important, as sex-effects may be differential at high or low BACs. Given the pharmacokinetic and pharmacodynamic sex differences in acute alcohol effects, sex differences in the presence and severity of hangover symptoms were also expected to be present during the hangover state.

2. Methods

Data from two previous surveys on alcohol hangovers were combined [66,67]. The combined dataset consisted of $n = 2446$ Dutch students (male = 50.7%, female = 49.35%), with an age range of 18 to 30 years old. For their past month's heaviest drinking occasion, they reported the number of alcoholic drinks they consumed and the start and stop time of drinking. The estimated peak blood alcohol concentration (eBAC) for this drinking occasion was calculated by applying an adapted Widmark equation [68]. As the eBAC formula includes data on weight, sex, duration of drinking, and the amount of alcohol consumed, eBAC can be regarded as the most overall alcohol consumption parameter. The severity of 22 individual hangover symptoms was scored on an 11 point scale ranging from 0 (absent) to 10 (extreme) [69]. The 22 individual hangover symptoms are a combination of those listed in the three most commonly used hangover symptom scales [70–72], and include headache, nausea, concentration problems, regret, sleepiness, heart pounding, heart racing, vomiting, tiredness, shivering, clumsiness, weakness, dizziness, apathy, sweating, stomach pain, confusion, sensitivity to light, thirst, anxiety, depression, and reduced appetite.

Statistical Analysis

Data were analyzed using SPSS version 25. Subjects who reported no hangover and those with an eBAC of 0.40% and higher were omitted from the analysis. The presence of each of the 22 hangover symptoms was calculated (i.e., the % of subjects with a score >0). Mean (SD) hangover symptom severity was calculated only for those subjects reporting a particular hangover symptom (i.e., excluding scores of 0). The presence and severity of individual hangover symptoms were compared between men and women using the nonparametric Chi-squared test (presence, comparing percentages) and the Mann–Whitney U test (severity, comparing means), respectively. A Bonferroni correction was applied to adjust for multiple comparisons. Sex differences were considered statistically significant if $p < 0.002$.

Since alcohol consumption levels may significantly differ between men and women, as well as the corresponding presence and severity of hangover symptoms, subjects were grouped according to their estimated blood alcohol concentration (eBAC): (a) $0\% \leq eBAC < 0.08\%$, (b) $0.08\% \leq eBAC < 0.11\%$, (c) $0.11\% \leq eBAC < 0.2\%$, (d) $0.2\% \leq eBAC < 0.3\%$ and (e) $0.3\% \leq eBAC < 0.4\%$. The lower cut-off values (a) correspond to common legal limits for driving, whereas $\geq 0.11\%$ (d) corresponds to the Alcohol Hangover Research Group proposed eBAC limit needed to provoke a hangover per se [73]. For the current study, participants with an eBAC of ≥ 0.4 were considered outliers (>2 SD from the mean) and excluded from the data analysis.

3. Results

From the $n = 2446$ subjects that completed the survey, those who reported no past month hangover were omitted from the analysis ($n = 681$, 27.6%). Thus, data on the presence and severity of hangover symptoms from $n = 1765$ subjects ($n = 895$ men and $n = 870$ women) were compared. Demographics of the sample are summarized in Table 1.

Table 1. Demographics.

Demographics	Overall ($n = 1765$)	Men ($n = 895$)	Women ($n = 870$)	p-Value
Age (years)	20.9 (2.3)	21 (2.4)	20.7 (2.2)	0.003 *
Body weight (kg)	71.9 (12.2)	77.4 (11.3)	66.3 (10.5)	0.000 *
Alcoholic drinks per week	13.2 (9.8)	16.6 (10.6)	9.7 (7.5)	0.000 *
Latest Past Month Drinking Session that Produced a Hangover				
Number of alcoholic drinks	12.1 (4.7)	14.1 (5)	10.0 (3.4)	0.000 *
Duration of the drinking session (h)	5.8 (2.0)	6.0 (2)	5.5 (1.9)	0.000 *
eBAC (%)	0.18 (0.08)	0.18 (0.08)	0.18 (0.08)	0.321
Overall hangover severity	6.1 (1.9)	6.1 (1.8)	6.2 (1.9)	0.367

Mean (SD) are shown. Differences are considered significant if $p < 0.05$ (indicated with *). Abbreviation: eBAC = estimated blood alcohol concentration.

Women reported consuming significantly less alcohol per week than men. Also, on the latest past month drinking occasion that resulted in a hangover, women reported consuming significantly fewer alcoholic drinks then men. Overall hangover severity and eBAC on the latest past month drinking occasion that resulted in a hangover did not significantly differ between men and women.

The overall results ($0 \leq$ eBAC $< 0.4\%$, see Table 2) show that women reported significantly higher hangover symptom scores than men for nausea, sleepiness, being tired, vomiting, weakness, sensitivity to light, and dizziness. Further, the presence of shivering was reported significantly more often by women, whereas the presence of heat racing, confusion, and sweating were reported significantly more frequently by men. Subjects were further categorized according to their eBAC. The results for the sex comparisons for each eBAC range are summarized in Tables 3–7.

Table 2. Presence and severity of individual hangover symptoms after alcohol consumption and reaching an eBAC $0\% \leq$ eBAC $< 0.4\%$.

Symptoms	Mean (SD) All	% All	Mean (SD) Men	% Men	Mean (SD) Women	% Women
Headache	5.7 (2.5)	91%	5.7 (2.4)	90.7%	5.8 (2.5)	91.3%
Nausea	5.6 (2.6)	86%	5.2 (2.5)	84.1%	5.9 (2.7) *	87.5%
Concentration problems	6.0 (2.4)	91.1%	6.1 (2.4)	92.2%	5.8 (2.5)	90.1%
Regret	4.5 (2.7)	58.9%	4.4 (2.7)	57.1%	4.6 (2.7)	60.2%
Sleepiness	6.6 (2.3)	94.9%	6.4 (2.3)	94.3%	6.8 (2.2) *	95.3%
Heart pounding	4 (2.4)	35.5%	4 (2.4)	35.7%	4.0 (2.5)	35.3%
Vomiting	5.3 (3.2)	26%	4.8 (3.1)	23.5%	5.6 (3.2) *	28.1%
Tired	6.9 (2.2)	97.7%	6.6 (2.2)	96.8%	7.1 (2.2) *	98.4%
Shivering	4.2 (2.5)	45.7%	4.2 (2.5)	41.4%	4.3 (2.5)	49.1%*
Clumsy	4.9 (2.4)	72.2%	4.8 (2.4)	72.4%	4.9 (2.4)	72.1%
Weakness	5.4 (2.4)	85.2%	5.1 (2.4)	83.4%	5.6 (2.4) *	86.7%
Dizziness	4.4 (2.6)	62.3%	4.1 (2.4)	61%	4.6 (2.7) *	63.3%
Apathy	5.8 (2.5)	80.1%	5.6 (2.5)	78.3%	5.9 (2.6)	81.5%
Sweating	4.3 (2.4)	54.8%	4.3 (2.4)	58.5%	4.3 (2.5)	51.7%*
Stomach Pain	4.4 (2.5)	57.3%	4.3 (2.4)	54.4%	4.5 (2.5)	59.7%
Confusion	4.2 (2.4)	46.3%	4.1 (2.4)	49.9%	4.2 (4.5)	43.4%*
Light sensitivity	4.1 (2.4)	52.2%	3.9 (2.3)	52.4%	4.3 (2.4) *	52%
Thirst	6.7 (2.4)	94.5%	6.7 (2.4)	95%	6.6 (2.4)	94.1%
Heart racing	4.2 (2.5)	41.6%	4.2 (2.5)	45.7%	4.1 (2.5)	38.3%*
Anxiety	3.2 (2.4)	18.3%	3.2 (2.4)	17.4%	3.2 (2.4)	19.1%
Depression	3.8 (2.5)	29.9%	3.9 (2.5)	28.9%	3.8 (2.5)	30.6%
Reduced appetite	5.5 (2.6)	70.1%	5.5 (2.6)	69.9%	5.5 (2.6)	70.3%

Significant differences ($p < 0.002$, after Bonferroni correction) between men and women are indicated with *.

Table 3. Presence and severity of individual hangover symptoms after alcohol consumption and reaching an eBAC 0% ≤ eBAC < 0.08%.

Symptoms	Mean (SD) All	% All	Mean (SD) Men	% Men	Mean (SD) Women	% Women
Headache	5.5 (2.3)	90.3%	5.7 (2.2)	89%	5.4 (2.5)	91.4%
Nausea	4.9 (2.6)	82.2%	4.9 (2.6)	80%	5 (2.6)	84.1%
Concentration problems	5.3 (2.4)	85.4%	5.4 (2.2)	86%	5.3 (2.6)	84.9%
Regret	4.1 (2.8)	50.0%	4.2 (2.8)	52.9%	4.1 (2.8)	47.4%
Sleepiness	6.2 (2.3)	92.7%	5.8 (2.3)	91.1%	6.5 (2.3)	94.1%
Heart pounding	3.8 (2.5)	30.2%	3.8 (2.7)	37.5%	3.8 (2.2)	23.7%
Vomiting	4.5 (3.1)	20.5%	4.4 (3.2)	22.1%	4.7 (3.1)	19.1%
Tired	6.3 (2.4)	97.2%	5.9 (2.3)	95.6%	6.6 (2.4)	98.7%
Shivering	3.8 (2.2)	34.4%	3.9 (2.4)	36%	3.6 (2.0)	32.9%
Clumsy	4.3 (2.3)	56.6%	4.3 (2)	56.6%	4.3 (2.5)	56.6%
Weakness	4.8 (2.4)	74.3%	4.7 (2.4)	72.8%	5 (2.4)	75.7%
Dizziness	3.9 (2.4)	52.8%	3.8 (2.3)	55.9%	4 (2.6)	50.0%
Apathy	5.5 (2.6)	74.2%	5.3 (2.6)	71.9%	5.8 (2.5)	76.3%
Sweating	4.1 (2.4)	44.6%	4.4 (2.4)	48.5%	3.8 (2.5)	41.1%
Stomach Pain	3.8 (2.2)	55.4%	3.8 (2.4)	53.7%	3.8 (2.1)	57.0%
Confusion	3.6 (2.5)	36.8%	3.8 (2.4)	41.2%	3.4 (2.6)	32.9%
Light sensitivity	4 (2.3)	42%	3.7 (2.3)	44.1%	4.3 (2.2)	40.1%
Thirst	6 (2.5)	90.9%	5.9 (2.4)	88.9%	6.1 (2.5)	92.7%
Heart racing	4 (2.6)	30.3%	4.2 (2.8)	36.3%	3.8 (2.4)	25%
Anxiety	3.1 (2.4)	12.9%	3.5 (2.8)	14.7%	2.6 (1.8)	11.3%
Depression	3.7 (2.3)	24.6%	3.9 (2.4)	25.9%	3.4 (2.1)	23.3%
Reduced appetite	4.6 (2.6)	60.4%	4.3 (2.5)	56.6%	4.8 (2.6)	63.8%

No significant differences ($p < 0.002$, after Bonferroni correction) were found between men and women.

Table 4. Presence and severity of individual hangover symptoms after alcohol consumption and reaching an eBAC of 0.08% ≤ eBAC < 0.11%.

Symptoms	Mean (SD) All	% All	Mean (SD) Men	% Men	Mean (SD) Women	% Women
Headache	5.6 (2.4)	88.8%	5.7 (2.3)	89.3%	5.4 (2.5)	88.4%
Nausea	5.3 (2.6)	83%	4.7 (2.5)	83.5%	5.8 (2.6) *	82.6%
Concentration problems	5.4 (2.4)	88.4%	5.9 (2.2)	87.6%	5.1 (2.5)	89%
Regret	4.4 (2.6)	59.8%	4.5 (2.7)	57%	4.3 (2.6)	61.9%
Sleepiness	6.4 (2.4)	92.8%	6.1 (2.4)	91.7%	6.5 (2.4)	93.5%
Heart pounding	3.5 (2.3)	32.6%	3.6 (2.4)	29.8%	3.4 (2.2)	34.8%
Vomiting	5.2 (3.1)	21.4%	4 (3.1)	15.7%	5.8 (3.0)	25.8%
Tired	6.7 (2.3)	96.4%	6.6 (2.2)	95%	6.8 (2.4)	97.4%
Shivering	4.2 (2.7)	42%	3.5 (2.5)	35.5%	4.6 (2.7)	47.1%
Clumsy	4.5 (2.3)	69.1%	4.6 (2.4)	68.6%	4.3 (2.3)	69.5%
Weakness	5.2 (2.3)	86.2%	4.9 (2.3)	84.3%	5.3 (2.3)	87.7%
Dizziness	4.3 (2.6)	58%	4.3 (2.3)	55.4%	4.2 (2.7)	60%
Apathy	5.5 (2.5)	77.5%	5.5 (2.3)	71.1%	5.5 (2.6)	82.5%
Sweating	3.9 (2.3)	48.6%	3.6 (2.2)	49.6%	4.2 (2.4)	47.7%
Stomach Pain	4.4 (2.6)	52.6%	4.1 (2.3)	53.3%	4.7 (2.8)	51.9%
Confusion	3.8 (2.2)	37.1%	3.7 (2.1)	40.5%	3.9 (2.4)	34.4%
Light sensitivity	4.1 (2.4)	50%	3.9 (2.3)	50.4%	4.2 (2.6)	49.7%
Thirst	6.4 (2.5)	93.1%	6.5 (2.5)	94.2%	6.3 (2.5)	92.3%
Heart racing	3.4 (2.2)	38.9%	3.4 (2.1)	43%	3.4 (2.3)	35.7%
Anxiety	2.9 (2.3)	16.4%	2.7 (1.9)	15.8%	3 (2.7)	16.9%
Depression	3.7 (2.5)	30.4%	3.4 (2.4)	25.6%	3.9 (2.5)	34.2%
Reduced appetite	5.6 (2.5)	62%	5.4 (2.6)	61.2%	5.8 (2.5)	62.6%

Significant differences ($p < 0.002$, after Bonferroni correction) between men and women are indicated with *.

Table 5. Presence and severity of individual hangover symptoms after alcohol consumption and reaching an eBAC of 0.11% ≤ eBAC < 0.2%.

Symptoms	Mean (SD) All	% All	Mean (SD) Men	% Men	Mean (SD) Women	% Women
Headache	5.8 (2.4)	91.1%	5.6 (2.5)	90.5%	5.9 (2.4)	91.7%
Nausea	5.7 (2.7)	86.4%	5.3 (2.7)	84.4%	6 (2.6) *	88.1%
Concentration problems	6.0 (2.4)	92.3%	6.1 (2.3)	94.6%	6 (2.4)	90.5%
Regret	4.5 (2.7)	61.1%	4.2 (2.6)	58.8%	4.7 (2.7)	62.9%
Sleepiness	6.6 (2.3)	95.7%	6.4 (2.3)	96.3%	6.8 (2.2)	95.2%
Heart pounding	3.9 (2.4)	36.3%	4 (2.4)	36.2%	3.9 (2.4)	36.3%
Vomiting	5.4 (3.1)	26%	5 (3.1)	23.5%	5.6 (3.1)	28.1%
Tired	7 (2.2)	97.3%	6.8 (2.2)	96.3%	7.2 (2.1) *	98.1%
Shivering	4.2 (2.6)	47.6%	4.3 (2.5)	42%	4.2 (2.7)	51.9%*
Clumsy	4.9 (2.4)	72.2%	5 (2.4)	73.7%	4.9 (2.5)	71.1%
Weakness	5.4 (2.4)	86.3%	5.1 (2.4)	83.7%	5.7 (2.4) *	88.4%
Dizziness	4.4 (2.5)	62.9%	4.0 (2.4)	62.4%	4.8 (2.6) *	63.4%
Apathy	5.8 (2.5)	81.3%	5.6 (2.5)	79%	6 (2.5)	83.1%
Sweating	4.2 (2.4)	55.4%	4.3 (2.4)	59.7%	4.2 (2.4)	52.1%
Stomach Pain	4.5 (2.5)	57.6%	4.3 (2.5)	54.8%	4.6 (2.5)	59.8%
Confusion	4 (2.4)	47.5%	3.9 (2.4)	53%	4.2 (2.4)	43.2%*
Light sensitivity	4.1 (2.5)	53.6%	3.8 (2.4)	55%	4.4 (2.5)	52.6%
Thirst	6.7 (2.3)	95.3%	6.8 (2.3)	95.9%	6.7 (2.3)	94.9%
Heart racing	4.2 (2.5)	41.2%	4.3 (2.5)	45.4%	4.2 (2.5)	37.9%
Anxiety	3.4 (2.4)	20%	3.3 (2.4)	19%	3.4 (2.4)	20.9%
Depression	3.8 (2.5)	31.5%	3.9 (2.5)	31.6%	3.8 (2.5)	31.5%
Reduced appetite	5.5 (2.6)	72.5%	5.5 (2.6)	73.1%	5.6 (2.6)	72.1%

Significant differences ($p < 0.002$, after Bonferroni correction) between men and women are indicated with *.

Table 6. Presence and severity of individual hangover symptoms after alcohol consumption and reaching an eBAC of 0.2% ≤ eBAC < 0.3%.

Symptoms	Mean (SD) All	% All	Mean (SD) Men	% Men	Mean (SD) Women	% Women
Headache	5.7 (2.6)	92.2%	5.8 (2.5)	92.8%	5.6 (2.6)	91.7%
Nausea	5.6 (2.6)	88.8%	5.2 (2.4)	86%	6 (2.7) *	91.1%
Concentration problems	6.1 (2.4)	93.1%	6.3 (2.4)	93.8%	6 (2.4)	92.5%
Regret	4.6 (2.7)	60.8%	4.5 (2.6)	59.9%	4.7 (2.8)	61.5%
Sleepiness	6.8 (2.1)	95.1%	6.7 (2.1)	93.4%	6.9 (2.1)	96.4%
Heart pounding	4.3 (2.5)	37.4%	4.4 (2.4)	35.4%	4.3 (2.5)	39%
Vomiting	5.5 (3.4)	27%	4.8 (3.1)	24%	6 (3.5)	29.5%
Tired	7 (2.1)	98.6%	6.7 (2.1)	98.3%	7.2 (2.1) *	98.9%
Shivering	4.4 (2.4)	48.6%	4.3 (2.5)	44.2%	4.4 (2.3)	52.2%
Clumsy	5.1 (2.4)	79%	4.9 (2.5)	78.9%	5.2 (2.4)	79.1%
Weakness	5.5 (2.4)	87.7%	5.3 (2.3)	86.9%	5.7 (2.4)	88.3%
Dizziness	4.4 (2.7)	66.7%	4.2 (2.5)	62.7%	4.6 (2.8)	69.9%
Apathy	5.7 (2.5)	81.9%	5.7 (2.4)	83.2%	5.8 (2.6)	80.8%
Sweating	4.5 (2.5)	58.4%	4.5 (2.4)	62.3%	4.5 (2.6)	55.2%
Stomach Pain	4.5 (2.5)	59.5%	4.4 (2.4)	54.6%	4.5 (2.6)	63.5%
Confusion	4.6 (2.5)	51.4%	4.5 (2.5)	52.6%	4.6 (2.6)	50.4%
Light sensitivity	4.2 (2.3)	55.2%	4.2 (2.2)	53.4%	4.3 (2.4)	56.7%
Thirst	6.8 (2.4)	95.5%	6.9 (2.3)	96.9%	6.8 (2.4)	94.4%
Heart racing	4.5 (2.5)	47.0%	4.6 (2.5)	49.5%	4.4 (2.6)	45%
Anxiety	3 (2.3)	19.3%	2.9 (2.3)	17.8%	3.1 (2.2)	20.6%
Depression	3.9 (2.4)	29.6%	4 (2.4)	29.8%	3.8 (2.5)	29.5%
Reduced appetite	5.5 (2.5)	73.3%	5.7 (2.4)	73.6%	5.4 (2.6)	73.1%

Significant differences ($p < 0.002$, after Bonferroni correction) between men and women are indicated with *.

Table 7. Presence and severity of individual hangover symptoms after alcohol consumption and reaching an eBAC of 0.3% ≤ eBAC < 0.4%.

Symptoms	Mean (SD) All	% All	Mean (SD) Men	% Men	Mean (SD) Women	% Women
Headache	5.9 (2.5)	90.6%	5.5 (2.5)	89.2%	6.3 (2.4)	91.8%
Nausea	6 (2.6)	84.3%	5.6 (2.5)	84%	6.4 (2.6)	84.5%
Concentration problems	6.4 (2.4)	89.8%	6.6 (2.4)	90%	6.3 (2.5)	89.7%
Regret	5.2 (2.8)	51.7%	5.4 (2.8)	45.1%	5.1 (2.8)	57.3%
Sleepiness	6.8 (2.4)	96.1%	6.4 (2.5)	95.2%	7 (2.3)	96.9%
Heart pounding	4.5 (2.6)	37.4%	4.1 (2.5)	40.2%	5 (2.6)	35.1%
Vomiting	5.3 (3.3)	38.5%	5.2 (3.2)	35.4%	5.5 (3.3)	41.2%
Tired	7 (2.4)	98.9%	6.7 (2.4)	98.8%	7.3 (2.3)	99%
Shivering	4.4 (2.7)	47.8%	4.3 (2.9)	46.3%	4.4 (2.6)	49%
Clumsy	5 (2.6)	77.7%	4.7 (2.7)	73.2%	5.2 (2.5)	81.4%
Weakness	5.8 (2.4)	85.4%	5.4 (2.3)	85.2%	6.2 (2.4)	85.6%
Dizziness	4.6 (2.5)	64.2%	4.3 (2.4)	64.6%	4.9 (2.6)	63.9%
Apathy	6.3 (2.5)	79.9%	6.3 (2.5)	78%	6.3 (2.5)	81.4%
Sweating	4.6 (2.4)	63.7%	4.7 (2.5)	68.3%	4.6 (2.3)	59.8%
Stomach Pain	4.7 (2.5)	58.2%	4.7 (2.5)	53.8%	4.7 (2.5)	61.9%
Confusion	4.6 (2.4)	50.3%	4.7 (2.3)	51.2%	4.6 (2.5)	49.5%
Light sensitivity	4.2 (2.3)	52.2%	3.6 (2.1)	50.6%	4.6 (2.3)	53.6%
Thirst	7 (2.3)	93.9%	7 (2.2)	94%	7 (2.4)	93.8%
Heart racing	4.4 (2.5)	46.9%	4 (2.4)	53.7%	4.9 (2.6)	41.2%
Anxiety	3.6 (2.5)	16.3%	3.3 (2.6)	13.6%	3.8 (2.5)	18.6%
Depression	4.1 (2.7)	28.5%	4.3 (3)	20.7%	4.1 (2.6)	35.1%
Reduced appetite	6.2 (2.5)	72.6%	6.2 (2.4)	73.2%	6.1 (2.5)	72.2%

No significant differences ($p < 0.002$, after Bonferroni correction) were found between men and women.

In the lowest eBAC group (0% ≤ eBAC < 0.08%, see Table 3) no significant differences were found between men ($n = 91$) and women ($n = 68$). In the subsequent eBAC group (0.08% ≤ eBAC < 0.11%, see Table 4), nausea was significantly more severe in women ($n = 95$) compared to men ($n = 93$). No other significant sex differences were found. Below an eBAC of 0.11%, the presence and severity of hangover symptoms were relatively low. This was expected given the relative low alcohol intake, and in line with the previous consensus paper of the Alcohol Hangover Research Group which stated that an eBAC < 0.11% is insufficient to produce a hangover per se. A minority of the total sample had an eBAC below 0.11%, which had an approximate equal sex distribution (18.7% of all women and 20.6% of all men).

Table 5 summarizes the results of the largest group of drinkers ($n = 403$ men and $n = 383$ women). Their eBAC ranged from 0.11% ≤ eBAC < 0.2%. In women, severity scores on nausea, tiredness, weakness, and dizziness were significantly higher than in men. In addition, confusion was significantly more often reported by men and shivering was significantly more often reported by women. All other sex comparisons did not reach statistical significance. Figure 1 gives an overview of the presence and severity of individual hangover symptoms for the eBAC range 0.11% ≤ eBAC < 0.2%. It is evident from Figure 1, that the observed significant sex differences are of modest magnitude. Except for dizziness, sex differences in symptom severity were seen only among symptoms that had high presence ratings. Figure 1 also reveals that some symptoms had a low severity and a low presence (e.g., anxiety and depression), whereas other symptoms had a high severity and high presence (e.g., sleepiness, headache, concentration problems, apathy). In this regard, vomiting was an interesting hangover symptom: vomiting had a low presence, but when it occurred its severity was relatively high.

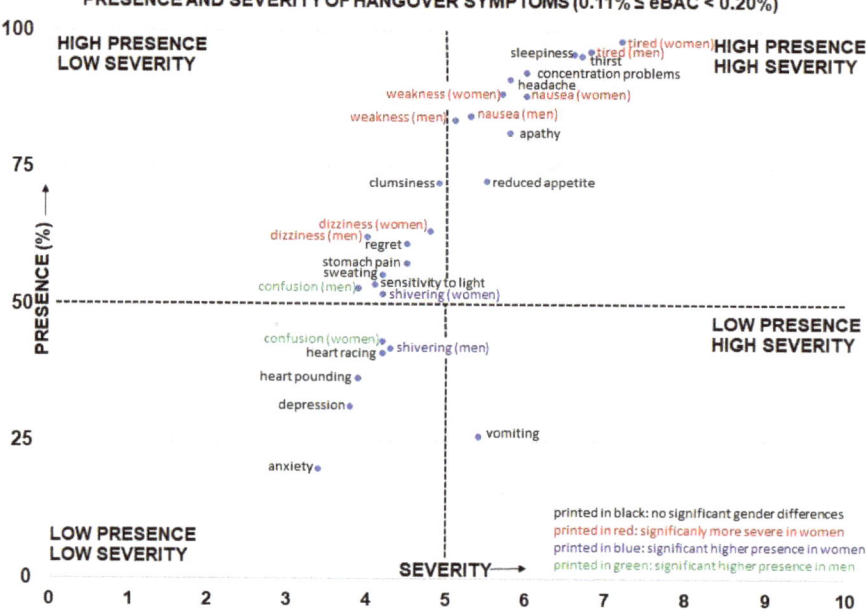

Figure 1. Presence and severity of individual hangover symptoms after alcohol consumption and reaching an eBAC of 0.11% ≤ eBAC < 0.2%. Sex differences are significant if $p < 0.002$. Abbreviation: eBAC = estimated blood alcohol concentration.

Table 6 summarizes data from subjects within the eBAC range 0.2% ≤ eBAC < 0.3% ($n = 244$ men and $n = 264$ women). In women, severity scores on nausea and tiredness were significantly higher than among men. All other sex comparisons did not reach statistical significance.

Finally, Table 7 summarizes data from subjects within the eBAC range 0.3% ≤ eBAC < 0.4% ($n = 62$ men and $n = 62$ women). For this eBAC range, no significant sex differences were observed.

4. Discussion

The current analyses showed that some sex differences in the presence and severity of hangover symptoms were observed. These were especially evident at the eBAC levels ranging from 0.11% ≤ eBAC < 0.2%. In the lowest eBAC group (0% ≤ eBAC < 0.08%), no significant sex differences were found. In the subsequent eBAC group (0.08% ≤ eBAC < 0.11%), severity of nausea was significantly higher in women than in men. In the third eBAC group (0.11% ≤ eBAC < 0.2%), women reported higher severity scores on nausea, tiredness, weakness, and dizziness than men. Men reported the presence of confusion significantly more often than women, and women reported the presence of shivering significantly more often than men. In the fourth eBAC group (0.2% ≤ eBAC < 0.3%), women reported higher severity scores on nausea and tiredness than men. The fifth eBAC group (0.3% ≤ eBAC < 0.4%), no significant sex differences were found. However, the observed sex differences were of modest magnitude and have little clinical relevance. This is illustrated in Figure 2 for the most frequently reported sex differences in hangover symptom severity, i.e., nausea and being tired. The magnitude of the differences in symptom severity scores for nausea and being tired, albeit statistically significant, across eBAC ranges were always below 1 on a scale ranging from 0 (absent) to 10 (extreme). Figure 2 further shows that symptom severity scores did somewhat increase when eBAC became higher. For example, the difference between the lowest eBAC group (<0.08%) and the highest eBAC group (0.3%–<0.4%) for nausea were +0.7 in men and +1.4 in women, and for being tired +0.8 in men and +0.7 in women. Thus, the present analyses suggest that men and women experience hangovers in comparable ways.

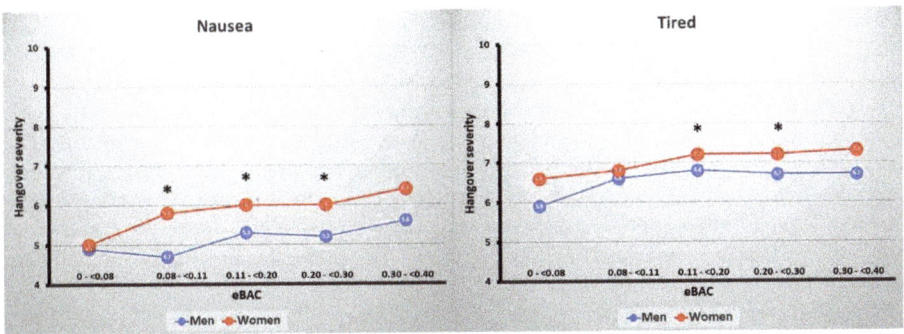

Figure 2. Severity scores for nausea and being tired for men and women across eBAC ranges. Significant differences ($p < 0.002$) are indicated by *. Abbreviation: eBAC = estimated blood alcohol concentration.

4.1. Alcohol Sensitivity during Intoxication and during the Hangover State

The absence of relevant sex differences in the presence and severity of hangover symptoms contrasts with the various sex differences that are commonly observed in pharmacokinetics and pharmacodynamics during the alcohol intoxication phase. Thus, there seems to be a discrepancy between sex differences observed during alcohol intoxication (in previous research) and the absence of relevant sex effects during the hangover state (found in the current study). Previous research has addressed this issue. For example, Piasecki and colleagues [74] reported that subjects with lower alcohol sensitivity appear to be more resistant to experiencing hangovers at a given number of drinks. However, as these drinkers often consume more alcohol than those with high alcohol sensitivity, they on average experience more hangovers during an arbitrary monitoring period. Unfortunately, the association with alcohol sensitivity and hangover severity was not assessed in this study.

Rohsenow et al. [75] did not administer the SRE but instead used an alcohol intoxication rating as a proxy measure for the sensitivity to acute alcohol effects. They found that hangover severity, as assessed with the AHS (Acute Hangover Scale), was positivity correlated with this intoxication rating. In this study, about one-third of the studied sample reported an absence of hangover after consuming alcohol to reach a peak BAC of 0.12%. As hangover insensitivity was associated with reports of being less intoxicated, the authors suggested that hangover sensitivity and intoxication sensitivity may be related to each other. In an earlier study, Ylikhari et al. [15] also found that, despite consuming the same amount of alcohol (1.5 g/kg), those who reported higher subjective intoxication levels also experienced more severe hangovers.

In a naturalistic study comprising of an alcohol and alcohol-free test day, Hogewoning et al. [69] administered the SRE and compared drinking behavior and hangover effects in both hangover-sensitive drinkers and hangover-resistant drinkers. Although both groups consumed a comparable amount of alcohol, achieving a peak BAC around 0.17%, the hangover resistant group reported not experiencing a hangover while the sensitive group did. Hogewoning et al. [69] found that the early life SRE score and total SRE score were significantly lower in women compared to men. However, in contrast to Piasecki et al. [74], SRE scores did not significantly differ between hangover-sensitive and hangover-resistant drinkers, and the SRE scores did not significantly correlate with hangover severity. Taken together, these findings suggest that sex differences in alcohol sensitivity seen during the intoxication phase may be unrelated to the sensitivity for alcohol effects expected during the hangover state. There may be several explanations for this observation.

Firstly, there is a time lag between the intoxication phase and the hangover phase. After stopping drinking, usually a period of sleep follows of which the quality and duration has shown to be associated with hangover severity [76,77]. Secondly, the acute alcohol effects are mainly caused by a direct presence of alcohol in the blood. Subsequently, alcohol can exert its effects virtually everywhere in the body, including the brain as it easily crosses the blood–brain barrier. In contrast, the hangover state starts when blood alcohol concentrations approach zero, and in many cases BAC readings are zero. Finally, although there is some overlap between the presence and severity of symptoms that are experienced during the intoxication and hangover phase, there are also several clear differences. Van Schrojenstein Lantman et al. [1] reported that the most common reported combination of symptoms describing the alcohol hangover comprised of the following: nausea, headache, tiredness, and apathy. These symptoms showed the highest associations between each other. Except for nausea, these are symptoms that are usually not reported to be present or have high severity ratings during the intoxication phase of drinking. Taken together during the intoxication phase, sex differences in the presence and severity of intoxication effects are common. However, in contrast, sex differences in symptomatology during the alcohol hangover are of limited relevance.

4.2. Implications

The current findings have several implications. Firstly, our findings do not suggest that results from studies conducted in men can be directly extrapolated to women, or that an equal sex balance would not be necessary. Although the presence and severity of hangover symptoms do not seem to vary much between men and women, their impact on mood and performance may be considerably different. Therefore, we advocate an equal sex balance in future hangover studies. Secondly, the current findings may have implications for drug development. In past research, the efficacy and safety of potential hangover treatments were often not investigated in women (e.g., [33,78–83]). Currently, most marketed hangover products lack sound scientific support that confirms their efficacy and safety [84,85], and no effective treatment has been developed yet. The current analyses suggest that in order to develop an effective and safe hangover treatment, similar symptoms should be targeted for both sexes, as they usually not differ in presence and severity. Of note, in both sexes, with increasing BAC levels the presence and severity of hangover symptoms increases to some extent. Therefore, dose-ranging studies to demonstrate efficacy and safety at various BAC levels (e.g., low, moderate, high intake) may be worthwhile to investigate. Given the potential negative impact of hangover symptoms on performance of daily activities such as driving and job performance [2–6], developing an effective hangover treatment could be useful. Previous research has shown that the vast majority of drinkers indicate that experiencing hangovers does not affect future drinking behavior [86], and that if an effective hangover treatment would be available this would not significantly alter their drinking behavior [87]. However, recent research did associate experiencing hangovers with having significantly more future drinking days and alcohol problems [88]. Therefore, more research should address the relationship between experiencing hangovers or their treatment with increased risk of developing alcohol use disorder.

As for all clinical trials investigating psychopharmacological treatments [89–91], it is also critical to examine the efficacy and safety of new hangover treatments in both men and women. Although the targeted hangover symptoms may have the same presence and severity, the safety, adverse effects, and efficacy of the hangover treatment itself may show significant sex differences. Even in the absence of a clear a priori explanation for expecting sex differences it is vital to examine possible dissociations in pharmacokinetics and pharmacodynamics of a hangover treatment.

From Figure 1, it is clear that some symptoms are high in both presence and severity. These can be regarded as core symptoms of the hangover state, and it is suggested that further research focuses on these symptoms. In this context, there are currently three hangover symptom scales in use that all comprise ratings of different hangover symptoms to sum up to an overall hangover score [70–72]. A new analysis, taking into account both the presence and severity of symptoms, and perhaps their impact on cognitive and physical functioning and mood [67], can be useful for the development of a new hangover scale. Finally, some symptoms such as anxiety and depression have both low presence and severity scores. This is a consistent finding in scientific literature, and although these symptoms are included in the Hangover Symptoms Scale (HSS) [71], it can be questioned if they are true hangover symptoms or should be considered as individual subject characteristics that are unrelated to alcohol consumption and its aftereffects.

4.3. Strengths and Limitations

Strengths of the current analysis include its large sample size, and the fact that subjects were stratified according to their estimated peak BAC level. The latter was critical to obtain a fair comparison between effects observed in men and women.

A limitation of the survey data consisted of retrospective ratings of hangover symptoms, experienced during their latest past month alcohol hangover. Therefore, in theory, impairments in accurate recall may have biased these results. However, there is also no reason to assume why possible recall bias would be different in men and women, or more pronounced for one hangover symptom over the others.

Second, the participants in the surveys consisted of students, aged 18 to 30 years old. This may limit the generalizability to other age groups or non-student samples. Hangover data on subjects aged 40 and beyond are virtually absent in the scientific literature. One study did report a shift in the presence and severity of hangover symptoms with progressing age [92]. This population-based Danish Health Examination Study (DANHES) gathered data from over 50,000 adults and showed that with increasing age, the presence of hangovers significantly decreased, and in a comparable rate for men and women. Both men and women reported fewer hangover symptoms (thirst, exhaustion, headache, dizziness, no appetite, stomach pain, nausea, heart racing, and vomiting) with increasing age. Although several reasons for this observation can be hypothesized (e.g., tolerance, changes in sensitivity, and alcohol metabolism), the most likely reason for the age-progressing decline in hangover frequency is the accompanying age-related reduction in the amount of (alcohol consumed on) heavy drinking sessions [93]. Although no formal testing was presented by Tolstrup et al. [92], inspection of the odds ratios for "(almost) always experiencing a hangover symptom" suggests that there are no clinically relevant sex differences across age groups (Tolstrup et al. [92], Table 3, page 469). Future research should further examine possible sex differences in hangover symptomatology in other age groups.

Third, we investigated the 22 hangover symptoms that are included in the three commonly used hangover scales [70–72]. Research has shown that there are several other hangover symptoms that we could have considered in the current study. For example, Penning et al. [66] identified 47 hangover symptoms in the literature. However, most of the symptoms that were omitted from the three hangover scales (e.g., nystagmus or tinnitus) are very infrequently reported and their severity scores are usually low. Hence, for the population as a whole these symptoms must not be regarded important contributors to the hangover state, and it is unlikely that important sex differences will be observed among these symptoms.

Fourth, there are several factors that contribute to hangover severity that were not included in the current analysis. These are discussed in various review articles on alcohol hangover [93–95], although no possible relationship with sex differences was suggested. Examples of factors that may impact the presence and severity of hangover symptoms are sleep duration and quality, physical activity while drinking, psychological state, congener content and type of drink, and having a family history of alcoholism. In the current analysis, these factors were not taken into account. Future research should examine the impact of factors that may aggravate hangover severity on possible sex differences. It may, for example, be possible that men may consume more congener rich drinks than women, or vice versa, which may explain observed gender differences. Alternatively, women may be more engaged in dancing during a drinking session than men or may consume more non-alcoholic drinks during a night out. Currently, this information is lacking from the scientific literature.

Finally, future research should investigate possible sex differences in memory, cognitive, and psychomotor functioning. The latter is important as the relative absence of important sex differences in the presence and severity of hangover symptoms does not imply that no sex differences may exist in performance outcomes and the conductance of daily life activities.

5. Conclusions

In conclusion, at the same BAC level, young adult male and female social drinkers experience hangover symptoms in comparable ways. No relevant sex differences were observed in the presence and severity of hangover symptoms across all eBAC levels. Future research should extend this investigation to other age groups.

Author Contributions: A.E.v.L.v.P. and J.C.V. made substantial contributions to conception and design, A.E.v.L.v.P., L.E.D., and J.C.V. analyzed the data, A.E.v.L.v.P. and J.C.V. drafted the manuscript, and all authors revised it critically for important intellectual content and approved the final version of the manuscript.

Conflicts of Interest: J.C.V. has received grants/research support from the Dutch Ministry of Infrastructure and the Environment, Janssen, Nutricia, Red Bull, Sequential, and Takeda, and has acted as a consultant for the Canadian Beverage Association, Centraal Bureau Drogisterijbedrijven, Clinilabs, Coleman Frost, Danone, Deenox, Eisai, Janssen, Jazz, More Labs, Purdue, Red Bull, Sanofi-Aventis, Sen-Jam Pharmaceutical, Sepracor, Takeda, Toast!, Transcept, Trimbos Institute, Vital Beverages, and ZBiotics. The other authors have no potential conflicts of interest to disclose.

References

1. Van Schrojenstein Lantman, M.; Mackus, M.; van de Loo, A.J.A.E.; Verster, J.C. Development of a definition for the alcohol hangover: Consumer descriptions and expert consensus. *Curr. Drug Abus. Rev.* **2016**, *9*, 148–154. [CrossRef] [PubMed]
2. Frone, M.R.; Verster, J.C. Alcohol hangover and the workplace: A need for research. *Curr. Drug Abus. Rev.* **2013**, *6*, 177–179. [CrossRef]
3. Gjerde, H.; Christophersen, A.S.; Moan, I.S.; Yttredal, B.; Walsh, J.M.; Normann, P.T.; Mørland, J. Use of alcohol and drugs by Norwegian employees: A pilot study using questionnaires and analysis of oral fluid. *J. Occup. Med. Toxicol.* **2010**, *15*, 13. [CrossRef] [PubMed]
4. Verster, J.C.; Bervoets, A.C.; de Klerk, S.; Vreman, R.A.; Olivier, B.; Roth, T.; Brookhuis, K.A. Effects of alcohol hangover on simulated highway driving performance. *Psychopharmacology* **2014**, *231*, 2999–3008. [CrossRef] [PubMed]
5. Verster, J.C.; Van Der Maarel, M.A.; McKinney, A.; Olivier, B.; De Haan, L. Driving during alcohol hangover among Dutch professional truck drivers. *Traffic Inj. Prev.* **2014**, *15*, 434–438. [CrossRef] [PubMed]
6. Hartung, B.; Schwender, H.; Mindiashvili, N.; Ritz-Timme, S.; Malczyk, A.; Daldrup, T. The effect of alcohol hangover on the ability to ride a bicycle. *Int. J. Legal Med.* **2015**, *129*, 751–758. [CrossRef] [PubMed]
7. Frezza, M.; di Padova, C.; Pozzato, G.; Terpin, M.; Baraona, E.; Lieber, C.S. High blood alcohol levels in women. The role of decreased gastric alcohol dehydrogenase activity and first-pass metabolism. *N. Engl. J. Med.* **1990**, *322*, 95–99. [CrossRef]
8. Mumenthaler, M.S.; Taylor, J.L.; O'Hara, R.; Yesavage, J.A. Gender differences in moderate drinking effects. *Alcohol Res. Health* **1999**, *23*, 55–64.
9. Jones, B.M.; Jones, M.K. Alcohol and memory impairment in male and female social drinkers. In *Alcohol and Human Memory*; Birnbaum, I.M., Parker, E.S., Eds.; Lawrence Erlbaum Associates: Hillsdale, NJ, USA, 1977; pp. 127–138.

10. Haut, J.S.; Beckwith, B.E.; Petros, T.V.; Russel, S. Gender differences in retrieval from long-term memory following acute intoxication with ethanol. *Physiol. Behav.* **1989**, *45*, 1161–1165. [CrossRef]
11. De Haan, L.; van der Palen, J.A.M.; Olivier, B.; Verster, J.C. Effects of consuming alcohol mixed with energy drinks versus consuming alcohol only on overall alcohol consumption and negative alcohol-related consequences. *Int. J. Gen. Med.* **2012**, *5*, 953–960.
12. Schuckit, M.A.; Smith, T.L.; Tipp, J.E. The self-rating of the effects of alcohol (SRE) form a retrospective measure of the risk for alcoholism. *Addiction* **1997**, *92*, 979–988. [CrossRef] [PubMed]
13. Van de Loo, A.J.A.E.; van Andel, N.; Verster, J.C. The relationship between self-Rating of the Effects of Alcohol (SRE) form scores and subjective intoxication and sleepiness scores after alcohol consumption. *J. Psychopharmacol.* **2014**, *28*, 19.
14. Myrsten, A.-L.; Post, B.; Frankenhaeuser, M. Catecholamine output during and after acute alcoholic intoxication. *Percept. Mot. Skills* **1971**, *33*, 652–654. [CrossRef] [PubMed]
15. Ylikahri, R.H.; Huttunen, M.O.; Eriksson, C.J.P.; Nikklä, E.A. Metabolic studies on the pathogenesis of hangover. *Eur. J. Clin. Investig.* **1974**, *4*, 93–100. [CrossRef]
16. Ylikahri, R.H.; Huttunen, M.O.; Härkönen, M. Effect of alcohol on anterior-pituitary secretion of trophic hormones. *Lancet* **1976**, *1*, 1353. [CrossRef]
17. Ylikahri, R.H.; Leino, T.; Huttunen, M.O.; Poso, A.R.; Eriksson, C.J.P.; Nikkilä, E.A. Effects of fructose and glucose on ethanol-induced metabolic changes and on the intensity of alcohol intoxication and hangover. *Eur. J. Clin. Investig.* **1974**, *6*, 93–102. [CrossRef]
18. Ylikahri, R.H.; Huttunen, M.O.; Härkönen, M.; Leino, T.; Helenius, T.; Liewendahl, K.; Karonen, S.-L. Acute effects of alcohol on anterior pituitary secretion of the tropic hormones. *J. Clin. Endocrinol. Metab.* **1978**, *46*, 715–720. [CrossRef] [PubMed]
19. Linkola, J.; Fyhrquist, F.; Nieminen, M.M.; Weber, T.H.; Tontti, K. Renin aldosterone axis in ethanol intoxication and hangover. *Eur. J. Clin. Investig.* **1976**, *6*, 191–194. [CrossRef]
20. Linkola, J.; Ylikahri, R.; Fyhrquist, F.; Wallenius, M. Plasma vasopressin in ethanol intoxication and hangover. *Acta Physiol. Scand.* **1978**, *104*, 180–187. [CrossRef]
21. Linkola, J.; Fyhrquist, F.; Ylikahri, R. Renin, aldoserone and cortisol during ethanol intoxication and hangover. *Acta Physiol. Scand.* **1979**, *106*, 75–82. [CrossRef]
22. Linkola, J.; Fyhrquist, F.; Ylikahri, R. Adenosine 3′, 5′cyclic monophosphate, calcium and magnesium excretion in ethanol intoxication and hangover. *Acta Physiol. Scand.* **1979**, *107*, 333–337. [CrossRef]
23. Heikkonen, E.; Mäki, T.; Kontula, K.; Ylikahri, R.; Härkönen, M. Effect of Acute Ethanol Intake and Hangover on the Levels of Plasma and Urinary Catecholamines and Lymphocytic β-Adrenergic Receptors. *Alcohol. Clin. Exp. Res.* **1989**, *13*, 20–24. [CrossRef]
24. Newlin, D.B.; Pretorius, M.B. Sons of alcoholics report greater hangover symptoms than sons of nonalcoholics: A pilot study. *Alcohol. Clin. Exp. Res.* **1990**, *14*, 713–716. [CrossRef] [PubMed]
25. Heikkonen, E.; Ylikahri, R.; Roine, R.; Välimäki, M.; Härkönen, M.; Salaspuro, M. The combined effect of alcohol and physical exercise on serum testosterone, luteinizing hormone, and cortisol in males. *Alcohol. Clin. Exp. Res.* **1996**, *20*, 711–716. [CrossRef] [PubMed]
26. Heikkonen, E.; Ylikahri, R.; Roine, R.; Välimäki, M.; Härkönen, M.; Salaspuro, M. Effect of alcohol on exercise-induced changes in serum glucose and serum free fatty acids. *Alcohol. Clin. Exp. Res.* **1998**, *22*, 437–443. [CrossRef]
27. Mäki, T.; Toivonen, L.; Koskinen, P.; Näveri, H.; Härkönen, M.; Leinonen, H. Effect of ethanol drinking, hangover, and exercise on adrenergic activity and heart rate variability in patients with a history of alcohol-induced atrial fibrillation. *Am. J. Cardiol.* **1998**, *82*, 317–322. [CrossRef]
28. Kim, D.J.; Kim, W.; Yoon, S.J.; Choi, B.M.; Kim, J.S.; Go, H.J.; Kim, Y.K.; Jeong, J. Effects of alcohol hangover on cytokine production in healthy subjects. *Alcohol* **2003**, *31*, 167–170. [CrossRef] [PubMed]
29. Carrolli, J.R.; Ashe, W.F.; Roberts, L.B. Influence of the aftereffects of alcohol combined with hypoxia on psychomotor performance. *Aerosp. Med. Hum. Perform.* **1964**, *35*, 990–993.
30. Ideström, C.M.; Cadenius, B. Time Relations of the Effects of Alcohol Compared to Placebo. *Psychopharmacologia* **1968**, *13*, 189–200. [CrossRef] [PubMed]
31. Dowd, P.J.; Wolfe, J.W.; Cramer, R.L. Aftereffects of alcohol on the perception and control pitch altitude during centripetal acceleration. *Aerosp. Med. Hum. Perform.* **1973**, *44*, 928–930.

32. Seppälä, T.; Leino, T.; Linnoila, M.; Huttunen, M.; Ylikahri, R. Effects of Hangover on Psychomotor Skills Related to Driving: Modification by Fructose and Glucose. *Basic Clin. Pharmacol. Toxicol.* **1976**, *38*, 209–218.
33. Myrsten, A.; Rydberg, U.; Lamble, R. Alcohol Intoxication and Hangover: Modification of Hangover by Chlormethiazole. *Psychopharmacology* **1980**, *125*, 117–125. [CrossRef]
34. Morrow, D.; Leirer, V.; Yesavage, J. The influence of alcohol and aging on radio communication during flight. *Aviat. Space Environ. Med.* **1990**, *61*, 12–20. [PubMed]
35. Roehrs, T.; Yoon, J.; Roth, T. Nocturnal and next-day effects of ethanol and basal level of sleepiness. *Hum. Psychopharmacol. Clin. Exp.* **1991**, *6*, 307–311. [CrossRef]
36. Lemon, J.; Chesher, G.; Fox, A.; Greeley, J.; Nabke, C. Investigation of the "Hangover" Effects of an Acute Dose of Alcohol on Psychomotor Performance. *Alcohol. Clin. Exp. Res.* **1993**, *17*, 665–668. [CrossRef]
37. Roehrs, T.; Beare, D.; Zorick, F.; Roth, T. Sleepiness and Ethanol Effects on Simulated Driving. *Alcohol. Clin. Exp. Res.* **1994**, *18*, 154–158. [CrossRef]
38. Streufert, S.; Pogash, R.; Braig, D.; Gingrich, D.; Kantner, A.; Landis, R.; Lonardi, L.; Roache, J.; Severs, W. Alcohol Hangover and Managerial Effectiveness. *Alcohol. Clin. Exp. Res.* **1995**, *19*, 1141–1146. [CrossRef] [PubMed]
39. Finnigan, F.; Hammersley, R.; Cooper, T. An examination of next-day hangover effects after a 100 mg/100 mL dose of alcohol in heavy social drinkers. *Addiction* **1998**, *93*, 1829–1838. [CrossRef] [PubMed]
40. Petros, T.; Bridewell, J.; Jensen, W.; Ferraro, F.R.; Bates, J.; Moulton, P.; Turnwell, S.; Rawley, D.; Howe, T.; Gorder, D. Postintoxication Effects of Alcohol on Flight Performance after Moderate and High Blood Alcohol Levels. *Int. J. Aviat. Psychol.* **2003**, *13*, 287–300. [CrossRef]
41. Kocher, H.M.; Warwick, J.; Al-Ghnaniem, R.; Patel, A.G. Surgical dexterity after a "night out on the town". *ANZ J. Surg.* **2006**, *76*, 110–112. [CrossRef] [PubMed]
42. Gallagher, A.G.; Boyle, E.; Toner, P.; Neary, P.C.; Andersen, D.K.; Satava, R.M.; Seymour, N.E. Persistent next-day effects of excessive alcohol consumption on laparoscopic surgical performance. *Arch. Surg.* **2011**, *146*, 419–426. [CrossRef] [PubMed]
43. Stock, A.K.; Hoffmann, S.; Beste, C. Effects of binge drinking and hangover on response selection sub-processes-a study using EEG and drift diffusion modeling. *Addict. Biol.* **2017**, *22*, 1355–1365. [CrossRef] [PubMed]
44. Wolff, N.; Gussek, P.; Stock, A.-K.A.-K.; Beste, C. Effects of high-dose ethanol intoxication and hangover on cognitive flexibility. *Addict. Biol.* **2016**, *23*, 503–514. [CrossRef] [PubMed]
45. Kruisselbrink, L.D.; Martin, K.L.; Megeney, M.; Fowles, J.R.; Murphy, R.J.L. Physical and psychomotor functioning of females the morning after consuming low to moderate quantities of beer. *J. Stud. Alcohol Drugs* **2006**, *67*, 416–420. [CrossRef]
46. Yesavage, J.A.; Dolhert, N.; Taylor, J.L. Flight Simulator Performance of Younger and Older Aircraft Pilots: Effects of Age and Alcohol. *J. Am. Geriatr. Soc.* **1994**, *42*, 577–582. [CrossRef]
47. Van Dyken, I.; Szlabick, R.E.; Sticca, R.P. Effect of alcohol on surgical dexterity after a night of moderate alcohol intake. *Am. J. Surg.* **2013**, *206*, 964–969. [CrossRef] [PubMed]
48. Anderson, S.; Dawson, J. Neuropsychological correlates of alcoholic hangover. *S. Afr. J. Sci.* **1999**, *95*, 145–147.
49. Chait, L.D.; Perry, J.L. Acute and residual effects of alcohol and marijuana, alone and in combination, on mood and performance. *Psychopharmacology* **1994**, *115*, 340–349. [CrossRef]
50. Grange, J.A.; Stephens, R.; Jones, K.; Owen, L. The effect of alcohol hangover on choice response time. *J. Psychopharmacol.* **2016**, *30*, 654–661. [CrossRef]
51. Howse, A.D.; Hassall, C.D.; Williams, C.C.; Hajcak, G.; Krigolson, O.E. Alcohol hangover impacts learning and reward processing within the medial-frontal cortex. *Psychophysiology* **2018**, *55*, e13081. [CrossRef]
52. McKinney, A.; Coyle, K. Next day effects of a normal night's drinking on memory and psychomotor performance. *Alcohol Alcohol.* **2004**, *39*, 509–513. [CrossRef] [PubMed]
53. McKinney, A.; Coyle, K. Next-day effects of alcohol and an additional stressor on memory and psychomotor performance. *J. Stud. Alcohol Drugs* **2007**, *68*, 446–454. [CrossRef] [PubMed]
54. Rohsenow, D.J.; Howland, J.; Minsky, S.J.; Arnedt, J.T. Effects of heavy drinking by maritime academy cadets on hangover, perceived sleep, and next-day ship power plant operation. *J. Stud. Alcohol Drugs* **2006**, *67*, 406–415. [CrossRef]
55. Chapman, L.F. Experimental induction of hangover. *Q. J. Stud. Alcohol* **1970**, *5*, 67–86.

56. Collins, W.E.; Chiles, W.D. Laboratory Performance during Acute Alcohol Intoxication and Hangover. *Hum. Factors* **1978**, *22*, 445–462. [CrossRef]
57. Collins, W.E. Performance effects of alcohol intoxication and hangover at ground level and at simulated altitude. *Aviat. Space Environ. Med.* **1980**, *51*, 327–335.
58. Verster, J.C.; van Duin, D.; Volkerts, E.; Schreuder, A.; Verbaten, M. Alcohol hangover effects on memory functioning and vigilance performance after an evening of binge drinking. *Neuropsychopharmacology* **2003**, *28*, 740–746. [CrossRef]
59. Howland, J.; Rohsenow, D.J.; Greece, J.A.; Littlefield, C.A.; Almeida, A.; Heeren, T.; Winter, M.; Bliss, C.A.; Hunt, S.; Herman, J. The effects of binge drinking on college students' next-day academic test-taking performance and mood state. *Addiction* **2010**, *105*, 655–665. [CrossRef]
60. Rohsenow, D.J.; Howland, J.; Arnedt, J.T.; Almeida, A.B.; Greece, J.; Minsky, S.; Kempler, C.S.; Sales, S. Intoxication with bourbon versus vodka: Effects on hangover, sleep, and next-day neurocognitive performance in young adults. *Alcohol: Clin. Exp. Res.* **2010**, *34*, 509–518. [CrossRef]
61. McKinney, A.; Coyle, K.; Penning, R.; Verster, J.C. Next day effects of naturalistic alcohol consumption on tasks of attention. *Hum. Psychopharmacol.* **2012**, *27*, 587–594. [CrossRef]
62. Laurell, H.; Tornros, J. Investigation of alcoholic hangover effects on driving performance. *Natl. Swed. Road Traffic Res. Inst.* **1983**, *581*, 489–499.
63. Howland, J.; Rohsenow, D.J.; Bliss, C.A.; Almeida, A.B.; Vehige Calise, T.; Heeren, T.; Winter, M. Hangover Predicts Residual Alcohol Effects on Psychomotor Vigilance the Morning After Intoxication. *J. Addict. Res. Ther.* **2010**, *23*, 101. [CrossRef] [PubMed]
64. Harburg, E.; Davis, D.; Cummings, K.M.; Gunn, R. Negative affect, alcohol consumption and hangover symptoms among normal drinkers in a small community. *J. Stud. Alcohol Drugs* **1981**, *42*, 998–1012. [CrossRef]
65. Harburg, E.; Gunn, R.; Gleiberman, L.; DiFranceisco, W.; Schork, A. Psychosocial factors, alcohol use, and hangover signs among social drinkers: A reappraisal. *J. Clin. Epidemiol.* **1993**, *46*, 413–422. [CrossRef]
66. Penning, R.; McKinney, A.; Verster, J.C. Alcohol hangover symptoms and their contribution to overall hangover severity. *Alcohol Alcohol.* **2012**, *47*, 248–252. [CrossRef] [PubMed]
67. Van Schrojenstein Lantman, M.; Mackus, M.; van de Loo, A.J.A.E.; Verster, J.C. The impact of alcohol hangover symptoms on cognitive and physical functioning, and mood. *Hum. Psychopharmacol. Clin. Exp.* **2017**, *32*. [CrossRef] [PubMed]
68. Watson, P.E.; Watson, I.D.; Batt, R.D. Prediction of blood alcohol concentrations in human subjects. Updating the Widmark Equation. *J. Stud. Alcohol Drugs* **1981**, *42*, 547–556. [CrossRef]
69. Hogewoning, A.; Van de Loo, A.J.A.E.; Mackus, M.; Raasveld, S.J.; De Zeeuw, R.; Bosma, E.R.; Bouwmeester, N.H.; Brookhuis, K.A.; Garssen, J.; Verster, J.C. Characteristics of social drinkers with and without a hangover after heavy alcohol consumption. *Subst. Abus. Rehabil.* **2016**, *7*, 161–167. [CrossRef]
70. Rohsenow, D.J.; Howland, J.; Minsky, S.J.; Greece, J.; Almeida, A.; Roehrs, T.A. The acute hangover scale: A new measure of immediate hangover symptoms. *Addict. Behav.* **2007**, *32*, 1314–1320. [CrossRef]
71. Slutske, W.S.; Piasecki, T.M.; Hunt-Carter, E.E. Development and initial validation of the Hangover Symptoms Scale: Prevalence and correlates of Hangover Symptoms in college students. *Alcohol. Clin. Exp. Res.* **2003**, *27*, 1442–1450. [CrossRef]
72. Penning, R.; McKinney, A.; Bus, L.D.; Olivier, B.; Slot, K.; Verster, J.C. Measurement of alcohol hangover severity: Development of the Alcohol Hangover Severity Scale (AHSS). *Psychopharmacology* **2013**, *225*, 803–810. [CrossRef] [PubMed]
73. Verster, J.C.; Stephens, R.; Penning, R.; Rohsenow, D.; McGeary, J.; Levy, D.; McKinney, A.; Finnigan, F.; Piasecki, T.M.; Adan, A.; et al. The alcohol hangover research group consensus statement on best practice in alcohol hangover research. *Curr. Drug Abus. Rev.* **2010**, *3*, 116–126. [CrossRef]
74. Piasecki, T.M.; Alley, K.J.; Slutske, W.S.; Wood, P.K.; Sher, K.J.; Shiffman, S.; Heath, A.C. Low Sensitivity to Alcohol: Relations with Hangover Occurrence and Susceptibility in an Ecological Momentary Assessment Investigation. *J. Stud. Alcohol Drugs* **2012**, *73*, 925–932. [CrossRef] [PubMed]
75. Rohsenow, D.J.; Howland, J.; Winter, M.; Bliss, C.A.; Littlefield, C.A.; Heeren, T.C.; Calise, T.V. Hangover sensitivity after controlled alcohol administration as predictor of post-college drinking. *J. Abnorm. Psychol.* **2012**, *121*, 270–275. [CrossRef] [PubMed]
76. Van Schrojenstein Lantman, M.; Mackus, M.; Roth, T.; Verster, J.C. Total sleep time, alcohol consumption and the duration and severity of alcohol hangover. *Nat. Sci. Sleep* **2017**, *9*, 181–186. [CrossRef] [PubMed]

77. Van Schrojenstein Lantman, M.; Roth, T.; Roehrs, T.; Verster, J.C. Alcohol hangover, sleep quality, and daytime sleepiness. *Sleep Vigil.* **2017**, *1*, 37–41. [CrossRef]
78. Bang, J.S.; Chung, Y.H.; Chung, S.J.; Lee, H.S.; Song, E.H.; Shin, Y.K.; Lee, Y.J.; Kim, H.C.; Nam, Y.; Jeong, J.H. Clinical effect of a polysaccharide-rich extract of Acanthopanax senticosus on alcohol hangover. *Pharmazie* **2015**, *70*, 269–273. [PubMed]
79. Chauhan, B.L.; Kulkarni, R.D. Effects of Liv.52, a herbal preparation, on absorption and metabolism of ethanol in humans. *Eur. J. Clin. Pharmacol.* **1991**, *40*, 189–191.
80. Lee, H.S.; Isse, T.; Kawamoto, T.; Baik, H.W.; Park, J.Y.; Yang, M. Effect of Korean pear (Pyruspyrifolia cv. Shingo) juice on hangover severity following alcohol consumption. *Food Chem. Toxocol.* **2013**, *58*, 101–106. [CrossRef]
81. Lee, M.H.; Kwak, J.H.; Jeon, G.; Lee, J.W.; Seo, J.H.; Lee, H.S. and Lee, J.H. Red ginseng relieves the effects of alcohol consumption and hangover symptoms in healthy men: A randomized crossover study. *Food Funct.* **2014**, *5*, 528–534. [CrossRef]
82. Muhonen, T.; Jokelainen, K.; Höök-Nikanne, J.; Methuen, T.; Salaspuro, M. Tropisetron and hangover. *Addict. Biol.* **1997**, *2*, 461–462. [CrossRef] [PubMed]
83. Noh, K.H.; Jang, J.H.; Kim, J.J.; Shin, J.H.; Kim, D.K.; Song, Y.S. Effect of dandelion juice supplementation on alcohol-induced oxidative stress and hangover in healthy male college students. *J. Korean Soc. Food Sci. Nutr.* **2009**, *38*, 683–693. [CrossRef]
84. Verster, J.C.; Penning, R. Treatment and prevention of alcohol hangover. *Curr. Drug Abus. Rev.* **2010**, *3*, 103–109. [CrossRef]
85. Jayawardena, R.; Thejani, T.; Ranasinghe, P.; Fernando, D.; Verster, J.C. Interventions for treatment and/or prevention of alcohol hangover: Systematic review. *Hum. Psychopharmacol. Clin. Exp.* **2017**, *32*, e2600. [CrossRef] [PubMed]
86. Mallett, K.A.; Lee, C.M.; Neighbors, C.; Larimer, M.E.; Turrisi, R. Do we learn from our mistakes? An examination of the impact of negative alcohol-related consequences on college students' drinking patterns and perceptions. *J. Stud. Alcohol* **2006**, *67*, 269–276. [CrossRef] [PubMed]
87. Mackus, M.; van Schrojenstein Lantman, M.; van de Loo, A.J.A.E.; Nutt, D.J.; Verster, J.C. An effective hangover treatment: Friend or foe? *Drug Sci. Policy Law* **2017**. [CrossRef]
88. Courtney, K.E.; Worley, M.; Castro, N.; Tapert, S.F. The effects of alcohol hangover on future drinking behavior and the development of alcohol problems. *Addict. Behav.* **2018**, *78*, 209–215. [CrossRef]
89. Greenblatt, D.J.; Harmatz, J.S.; Roth, T. Zolpidem and gender: Are women really at risk? *J. Clin. Psychopharmacol.* **2019**, *39*, 189–199. [CrossRef]
90. Food and Drug Administration (FDA). Guideline for the study and evaluation of gender differences in the clinical evaluation of drugs. *Fed. Regist.* **1993**, *58*, 39406–39416.
91. European Medicines Agency (EMA). Gender Considerations in the Conduct of Clinical Trials. Available online: https://www.ema.europa.eu/en/documents/scientific-guideline/ich-gender-considerations-conduct-clinical-trials-step-5_en.pdf (accessed on 18 May 2019).
92. Tolstrup, J.S.; Stephens, R.; Grønbaek, M. Does the severity of hangovers decline with age? Survey of the incidence of hangover in different age groups. *Alcohol. Clin. Exp. Res.* **2014**, *38*, 466–470. [CrossRef]
93. Kanny, D.; Liu, Y.; Brewer, R.D. Binge drinking-United States, 2009. *MMWR Surveill. Summ.* **2011**, *60*, 101–104.
94. Prat, G.; Adan, A.; Sánchez-Turet, M. Alcohol hangover: A critical review of explanatory factors. *Hum. Psychopharmacol.* **2009**, *24*, 259–267. [CrossRef] [PubMed]
95. Tipple, C.T.; Benson, S.; Scholey, A. A Review of the Physiological Factors Associated with Alcohol Hangover. *Curr. Drug Abus. Rev.* **2016**, *9*, 93–98. [CrossRef] [PubMed]

© 2019 by the authors. Licensee MDPI, Basel, Switzerland. This article is an open access article distributed under the terms and conditions of the Creative Commons Attribution (CC BY) license (http://creativecommons.org/licenses/by/4.0/).

Article

The Impact of Mood and Subjective Intoxication on Hangover Severity

Joris C. Verster [1,2,3], Lizanne Arnoldy [1,3], Aurora J.A.E. van de Loo [1,2], Sarah Benson [3], Andrew Scholey [3] and Ann-Kathrin Stock [4,5,*]

1. Division of Pharmacology, Utrecht Institute for Pharmaceutical Sciences (UIPS), Utrecht University, 3584CG Utrecht, The Netherlands; J.C.Verster@uu.nl (J.C.V.); larnoldy@swin.edu.au (L.A.); a.j.a.e.vandeloo@uu.nl (A.J.A.E.v.d.L.)
2. Institute for Risk Assessment Sciences (IRAS), Utrecht University, 3584CM Utrecht, The Netherlands
3. Centre for Human Psychopharmacology, Swinburne University, Melbourne 3122, Australia; sarahmichellebenson@gmail.com (S.B.); andrew@scholeylab.com (A.S.)
4. Cognitive Neurophysiology, Department of Child and Adolescent Psychiatry, Faculty of Medicine, TU Dresden, Fetscherstr. 74, 01307 Dresden, Germany
5. Biopsychology, Department of Psychology, School of Science, TU Dresden, Zellescher Weg 19, 01069 Dresden, Germany
* Correspondence: Ann-Kathrin.Stock@uniklinikum-dresden.de

Received: 4 July 2020; Accepted: 27 July 2020; Published: 1 August 2020

Abstract: The aim of this study was to investigate whether baseline mood and/or mood while drinking have an impact on alcohol hangover severity. A survey was held among $N = 331$ young adults (mean age = 23.6 years, range = 18–35 years). Demographics, alcohol consumption, subjective intoxication, and hangover severity were assessed for the past three days. In addition, mood (baseline, while drinking, and during hangover) was also assessed. $N = 143$ participants reported to be hungover on the day of assessment, $N = 122$ participants reported to have been hungover the previous day ('yesterday'), and $N = 87$ participants reported to have been hungover two days before the assessment ('2 days ago'). The analyses revealed that baseline mood and mood while drinking had no relevant effect on the amount of consumed alcohol and did not significantly contribute to hangover severity. However, hangover severity was associated with significantly increased negative affect, particularly with higher levels of subjective stress on the day of the hangover.

Keywords: alcohol; hangover; mood; subjective intoxication; stress; neuroticism

1. Introduction

Negative baseline mood and mood while drinking have been reported to influence alcohol consumption. This is reflected by reports that situational factors (experiencing negative life events and stress) and intrinsic factors (e.g., neuroticism) can be associated with negative mood [1,2]. Subsequently, certain individuals are more likely to develop negative coping styles, including increased alcohol use [3]. Despite their conjoint effect on mood, it is important to note that trait-like, rather stable, baseline mood may be different from the current affective state while drinking (i.e., state mood) [4], so both should be considered when investigating associations between mood and drinking behavior. Additionally, sex may play a role as women are usually more likely than men to increase their drinking behavior in response to negative affect and stress and to report greater stress relief by drinking [5], but they also suffer more negative affective (long-term) consequences of excessive alcohol consumption [6,7]. Finally, it should be kept in mind that while alcohol intake may be motivated by the wish for relief of negative affect, it may also enhance trait or state mood and emotions, thus constituting a bidirectional association [8].

The functional link between stress and alcohol consumption has repeatedly been demonstrated via changes in the hypothalamic–pituitary-adrenal (HPA) axis, especially with respect to the corticotropin-releasing factor (CRF) [6,9]. Specifically, it has been found that higher levels of stress as well as reduced sensitivity of the HPA axis escalate alcohol consumption and increase the likelihood of binge drinking [10]. Lastly, a genetic study conducted in *Drosophila* has suggested that alcohol hangover/tolerance development and the response to stress may be partly mediated by the same molecular pathway [11].

Due to the phenomenological (affective) and functional (neurobiochemical/neuroendocrinological) overlap between mood, stress, and hangover, mood and subjective stress may be predictors of hangover severity. Despite the close link between mood and alcohol consumption, only very few studies have investigated the effects of baseline and/or current mood on experiencing hangovers.

Hangover is the most commonly reported negative consequence of alcohol consumption [12] and has been defined as the combination of negative mental and physical symptoms which may be experienced after a single episode of alcohol consumption, starting when blood alcohol concentration (BAC) approaches zero [13,14]. In a sample of male alcohol use disorder patients and social drinkers, Gunn [15] reported that negative attitudes towards drinking alcohol and feeling guilty about drinking were associated with experiencing more severe hangovers, but this association does not allow for any conclusions on causality. More recently, a regression analysis by Piasecki et al. [16] revealed that experiencing depressive symptoms was associated with both current and future hangover susceptibility, and Royle et al. [17] found that drinkers who had higher levels of pain catastrophizing reported experiencing more severe hangovers. However, research in the area of alcohol hangover is still severely limited. Hence, more studies need to be conducted to elucidate to what extent personality aspects and baseline mood have an impact on the susceptibility to the occurrence and severity of hangovers.

While it is commonly reported that mood and emotions are negatively affected during the hangover state [18–20], the extent to which mood while drinking impacts hangover severity has largely been neglected in hangover literature. In fact, only two studies investigated the direct impact of mood during drinking on next-day hangover in the same sample [21,22]. The first article [21], which included $N = 1266$ subjects (randomly selected from the Tecumseh Community Health Study), reported positive correlations between hangover symptom frequency and psychosocial factors, including negative life events, neuroticism, guilt about drinking, feeling depressed while drinking, and being angry while drinking. In the second paper, Harburg et al. [22] excluded all sober subjects from their dataset, i.e., subjects who reported 'never' being 'tipsy, high, or drunk'. When these were removed and the data of the remaining $N = 1104$ subjects were re-analyzed, the correlations between hangover symptom frequency and mood while drinking were less pronounced. Significant correlations were reported for neuroticism, guilt about drinking, drinking to escape, negative life events, and feelings of depression and anger while drinking. A stepwise linear regression analysis including all of the assessed variables was performed separately for men and women. Both analyses yielded a model with only modest predictive validity (19% in men and 21% in women) for the variance in the reported frequency of hangover symptoms. The analyses showed that with regard to the contribution of individual variables to the explained variance, guilt about drinking was the strongest predictor of hangover severity (9% in men and 11% in women), followed by neuroticism (4% in both men and women), being angry when high/drunk (3% in men and 2% in women), and negative life events (2% in men and 1% in women). In men, being depressed when high/drunk (1%) and the amount of consumed ethanol (<1%) further contributed to the model. In women, additional significant predictors were being younger (2%) and having experienced being drunk for the first time at a younger age (1%). Together, these findings suggest that mood during drinking has relatively little impact on experiencing hangover symptoms.

However, there are several issues that complicate the interpretation of the results presented by Harburg et al. [21,22]. For example, while they excluded subjects who reported not being drunk (but could still have had a hangover), 23% of the subjects who reported experiencing no hangover remained in the sample that underwent statistical analyses. As a consequence, the Pearson's correlations and forward stepwise multiple regression used to examine the data may have produced less pronounced associations in this zero-inflated sample [23]. Moreover, the authors assumed that they measured hangover severity, but the Hangover Sign Index (HSI) actually assesses the frequency with which eight different hangover symptoms ('headache or hangover', 'loss of appetite', 'diarrhea', 'stomach pains', 'anxiety', 'blackout or loss of memory', 'tremors or hand shaking', and 'thoughts of suicide') occur after drinking, rather than their severity. In the study by Harburg et al. [21,22], subjects indicated whether or not they experienced these symptoms on the day following the latest occasion when they were drunk. The specific combination of experienced hangover symptoms was used to estimate hangover severity. The six levels of hangover severity distinguished by Harburg et al. included (1) 'no signs' (gets drunk, but reports no hangover signs), (2) weak (any or all of these three symptoms: headache, diarrhea, and loss of appetite), (3) 'mild' (anxiety and/or stomach pains), (4) strong (any one of blackout, tremor, and thoughts of suicide), (5) 'very strong' (anxiety plus any one of blackout, tremor, and thoughts of suicide), and (6) 'severe hangover' (two or more of blackout, tremor, and suicidal thoughts). The validity of the HSI to reliably measure hangover severity can be questioned in several ways. First, the frequency of symptom occurrence does not tell us anything about their severity. Second, the HSI contains items that are not hangover signs, but signs of intoxication (e.g., blackouts). Third, it omits several core symptoms of the hangover state (e.g., fatigue or nausea), while including other signs such as 'thoughts of suicide', which are seldom regarded as hangover symptoms in the scientific literature [19,20]. There is great variability with regard to the presence and severity of hangover symptoms [19,24] and a recent study suggested that composite hangover scales are less accurate as they may over- or under-represent core symptoms and/or hangover-irrelevant symptoms [24]. In addition, the HSI score does not account for the impact of the experienced symptoms.

Taken together, it is important to replicate and improve the study by Harburg et al. by including a valid assessment of hangover severity to infer whether mood while drinking has an impact on the presence and severity of next-day alcohol hangover. Therefore, the current study aimed to verify and extend the observations by Harburg et al. in an international sample of young adults by applying a 1-item overall hangover severity scale, which is regarded to be superior to composite symptom scores [24]. Specifically, we investigated whether baseline mood, mood while drinking, and mood during hangover were associated with and/or predicting current and retrospective hangover severity.

2. Methods

In August 2018, a survey was conducted among an international sample of young adults who came to Fiji either for work or holidays. Both men and women within the age range of 18–35 years were included. The young adults were approached at Wailoaloa Beach and asked to complete a survey.

Subjects who were willing to participate and sufficiently understood the English language completed the survey on location. The location was chosen because a relatively large number of young adults congregated here to spend a holiday or relax after work.

The survey was anonymous and subjects did not receive an incentive for completing the survey. The investigator was present to clarify any issues arising from English not being the participants' mother tongue. The study was conducted by Utrecht University, informed consent was obtained from all subjects, and the Ethics Committee of the Faculty of Social and Behavioral Sciences of Utrecht University granted ethical approval (approval code FETC17-061).

The survey collected demographic information, including age, gender, height, and weight to compute body mass index (BMI), and usual weekly alcohol intake. The survey contained guidance about standard drinking sizes, and how to convert, for example, a bottle of wine into standardized alcohol units, which contain 10 g of alcohol each.

To assess the past year's immune status, the Immune Status Questionnaire (ISQ) was completed [25]. Current perceived immune fitness was assessed using a 1-item scale ranging from 0 (very poor) to 10 (excellent) [25,26]. The scale has a Cronbach's alpha of 0.80 [25] and was included as previous research found an association between having hangovers and immune status [27,28]. Additionally, a short scale was used to assess baseline mood. The six items reflected the subscales of the short version of the Profiles of Mood States (POMS-SF) [29], and included tension/stress, anxiety, depression, being active, fatigue, and anger/hostility. The items were scored on a scale ranging from 0 (absent) to 10 (extreme). The 11-point scale has successfully been used in previous research [30,31], which showed that single item visual analog scales are just as sensitive and reliable as full-scale construct assessments of mood states like depression [32], fatigue [33], or quality of life [34].

Neuroticism was assessed with the neuroticism scale of the Eysenck Personality Questionnaire-Revised Short Scale (EPQ-RSS) [35,36]. The neuroticism scale consists of 12 items that can be answered with 'yes' or 'no', which correspond to the values of 1 and 0, respectively. The sum score of items ranges from 0 to 12, with higher scores implying more neuroticism. Cronbach's alpha of the neuroticism scale is 0.82 [36].

In addition to demographics and baseline mood, various other assessments regarding alcohol consumption and mood were made for the past three days (referred to as 'today', 'yesterday', and '2 days ago'). For each of these days, subjects reported their alcohol consumption. Both the number of alcoholic drinks and the time frame of consumption were assessed. The estimated BAC was computed with a modified Widmark equation [37]. Subjective intoxication was rated on a scale ranging from 0 (sober) to 10 (very drunk) [38]. To assess the current mood while drinking, participants rated their mood state while drinking, including being 'angry/hostile/irritable' and being 'depressed/sad' on scales ranging from 0 (absent) to 10 (extreme). Total sleep time was assessed and subjects rated their sleep quality on a scale ranging from 0 to 10 [39,40]. Regarding next-day effects, hangover severity was scored with a 1-item severity score, ranging from 0 (absent) to 10 (extreme) [24]. Using the same 0–10 scale, 'fatigue, sleepiness', 'stress', and 'guilt about drinking' were also assessed as measures of current (hangover) mood.

Statistical analyses were conducted with SPSS (IBM SPSS Statistics for Windows, version 25.0, released in 2013; IBM Corp., Armonk, NY, USA). Mean and standard deviation (SD) were computed for each variable. Outlier data (alcohol intake on evening > +3SD of group average) were omitted from the analyses.

For each test day, participants were independently allocated to the 'no hangover' or 'hangover' group. This was based on the reported absence (score 0) or presence (score 1–10) of a hangover for that particular day. Thus, group sizes differed between the three days, and individual subjects could be allocated to the hangover group on one day, but to the no hangover group on another day, depending on the reported presence and absence of hangover for that particular day.

All statistical analyses were conducted separately for each of the three days. Most study outcome variables did not follow a normal distribution. Therefore, nonparametric statistics were used to analyze the data. To compare demographics and baseline mood between the hangover and the no hangover group, independent-samples Mann–Whitney U tests were used.

Spearman's rho correlations were computed between drinking variables and mood outcomes. Results were considered significant if $p < 0.05$. Linear stepwise regression analyses (for which independent variables do not need to be normally distributed or continuous) were conducted to determine which variables (i.e., demographics, mood, and drinking variables) were significant predictors of (a) having a hangover (yes/no) and of (b) hangover severity (1–10 score on the single-item hangover severity assessment). Further, linear stepwise regression analyses were conducted to determine which of the assessed variables were significant predictors of (c) the amount of consumed alcohol and of (d) subjective intoxication on the evenings preceding the next-day alcohol hangover. Analyses were conducted for the whole sample, and for men and women separately.

3. Results

The survey was completed by $N = 331$ subjects. Their demographics and baseline mood are summarized in Table 1. A total of $N = 143$ subjects (43.2% of the sample) reported having a hangover on the day of the assessment (referred to as 'today' throughout the article). Table 1 contrasts their demographics and past day drinking behaviors with the N = 188 subjects who did not report having a hangover. The comparisons revealed that subjects with a hangover scored significantly higher on some of the baseline mood scales and perceived immune fitness compared to subjects who reported no hangover. However, it should be noted that the magnitudes of the observed differences were small (<1 on 11-point scales).

Table 1. Demographics and baseline mood.

	Total Sample	Hangover	No Hangover
Demographics			
N (%)	331 (100.0%)	143 (43.2%)	188 (56.8%)
Age (years)	23.6 (4.2)	23.5 (4.3)	23.6 (4.1)
Sex (m/f)	143/188	81/62	63/125
BMI (kg/m^2)	23.5 (3.9)	23.8 (4.4)	23.2 (3.5)
Usual weekly alcohol consumption (units)	11.5 (11.0)	13.3 (12.1) *	10.2 (9.8)
Past year's immune fitness (ISQ)	7.0 (2.3)	6.9 (2.2)	7.1 (2.3)
Perceived immune fitness	8.0 (1.6)	7.8 (1.6) *	8.1 (1.5)
Baseline Mood Ratings			
Tension, stress	1.8 (1.8)	1.9 (1.8)	1.7 (1.7)
Anxiety	1.5 (1.9)	1.5 (1.9)	1.5 (1.9)
Depression	0.7 (1.5)	0.8 (1.7) *	0.6 (1.3)
Being active	5.4 (2.8)	5.8 (2.6) *	5.1 (2.9)
Fatigue	3.5 (2.6)	4.1 (2.6) *	3.1 (2.6)
Anger, hostility	0.9 (1.6)	1.1 (1.8) *	0.7 (1.4)
Neuroticism	2.1 (2.2)	2.2 (2.2)	2.0 (2.2)

Mean and standard deviation (between brackets) are shown. Significant differences ($p < 0.05$) between the hangover and the no hangover group are indicated by *. No significant partial correlations with hangover severity were found ($p < 0.05$), controlling for estimated BAC. Abbreviations: BMI = body mass index, ISQ = Immune Status Questionnaire, BAC = blood alcohol concentration.

Variables related to alcohol consumption and mood (rated separately for mood during drinking and mood while hungover) are summarized in Table 2. For subjects with a hangover, the partial correlation with hangover severity (controlled for estimated BAC) is also indicated.

Table 2. Study outcomes and their association with hangover severity.

	Today		Yesterday		Two Days Ago	
	Hangover	No Hangover	Hangover	No Hangover	Hangover	No Hangover
n (%)	143	188	122	208	87	243
Drinking variables						
Alcohol consumed (units)	12.3 (7.3) *	4.0 (5.3)	11.4 (7.3) *†	3.2 (4.6)	9.9 (7.4) *†	1.9 (3.6)
Time spent drinking (h)	6.9 (3.5) *	3.0 (3.8)	6.6 (4.1) *†	4.3 (4.5)	6.5 (4.6) *†	3.9 (4.1)
Estimated BAC (%)	0.16 (0.1) *†	0.08 (0.1)	0.17 (0.2) *†	0.06 (0.1)	0.14 (0.1) *†	0.05 (0.1)
Subjective intoxication	6.2 (2.5) *†	1.7 (2.4)	6.1 (2.7) *†	1.7 (2.3)	5.6 (3.1) *†	1.0 (2.1)
Cigarettes smoked	3.7 (6.4) *	1.4 (3.8)	2.8 (5.4) *	1.2 (3.1)	2.8 (5.3) *	1.1 (3.0)
Mood during drinking						
Angry, hostile, irritable	0.7 (1.6) *	0.4 (1.2)	0.9 (2.0) *†	0.4 (1.3)	0.6 (1.2) *	0.3 (1.1)
Depressed, sad	0.6 (1.5) *	0.3 (1.2)	0.9 (2.0) *	0.4 (1.3)	0.9 (1.7) *	0.5 (1.7)
Sleep						
Total sleep time (h)	6.2 (2.0) †	7.3 (2.1)	6.5 (1.8) *†	7.2 (1.9)	6.6 (1.9) *†	6.9 (2.5)
Sleep quality	6.0 (2.5) *†	6.7 (2.3)	6.7 (2.3) †	6.7 (2.3)	6.4 (2.5) †	6.2 (2.7)
Next-day mood						
Hangover severity	3.5 (2.5) *	0.0 (0.0)	3.7 (2.7) *	0.0 (0.0)	3.2 (2.3) *	0.0 (0.0)
Fatigue, sleepiness	4.7 (2.9) *†	1.7 (2.4)	5.1 (2.6) *†	1.5 (2.3)	4.5 (3.0) *†	1.7 (2.6)
Stress	1.3 (2.1) *†	0.4 (1.3)	1.7 (2.5) *†	0.4 (1.2)	1.8 (2.6) *†	0.3 (1.0)
Guilt about drinking	1.4 (2.3) *†	0.2 (0.9)	1.4 (2.3) *†	0.1 (0.6)	1.5 (2.4) *†	0.1 (0.7)

Significant differences ($p < 0.05$) between the hangover and the no hangover group are indicated by *. Significant partial correlations ($p < 0.05$), controlling for estimated BAC, with hangover severity are indicated by †. Abbreviation: BAC = blood alcohol concentration.

As can be seen in Table 2, the hangover group significantly differed from the no hangover group in almost all of the drinking-, sleep-, and current mood-associated variables. Significant partial correlations (controlling for estimated BAC) were found between hangover severity and subjective intoxication (being drunk) ($r = 0.453$, $p < 0.0001$), between hangover severity and total sleep time ($r = -0.226$, $p = 0.009$), and between hangover severity and sleep quality ($r = -0.183$, $p = 0.036$). There were no significant correlations between hangover severity and baseline mood or neuroticism. Ratings of mood while drinking did not significantly correlate with hangover severity. However, significant correlations were found between hangover severity and 'fatigue, sleepiness' experienced during hangover ($r = 0.514$, $p < 0.0001$), between hangover severity and 'stress' experienced during hangover ($r = 0.423$, $p < 0.0001$), and between hangover severity and 'guilt about drinking' experienced during hangover ($r = 0.361$, $p < 0.0001$). A similar pattern of outcomes was seen for the other two days that were assessed (Table 2).

Tables 3 and 4 present the results of stepwise linear regression analyses including all the variables summarized in Tables 1 and 2. Tables 3 and 4 list those predictors that significantly contributed to each regression model (once while excluding and once while including "next-day" variables). The percentage that each particular variable contributed to the model (R^2) and the beta coefficient (β) are also included.

Table 3. Significant predictors of hangover severity (excluding next-day variables).

Full Sample	Today ($n = 313$) Model: $R^2 = 45.2\%$	Yesterday ($n = 243$) Model: $R^2 = 42.8\%$	Two days ago ($n = 175$) Model: $R^2 = 43.1\%$
Contributing variables	Subjective intoxication ($R^2 = 43.0\%$) ($\beta = 0.644$, $p < 0.0001$) Baseline fatigue ($R^2 = 1.5\%$) ($\beta = 0.117$, $p = 0.006$) Sleep quality ($R^2 = 0.7\%$) ($\beta = -0.096$, $p = 0.024$)	Subjective intoxication ($R^2 = 38.9\%$) ($\beta = 0.524$, $p < 0.0001$) ISQ ($R^2 = 2.1\%$) ($\beta = -0.165$, $p = 0.001$) Estimated BAC ($R^2 = 1.8\%$) ($\beta = 0.167$, $p = 0.004$)	Subjective intoxication ($R^2 = 37.0\%$) ($\beta = 0.561$, $p < 0.0001$) Estimated BAC ($R^2 = 3.9\%$) ($\beta = 0.420$, $p < 0.0001$) Alcohol intake evening ($R^2 = 1.2\%$) ($\beta = -0.288$, $p = 0.022$) Sleep quality ($R^2 = 1.0\%$) ($\beta = -0.115$, $p = 0.049$)
Men only	Today ($n = 120$) Model: $R^2 = 36.2\%$	Yesterday ($n = 110$) Model: $R^2 = 60.5\%$	Two days ago ($n = 91$) Model: $R^2 = 44.3\%$
Contributing variables	Subjective intoxication ($R^2 = 33.5\%$) ($\beta = 0.570$, $p < 0.0001$) Baseline anger, hostility ($R^2 = 2.7\%$) ($\beta = 0.181$, $p = 0.015$)	Subjective intoxication ($R^2 = 38.9\%$) ($\beta = 0.485$, $p < 0.0001$) Stress while hangover ($R^2 = 12.3\%$) ($\beta = 0.293$, $p < 0.0001$) Fatigue while hangover ($R^2 = 3.6\%$) ($\beta = 0.179$, $p = 0.019$) Estimated BAC ($R^2 = 3.1\%$) ($\beta = 0.495$, $p = 0.002$) Weekly alcohol intake ($R^2 = 1.2\%$) ($\beta = 0.159$, $p = 0.019$) Alcohol intake evening ($R^2 = 1.4\%$) ($\beta = -0.404$, $p = 0.030$)	Subjective intoxication ($R^2 = 41.0\%$) ($\beta = 0.818$, $p < 0.0001$) Drinking time ($R^2 = 3.3\%$) ($\beta = -0.263$, $p = 0.013$)

Table 3. *Cont.*

Women only	Today (n = 133) Model: R^2 = 41.7%	Yesterday (n = 132) Model: R^2 = 46.2%	Two days ago (n = 83) Model: R^2 = 44.0%
Contributing variables	Subjective intoxication (R^2 = 38.9%) (β = 0.580, p < 0.0001) Total sleep time (R^2 = 2.8%) (β = −0.183, p = 0.009)	Subjective intoxication (R^2 = 38.9%) (β = 0.560, p < 0.0001) Weekly alcohol intake (R^2 = 3.2%) (β = −0.221, p = 0.001) ISQ (R^2 = 2.4%) (β = −0.172, p = 0.010) Estimated BAC (R^2 = 1.7%) (β = 0.207, p = 0.008)	Subjective intoxication (R^2 = 32.5%) (β = 0.392, p = 0.001) Current immune fitness (R^2 = 9.1%) (β = −0.279, p = 0.001) Estimated BAC (R^2 = 2.1%) (β = 0.235, p = 0.036)

Linear stepwise regression analyses were conducted on the data of participants who reported having a hangover. The included variables were demographics, baseline mood, neuroticism, alcohol consumption variables, and sleep outcomes. The percentage of variance explained (adjusted R^2), the unadjusted beta coefficient (β), and standard error (SE) are provided. Abbreviation: BAC = estimated blood alcohol concentration, ISQ = Immune Status Questionnaire.

The analysis of the current day revealed that three variables accounted for 45.2% of the variance in overall hangover severity. With regard to the variance explained by individual variables, subjective intoxication was the strongest predictor of hangover severity (43.0%), followed by baseline fatigue (1.5%) and sleep quality (0.7%). The addition of the 'next day' variables mood and guilt experienced while hungover yielded a model where four variables accounted for 56.1% of the variance in overall hangover severity. Subjective intoxication was again the strongest predictor of hangover severity (43.0%), followed by fatigue while hungover (8.7%), guilt about drinking while hungover (4.0%), and stress while hungover (0.4%). Regression analyses for the other two days of assessments yielded comparable results, as subjective intoxication was always the best predictor (Tables 3 and 4). Tables 3 and 4 also show the outcomes of separate regression analyses for men only and women only. These analyses again yielded comparable results as subjective intoxication was the most important predictor of hangover severity. In contrast to subjective intoxication, the amount of consumed alcohol and estimated BAC only had a marginal impact on hangover severity across all models. The demographic and mood variables that affected alcohol consumption are summarized in Table 5. For the 'today' data, a stepwise regression analysis revealed that three variables accounted for 25.0% of the variance in the amount of alcohol consumed (Table 5). The analysis showed that the number of smoked cigarettes was the strongest predictor and explained the most variance in the amount of consumed alcohol (16.4%). This was followed by sex (6.2%) and weekly alcohol consumption (2.4%). The data for the other two days yielded similar results, as the number of smoked cigarettes was always a relevant predictor of the amount of consumed alcohol.

The demographic and mood variables that affected alcohol consumption are summarized in Table 6. Three variables accounted for 41.0% of the variance in subjective intoxication (drunkenness) (Table 6). The analysis showed that the amount of consumed alcohol was the strongest predictor of subjective intoxication (37.9%), followed by feeling angry while drinking (2.1%) and age (1.0%). The data for the other two days (Table 6) yielded similar results: alcohol intake on the respective evening was always the most important predictor of drunkenness. Taken together, baseline mood and feeling more 'angry/hostile' or 'depressed' while drinking had only marginal effects on the amount of consumed alcohol (<5%). The overall variance of alcohol consumption explained across the models was low, with the number of cigarettes smoked being the most important predictor. This suggests that, instead of mood, other (not assessed) variables are more important predictors of the amount of alcohol consumption. The models for subjective intoxication were more robust, with the amount of consumed alcohol being the best predictor of subjective intoxication. Baseline mood and feeling angry/hostile or depressed while drinking had only small effects on subjective intoxication.

Table 4. Significant predictors of hangover severity (including next-day variables).

Full sample	Today (n = 313) Model: R^2 = 56.1%	Yesterday (n = 243) Model: R^2 = 58.4%	Two days ago (n = 175) Model: R^2 = 56.7%
Contributing variables	Subjective intoxication (R^2 = 43.0%) (β = 0.472, p < 0.0001) Fatigue while hungover (R^2 = 8.7%) (β = 0.237, p < 0.0001) Guilt about drinking (R^2 = 4%) (β = 0.186, p < 0.0001) Stress while hungover (R^2 = 0.4%) (β = 0.093, p = 0.048)	Subjective intoxication (R^2 = 38.9%) (β = 0.331, p < 0.0001) Stress while hungover (R^2 = 13.0%) (β = 0.220, p < 0.0001) Fatigue while hungover (R^2 = 3.9%) (β = 0.227, p < 0.0001) Guilt about drinking (R^2 = 2.0%) (β = 0.171, p = 0.001) Estimated BAC (R^2 = 0.6%) (β = 0.107, p = 0.028)	Subjective intoxication (R^2 = 37.0%) (β = 0.319, p < 0.0001) Stress while hungover (R^2 = 11.8%) (β = 0.351, p = 0.0001) Estimated BAC (R^2 = 4.5%) (β = 0.227, p < 0.0001) Fatigue while hungover (R^2 = 2.2%) (β = 0.188, p = 0.001) Angry while drinking (R^2 = 1.6%) (β = −0.142, p = 0.006)
Men only	Today (n = 120) Model: R^2 = 50.3%	Yesterday (n = 110) Model: R^2 = 50.8%	Two days ago (n = 91) Model: R^2 = 66.5%
Contributing variables	Subjective intoxication (R^2 = 33.5%) (β = 0.423, p < 0.0001) Fatigue while hungover (R^2 = 12.5%) (β = 0.306, p < 0.0001) Guilt about drinking (R^2 = 4.3%) (β = 0.234, p = 0.001)	Subjective intoxication (R^2 = 38.9%) (β = 0.578, p < 0.0001) Estimated BAC (R^2 = 4.8%) (β = 0.657, p < 0.0001) Baseline anger, hostility (R^2 = 3.2%) (β = 0.226, p = 0.001) Alcohol intake evening (R^2 = 2.1%) (β = −0.551, p = 0.008) Weekly alcohol intake (R^2 = 1.8%) (β = 0.155, p = 0.033)	Subjective intoxication (R^2 = 41.0%) (β = 0.618, p < 0.0001) Stress while hungover (R^2 = 17.1%) (β = 0.426, p < 0.0001) Estimated BAC (R^2 = 4.1%) (β = 0.522, p < 0.0001) Alcohol intake evening (R^2 = 2.1%) (β = −0.437, p = 0.009) Angry while drinking (R^2 = 1.7%) (β = −0.143, p = 0.024)
Women only	Today (n = 133) Model: R^2 = 54.4%	Yesterday (n = 132) Model: R^2 = 61.4%	Two days ago (n = 83) Model: R^2 = 53.1%
Contributing variables	Subjective intoxication (R^2 = 38.9%) (β = 0.419, p < 0.0001) Fatigue while hungover (R^2 = 9.8%) (β = 0.322, p < 0.0001) Guilt about drinking (R^2 = 4.1%) (β = 0.239, p < 0.0001) Baseline anxiety (R^2 = 1.6%) (β = −0.139, p = 0.021)	Fatigue while drinking (R^2 = 39.6%) (β = 0.226, p = 0.002) Guilt about drinking (R^2 = 12.4%) (β = 0.250, p < 0.0001) Subjective intoxication (R^2 = 5.5%) (β = 0.343, p < 0.0001) Stress while drinking (R^2 = 2.3%) (β = 0.196, p = 0.005) Weekly alcohol intake (R^2 = 1.8%) (β = −0.147, p = 0.010)	Subjective intoxication (R^2 = 32.5%) (β = 0.358, p < 0.0001) Fatigue while hungover (R^2 = 9.1%) (β = 0.325, p < 0.0001) Stress while hungover (R^2 = 5.9%) (β = 0.224, p = 0.012) Baseline fatigue (R^2 = 3.3%) (β = −0.220, p = 0.006) Current immune fitness (R^2 = 2.3%) (β = −0.189, p = 0.031)

Linear stepwise regression analyses were conducted on the data of participants who reported having a hangover. The included variables were demographics, baseline mood, neuroticism, alcohol consumption variables, sleep outcomes, and next-day variables on mood while hungover, as well as guilt about drinking. The percentage of variance explained (adjusted R^2), the standardized beta coefficient (β), and p-value are given.

Table 5. Significant predictors of alcohol consumption.

Full sample	Today (n = 318) Model: R^2 = 25.0%	Yesterday (n = 315) Model: R^2 = 20.7%	Two days ago (n = 317) Model: R^2 = 26.1%
Contributing variables	Cigarettes smoked (R^2 = 16.4%) (β = 0.320, p < 0.0001) Sex (R^2 = 6.2%) (β = −0.237, p < 0.0001) Weekly alcohol intake (R^2 = 2.4%) (β = 0.167, p = 0.001)	Cigarettes smoked (R^2 = 14.0%) (β = 0.293, p < 0.0001) Weekly alcohol intake (R^2 = 6.7%) (β = 0.275, p < 0.0001)	Cigarettes smoked (R^2 = 16.6%) (β = 0.353, p < 0.0001) Weekly alcohol intake (R^2 = 3.6%) (β = 0.165, p = 0.001) Angry while drinking (R^2 = 3.1%) (β = 0.194, p < 0.0001) Baseline anger, hostility (R^2 = 1.6%) (β = −0.148, p = 0.003) Sex (R^2 = 1.4%) (β = −0.132, p = 0.009)
Men only	Today (n = 140) Model: R^2 = 12.1%	Yesterday (n = 139) Model: R^2 = 23.3%	Two days ago (n = 140) Model: R^2 = 22.2%
Contributing variables	Cigarettes smoked (R^2 = 8.8%) (β = 0.361, p < 0.0001) Baseline anxiety (R^2 = 3.3%) (β = −0.204, p = 0.014)	Cigarettes smoked (R^2 = 17.1%) (β = 0.318, p < 0.0001) Weekly alcohol intake (R^2 = 6.2%) (β = 0.279, p = 0.001)	Cigarettes smoked (R^2 = 12.9%) (β = 0.384, p < 0.0001) Angry while drinking (R^2 = 6.9%) (β = 0.304, p < 0.0001) Baseline anger, hostility (R^2 = 2.4%) (β = −0.175, p = 0.025)
Women only	Today (n = 177) Model: R^2 = 24.6%	Yesterday (n = 175) Model: R^2 = 12.3%	Two days ago (n = 176) Model: R^2 = 22.4%
Contributing variables	Cigarettes smoked (R^2 = 22.1%) (β = 0.442, p < 0.0001) Weekly alcohol intake (R^2 = 2.5%) (β = 0.174, p = 0.010)	Weekly alcohol intake (R^2 = 8.0%) (β = 0.261, p < 0.0001) Cigarettes smoked (R^2 = 4.3%) (β = 0.219, p = 0.003)	Cigarettes smoked (R^2 = 16.6%) (β = 0.369, p < 0.0001) Weekly alcohol intake (R^2 = 5.8%) (β = 0.253, p < 0.0001)

Linear stepwise regression analyses were conducted on the data of participants who reported having a hangover. The included variables were demographics, baseline mood, neuroticism, and mood while drinking. The percentage of variance explained (adjusted R^2), the standardized beta coefficient (β), and p-value are given.

A summary of all findings is presented in Figure 1.

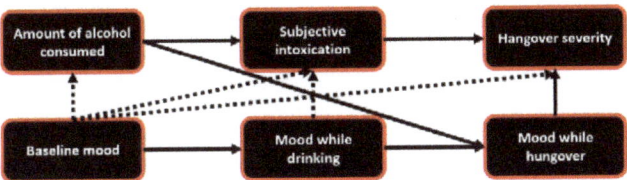

Figure 1. Associations between drinking variables, mood, and hangover severity. Lines represent significantly contributing variables to the regression analyses. Dashed lines connect variables that contributed less than 5% to the associations. The actual percentages are listed in Tables 3–6.

Table 6. Significant predictors of subjective intoxication.

Full sample	Today (n = 254) Model: R^2 = 41.0%	Yesterday (n = 244) Model: R^2 = 45.5%	Two days ago (n = 175) Model: R^2 = 54.5%
Contributing variables	Alcohol intake evening (R^2 = 37.9%) (β = 0.617, p < 0.0001) Angry while drinking (R^2 = 2.1%) (β = 0.144, p = 0.003) Age (R^2 = 1.0%) (β = −0.112, p = 0.021)	Alcohol intake evening (R^2 = 39.8%) (β = 0.623, p < 0.0001) Angry while drinking (R^2 = 3.8%) (β = 0.196, p < 0.0001) Age (R^2 = 1.9%) (β = −0.145, p = 0.002)	Alcohol intake evening (R^2 = 52.2%) (β = 0.735, p < 0.0001) Baseline fatigue (R^2 = 2.3%) (β = 0.160, p = 0.002)
Men only	Today (n = 120) Model: R^2 = 49.3%	Yesterday (n = 110) Model: R^2 = 58.4%	Two days ago (n = 91) Model: R^2 = 67.8%
Contributing variables	Alcohol intake evening (R^2 = 44.3%) (β = 1.060, p < 0.0001) Estimated BAC (R^2 = 3.1%) (β = −0.438, p = 0.006) Baseline being active (R^2 = 1.9%) (β = 0.151, p = 0.022)	Alcohol intake evening (R^2 = 48.4%) (β = 1.033, p < 0.0001) Angry while drinking (R^2 = 8.0%) (β = 0.252, p < 0.0001) Estimated BAC (R^2 = 1.8%) (β = −0.372, p = 0.018)	Alcohol intake evening (R^2 = 55.1%) (β = 1.113, p < 0.0001) Estimated BAC (R^2 = 5.7%) (β = −0.467, p < 0.0001) Baseline being active (R^2 = 3.6%) (β = 0.148, p = 0.019) Baseline neuroticism (R^2 = 1.9%) (β = 0.177, p = 0.006) Age (R^2 = 1.5%) (β = −0.133, p = 0.031)
Women only	Today (n = 133) Model: R^2 = 33.0%	Yesterday (n = 133) Model: R^2 = 44.3%	Two days ago (n = 83) Model: R^2 = 49.5%
Contributing variables	Alcohol intake evening (R^2 = 27.8%) (β = 0.522, p < 0.0001) Angry while drinking (R^2 = 3.5%) (β = 0.197, p = 0.006) Age (R^2 = 1.7%) (β = −0.147, p = 0.040)	Alcohol intake evening (R^2 = 34.7%) (β = 0.908, p < 0.0001) Age (R^2 = 4.7%) (β = −0.222, p = 0.001) Estimated BAC (R^2 = 2.7%) (β = −0.386, p = 0.006) Depressed while drinking (R^2 = 2.2%) (β = 0.160, p = 0.016)	Alcohol intake evening (R^2 = 49.5%) (β = 0.708, p < 0.0001)

Linear stepwise regression analyses were conducted on the data of participants who reported having a hangover. The included variables were demographics, baseline mood, neuroticism, mood while drinking, and amount of alcohol consumed (alcohol intake during the evening). The percentage variance explained (adjusted R^2), the standardized beta coefficient (β), and p-value are given.

4. Discussion

The current study aimed to verify and extend the observations by Harburg et al. in an international sample of young adults by applying a 1-item overall hangover severity scale to investigate whether baseline mood, mood while drinking, as well as mood during hangover, were associated with and/or predicted current and retrospective hangover severity. In contrast to previous reports [15,21,22], the findings of our study suggest that even though mood while drinking seemed to differ between hungover and non-hungover populations, this factor has a rather negligible impact on hangover severity. Instead, variables related to alcohol intake (in particular, subjective intoxication and estimated BAC) and sleep (in particular, sleep quality) were much more strongly related to hangover severity. Feeling stressed and fatigued during hangover were also significantly associated with hangover severity, confirming that mood changes accompany alcohol hangover. Finally, guilt was experienced most frequently by drinkers in the hangover group. Guilt about drinking significantly correlated with both the amount of alcohol consumed and with hangover severity.

The confirmatory regression analyses further supported our conclusions. The obtained models revealed that subjective intoxication (drunkenness) was the most important contributor to hangover severity. In comparison to that, mood while drinking had no relevant impact on hangover severity. Mood while hungover and guilt about drinking while hungover significantly contributed to the model predicting hangover severity, but it should also be noted that stress, fatigue, and guilt during hangover are most likely the consequences, rather than the cause of hangover severity.

Regression models predicting subjective intoxication revealed that the most important contributing factor was the amount of consumed alcohol. For both subjective intoxication and the amount of alcohol consumed, the regression models revealed that baseline mood and mood during drinking only had a small contribution to the models, if any (usually < 5%). At first sight, this might be regarded as comparable to the findings reported by Harburg et al. [22], who reported that being angry when high/drunk accounted for 3% (2%) of the observed hangover severity variance in men (women) and that being depressed when high/drunk accounted for another 1% of the variance in hangover severity observed for men when using composite HSI scores. Yet, our findings need to be interpreted within the context that out of all the assessed factors, the mood variables tended to explain the least variance, thus being the least suitable predictors of hangover severity.

The observation that subjective intoxication was the most important predictor of hangover severity is in line with results of previous studies [41–43]. One of these studies suggested that 'consuming more alcohol than usual' was an even better predictor of hangover severity than subjective intoxication [43]. Unfortunately, this variable was not included in the current study. We therefore recommend assessing how much alcohol is typically consumed at an average drinking occasion. This might be done either with individualized questions or with the help of (semi)structured clinical interview tools. Furthermore, it might also be beneficial to include measures of overall alcohol sensitivity, such as the Self-Rating of the Effects of Alcohol (SRE) form [44] or the Alcohol Sensitivity Questionnaire (ASQ) [45]. Another potential limitation of the current study was the relatively young sample, which makes it unclear to what extent the results can be generalized to other age groups. We therefore recommend assessing samples that cover wide age ranges, whenever possible. The data also relied on retrospective self-reports, which might have suffered from recall bias in some participants and which might potentially have led to the smaller number of hangovers reported with increasing recall period. Therefore, retrospective hangover assessments should ideally not be averaged over days that differ in recall period. Retrospective assessments also mean that BAC was not assessed while drinking, but instead calculated using the Widmark formula [37]. Given the possibility of recall bias and individual differences, the BAC was therefore reported as an estimate throughout this article. If possible, the BAC should ideally be determined on the night of drinking, but it should also be kept in mind that the measurement itself and the presence of investigators might induce bias, thus making drinking behavior less naturalistic [46].

Research on the relationship between smoking and the presence and severity of hangover is limited, and this is an important topic for future research. Tables 3 and 4 show that both cigarette smoking and drug use were not significant predictors of hangover severity or subjective intoxication. However, this observation is in contrast to previous research that found smoking to significantly increase the odds of hangover incidence and hangover severity [47]. Our analysis did reveal that the number of cigarettes smoked was the strongest predictor of the amount of alcohol consumed. This observation is in line with other research showing that drinking and smoking often go hand in hand [48].

Finally, we examined possible sex differences in variables contributing to hangover severity. Previous research showed that the presence and severity of hangover symptoms did not relevantly differ between men and women at comparable BAC levels [49,50]. In the current study, conducting the statistical analysis separately for men and women revealed that men consumed significantly more alcohol than women, but we found no important sex differences in which variables significantly contributed to hangover severity. Across all analyses, subjective intoxication was the most important predictor of hangover severity. While mood during drinking had no relevant impact, mood during hangover was clearly associated with hangover severity.

Author Contributions: Conceptualization and design of the study, L.A., A.-K.S., J.C.V., A.J.A.E.v.d.L., S.B., and A.S.; data acquisition, L.A.; data analysis and interpretation, J.C.V.; writing—original draft, J.C.V. All authors have read and agreed to the published version of the manuscript.

Funding: This research received no external funding.

Conflicts of Interest: S.B. has received funding from Red Bull GmbH, Kemin Foods, Sanofi Aventis, Phoenix Pharmaceutical, BioRevive, and GlaxoSmithKline. Over the past 36 months, A.S. has held research grants from Abbott Nutrition, Arla Foods, Bayer, BioRevive, DuPont, Fonterra, Kemin Foods, Nestlé, Nutricia-Danone, and Verdure Sciences. He has acted as a consultant/expert advisor to Bayer, Danone, Naturex, Nestlé, Pfizer, Sanofi, and Sen-Jam Pharmaceutical, and has received travel/hospitality/speaker fees from Bayer, Sanofi, and Verdure Sciences. Over the past 36 months, J.C.V. has held grants from Janssen, Nutricia, and Sequential Medicine, and acted as a consultant/expert advisor to Clinilabs, More Labs, Sen-Jam Pharmaceutical, Toast!, and ZBiotics. A.-K.S. has received funding from the Daimler and Benz Foundation. A.J.A.E.v.d.L. and L.A. have no conflict of interest to declare.

References

1. Clark, L.A.; Watson, D.; Mineka, S. Temperament, personality, and the mood and anxiety disorders. *J. Abnorm. Psychol.* **1994**, *103*, 103–116. [CrossRef] [PubMed]
2. Pemberton, R.; Fuller Tyszkiewicz, M.D. Factors contributing to depressive mood states in everyday life: A systematic review. *J. Affect. Disord.* **2016**, *200*, 103–110. [CrossRef] [PubMed]
3. Hogarth, L.; Hardy, L.; Mathew, A.R.; Hitsman, B. Negative mood-induced alcohol-seeking is greater in young adults who report depression symptoms, drinking to cope, and subjective reactivity. *Exp. Clin. Psychopharmacol.* **2018**, *26*, 138–146. [CrossRef] [PubMed]
4. Lac, A.; Donaldson, C.D. Validation and psychometric properties of the alcohol positive and negative affect schedule: Are drinking emotions distinct from general emotions? *Psychol. Addict. Behav.* **2018**, *32*, 40–51. [CrossRef]
5. Abulseoud, O.A.; Karpyak, V.M.; Schneekloth, T. A retrospective study of gender differences in depressive symptoms and risk of relapse in patients with alcohol dependence. *Am. J. Addict.* **2013**, *22*, 437–442. [CrossRef]
6. Logrip, M.L.; Milivojevic, V.; Bertholomey, M.L.; Torregrossa, M.M. Sexual dimorphism in the neural impact of stress and alcohol. *Alcohol* **2018**, *72*, 49–59. [CrossRef]
7. Peltier, M.R.; Verplaetse, T.L.; Mineur, Y.S.; Petrakis, I.L.; Cosgrove, K.P.; Picciotto, M.R.; McKee, S.A. Sex differences in stress-related alcohol use. *Neurobiol. Stress* **2019**, *10*, 100149. [CrossRef]
8. Sayette, M.A. The effects of alcohol on emotion in social drinkers. *Behav. Res. Ther.* **2017**, *88*, 76–89. [CrossRef]
9. Schreiber, A.L.; Gilpin, N.W. Corticotropin-Releasing Factor (CRF) Neurocircuitry and Neuropharmacology in Alcohol Drinking. *Handb. Exp. Pharmacol.* **2018**, *248*, 435–471.
10. Blaine, S.K.; Nautiyal, N.; Hart, R.; Guarnaccia, J.B.; Sinha, R. Craving, cortisol and behavioral alcohol motivation responses to stress and alcohol cue contexts and discrete cues in binge and non-binge drinkers. *Addict. Biol.* **2019**, *24*, 1096–1108. [CrossRef]
11. Scholz, H.; Franz, M.; Heberlein, U. The hangover gene defines a stress pathway required for ethanol tolerance development. *Nature* **2005**, *436*, 845–847. [CrossRef] [PubMed]
12. Verster, J.C.; van Herwijnen, J.; Olivier, B.; Kahler, C.W. Validation of the Dutch Brief Young Adult Alcohol Consequences Questionnaire (B-YAACQ). *Addict. Behav.* **2009**, *34*, 411–414. [CrossRef] [PubMed]
13. Van Schrojenstein Lantman, M.; van de Loo, A.J.; Mackus, M.; Verster, J.C. Development of a definition for the alcohol hangover: Consumer descriptions and expert consensus. *Curr. Drug Abuse Rev.* **2016**, *9*, 148–154. [CrossRef] [PubMed]
14. Verster, J.C.; Scholey, A.; van de Loo, A.J.A.E.; Benson, S.; Stock, A.-K. Updating the definition of the alcohol hangover. *J. Clin. Med.* **2020**, *9*, 823. [CrossRef] [PubMed]
15. Gunn, R.C. Hangovers and attitudes toward drinking. *Q. J. Stud. Alcohol* **1973**, *34*, 194–198. [CrossRef]

16. Piasecki, T.M.; Trela, C.J.; Mermelstein, R.J. Hangover symptoms, heavy episodic drinking, and depression in young adults: A cross-lagged analysis. *J. Stud. Alcohol Drugs* **2017**, *78*, 580–587. [CrossRef]
17. Royle, S.; Owen, L.; Roberts, D.; Marrow, L. Pain catastrophising predicts alcohol hangover severity and symptoms. *J. Clin. Med.* **2020**, *9*, E280. [CrossRef]
18. McKinney, A. A review of the next day effects of alcohol on subjective mood ratings. *Curr. Drug Abuse Rev.* **2010**, *3*, 88–91. [CrossRef]
19. Penning, R.; McKinney, A.; Verster, J.C. Alcohol hangover symptoms and their contribution to overall hangover severity. *Alcohol Alcohol.* **2012**, *47*, 248–252. [CrossRef]
20. Van Schrojenstein Lantman, M.; Mackus, M.; van de Loo, A.J.A.E.; Verster, J.C. The impact of alcohol hangover symptoms on cognitive and physical functioning, and mood. *Hum. Psychopharmacol.* **2017**, *32*, e2623. [CrossRef]
21. Harburg, E.; Davis, D.; Cummings, K.M.; Gunn, R. Negative affect, alcohol consumption and hangover symptoms among normal drinkers in a small community. *J. Stud. Alcohol.* **1981**, *42*, 998–1012. [CrossRef]
22. Harburg, E.; Gunn, R.; Gleiberman, L.; DiFranceisco, W.; Schork, A. Psychosocial factors, alcohol use, and hangover signs among social drinkers: A reappraisal. *J. Clin. Epidemiol.* **1993**, *46*, 413–422. [CrossRef]
23. Huson, L.W. Performance of some correlation coefficients when applied to zero-clustered data. *J. Mod. Appl. Stat. Methods* **2007**, *6*, 530–536. [CrossRef]
24. Verster, J.C.; van de Loo, A.J.A.E.; Benson, S.; Scholey, A.; Stock, A.K. The assessment of overall hangover severity. *J. Clin. Med.* **2020**, *9*, 786. [CrossRef] [PubMed]
25. Wilod Versprille, L.J.F.; van de Loo, A.J.A.E.; Mackus, M.; Arnoldy, L.; Sulzer, T.A.L.; Vermeulen, S.A.; Abdulahad, S.; Huls, H.; Baars, T.; Kraneveld, A.D.; et al. Development and validation of the Immune Status Questionnaire (ISQ). *Int. J. Environ. Res. Public Health* **2019**, *16*, E4743. [CrossRef] [PubMed]
26. Van Schrojenstein Lantman, M.; Otten, L.S.; Mackus, M.; de Kruijff, D.; van de Loo, A.J.A.E.; Kraneveld, A.D.; Garssen, J.; Verster, J.C. Mental resilience, perceived immune functioning, and health. *J. Multidiscip. Healthc.* **2017**, *10*, 107–112. [CrossRef] [PubMed]
27. Van de Loo, A.J.A.E.; Mackus, M.; van Schrojenstein Lantman, M.; Kraneveld, A.D.; Garssen, J.; Scholey, A.; Verster, J.C. Susceptibility to alcohol hangovers: The association with self-reported immune status. *Int. J. Environ. Res. Public Health* **2018**, *15*, 1286. [CrossRef]
28. van de Loo, A.J.A.E.; van Schrojenstein Lantman, M.; Mackus, M.; Scholey, A.; Verster, J.C. Impact of mental resilience and perceived immune functioning on the severity of alcohol hangover. *BMC Res. Notes* **2018**, *11*, 526. [CrossRef] [PubMed]
29. McNair, D.M.; Lorr, M.; Droppleman, L.F. *Manual for the Profile of Mood States*; Educational and Industrial Testing Service: San Diego, CA, USA, 1971.
30. Baars, T.; Berge, C.; Garssen, J.; Verster, J.C. Effect of raw milk consumption on perceived health, mood and immune functioning among US adults with a poor and normal health: A retrospective questionnaire based study. *Complement. Ther. Med.* **2019**, *47*, 102196. [CrossRef]
31. Baars, T.; Berge, C.; Garssen, J.; Verster, J.C. The impact of raw fermented milk products on perceived health and mood among Dutch adults. *Nutr. Food Sci.* **2019**, *49*, 1195–1206. [CrossRef]
32. Killgore, W.D. The visual analogue mood scale: Can a single-item scale accurately classify depressive mood state? *Psychol. Rep.* **1999**, *85*, 1238–1243. [CrossRef] [PubMed]
33. Wolfe, F. Fatigue assessments in rheumatoid arthritis: Comparative performance of visual analog scales and longer fatigue questionnaires in 7760 patients. *J. Rheumatol.* **2004**, *31*, 1896–1902. [PubMed]
34. De Boer, A.G.; van Lanschot, J.J.; Stalmeier, P.F.; van Sandick, J.W.; Hulscher, J.B.; de Haes, J.C.; Sprangers, M.A. Is a single-item visual analogue scale as valid, reliable and responsive as multi-item scales in measuring quality of life? *Qual. Life Res.* **2004**, *13*, 311–320. [CrossRef] [PubMed]
35. Eysenck, H.J.; Eysenck, S.B. *Manual of the Eysenck Personality Questionnaire: (EPQ-R Adult)*; EdITS/Educational and Industrial Testing Service: San Diego, CA, USA, 1994.
36. Eysenck, S.B.; Eysenck, H.J.; Barrett, P. A revised version of the psychoticism scale. *Personal. Individ. Differ.* **1985**, *6*, 21–29. [CrossRef]
37. Watson, P.E.; Watson, I.D.; Batt, R.D. Prediction of blood alcohol concentrations in human subjects. Updating the Widmark Equation. *J. Stud. Alcohol Drugs* **1981**, *42*, 547–556. [CrossRef] [PubMed]

38. van de Loo, A.J.A.E.; van Andel, N.; van Gelder, C.A.G.H.; Janssen, B.S.G.; Titulaer, J.; Jansen, J.; Verster, J.C. The effects of alcohol mixed with energy drink (AMED) on subjective intoxication and alertness: Results from a double-blind placebo-controlled clinical trial. *Hum. Psychopharmacol.* **2016**, *31*, 200–205. [CrossRef]
39. Donners, A.A.M.T.; Tromp, M.D.P.; Garssen, J.; Roth, T.; Verster, J.C. Perceived immune status and sleep: A survey among Dutch students. *Sleep Disord.* **2015**, *2015*, 721607. [CrossRef]
40. Tromp, M.D.P.; Donners, A.A.M.T.; Garssen, J.; Verster, J.C. Sleep, eating disorder symptoms, and daytime functioning. *Nat. Sci. Sleep* **2016**, *8*, 35–40.
41. Rohsenow, D.J.; Howland, J.; Winter, M.; Bliss, C.A.; Littlefield, C.A.; Heeren, T.C.; Calise, T.V. Hangover sensitivity after controlled alcohol administration as predictor of post-college drinking. *J. Abnorm. Psychol.* **2012**, *121*, 270–275. [CrossRef]
42. Piasecki, T.M.; Alley, K.J.; Slutske, W.S.; Wood, P.K.; Sher, K.J.; Shiffman, S.; Heath, A.C. Low sensitivity to alcohol: Relations with hangover occurrence and susceptibility in an ecological momentary assessment investigation. *J. Stud. Alcohol Drugs* **2012**, *73*, 925–932. [CrossRef]
43. Verster, J.C.; Kruisselbrink, L.D.; Slot, K.A.; Anogeianaki, A.; Adams, S.; Alford, C.; Arnoldy, L.; Ayre, E.; Balikji, S.; Benson, S.; et al. On behalf of the Alcohol Hangover Research Group. Sensitivity to experiencing alcohol hangovers: Reconsideration of the 0.11% blood alcohol concentration (BAC) threshold for having a hangover. *J. Clin. Med.* **2020**, *9*, 179. [CrossRef] [PubMed]
44. Schuckit, M.A.; Smith, T.L.; Tipp, J.E. The Self-Rating of the Effects of Alcohol (SRE) form as a retrospective measure of the risk for alcoholism. *Addiction* **1997**, *92*, 979–988. [CrossRef] [PubMed]
45. Fleming, K.A.; Bartholow, B.D.; Hilgard, J.; McCarthy, D.M.; O'Neill, S.E.; Steinley, D.; Sher, K.J. The alcohol sensitivity questionnaire: Evidence for construct validity. *Alcohol. Clin. Exp. Res.* **2016**, *40*, 880–888. [CrossRef] [PubMed]
46. Verster, J.C.; van de Loo, A.J.A.E.; Adams, S.; Stock, A.-K.; Benson, S.; Alford, C.; Scholey, A.; Bruce, G. Advantages and limitations of naturalistic study designs and their implementation in alcohol hangover research. *J. Clin. Med.* **2019**, *8*, 2160. [CrossRef] [PubMed]
47. Jackson, K.M.; Rohsenow, D.J.; Piasecki, T.M.; Howland, J.; Richardson, A.E. Role of tobacco smoking in hangover symptoms among university students. *J. Stud. Alcohol Drugs* **2013**, *74*, 41–49. [CrossRef]
48. Verster, J.C. Smoking and drinking go hand in hand. *Curr. Drug Abuse Rev.* **2008**, *1*, 112–113. [CrossRef]
49. van Lawick van Pabst, A.E.; Devenney, L.E.; Verster, J.C. Sex differences in the presence and severity of alcohol hangover symptoms. *J. Clin. Med.* **2019**, *8*, E867. [CrossRef]
50. van Lawick van Pabst, A.E.; Devenney, L.E.; Verster, J.C. Correction: Van Lawick van Pabst et al. Sex differences in the presence and severity of alcohol hangover symptoms. *Journal of Clinical Medicine* **2019**, *8*, 867. *J. Clin. Med.* **2019**, *8*, E1308. [CrossRef]

© 2020 by the authors. Licensee MDPI, Basel, Switzerland. This article is an open access article distributed under the terms and conditions of the Creative Commons Attribution (CC BY) license (http://creativecommons.org/licenses/by/4.0/).

Article

The Effects of Alcohol Hangover on Mood and Performance Assessed at Home

Chris Alford [1,*], Zuzana Martinkova [1], Brian Tiplady [2], Rebecca Reece [1] and Joris C. Verster [3,4]

1 Department of Health and Social Sciences, University of the West of England, Bristol BS16 1QY, UK
2 Department of Anaesthesia, Critical Care & Pain Medicine, University of Edinburgh, Edinburgh EH8 9YL, UK
3 Division of Pharmacology, Utrecht Institute for Pharmaceutical Sciences (UIPS), Utrecht University, 3584CG Utrecht, The Netherlands
4 Centre for Human Psychopharmacology, Swinburne University, Melbourne, VIC 3122, Australia
* Correspondence: chris.alford@uwe.ac.uk

Received: 24 March 2020; Accepted: 7 April 2020; Published: 9 April 2020

Abstract: The current study evaluated the next day consequences of a social night of drinking compared to a no alcohol night, with standardised mood and portable screen-based performance measures assessed in the morning at participants' homes, and a breathalyser screen for zero alcohol. A mixed sex group ($n = 20$) took part in the study. Participants reported consuming on average 16.9 units (135 g) alcohol, resulting in a hangover rating of 60 (out of 100) compared to 0.3 following the no alcohol night. Statistical significance comparisons contrasting the hangover with the no alcohol condition revealed an increase in negative mood and irritability during hangover and an (unexpected) increase in risk and thrill seeking. Performance scores showed an overall slowing of responses across measures, but with less impact on errors. The results support the description of hangover as a general state of cognitive impairment, reflected in slower responses and reduced accuracy across a variety of measures of cognitive function. This suggests a general level of impairment due to hangover, as well as increased negative mood. The use of a naturalistic design enabled the impact of more typical levels of alcohol associated with real life social consumption to be assessed, revealing wide ranging neurocognitive impairment with these higher doses. This study has successfully demonstrated the sensitivity of home-based assessment of the impact of alcohol hangover on a range of subjective and objective measures. The observed impairments, which may significantly impair daily activities such as driving a car or job performance, should be further investigated and taken into account by policy makers.

Keywords: alcohol; hangover; mood; performance; assessment at home; mobile testing

1. Introduction

The negative consequences of alcohol consumption on safety and productivity are well known, but the separate impact of alcohol hangover has historically received less attention. This is changing, and alcohol hangover is now being recognised as important in its own right. Around 9% of workers in the USA report having a hangover whilst at work, which can impair performance both through absenteeism and "presenteeism"—attending but unable to work effectively [1]. There are also wider implications for health and safety [2–4]. Safety critical daily activities such as driving have shown impairment with hangover [5,6]. The economic costs of hangover in terms of absenteeism and presenteeism are estimated at 4 billion GBP annually [7].

The alcohol hangover refers to the combination of negative mental and physical symptoms which can be experienced after a single episode of alcohol consumption, starting when blood alcohol concentration (BAC) approaches zero [8,9]. Symptoms reflect a general state of malaise described medically as veisalgia. The most common symptoms may be grouped as follows: (1) drowsiness, including fatigue, sleepiness and weakness; and (2) cognitive problems, including reduced alertness and difficulties with memory and concentration. Other symptoms, such as disturbed water balance, contribute less to the overall hangover [10,11].

A simple "culprit" responsible for this range of symptoms has yet to be identified. The toxic metabolite acetaldehyde has largely gone from the system when hangover remains, although research suggests that an inflammatory response and cytokine elevation may account for some symptoms [12–15]. Changes in neurotransmitters and mitochondrial dysfunction, as well as the congeners added to different types of drink may also have a role [16]. The presence and severity of a hangover have been linked to the level of prior consumption in some studies, with the proportion of hangover resistant participants reducing to closer to zero as consumption levels achieve 0.3–0.4% estimated BAC [17,18]. However, individual differences affecting metabolism, such as genetic variation in alcohol dehydrogenase, as well as personality and health status, reflect a wide range of reported symptoms and overall severity [13,16]. Sex differences in alcohol metabolism are well known and may also reflect variability of the presence and severity of hangover symptoms in men and women, and their impact on cognitive and psychomotor functioning and mood [19,20]. Surveys generally indicate higher social consumption levels by males, although gender differences in consequential hangover severity have not been consistently reported, and may be more apparent in women at intermediate BAC levels (0.1–0.3%) [21,22].

Acute alcohol consumption tends to lead to an increase in errors, with less effect on speed of performance, possibly reflecting a "risky shift" [23,24]. In contrast, slowed responses seems a more general consequence of alcohol hangover and may reflect the distinct neurocognitive effects when contrasted with alcohol intoxication [25,26]. A key problem facing alcohol hangover research is the need for participants to achieve sufficient levels of alcohol consumption, in order to produce a measurable hangover for impairment studies to be effective. Due to ethics constraints, sufficient alcohol dosing required to achieve higher real-life BACs cannot always be administered in experimental settings; or requires stringent assessment and prolonged post dose participant support. For example, supervised overnight stays at the research facility are required to monitor participants' wellbeing. This makes these studies resource intensive and demanding for participants [27]. In addition, social drinking frequently includes a mixture of alcoholic beverages being consumed in a single drinking occasion which cannot easily be reproduced in a laboratory. This has led to a number of naturalistic studies where typically residual effects next day are assessed, contrasting a regular drinking night as part of normal social life with a no alcohol consumption night [27]. The resulting hangover assessment can then still include the same validated measures as used in RCTs [27,28]. However, the hangover state in naturalistic studies reflects a real-life hangover experience significantly better than studies that involve lab-based alcohol administration [26,27,29].

The following investigation was based on a naturalistic study design with participants assessed within their own homes in order to maximise their safety and comfort after a potentially heavy social drinking session. The aim of using the naturalistic study design was to mimic real-life as closely as possible and as such is characterised by a minimum of lifestyle rules for participants. The investigators did not (actively) interfere with their activities, and behaviours and activities of the participants were neither standardised nor regulated by the study protocol. All assessments were undertaken whilst participants remained in their usual environment (i.e., at home), which was less demanding for participants and the results may more accurately reflect real life. Balanced gender groups were included to see if there were any marked differences. The principal research questions were to determine whether subjective state and objective performance measures differed when assessed at home the morning after a night of social drinking compared to a night without alcohol. Based on previous research, it was hypothesised that both performance and mood are impaired during alcohol hangover. The secondary exploratory research question was to evaluate possible relationships between measures including reported alcohol consumption and subsequent subjective state and performance.

2. Methods

2.1. Design

The study was a two-period comparison of performance and subjective state between days following an evening of drinking or an evening with no alcohol. One day in each condition was assessed for each subject. The study was unblinded, and the order of conditions was not specified in advance. The study was approved by the University of the West of England Faculty Ethics Committee (Approval number: HLS130235, 1 March 2013), informed consent was obtained from all participants and the study was undertaken in accordance with the British Psychological Society Code of Ethics and Conduct (2009).

2.2. Participants

Participants were social drinkers, with no specific criterion of hangover frequency. They were recruited using the snowball technique from acquaintances of people known personally to the principal experimenter, rather than the direct friends of the experimenter. This approach was used to minimise the impact of personal relationships, whilst optimising researcher safety as well as participant comfort for home-based assessment. Exclusion criteria were frequent illicit drug use, or use prior to testing, any health issues, including those adversely affected by alcohol, breast feeding, pregnancy and those who may be pregnant. No financial payment was made for participation.

Most participants were (friends of) students from the University of the West of England, providing an age range of 18–33 years old. The sample consisted of twenty healthy Caucasian participants, ten female (mean ages of 28.8 years old). Thirty-five per cent were smokers (mean 4–5 cigarettes per day), with 65% in full employment and 5% students. They reported having a mean (SD) of 2.8 (1.3) (range 1–6) hangovers per month.

2.3. Procedure

Participants were given information sheets describing the study as well as inclusion and exclusion criteria. Those eligible to be enrolled then provided signed informed consent and were advised of their rights to withdraw. Enrolled participants completed a demographic questionnaire and assigned their unique participant code facilitating anonymity, and reminded that participation and all information was confidential. Participants were reminded not to consume any stimulants, e.g., coffee, tea and chocolate, or smoke an hour before assessment, as well as to make a note of the number and type of drinks they consumed the previous evening for their residual alcohol assessment days.

Participants contacted the investigator to arrange a suitable time for an assessment visit and were tested at their homes in the morning (08:00–12:00) with the experimenter accompanied by a research buddy who remained in the car outside during the visit. Assessment day and treatment order selected by participants resulting in 65% undertaking the hangover condition first. An initial positive breath alcohol test resulted in the schedule start being moved back enabling a zero reading to be obtained, or testing rearranged, so that all assessments were completed with zero BAC% following drinking, as well as in the control condition. Subjective assessments were followed by performance tests with the total assessment period completed within a 45 min period. There was an average of 5 weeks between treatments and minimum washout period of a week following the hangover assessment. Participants were given a debrief sheet including contacts for support and advice about excessive drinking after completing both assessments, and being reminded of their rights to withdraw their data.

2.4. Assessments

Breath Alcohol Concentration (BAC%) was assessed with a Lion Alcolmeter 400 (Lion Laboratories, Barry, South Glamorgan, UK) by the experimenter who kept their hand over the visual display so that participants were blind to readings.

Subjective Assessments comprised paper questionnaires including general mood assessment dimensions, namely "alert", "calm" and "content", assessed with 100 mm Visual Analog Scales (VAS) [30]; an Irritability Questionnaire based on a 4-point Likert scale [31]; risk taking comprising a 100 mm VAS [32]; the Alcohol Hangover Severity Scale (AHSS) with an 11-point scale, 0–10 [28,33]; and a 1-item overall hangover severity score (a single item, rated on a 100 mm VAS, ranging from "not at all" to "worst ever hangover") [28]. Participants were also required to record their alcohol consumption on their social nights out with the aid of a retrospective diary when assessed in their homes.

Performance Tests were selected from the Penscreen Test Battery (www.MobileCognition.com) presented on ARNOVA 7 G2 android 7 Inch/17.8 cm Screens. The five a priori selected tests (Arrow Flankers, Continuous Performance, Four Choice Reaction Time, Serial Sevens and Stop Signal Task) each included a single measure of response time and a single error measure for each test to enable a speed–accuracy trade-off evaluation to be undertaken, and with a view to limiting type I errors in the subsequent statistical analyses. Each test was explained to the participant and performed by the researcher where necessary, before the participants had an initial practice, as well as being provided with a printed sheet describing each test including the stimulus display and the required responses. Participants were asked to respond as quickly and accurately as possible, with a practice block preceding each test which took around 5 min to complete with derived test variables including response speed (reaction time) and number of errors, and the total assessment period taking up to 45 min.

Arrow Flankers [34,35] measures divided attention and inhibition. Sets of five symbols appear on the screen one set at a time. The central symbol (target) is always an arrow, pointing either to the right or the left. The other four symbols (flankers) are either congruent (arrows pointing in the same direction as the target); incongruent (arrows pointing in the opposite direction to the target); neutral (squares); or suppressors (crosses). The task is to press a left or right button, corresponding to the central arrow, unless the flankers are crosses, in which case no response should be made. Outcome measures are the mean overall reaction time and the total number of errors.

Continuous Performance [36,37] measures sustained attention. The A–X version of the test was used. Letters appear on the screen one at a time. The task is to respond when X appears immediately following A. Outcome measures are the mean reaction time for correct responses and the number of missed targets.

Four Choice Reaction Time [38,39] is a measure of psychomotor performance. An array of four "lights" on the screen corresponds to four buttons below. Each of the lights is highlighted in turn, and the participant responds by tapping the corresponding button as quickly as possible. Outcome measures are the mean reaction time for correct responses and the number of incorrect responses.

Serial Sevens [39,40] assesses arithmetic and working memory. A starter number in the range 800–899 is presented on the screen followed by a series of descending 3-digit numbers. The task is to tap a Yes key if the number is seven less than the previous number shown. In other cases, participants press "No". Outcome measures are the mean reaction time for correct responses and the number of incorrect responses.

Stop Signal Task [41,42] is a measure of impulsivity and inhibition. A letter stimulus which is either an O or an X is displayed, together with two on-screen buttons. The task is to tap the left button for an X, the right button for an O, as quickly as possible. In one trial in four, a stop-signal is presented after the onset of the letter, consisting of two horizontal red lines superimposed on the letter display. When a stop-signal appears, the user should not respond to the letter stimulus. Outcome measures are the mean reaction time for correct responses and the number of missed stop signal items (i.e., response not withheld).

2.5. Statistical Analyses

Statistical analyses were undertaken using IBM SPSS Version 25 (IBM SPSS Statistics 2015, IBM Corp, Armonk, NY, USA). In line with the research questions, statistical analyses were based on contrasting data from the residual alcohol (alcohol hangover) with the no alcohol condition, as well investigating possible relationships among subjective hangover, the amount of alcohol consumed the previous night, and performance and other subjective measures. Separate MANOVAS were first performed to limit type I errors for the subjective and performance data, including independent variables within groups: measure (performance tests or questionnaires) and alcohol (post-alcohol or no alcohol); and between groups factors: assessment order (post-alcohol or no alcohol first) and gender (male or female). They were rerun as a reduced model (questionnaires) with nonsignificant factors removed (assessment order and gender). Where MANOVA indicated an interaction between alcohol and gender (performance tests), ANOVAS were run including alcohol and gender as factors. Overall significant effects for alcohol enabled subsequent paired comparisons to be run for individual variables (each performance test, or each questionnaire) contrasting no alcohol with the post-alcohol condition. Nonparametric paired comparisons were run as confirmatory tests as a control for departures from normality amongst test variable data and identified the same significant paired contrasts as the parametric analyses. Exact p values are reported. Tests of association were run as partial correlations with hangover frequency (monthly) as a control for possible tolerance effects [43] as well as smoking as a control, which may impact hangover [44]. Two-tailed significance levels ($p < 0.05$) were used for all analyses.

3. Results

Twenty participants successfully completed the experimental protocol undertaking assessments in their own homes after a night of social drinking (alcohol night) and a night without alcohol consumption (no alcohol). All participants had zero BAC confirmed with the breathalyser prior to their assessments.

3.1. Alcohol Consumption

All participants reported zero alcohol consumption on their no alcohol nights before testing, and they reported consuming a mean (SD) of 16.9 (4.2) units (range 8–26), calculated using the UK National Health Service (NHS) website tool (available at: www.nhs.uk) to convert recorded drinks into standard UK units (1 unit = 10 mL, 8 g pure alcohol). This level of consumption is similar to those reported elsewhere including 13.5 standard drinks (10 g pure alcohol) reported in a recent Australian study [26] averaging 135 g pure alcohol, whilst for this study 16.9 units of 8 g/10 mL is also equivalent to averaging 135 g pure alcohol based on the UK definition. Most participants mixed their drinks, with two participants having only one type of drink, most 2–3, and two of the 20 participants mixing four different drinks, typical of a naturalistic study.

3.2. Subjective Assessments

Multivariate analysis failed to reveal significant effects for treatment order ($\eta^2 p = 0.00$) or gender ($\eta^2 p = 0.08$), including interactions; thus, these factors were removed and data reanalysed with the reduced model where alcohol remained significant ($F_{(1,19)} = 17.98$, $p < 0.0001$, $\eta^2 p = 0.49$), reflecting an overall difference in subjective ratings between the alcohol and no alcohol conditions and enabling paired comparisons to be undertaken for the individual measures. The subjective assessments and results of the pairwise statistical analyses are summarised in Table 1.

Table 1. Subjective assessments of mood and hangover.

Measure	Control	Alcohol
Alert VAS (0–100)	51.2 (4.9)	44.1 (6.8) *,†
Content VAS (0–100)	54.0 (6.7)	50.0 (9.1)
Calm VAS (0–100)	70.0 (17.7)	53.4 (12.2) *,†
Irritability (0–3)	0.98 (0.29)	1.56 (0.52) *,†
Overall risk taking VAS (0–100)	53.1 (4.4)	53.3 (4.2)
Risk and thrill seeking VAS (0–100)	50.6 (6.9)	54.1 (6.0) *,†
Self-confidence EVAR (0–100)	54.2 (6.2)	51.0 (8.6)
Need for control EVAR (0–100)	54.5 (9.6)	54.9 (10.1)
Alcohol Hangover Severity Scale (0–10)	0.6 (0.4)	3.8 (1.0) *,†
One-item hangover severity (0–100)	0.3 (1.1)	60.3 (20.0) *,†

Mean and standard deviation (SD) are shown. Abbreviations: VAS, visual analogue scale; EVAR, Evaluation of Risks Scale. Significant differences ($p < 0.05$) are indicated by * (parametric) and † (nonparametric).

The parametric analyses provided the same significant contrasts as the nonparametric confirmatory analysis. Significant increases ($p < 0.0001$) were seen following alcohol for both the AHSS and the one-item hangover severity VAS, with both showing close to zero symptom ratings after no alcohol but notable increases after alcohol (see Table 1). For example, a mean (SD) one-item hangover severity VAS scores of 0.3 (1.1) (range 0–100) was reported following no alcohol rising to 60.3 (20.0) (range 15–90) the day after alcohol consumption, confirming overall hangover status amongst participants.

The VAS mood scale revealed a significant reduction in measures of alertness ($p = 0.002$) and calmness ($p = 0.005$) in the hangover condition and a trend for a reduction in contentedness ($p = 0.075$). A significant increase was also recorded with the irritability scale ($p < 0.0001$), reflecting an overall reduction in positive mood and increased negative state with alcohol hangover compared to the no hangover condition. Although no overall change in subjective risk taking was recorded with EVAR, a modest increase was seen in risk and thrill seeking ($p = 0.032$), although no change in need for control.

3.3. Performance Tests

Multivariate analysis again failed to find overall significant effects for either treatment order ($\eta^2 p = 0.00$) or gender ($\eta^2 p = 0.10$), but did find a significant alcohol × gender interaction ($F_{(1,16)} = 8.1$, $p = 0.012$, $\eta^2 p = 0.34$), which was explored with ANOVAs run for each variable and including alcohol and gender as factors. Alcohol × gender interaction was found for serial sevens response time ($F_{(1,18)} = 7.1$, $p = 0.016$, $\eta^2 p = 0.28$) and errors ($F_{(1,18)} = 8.1$, $p = 0.011$, $\eta^2 p = 0.31$) where males had significantly slower responses (1712.0 (699.9) versus 1005.3 (332.9), $p < 0.0001$) and more errors (6.2 (5.9) versus 3.1 (4.6), $p = 0.011$) after alcohol compared to no alcohol, although no significant effects were found for females. There were no other significant interactions or gender differences. An overall effect for alcohol was seen with response times for arrow flankers ($F_{(1,18)} = 5.1$, $p = 0.036$, $\eta^2 p = 0.22$), continuous performance ($F_{(1,18)} = 8.1$, $p = 0.011$, $\eta^2 p = 0.31$) and choice reaction time ($F_{(1,18)} = 11.5$, $p = 0.003$, $\eta^2 p = 0.39$), and for errors with arrow flankers alone ($F_{(1,18)} = 10.1$, $p = 0.005$, $\eta^2 p = 0.36$). These differences were further examined with paired comparisons. Results for the cognitive tests and the pairwise comparisons are summarised in Table 2.

The parametric paired contrasts reflected the nonparametric analyses (see Table 2). Responses were significantly slower with arrow flankers ($p = 0.036$), continuous performance ($p = 0.011$) and choice reaction time ($p = 0.005$). Errors were significantly increased in the hangover condition for arrow flankers ($p = 0.004$).

Table 2. Performance outcomes.

Test	Measure	Control	Alcohol
Arrow Flankers	RT (ms)	625.8 (127.8)	693.9 (143.5) *,†
	Errors (*n*)	3.4 (6.6)	13.9 (16.5) *,†
Serial Sevens	RT (ms)	1948.9 (1938.2)	2055.6 (1339.3)
	Errors (*n*)	3.3 (3.7)	4.4 (4.6)
Continuous Performance	RT (ms)	468.5 (67.5)	506.0 (69.0) *,†
	Errors (*n*)	2.4 (3.0)	2.6 (2.5)
Choice Reaction Time	RT (ms)	597.9 (263.0)	705.0 (245.2) *,†
	Errors (*n*)	2.8 (2.1)	4.5 (10.6)
Stop Signal Task	RT (ms)	797.3 (115.1)	828.3 (120.6)
	Errors (*n*)	4.2 (1.5)	3.7 (1.3)

Mean and standard deviation (SD) are shown. Abbreviations: RT, reaction time; ms, milliseconds; *n*, number. Significant differences ($p < 0.05$) are indicated by * (parametric) and † (nonparametric).

3.4. Associations between Measures

The secondary research question was concerned with possible associations between subjective state including perceived hangover, performance and other variables. Statistical analysis employed partial correlations with monthly hangover frequency to take into account possible tolerance effects [43] and smoking habit (categorised as yes, occasional, no) [44], included as control variables which may impact scores in the assessed hangover condition).

The one-item overall hangover severity score was selected as an overall measure of subjective hangover and was found to be more sensitive than the AHSS [28], with a relatively weak association between the two scales ($r = 0.439$, $p = 0.068$). Overall hangover severity (one-item) correlated significantly with the recalled amount of alcohol consumed the night before ($r = 0.548$, $p = 0.019$), with the more alcohol consumed the previous evening the greater the subjective hangover the following day. The amount of alcohol consumed the previous evening failed to show any significant associations with performance or other subjective variables. However, overall hangover severity was also correlated with other performance and subjective data collected in the hangover condition.

With the VAS, increased one-item overall hangover severity was associated with participants feeling less content ($r = -0.524$, $p = 0.026$). Risk taking (EVAR) indicated a reduction in the self-confidence subscale with increased hangover ($r = -0.493$, $p = 0.037$), but an unexpected increase in the risk and thrill seeking subscale ($r = 0.479$, $p = 0.044$).

There was a significant association between the 1-item overall hangover severity score and cognitive performance. With serial sevens (continuous subtraction), increased hangover was unexpectedly associated with faster responses ($r = -0.623$, $p = 0.006$) and fewer errors ($r = -0.541$, $p = 0.020$), although there were no other significant associations. In contrast to the one-item overall hangover severity score, none of the associations among mood, performance and AHSS scores were significant.

4. Discussion

The main aim of the study was to investigate differences in both subjective state and performance following a night of social drinking that may result in a hangover next day when assessed at home and contrasted with a night without alcohol consumption using a naturalistic design. A relatively high mean (SD) alcohol consumption was reported of 17 (4.2) units (range 8–26) based on primed recall next day, and with most participants consuming two or more types of drinks. Mixing drinks, or the order in which they are consumed, has not in itself shown an impact on hangover [45], although congeners may have a role in determining hangover severity [46]. The mean (SD) hangover severity score of 60.3 (20.0) on a one-item scale ranging from "not at all" (0) to "worst ever" (100) reflects a clearly demonstrable hangover for this group overall, although BAC was not directly measured. Significant differences in the present study were associated with overall effect sizes ($\eta^2 p$) of around of 0.5 for subjective measures and 0.2–0.4 for performance measures. These are comparable with those reported in compatible repeated measures studies [47], and supported by impairments reported in other measures such as driving [5], suggesting the real life consequences of alcohol hangover after social drinking occasions in everyday life.

4.1. Subjective Measures

Six of the 10 subjective measures returned significant differences between the alcohol hangover and no hangover conditions (see Table 1) showing that participants were clearly experiencing different subjective states, and the reduction in alertness is consistent with the study of 1400 participants identifying key factors of the subjective hangover state [11]. The reduction in positive mood measured with the VAS and increase in irritability supports the descriptions of hangover as a state of malaise and supported by significant increases in both specific hangover symptoms measured with the AHSS as well as the overall one-item hangover severity score.

The finding of a significant positive association between recalled consumption the previous night and one-item hangover severity assessed next day ($r = 0.548$, $p = 019$) has been found in some other studies but not all, and thus the relatively small sample size included here provides limited support for this [26]. Given the wide range of hangover symptoms including CNS, gastrointestinal and more general physiological effects [11,48], as well as individual differences in alcohol metabolism that may impact hangover including reported immunity, larger samples are required to establish this association, possibly with selected cohorts where distinct groups of phenotypes are found. The average change in subjective hangover (one-item severity score) ratings from less than 1 on the control day to 60 (out of 100) on the hangover day indicates a notable hangover for the majority of participants. The AHSS also increased significantly, but a relative lack of sensitivity for the AHSS when correlated with other measures may reflect its component nature, where a restricted number of individual symptoms may not reflect the complete hangover experience and its impact accurately [28]. The decrease in positive mood with the VAS supports the hangover ratings, with significant reductions in feeling alert and calm, and a trend for a reduction in feeling content in the hangover condition compared to no alcohol, and supports the reliability of the subjective data.

The EVAR scale has been widely used in assessing risk taking and found to be a sensitive and reliable measure [32]. The component scores showed a differential effect when contrasting hangover with no alcohol conditions. Overall risk taking, need for control and self-confidence were relatively unchanged. Interestingly, risk and thrill seeking were increased which would not be predicted for participants in a hangover state that may be associated with increased feelings of fragility and vulnerability. However, reduced executive function [25] may result in disinhibition during hangover that may then be reflected in increased risk and thrill seeking ratings. This has been seen alongside increased impulsivity with alcohol and driving [49], although the stop signal task failed to show significant changes in the present study, suggesting a different impact profile for neurocognitive domains linked to risk taking and impulsivity behaviour with hangover.

Although recalled alcohol consumption was associated with one-item overall hangover severity, other measures were not, suggesting a lack of direct changes in mood states with quantitative changes in alcohol consumption and resulting hangover status. However, hangover severity did correlate significantly with decreases in self-confidence and contentedness, as well as increased risk and thrill seeking, suggesting the sensitivity of subjective awareness of a hangover state. These results contrast with the more limited associations found between subjective hangover and performance measures.

4.2. Performance Tests

The five performance tests assessed a range of cognitive functions including psychomotor performance (choice reaction time), sustained attention (continuous performance) and executive functions. All tests included a response time measure as well as errors enabling a potential speed—accuracy trade-off to be evaluated. Executive function measures included working memory (serial sevens) as well as inhibition (arrow flankers and stop signal task). In contrasting the no alcohol and hangover conditions, four of the five tests indicated a significant contrast (arrow flankers, choice reaction time, continuous performance and (for males only) serial sevens), reflecting impaired performance with slower responses with hangover. Significantly increased errors were only seen with arrow flankers and (for males only) serial sevens, although more errors were generally seen in the hangover condition. The stop signal task alone failed to show differences (see Table 2).

Slower response speeds were evident for all but the stop signal task, resulting in significant slowing with hangover across test averages, although for serial sevens the difference was only significant for male participants. Increased errors only achieved significance with arrow flankers and (for males only) serial sevens, but all except the stop signal task showed an increased number of errors with hangover. These results are in opposition to the speed–accuracy trade-off that has been reported for alcohol, where response speeds are maintained but with an increase in errors [23]. This may be associated with a "risky shift" after acute alcohol administration and observed in the field [50] and characteristic of disinhibition. The speed—accuracy trade-off has not always been observed and error rates can also increase with alcohol induced impairment [51]. However, based on the overall profile of significant results observed in this study, response slowing is a more consistent feature of alcohol hangover induced impairment than increased errors and this has been observed in other studies [26,52,53]. The greater impact on response slowing is therefore an emerging important factor in profiling alcohol hangover induced impairment in contrast to the effects of acute alcohol intoxication.

The overall results of slower response and more errors for all except for the stop signal task, supports alcohol hangover being associated with impairment and slowing of cognitive functions. The apparently contradictory finding of a significant association with serial sevens (assessing mental arithmetic, working memory) and increased hangover severity with faster and more accurate responses, albeit within a general slowing of responses of 100 ms between the no alcohol and hangover conditions, was unexpected. This might reflect an increased situational awareness [54] of being in a sensitive state with increased perception of hangover and possibly increased effort as a result, but was found for this performance measure alone and requires replication.

4.3. Limitations and Strengths

Study limitations included the limited number of participants ($n = 20$) and, although this was sufficient to show both differences between the no alcohol and alcohol hangover conditions, which has been supported by a partial replication, as well as some associations between measures, it was underpowered for gender comparisons and a more reliable investigation of associations between measures. The age range of participants was 18–33 years old. Although high level drinking and hangovers are frequently reported by this age range, future studies should examine to what extent the current findings can be generalised to other age groups. Especially, hangover data of elderly are lacking. Whereas declines are observed in overall drinking in the last 15 years [55] and a growing number of British young adults routinely do not drink alcohol [56], reported alcohol use and corresponding problems in older age groups are reported to be on the increase [57,58].

The naturalistic design did enable the impact of higher alcohol consumption levels (17 units) to be assessed reflecting real life. This enabled a number of differences in both subjective and performance measures to be found between the hangover and no alcohol conditions, suggesting an overall successful investigation in relation to the primary research question. The partial replication also supports these general findings. Assessment at home provided a more natural environment to record actual subjective state and functionality, as well as improving participant safety if travelling in a potentially vulnerable condition. In addition, where participants initially showed a small positive BAC with breathalyser the morning after drinking, assessments could be comfortably delayed until a zero BAC was recorded. These attributes of home-based assessment enabled all participants to complete the study protocol without dropping out, suggesting that this approach places less demands on participants and is therefore a further benefit of naturalistic designs in alcohol hangover research [27]. These differences may account for some of the observed differences in alcohol hangover studies when comparing lab with naturalistic studies. However, one study which directly compared the effects of acute alcohol consumption in the laboratory and in a naturalistic setting showed similar patterns of impairment by alcohol in both settings [59].

5. Conclusions

This study employed a naturalistic design so that participants were able to reflect a typical hangover state after a night out of social drinking with higher consumption, unrestrained from laboratory protocol and ethical limitations on alcohol administration. The use of home-based assessment was successful, demonstrating significant effects on both mood and performance. The pattern of impairment found here supports other findings of alcohol hangover as inducing a state of malaise or veisalgia which negatively impacts a range of cognitive abilities including executive functions, and in addition to more negative mood and experience of hangover symptoms. The findings support others showing both mood and performance impairment with alcohol hangover, and a pattern of general impairment is emerging that is distinct to that found in acute alcohol administration studies. Future studies should determine how the observed impairments may have negative consequences in everyday life, work performance and daily activities such as driving.

Author Contributions: Conceptualisation, C.A., Z.M. and J.C.V.; Software, B.T.; data collection, Z.M.; data analysis, C.A., Z.M. and R.R.; writing—original draft preparation, C.A.; and writing—review and editing, C.A., J.C.V. and B.T. All authors have read and agreed to the published version of the manuscript.

Funding: This research received no external funding.

Conflicts of Interest: C.A. has undertaken sponsored research, or provided consultancy, for a number of companies and organisations including Airbus Group Industries, Astra, British Aerospace/BAE Systems, UK Civil Aviation Authority, Duphar, Farmitalia Carlo Erba, Ford Motor Company, ICI, Innovate UK, Janssen, LERS Synthélabo, Lilly, Lorex/Searle, UK Ministry of Defence, More Labs, Quest International, Red Bull GmbH, Rhone-Poulenc Rorer, Sanofi Aventis and Vital Beverages. B.T. is owner and director of Mobile Cognition Ltd., which provides research consultancy services and develops and markets the PenScreeenSix software. Over the past three years, J.C.V. has received grants/research support from the Dutch Ministry of Infrastructure and the Environment, Janssen Research and Development and Sequential, and has acted as a consultant/advisor for Clinilabs, More Labs, Red Bull GmbH, Sen-Jam Pharmaceutical, Toast! and ZBiotics. The other authors have no potential conflicts of interest to disclose.

References

1. Thorrisen, M.M.; Bonsaksen, T.; Hashemi, N.; Kjeken, I.; van Mechelen, W.; Aas, R.W. Association between alcohol consumption and impaired work performance (presenteeism): A systematic review. *BMJ Open* **2019**, *9*, e029184. [CrossRef] [PubMed]
2. Frone, M.R. Prevalence and distribution of alcohol use and impairment in the workplace: A U.S. national survey. *J. Stud. Alcohol* **2006**, *67*, 147–156. [CrossRef]
3. Frone, M.R.; Verster, J.C. Alcohol hangover and the workplace: A need for research. *Curr. Drug Abus. Rev.* **2013**, *6*, 177–179. [CrossRef]
4. Verster, J.C.; Stephens, R.; Penning, R.; Rohsenow, D. On behalf of the Alcohol Hangover Research Group. The Alcohol Hangover Research Group. Consensus statement on best practice in alcohol hangover research. *Curr. Drug Abus. Rev.* **2010**, *3*, 116–127. [CrossRef] [PubMed]
5. Verster, J.C.; Bervoets, A.C.; de Klerk, S.; Vreman, R.A.; Olivier, B.; Roth, T.; Brookhuis, K.A. Effects of alcohol hangover on simulated highway driving performance. *Psychopharmacology* **2014**, *231*, 2999–3008. [CrossRef] [PubMed]
6. Verster, J.C.; Van Der Maarel, M.A.; McKinney, A.; Olivier, B.; De Haan, L. Driving during alcohol hangover among Dutch professional truck drivers. *Traffic Inj. Prev.* **2014**, *15*, 434–438. [CrossRef]
7. Bhattacharya, A. *Financial Headache. The Cost of Workplace Hangovers and Intoxication to the UK Economy*; Alliance House: London, UK, 2019.
8. Van Schrojenstein Lantman, M.; van de Loo, A.J.; Mackus, M.; Verster, J.C. Development of a definition for the alcohol hangover: Consumer descriptions and expert consensus. *Curr. Drug Abus. Rev.* **2016**, *9*, 148–154. [CrossRef]

9. Verster, J.C.; Scholey, A.; van de Loo, A.J.A.E.; Benson, S.; Stock, A.-K. Updating the definition of the alcohol hangover. *J. Clin. Med.* **2020**, *9*, 823. [CrossRef]
10. Penning, R.; van Nuland, M.; Fliervoet, L.A.L.; Olivier, B.; Verster, J.C. The pathology of alcohol hangover. *Curr. Drug Abus. Rev.* **2010**, *3*, 68–75. [CrossRef]
11. Penning, R.; McKinney, A.; Verster, J.C. Alcohol hangover symptoms and their contribution to overall hangover severity. *Alcohol Alcohol.* **2012**, *47*, 248–252. [CrossRef]
12. Prat, G.; Adan, A.; Sánchez-Turet, M. Alcohol hangover: A critical review of explanatory factors. *Hum. Psychopharmacol.* **2009**, *24*, 259–267. [CrossRef] [PubMed]
13. Tipple, C.T.; Benson, S.; Scholey, A. A review of the physiological factors associated with alcohol hangover. *Curr. Drug Abus. Rev.* **2016**, *9*, 93–98. [CrossRef] [PubMed]
14. Van De Loo, A.J.A.E.; Slot, K.A.; Kleinjan, M.; Knipping, K.; Garssen, J.; Verster, J.C. Time-dependent changes in saliva cytokine concentrations during alcohol hangover: A comparison of two naturalistic studies. *Alcohol. Clin. Exp. Res.* **2016**, *40*, 95A.
15. Van de Loo, A.J.A.E.; van Schrojenstein Lantman, M.; Mackus, M.; Scholey, A.; Verster, J.C. Impact of mental resilience and perceived immune functioning on the severity of alcohol hangover. *BMC Res. Notes* **2018**, *11*, 526. [CrossRef] [PubMed]
16. Palmer, E.; Tyacke, R.; Sastre, M.; Lingford-Hughes, A.; Nutt, D.; Ward, R.J. Alcohol Hangover: Underlying Biochemical, Inflammatory and Neurochemical Mechanisms. *Alcohol Alcohol.* **2019**, *54*, 196–203. [CrossRef]
17. Kruisselbrink, L.D.; Bervoets, A.C.; de Klerk, S.; van de Loo, A.J.A.E.; Verster, J.C. Hangover resistance in a Canadian university student population. *Addict. Behav. Rep.* **2017**, *5*, 14–18. [CrossRef]
18. Verster, J.C.; de Klerk, S.; Bervoets, A.C.; Kruisselbrink, L.D. Can hangover immunity really be claimed? *Curr. Drug Abus. Rev.* **2013**, *6*, 253–254. [CrossRef]
19. Frezza, M.; di Padova, C.; Pozzato, G.; Terpin, M.; Baraona, E.; Lieber, C.S. High blood alcohol levels in women. The role of decreased gastric alcohol dehydrogenase activity and first-pass metabolism. *N. Engl. J. Med.* **1990**, *322*, 95–99. [CrossRef]
20. Mumenthaler, M.S.; Taylor, J.L.; O'Hara, R.; Yesavage, J.A. Gender differences in moderate drinking effects. *Alcohol Res. Health* **1999**, *23*, 55–64.
21. Van Lawick van Pabst, A.E.; Devenney, L.E.; Verster, J.C. Sex differences in the presence and severity of alcohol hangover symptoms. *J. Clin. Med.* **2019**, *8*, 867. [CrossRef]
22. Van Lawick van Pabst, A.E.; Devenney, L.E.; Verster, J.C. Correction: Van Lawick van Pabst et al. Sex differences in the presence and severity of alcohol hangover symptoms. Journal of Clinical Medicine 2019, 8, 867. *J. Clin. Med.* **2019**, *8*, 1308. [CrossRef] [PubMed]
23. Tiplady, B.; Drummond, G.; Cameron, E.; Gray, E.; Hendry, J.; Sinclair, W.; Wright, P. Ethanol, errors, and the speed-accuracy trade-off. *Pharmacol. Biochem. Behav.* **2001**, *69*, 635–641. [CrossRef]
24. Tiplady, B.; Franklin, N.; Scholey, A. Effect of ethanol on judgments of performance. *Br. J. Psychol.* **2004**, *95*, 105–118. [CrossRef]
25. Gunn, C.; Mackus, M.; Griffin, C.; Munafò, M.R.; Adams, S. A systematic review of the next-day effects of heavy alcohol consumption on cognitive performance. *Addiction* **2018**, *113*, 2182–2193. [CrossRef] [PubMed]
26. Scholey, A.; Benson, S.; Kaufman, J.; Terpstra, C.; Ayre, E.; Verster, J.C.; Allen, C.; Devilly, G. Effects of alcohol hangover on cognitive performance: A field/internet mixed methodology approach. *J. Clin. Med.* **2019**, *8*, 440. [CrossRef] [PubMed]
27. Verster, J.C.; van de Loo, A.J.A.E.; Adams, S.; Stock, A.-K.; Benson, S.; Alford, C.; Scholey, A.; Bruce, G. Advantages and limitations of naturalistic study designs and their implementation in alcohol hangover research. *J. Clin. Med.* **2019**, *8*, 2160. [CrossRef] [PubMed]
28. Verster, J.C.; van de Loo, A.J.A.E.; Benson, S.; Scholey, A.; Stock, A.-K. The assessment of overall hangover severity. *J. Clin. Med.* **2020**, *9*, 786. [CrossRef]
29. Stephens, R.; Grange, J.A.; Jones, K.; Owen, L. A critical analysis of alcohol hangover research methodology for surveys or studies of effects on cognition. *Psychopharmacology* **2014**, *231*, 2223–2236. [CrossRef]
30. Bond, A.; Lader, M. The use of analogue rating scales in rating subjective feelings. *Br. J. Med. Psychol.* **1974**, *47*, 211–218. [CrossRef]
31. Craig, K.; Hietenan, H.; Marova, I.; Berrios, G. The irritability questionnaire: A new scale for measurement of irritability. *Psychiatry Res.* **2008**, *159*, 367–375. [CrossRef]

32. Killgore, W.D.S.; Vo, A.H.; Castro, C.A.; Hoge, C.W. Assessing Risk Propensity in American Soldiers: Preliminary Reliability and Validity of the Evaluation of Risks (EVAR) Scale—English Version. *Mil. Med.* **2006**, *171*, 233–239. [CrossRef]
33. Penning, R.; McKinney, A.; Bus, L.D.; Olivier, B.; Slot, K.; Verster, J.C. Measurement of alcohol hangover severity: Development of the Alcohol Hangover Severity Scale (AHSS). *Psychopharmacology* **2013**, *225*, 803–810. [CrossRef] [PubMed]
34. Tiplady, B.; Bowness, E.; Stien, L.; Drummond, G. Selective effects of clonidine and temazepam on attention and memory. *J. Psychopharmacol.* **2005**, *19*, 259–265. [CrossRef] [PubMed]
35. Sanders, L.M.J.; Hortobágyi, T.; Balasingham, M.; Van der Zee, E.A.; van Heuvelen, M.J.G. Psychometric Properties of a Flanker Task in a Sample of Patients with Dementia: A Pilot Study. *Dement. Geriatr. Cogn. Dis. Extra* **2018**, *8*, 382–392. [CrossRef] [PubMed]
36. Rosvold, H.E.; Mirsky, A.F.; Sarason, I.; Bransome, E.D., Jr.; Beck, L.H. A continuous performance test of brain damage. *J. Consult. Psychol.* **1956**, *20*, 343. [CrossRef] [PubMed]
37. Smid, H.G.; de Witte, M.R.; Homminga, I.; van den Bosch, R.J. Sustained and transient attention in the continuous performance task. *J. Clin. Exp. Neuropsychol.* **2006**, *28*, 859–883. [CrossRef]
38. Wilkinson, R.T.; Houghton, D. Portable four-choice reaction time test with magnetic tape memory. *Behav. Res. Methods Instrum.* **1975**, *7*, 441–446. [CrossRef]
39. Benson, S.; Tiplady, B.; Scholey, A. Attentional and working memory performance following alcohol and energy drink: A randomised, double-blind, placebo-controlled, factorial design laboratory study. *PLoS ONE* **2019**, *14*, e0209239. [CrossRef]
40. Scholey, A.B.; Harper, S.; Kennedy, D.O. Cognitive demand and blood glucose. *Physiol. Behav.* **2001**, *73*, 585–592. [CrossRef]
41. Logan, G.D.; Schachar, R.J.; Tannock, R. Impulsivity and Inhibitory Control. *Psychol. Sci.* **1997**, *8*, 60–64. [CrossRef]
42. Jones, A.; Tiplady, B.; Houben, K.; Nederkoorn, C.; Field, M. Do daily fluctuations in inhibitory control predict alcohol consumption? An ecological momentary assessment study. *Psychopharmacology* **2018**, *235*, 1487–1496. [PubMed]
43. Verster, J.C.; Slot, K.A.; Arnoldy, L.; Van Lawick van Pabst, A.E.; van de Loo, A.J.A.E.; Benson, S.; Scholey, A. The association between alcohol hangover frequency and severity: Evidence for reverse tolerance? *J. Clin. Med.* **2019**, *8*, 1520. [CrossRef] [PubMed]
44. Jackson, K.M.; Rohsenow, D.J.; Piasecki, T.M.; Howland, J.; Richardson, A.E. Role of tobacco smoking in hangover symptoms among university students. *J. Stud. Alcohol Drugs* **2013**, *74*, 41–49. [CrossRef]
45. Köchling, J.; Geis, B.; Wirth, S.; Hensel, K.O. Grape or grain but never the twain? A randomized controlled multiarm matched-triplet crossover trial of beer and wine. *Am. J. Clin. Nutr.* **2019**, *109*, 345–352. [CrossRef]
46. Rohsenow, D.J.; Howland, J. The role of beverage congeners in hangover and other residual effects of alcohol intoxication: A review. *Curr. Drug Abus. Rev.* **2010**, *3*, 76–79. [CrossRef] [PubMed]
47. Lakens, D. Calculating and reporting effect sizes to facilitate cumulative science: A practical primer for t-tests and ANOVAs. *Front. Psychol.* **2013**, *4*, 863. [CrossRef] [PubMed]
48. Van Schrojenstein Lantman, M.; Mackus, M.; van de Loo, A.J.A.E.; Verster, J.C. The impact of alcohol hangover symptoms on cognitive and physical functioning, and mood. *Hum. Psychopharmacol.* **2017**, *32*. [CrossRef] [PubMed]
49. Van Dyke, N.A.; Fillmore, M.T. Alcohol Effects on Simulated Driving Performance. *Exp. Clin. Psychopharmacol.* **2014**, *22*, 484–493. [CrossRef]
50. Alford, C.; Hamilton-Morris, J.; Verster, J.C. The effect of energy drink in Combination with alcohol on performance and subjective awareness. *Psychopharmacology* **2012**, *222*, 519–532. [CrossRef]
51. Mackay, M.; Tiplady, B.; Scholey, A.B. Interactions between alcohol and caffeine in relation to psychomotor speed and accuracy. *Hum. Psychopharmacol.* **2002**, *17*, 151–156. [CrossRef]
52. Grange, J.A.; Stephens, R.; Jones, K.; Owen, L. The effect of alcohol hangover on choice response time. *J. Psychopharmacol.* **2016**, *30*, 654–661. [CrossRef] [PubMed]
53. Devenney, L.E.; Coyle, K.B.; Verster, J.C. Memory and attention during an alcohol hangover. *Hum. Psychopharmacol.* **2019**, *34*, e2701. [CrossRef] [PubMed]
54. Wickens, C.D. Situation awareness: Review of mica Endsley's 1995 articles on situation awareness theory and measurement. *Hum. Factors* **2008**, *50*, 397–403. [CrossRef]

55. Svensson, J.; Andersson, D.E. What role do changes in the demographic composition play in the declining trends in alcohol consumption and the increase of non-drinkers among Swedish youth? A time-series analysis of trends in non-drinking and region of origin 1971–2012. *Alcohol Alcohol.* **2016**, *51*, 172–176. [CrossRef] [PubMed]
56. Conroy, D.; de Visser, R.O. Benefits and drawbacks of social non-drinking identified by British university students. *Drug Alcohol Rev.* **2017**, *37*, S89–S97. [CrossRef] [PubMed]
57. Dar, K. Alcohol use disorders in elderly people: Fact or fiction? *Adv. Psychiat. Treat.* **2006**, *12*, 173–181. [CrossRef]
58. Han, B.H.; Moore, A.A.; Ferris, R.; Palamar, J.J. Binge drinking among older aldults in the United States, 2015 to 2017. *J. Am. Geriatr. Soc.* **2019**, *67*, 2139–2144. [CrossRef]
59. Tiplady, B.; Oshinowo, B.; Thomson, J.; Drummond, G.B. Alcohol and cognitive function: Assessment in everyday life and laboratory settings using mobile phones. *Alcohol. Clin. Exp. Res.* **2009**, *33*, 2094–2102. [CrossRef]

© 2020 by the authors. Licensee MDPI, Basel, Switzerland. This article is an open access article distributed under the terms and conditions of the Creative Commons Attribution (CC BY) license (http://creativecommons.org/licenses/by/4.0/).

Article

Effects of Alcohol Hangover on Cognitive Performance: Findings from a Field/Internet Mixed Methodology Study

Andrew Scholey [1,*], Sarah Benson [1], Jordy Kaufman [2], Chantal Terpstra [1], Elizabeth Ayre [1], Joris C. Verster [1,3], Cory Allen [4] and Grant J. Devilly [5]

1 Centre for Human Psychopharmacology, Swinburne University, Melbourne, VIC 3122, Australia; sarahbenson@swin.edu.au (S.B.); cterpstra@swin.edu.au (C.T.); eayre@swin.edu.au (E.A.)
2 Swinburne BabyLab, Swinburne University, Melbourne, VIC 3122, Australia; jkaufman@swin.edu.au
3 Division of Pharmacology, Utrecht University, 3584 CG Utrecht, The Netherlands; j.c.verster@uu.nl
4 Queensland Police Service Academy, GPO Box 1110, Archerfield, QLD 4108, Australia; allencorey1966@hotmail.com
5 School of Applied Psychology and Griffith Criminology Institute, Griffith University (Mt Gravatt Campus), Mt Gravatt, QLD 4122, Australia; grant@devilly.org
* Correspondence: andrew@scholeylab.com; Tel.: +44-(3)92148932

Received: 11 February 2019; Accepted: 21 March 2019; Published: 30 March 2019

Abstract: Results from studies into the cognitive effects of alcohol hangover have been mixed. They also present methodological challenges, often relying on self-reports of alcohol consumption leading to hangover. The current study measured Breath Alcohol Concentration (BAC, which was obtained via breathalyzer) and self-reported drinking behavior during a night out. These were then related to hangover severity and cognitive function, measured over the internet in the same subjects, the following morning. Volunteers were breathalyzed and interviewed as they left the central entertainment district of an Australian state capital. They were provided with a unique identifier and, the following morning, logged on to a website. They completed a number of measures including an online version of the Alcohol Hangover Severity Scale (AHSS), questions regarding number and type of drinks consumed the previous night, and the eTMT-B-a validated, online analogue of the Trail Making Test B (TMT-B) of executive function and working memory. Hangover severity was significantly correlated with one measure only, namely the previous night's Breath Alcohol Concentration ($r = 0.228$, $p = 0.019$). Completion time on the eTMT-B was significantly correlated with hangover severity ($r = 0.245$, $p = 0.012$), previous night's BAC ($r = 0.197$, $p = 0.041$), and time spent dinking ($r = 0.376$, $p < 0.001$). These findings confirm that alcohol hangover negatively affects cognitive functioning and that poorer working memory and executive performance correlate with hangover severity. The results also support the utility and certain advantages of using online measures in hangover research.

Keywords: hangover; alcohol; internet; attention; executive function; working memory

1. Introduction

The alcohol hangover (AH) is defined as "the combination of mental and physical symptoms, experienced the day after a single episode of heavy drinking, starting when blood alcohol concentration approaches zero" [1]. It describes a feeling of malaise that follows a bout of drinking when blood alcohol levels are at or returning to zero [2–7]. The AH is variously characterized by somatic and behavioral symptoms including headache, thirst, stomach upsets, negative mood, and cognitive problems.

There can be considerable inter-individual variability in the pattern, severity, and temporal characteristics of hangover symptoms [8], with no clear relationship between AH severity and any single physiological process (although cytokine response to alcohol is emerging as a possible key factor [9,10]). Other mechanisms that may contribute to AH include, but are not limited to, gut dysbiosis (including ghrelin-mediated), decreased blood glucose concentrations, poor sleep architecture, dehydration (and concomitant electrolyte imbalances), oxidative stress, and inflammatory responses [2,11,12]. These last two may in part be elevated in response to circulating ethanol metabolites.

The majority of previous research exploring the cognitive effects of alcohol has focused on acute intoxication, and the long-term neurocognitive consequences of alcohol dependence [7,13,14]. Acute intoxication impairs aspects of memory, attention, and psychomotor performance [15–20]. Alcohol also produces a characteristic shift in the speed/accuracy trade-off. Unlike other impairing drugs which tend to slow responding, alcohol typically increases error rates with little effect on response speed [16,17,21].

Compared with alcohol intoxication, relatively little research has been directed at the specific cognitive effects of AH [22,23]. A recent meta-analysis of next-day cognitive effects of heavy alcohol consumption included 19 studies published in 11 articles since 1970 [24]. It concluded that sustained attention, short- and long-term memory, and psychomotor speed are the cognitive domains most susceptible to hangover.

Any hangover-related cognitive impairment could have major implications for everyday activities. For example, hangover impairments to driving ability were similar to those observed at Breath Alcohol Concentrations (BAC) of 0.05–0.08% [25], that is similar to those. Such impairments have clear ramifications for safety-sensitive occupations, but also negatively impact on those that continue to engage in everyday activities while in a hangover state. In the context of absenteeism and presenteeism, it has been estimated that alcohol hangover costs the US economy 179 billion annually in lost productivity [26].

There are a number of methodological approaches to the study of hangover effects on cognition (see Stephens et al., 2014 for a critical review) [2]. These include laboratory studies where controlled doses of alcohol are administered, usually in a relatively pure form (typically vodka), and cognitive outcomes are measured once BACs have returned to zero. Alcohol is either administered at fixed doses or titrated to reach a target BAC. This approach has the advantage of providing relatively high levels of control, particularly regarding the timing of alcohol administration and measurement of physiological and functional endpoints. On the other hand, it may not have high ecological validity. For example, in real-life drinking situations individuals may consume a variety of beverages over different lengths of time. Secondly, because Ethics Boards tend to err on the side of caution, laboratory studies typically use lower levels of alcohol than those observed in the field. Even when bar-like settings are simulated in the laboratory there tends to be a limit on target alcohol levels.

An alternative methodology in AH research is to use a so-called 'naturalistic' design. Here participants visit the laboratory on two mornings, one after a night's drinking and another after a sober night (with order counterbalanced across participants) [27]. This method has the advantage of ecological validity by not limiting participants' drinking. On the other hand, the approach relies on recollection of levels of alcohol consumed to generate an estimated Blood Alcohol Level (eBAC). Given that alcohol intoxication is associated with memory problems (as may be hangover), this is problematic when trying to explore the relationship between alcohol consumed and functional consequences of AH.

The relative utility of these approaches is illustrated by disparate findings regarding cognitive impairments associated with hangover [28]. For example, psychomotor deficits associated with AH were observed in naturalistic studies [23,29,30] but not in laboratory settings [31–33].

The current study took a somewhat different approach. Over the past decade or so, internet studies have been increasingly used in psychological research, including to evaluate the impact of substance use and complement field and laboratory studies on the effects of recreational drugs and alcohol [14,16,17,34–42]. Here, we employed a mixed methodology involving a subgroup of individuals who were taking part in ongoing field studies, SmartStart [43] and Last Drinks [44], into patterns and consequences of alcohol consumption. These studies included a cohort of patrons who were breathalyzed while leaving the entertainment area of an Australian capital city and agreed to be contacted the following morning to complete an internet study of hangover. This included a version of the Alcohol Hangover Severity Scale (AHSS [28]) and an online analogue of the Trail Making Test B (TMT-B), named the eTMT-B-a test of psychomotor function, working memory, and executive function. This task was chosen as it is relatively brief, the task demands are easily understood, and it captures elements of the major cognitive domains (psychomotor function, attention, and executive function) affected by AH [24,45].

This approach has several advantages. Next-day symptomatology and performance can be linked to measured alcohol levels and drinking characteristics, collecting data via the internet is relatively convenient in that it does not require travelling to a testing location, and the method can address certain methodological questions—for example what factors contribute to attrition in hangover studies.

We hypothesized that previous night BAC would be related to both worse hangover severity and cognitive performance as measured using the eTMT-B. Further analyses explored the influence of beverage types and patterns of drinking on hangover severity and cognitive performance.

2. Experimental Section

2.1. Pilot Study

A pilot study was conducted to evaluate the validity of the online eTMT-B. Twenty-four young adult volunteers (17 female, mean age 30.29 years, SD 5.03) completed both the pencil-and-paper and online versions of the test (with order counterbalanced across participants).

The traditional TMT-B requires participants to draw lines connecting 25 circles distributed on a page containing single digits and letters. Correct completion involves joining the stimuli alternating between ascending numbers and letters (e.g., 1-A-2-B-3-C ..., etc.), with completion time as the main outcome [29]. The TMT-B was administered according to the published protocol [46], with the exception that errors were not corrected (as this would not be possible in the online version).

The eTMT-B was designed as an online analogue of the Pencil-and-Paper TMT-B. It consisted of a 5 × 5 grid of rectangular panels each labeled with a digit or number. Similar to the paper version, the task involved clicking on alternating ascending numbers and digits starting with 1-A and so on (see Figure 1). Once a panel was pressed it changed color from white to grey, after which clicking on it had no effect. If a participant attempted to press an incorrect button, no response could be made and an incorrect response was recorded. Both versions were scored for errors and completion time.

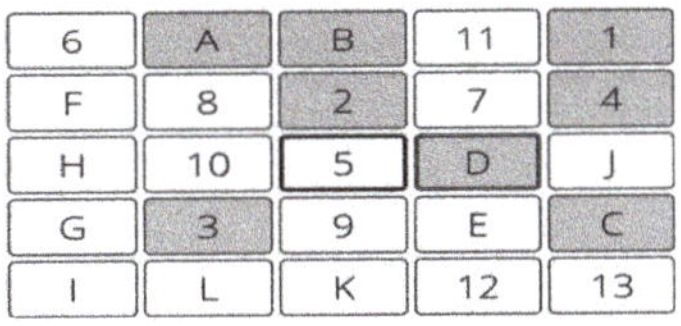

6	A	B	11	1
F	8	2	7	4
H	10	5	D	J
G	3	9	E	C
I	L	K	12	13

Figure 1. Layout of the online analogue of the Trail Making Test B, eTMT-B. The task requires participants to click ascending numbers and letters alternating between numbers and figures (i.e., 1-A-2-B-3-C, etc.).

2.2. Main Study

2.2.1. Design

This study formed part of a larger series of studies aimed at determining patterns and drivers for drinking in and around an Australian state capital. This part of the study employed a mixed methodology approach, whereby individuals were approached as they left the Central Entertainment District (CED) and then contacted the following morning to conduct an internet study.

The study was approved by the ethics committees of both Griffith University (2015/704) and Swinburne University (2016/167).

When first engaged during the CED phase, participants were presented verbally with a summary of the nature of the study and what their involvement would entail. Survey and breathalyzer measures were completed only if the participant provided verbal consent for both the concurrent and next morning measures (as approved by the ethics committee) as is usual for this kind of research. Each participant was given a unique identifier card with a link to the study information sheet (www.last-drinks.com.au). They were informed that they would be allowed to withdraw their data (using the anonymous ID number given on the card) and obtain further information regarding the study if they so wished.

2.2.2. Participants

One hundred and five participants provided usable datasets. They were recruited at and around exit points (taxi ranks and train station) in the central entertainment area of Brisbane, Queensland, Australia.

2.2.3. Breath Alcohol

Breath alcohol levels were measured using an Alcolizer LE5 (Alcolizer Pty Ltd, Australia). This device is used by law enforcement agencies throughout Australia and South East Asia, is Australia Standard 3547 certified, and has an accuracy of greater than 0.005 at 0.100 BAC g/100 mL. It has been demonstrated to have high reliability and validity for measuring intoxication in a sample of people attending nighttime entertainment districts [47].

2.2.4. Online Measures

A website survey was constructed which included questions collecting demographic and morphometric information (birth year, gender, weight, and height). Participants were asked specifically about the number of beers/ciders, glasses of wine, shots (unmixed), and alcohol mixed with either energy drinks or other beverages they had consumed the previous night and on a typical night out. This was followed by an 11-item version of the Alcohol Hangover Severity Scale (AHSS [28]), comprising of 11-point scales with endpoints (0 and 10) labelled as 'absent' and 'extreme'. Individual items were 'sweating', 'confusion', 'thirst', 'nausea', 'fatigue (being tired)', 'heart pounding', 'dizziness', 'shivering', 'clumsiness', 'apathy (lack of interest/concern)', and 'stomach pain'. Note that the 'difficulty concentrating' item was omitted to reduce expectancy effects while engaging in a task requiring an element of focused attention.

The next sections asked questions regarding the previous night's alcohol consumption (specifically numbers and types of beverages), sleep, and qualitative hangover data (these will be reported elsewhere). Alcohol intake data were collected according to the number of standard drinks consumed for each drink type (an Australian standard drink contains 10 grams of alcohol). Participants were provided with a link to the Better Health Channel online drink calculator where they were shown images of glasses containing an Australian standard drink. Participants were able to compute the number of standard drinks they had consumed by moving a slider to add or remove alcohol from the glass. Calculations were based on the average alcohol content for each drink type. For example, red wine was calculated at 13% alcohol, white wine at 11.5% alcohol, and full-strength beer at 4.8% alcohol. The full list of average alcohol contents by drink type can be found at http://mapi.betterhealth.vic.gov.au/saywhen/my-drinking/calculator.

Participants were then directed to a website with the eTMT-B. Following completion of the online test, a debriefing page was displayed, and participants were given details for entry to an online competition to win an iPad. They were also compensated with a $15 AUD iTunes voucher if they provided an email address.

2.2.5. Procedure

Individuals were recruited by being approached as they left the central entertainment area of Brisbane, Queensland. If willing to be interviewed, they provided consent, were breathalyzed, and gave details regarding their consumption of alcohol (including time drinking and number of drinks) that night. Those with BACs around 0.05% or above were asked if they agreed to be contacted the following morning. Those consenting (N = 346) provided their contact details and estimated bedtime for that night. Participants were given a card with a unique identifier which allowed anonymous linking of their BAC and drinking data to the data collected over the internet.

The following morning individuals were contacted by text message 8 h following their estimated bedtime and again 6 h later if they had not completed the survey. If participating, they entered their unique identifier and then completed the webpages as described above.

2.2.6. Statistics and Analysis

Number of drinks recorded were converted to standard drinks (one Australian standard drink is equivalent to 10 g alcohol). Initial analyses involved exploring the relationship between BAC, hangover severity, and cognitive performance with various drinking factors. These included number of standard drinks, time drinking, and number of specific types of drink. The relationships between these factors were analyzed using Pearson's correlations. To determine whether BAC affected participation in the hangover part of the study, t-tests compared BACs of those who did and did not participate in the next-day online phase of the study.

3. Results

3.1. Pilot Study

Response times were similar between the paper-and pencil and internet platforms. With a mean completion time of 40.27 s (range 14.71–71.59) for the paper version of the TMT-B and 41.49 s (23.02–75.10) for the eTMT-B (t(23) = 0.116, p = 0.909). These were significantly correlated between individuals completing the two platforms (r = 0.499, p = 0.013).

Published norms for completion of the paper version of the task are typically around 50 s [44,48]. Considering our sample were predominantly students and the paper version involves error correction, our figures are in line with normative data. Errors were rare on both platforms with 4 people making errors on the paper version and 10 people on the eTMT-B. The vast majority of these were single errors.

3.2. Main Study

Of 346 participants who were breathalyzed and indicated that they may be prepared to participate in the next-day phase of the study, 105 provided complete online datasets. Sample characteristics of this cohort are presented in Table 1.

Table 1. Demographic and morphometric data of sample. Apart from gender figures are means with range in parentheses.

N	105
Males/Females (%)	51.4/48.6
Age	24.7 (17–49) years
Weight	74.0 (43–115) kg
BMI	24.15 (16.58–40.28)

Drinking characteristics of the cohort are presented in Table 2. The sample had a mean BAC of 0.11% (SD ± 0.40). They reported drinking for an average of 7.45 hours and had consumed a mean number of 13.5 standard drinks. Analyses of next-day reports revealed that the most consumed drink was alcohol mixed with non-energy drink mixers (N = 83 drinkers, consuming a mean of 7.27 drinks) followed by beer/cider (N = 59, mean = 7.16), shots (N = 43, mean = 3.51), wine (N = 31, mean = 5.68), with alcohol mixed with energy drinks being the least consumed beverage both in terms of number of drinkers and average number of beverages (N = 30, mean = 2.87).

Table 2. Reported drinking characteristics, including types of drinks consumed on the night of data collection, Breath Alcohol Concentration (BAC), and hours drinking.

	N	Mean	SD	Max
Drinks consumed				
Beer/cider	59	7.16	5.86	30
Wine	31	5.68	4.56	20
Shots (unmixed)	43	3.51	3.21	16
Alcohol mixed with Energy Drink	30	2.87	1.98	8
Alcohol mixed with Other Beverage	83	7.27	5.80	40
Total	105	13.48	5.94	35
Drinking measures				
BAC (%)	-	0.110	0.040	0.25
Hours drinking	-	7.45	4.09	17

The main focus of the study was to examine factors associated with hangover severity and cognitive performance (see Figure 2).

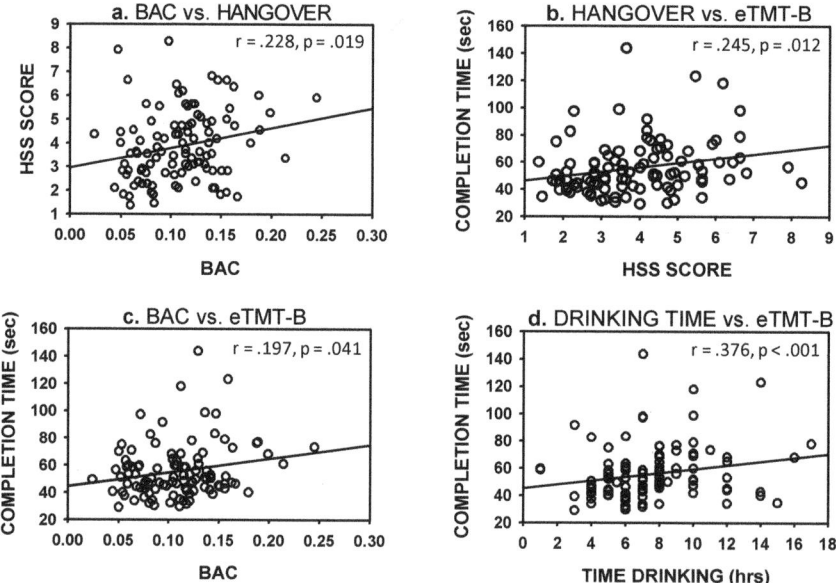

Figure 2. Graphs depicting significant associations between (**a**). previous night's breath alcohol content (BAC) and hangover severity (Hangover Severity Scale (HSS) score), (**b**). hangover severity and cognitive performance (eTMT-B), (**c**). BAC and cognitive performance (**d**). drinking time and cognitive performance.

Hangover severity was significantly related to one measure only, namely BAC ($r = 0.228$, $p = 0.019$). The correlation between total number of standard drinks and HSS gave a value of $r = 0.184$, $p = 0.064$. Speed of completion of eTMT-B correlated with self-rated hangover severity ($r = 0.245$, $p = 0.0120$, previous night's BAC ($r = 0.197$, $p = 0.04$), and drinking time ($r = 0.376$, $p < 0.001$).

Further analyses revealed that BAC correlated significantly with number of standard drinks consumed ($r = 0.486$, $p < 0.001$) and time drinking ($r = 0.376$, $p < 0.001$). Examination of the relationship between individual drinks and alcohol levels revealed significant correlations between BAC and amount of alcohol consumed as beer/cider ($r = 0.361$, $p = 0.005$), wine ($r = 0.398$, $p = 0.026$), and alcohol mixed with other beverages ($r = 0.228$, $p = 0.038$) but not between BAC and shots alone or alcohol mixed with energy drink (Table 3).

Table 3. Correlations between previous night's BAC, hangover severity, and drinking characteristics.

	BAC (%)	Time Drinking (h)	HSS Score	Standard Drinks (N)	Beer/Cider (N)	Wine (N)	Shots (N)	AMED (N)	AMOB (N)
N	105	105	105	102	59	31	43	30	83
BAC (%)	-	0.376 ***	0.228 *	0.486 ***	0.361 **	0.398 *	−0.028	−0.099	0.228 *
Time drinking (h)		-	0.148	0.633 ***	0.503 ***	0.223	0.166	0.068	0.257 *
HSS score			-	0.184	0.171	0.293	−0.010	0.019	0.018
Standard drinks (N)				-	0.623 ***	0.437 *	0.249	0.360	0.369 **
Beer/cider (N)	-	-	-	-	-	−0.388	−0.047	0.033	−0.255
Wine (N)	-	-	-	-	-	-	−0.131	0.162	−0.284
Shots (N)	-	-	-	-	-	-	-	0.102	−0.067
AMED (N)	-	-	-	-	-	-	-	-	−0.023

BAC = Blood Alcohol Content, HSS = Hangover Severity Scale, AMED = alcohol mixed with energy drinks, AMOB = alcohol mixed with other beverage. Significant correlations are indicated in bold (*, $p < 0.05$; **, $p < 0.01$; ***, $p < 0.0001$). Drinking characteristics including types of drinks consumed reported on the night of data collection, BAC, and hours drinking.

Time drinking significantly correlated with standard drinks consumed in total ($r = 0.633$, $p < 0.001$), and as beer/cider ($r = 0.503$, $p < 0.001$) and alcohol mixed with other beverage (AMOB, $r = 0.257$, $p = 0.019$). Standard drinks significantly correlated with drinks consumed as beer/cider ($r = 0.623$, $p < 0.001$), wine ($r = 0.437$, $p = 0.014$), and AMOB ($r = 0.369$, $p = 0.001$). There were no other significant correlations.

There was no difference in the BACs of those who did ($x = 0.11\% \pm SD\ 0.0408$) and did not ($x = 0.11\% \pm SD\ 0.0405$) complete the online hangover phase of the study ($t(345) = 0.240$, $p = 0.81$).

4. Discussion

Using mixed field and internet methodology, we found that hangover severity is significantly related to BAC and both are associated with worse performance on the eTMT-B test of attention and executive function. Drinking time was also associated with BAC and with worse performance on the task (though not with hangover score).

The current methodology allowed the measurement of BACs and next morning collection of data regarding types of alcohol consumed, evaluation of severity of alcohol hangover, and cognitive functioning. The study confirmed that previous night's BAC was significantly associated with hangover severity. No other measure was significantly correlated with hangover scores despite some intercorrelations with other measures and BAC including type of beverage consumed.

It has been suggested that different types of alcoholic beverage may influence hangover severity. In particular it has been suggested that non-alcohol constituents of drinks, known as congeners, may differentially affect AH [49]. In particular, it has been suggested that congener-rich drinks such as whiskey produce worse hangover symptoms than beverages with essentially no congeners, such as vodka, although there is little systematic research in this area. Our (albeit limited) data do not suggest that different alcohol types contribute differentially to AH symptomatology. Similarly, different types of mixers had no differential effect on hangover (symptom) severity.

Conversely there was some evidence that certain drink types were more closely associated with BAC. As well as correlating with amount of time drinking, BAC was related to the amount of beer/cider, wine, and AMOBs reportedly consumed (but not shots, or alcohol mixed with energy drinks—AMED). It seems unlikely that the nature of beverage consumed could differentially contribute to BAC. AMOBs were the most commonly consumed drinks (by $N = 83$ people, consuming $\bar{x} = 7.27$ standard drinks), followed by beer/cider ($N = 59$, $\bar{x} = 7.16$), and shots ($N = 43$, $\bar{x} = 3.51$). AMEDs were the least consumed beverages ($N = 30$, $\bar{x} = 2.87$), and although wine was consumed by a similar number of people ($N = 31$), on average it was consumed at a higher level ($\bar{x} = 7.16$ drinks). Thus, our data confirm other findings using other methodology [50] showing that irrespective of the type of alcoholic drink or mixer, the most meaningful association was between the number of drinks consumed and BAC.

The error rate on the eTMT-B was generally low, with more than half of the sample (54%) making no errors, and around one quarter (26%) making one or two errors only. This suggests that participants engaged with the eTMT-B and understood task demands of the online version. This confirms the results of the pilot study where there was high correspondence between the paper and online versions of the task. Thus AH-related impairment was largely manifest by slowed function. A pattern of slower reaction times with increasing hangover severity would differentiate alcohol hangover from alcohol intoxication. The latter is typified by a characteristic shift in the speed/accuracy trade-off (SATO) with intoxication leading to more errors while having relatively little effect on reaction times [16,17,21,51]. Since few errors were made on either version of the TMT-B however, we cannot draw strong conclusions from this limited dataset. Nevertheless, the literature does suggest that slowing of response times during AH is more robust than increased errors [24,52]. One focus of this study was to implement a cognitive task that was sensitive to impairment and could be used online (and thus be relatively brief with clear task demands). Clearly, future studies would benefit from a more comprehensive assessment of working memory and executive functions as well as other cognitive domains.

The current study had certain advantages over previous studies applying naturalistic methodology. For example, rather than relying on next-day recall of previous night's drinks, we had an objective measure of BACs. Additionally, the number of alcoholic beverages consumed and length of time drinking were recorded on the night, making these data less susceptible (though not invulnerable) to recall bias.

While we can be confident that measured BACs were accurate [47], we do not know which phase of drinking participants were in when breathalyzed. Specifically, we do not know if BACs were measured at peak alcohol levels or during the rising or falling limb of the blood alcohol curve. Further, although the BAC measurement occurred at Central Entertainment District exit points (taxi ranks and train stations), we cannot preclude the possibility that some participants continued to consume alcohol, which would affect hangover and related functional consequences.

Another issue is that approximately one third of the sample did not attempt the next-day measures. Drop-out rates in hangover studies are typically high (e.g., 70 % attrition in Grange et al. [53]), suggesting that the current methodology is as viable as others in this respect. One possibility is that those subjects who consume the most alcohol and/or have the worse hangover symptoms are less likely to complete the next-day measures. The current methodology allowed us to address this directly by comparing the BACs of those who did and did not complete the hangover part of the study. This showed that the BACs were more-or-less identical. Furthermore, the sample who completed the next-day measures had a large range of BACs (up to 0.245%), suggesting that reaching a relatively high BAC did not affect the ability to complete the next-day measures. This may be an advantage of the current methodology, as next-day measures were completed online and so were relatively convenient.

The similarity of BACs between the next-day completers and non-completers also partially addresses a potential ethical issue in this type of research, specifically pertaining to consent from intoxicated individuals. It is reasonable to assume that any individual who consented in error would simply not enter into the next-day data collection phase. The fact that previous night's BACs were statistically similar between those who did and did not complete next-day measures strongly suggests that level of intoxication did not affect consent to the extent that the sample were 'self-selecting' in this context.

5. Conclusions

In conclusion, using novel methodology, this study has confirmed that higher BACs and associated measures result in worse hangover symptoms and poorer performance on a newly-validated online measure of working memory and executive function—the eTMT-B. Such hangover-related impairments are likely to have clear, real-life ramifications. For example, they would impact on the ability to engage in complex behaviors and may explain some of the hangover-associated impairment of driving [25]. They are also likely to impinge on fundamental aspects of cognitive functioning. Moreover, our findings suggest that a mixed field/internet approach provides a novel and viable methodology for hangover research.

Author Contributions: Conceptualization, A.S., S.B., and G.J.D.; data curation, S.B., C.T., and E.A.; formal analysis, A.S., E.A., and G.J.D.; funding acquisition, G.J.D.; methodology, J.C.V.; project administration, S.B. and C.A.; resources, C.A.; software, J.K.; supervision, J.C.V. and G.J.D.; validation, J.K. and C.T.; visualization, A.S.; writing—original draft, A.S.; writing—review and editing, S.B., J.K., C.T., E.A., J.C.V., C.A., and G.J.D.

Funding: These data were drawn from a sample of a larger study funded by a grant to G.D. from the National Drug Strategy Law Enforcement Funding Committee (# 1314004). The funders had no role in the design of the study (in the collection, analyses, or interpretation of data), in the writing of the manuscript, or in the decision to publish the results.

Conflicts of Interest: The authors declare no conflict of interest.

References

1. Van Schrojenstein Lantman, M.; van de Loo, A.J.; Mackus, M.; Verster, J.C. Development of a definition for the alcohol hangover: Consumer descriptions and expert consensus. *Curr. Drug Abuse Rev.* **2016**, *9*, 148–154. [CrossRef] [PubMed]
2. Stephens, R.; Grange, J.A.; Jones, K.; Owen, L. A critical analysis of alcohol hangover research methodology for surveys or studies of effects on cognition. *Psychopharmacology* **2014**, *231*, 2223–2236. [CrossRef]
3. Prat, G.; Adan, A.; Sánchez-Turet, M. Alcohol hangover: A critical review of explanatory factors. *Hum. Psychopharmacol.* **2009**, *24*, 259–267. [CrossRef]
4. Penning, R.; van Nuland, M.; Fliervoet, L.A.; Olivier, B.; Verster, J.C. The pathology of alcohol hangover. *Curr. Drug Abuse Rev.* **2010**, *3*, 68–75. [CrossRef]
5. Swift, R.; Davidson, D. Alcohol hangover: Mechanisms and mediators. *Alcohol Health Res. World* **1998**, *22*, 54–60. [PubMed]
6. Verster, J.C. The alcohol hangover—A puzzling phenomenon. *Alcohol Alcohol.* **2008**, *43*, 124–126. [CrossRef] [PubMed]
7. Wiese, J.G.; Shlipak, M.G.; Browner, W.S. The alcohol hangover. *Ann. Int. Med.* **2000**, *132*, 897–902. [CrossRef] [PubMed]
8. Verster, J.C.; van Schrojenstein Lantman, M.; Mackus, M.; van de Loo, A.J.A.E.; Garssen, J.; Scholey, A. Differences in the Temporal Typology of Alcohol Hangover. *Alcohol. Clin. Exp. Res.* **2018**, *42*, 691–697. [CrossRef] [PubMed]
9. Van De Loo, A.J.A.E.; Slot, K.A.; Kleinjan, M.; Knipping, K.; Garssen, J.; Verster, J.C. Time-dependent changes in saliva cytokine concentrations during alcohol hangover: A comparison of two naturalistic studies. *Alcohol. Clin. Exp. Res.* **2016**, *40*, 95A.
10. Van de Loo, A.J.A.E.; van Schrojenstein Lantman, M.; Mackus, M.; Scholey, A.; Verster, J.C. Impact of mental resilience and perceived immune functioning on the severity of alcohol hangover. *BMC Res. Notes* **2018**, *11*, 526.
11. Tipple, C.T.; Benson, S.; Scholey, A. A review of the physiological factors associated with alcohol hangover. *Curr. Drug Abuse Rev.* **2016**, *9*, 93–98. [CrossRef] [PubMed]
12. Farokhnia, M.; Lee, M.R.; Farinelli, L.A.; Ramchandani, V.A.; Akhlaghi, F.; Leggio, L. Pharmacological manipulation of the ghrelin system and alcohol hangover symptoms in heavy drinking individuals: Is there a link? *Pharmacol. Biochem. Behav.* **2018**, *172*, 39–49. [CrossRef] [PubMed]
13. Brust, J. Ethanol and cognition: Indirect effects, neurotoxicity and neuroprotection: A review. *Int. J. Environ. Res. Public Health* **2010**, *7*, 1540–1557. [CrossRef] [PubMed]

14. Ling, J.; Heffernan, T.M.; Buchanan, T.; Rodgers, J.; Scholey, A.B.; Parrott, A.C. Effects of alcohol on subjective ratings of prospective and everyday memory deficits. *Alcohol. Clin. Exp. Res.* **2003**, *27*, 970–974. [CrossRef]
15. Ogden, E.J.; Moskowitz, H. Effects of alcohol and other drugs on driver performance. *Traffic Inj. Prev.* **2004**, *5*, 185–198. [CrossRef]
16. Scholey, A.B.; Benson, S.; Neale, C.; Owen, L.; Tiplady, B. Neurocognitive and mood effects of alcohol in a naturalistic setting. *Hum. Psychopharmacol.* **2012**, *27*, 514–516. [CrossRef]
17. Tiplady, B.; Franklin, N.; Scholey, A. Effect of ethanol on judgments of performance. *Br. J. Psychol.* **2004**, *95*, 105–118. [CrossRef]
18. Parrott, A.; Morinan, A.; Moss, M.; Scholey, A. *Understanding Drugs and Behaviour*; John Wiley & Sons Ltd.: Chichester, West Sussex, UK, 2005.
19. Farquhar, K.; Lambert, K.; Drummond, G.B.; Tiplady, B.; Wright, P. Effect of ethanol on psychomotor performance and on risk taking behaviour. *J. Psychopharmacol.* **2002**, *16*, 379–384. [CrossRef] [PubMed]
20. Maylor, E.A.; Rabbitt, P. Alcohol, reaction time and memory: A meta-anlysis. *Br. J. Psychol.* **1993**, *84*, 301–317. [CrossRef] [PubMed]
21. Mackay, M.; Tiplady, B.; Scholey, A.B. Interactions between alcohol and caffeine in relation to psychomotor speed and accuracy. *Hum. Psychopharmacol.* **2002**, *17*, 151–156. [CrossRef]
22. Van Schrojenstein Lantman, M.; Mackus, M.; van de Loo, A.J.A.E.; Verster, J.C. The impact of alcohol hangover symptoms on cognitive and physical functioning, and mood. *Hum. Psychopharmacol.* **2017**, *32*. [CrossRef] [PubMed]
23. Verster, J.C.; van Duin, D.; Volkerts, E.R.; Schreuder, A.H.; Verbaten, M.N. Alcohol hangover effects on memory functioning and vigilance performance after an evening of binge drinking. *Neuropsychopharmacology* **2003**, *28*, 740–746. [CrossRef] [PubMed]
24. Gunn, C.; Mackus, M.; Griffin, C.; Munafò, M.R.; Adams, S. A systematic review of the next-day effects of heavy alcohol consumption on cognitive performance. *Addiction* **2018**, *113*, 2182–2193. [CrossRef] [PubMed]
25. Verster, J.C.; Bervoets, A.C.; de Klerk, S.; Vreman, R.A.; Olivier, B.; Roth, T.; Brookhuis, K.A. Effects of alcohol hangover on simulated highway driving performance. *Psychopharmacology* **2014**, *231*, 2999–3008. [CrossRef]
26. Centers for Disease Control and Prevention. The Cost of Excessive Alcohol Use. 2015. Available online: www.cdc.gov/alcohol/onlinemedia/infographics/cost-excessive-alcohol-use.html (accessed on 10 February 2019).
27. McKinney, A.; Coyle, K.; Penning, R.; Verster, J.C. Next day effects of naturalistic alcohol consumption on tasks of attention. *Hum. Psychopharmacol.* **2012**, *27*, 587–594. [CrossRef]
28. Penning, R.; McKinney, A.; Bus, L.D.; Olivier, B.; Slot, K.; Verster, J.C. Measurement of alcohol hangover severity: Development of the Alcohol Hangover Severity Scale (AHSS). *Psychopharmacology* **2013**, *225*, 803–810. [CrossRef] [PubMed]
29. McKinney, A.; Coyle, K. Next-day effects of alcohol and an additional stressor on memory and psychomotor performance. *J. Stud Alcohol Drugs* **2007**, *68*, 446–454. [CrossRef] [PubMed]
30. McKinney, A.; Coyle, K. Next day effects of a normal night's drinking on memory and psychomotor performance. *Alcohol Alcohol.* **2004**, *39*, 509–513. [CrossRef]
31. Collins, W.E.; Chiles, W.D. Laboratory performance during acute alcohol intoxication and hangover. *Hum. Factors* **1980**, *22*, 445–462. [CrossRef]
32. Kelly, M.; Myrsten, A.-L.; Neri, A.; Rydberg, U. Effects and after-effects of alcohol on psychological functions in man—A controlled study. *Blutalkohol* **1970**, *7*, 422–436.
33. Kruisselbrink, L.D.; Martin, K.L.; Megeney, M.; Fowles, J.R.; Murphy, R.J. Physical and psychomotor functioning of females the morning after consuming low to moderate quantities of beer. *J. Stud. Alcohol* **2006**, *7*, 416–420. [CrossRef]
34. Heffernan, T.M.; Jarvis, H.; Rodgers, J.; Scholey, A.B.; Ling, J. Prospective memory, everyday cognitive failure and central executive function in recreational users of Ecstasy. *Hum. Psychopharmacol.* **2001**, *16*, 607–612. [CrossRef] [PubMed]
35. Parrott, A.C.; Buchanan, T.; Scholey, A.B.; Heffernan, T.; Ling, J.; Rodgers, J. Ecstasy/MDMA attributed problems reported by novice, moderate and heavy recreational users. *Hum. Psychopharmacol.* **2002**, *17*, 309–312. [CrossRef] [PubMed]
36. Parrott, A.C.; Rodgers, J.; Buchanan, T.; Ling, J.; Heffernan, T.; Scholey, A.B. Dancing hot on Ecstasy: Physical activity and thermal comfort ratings are associated with the memory and other psychobiological problems reported by recreational MDMA users. *Hum. Psychopharmacol.* **2006**, *21*, 285–298. [CrossRef]

37. Rodgers, J.; Buchanan, T.; Scholey, A.B.; Heffernan, T.M.; Ling, J.; Parrott, A. Differential effects of Ecstasy and cannabis on self-reports of memory ability: A web-based study. *Hum. Psychopharmacol.* **2001**, *16*, 619–625. [CrossRef]
38. Rodgers, J.; Buchanan, T.; Scholey, A.B.; Heffernan, T.M.; Ling, J.; Parrott, A.C. Patterns of drug use and the influence of gender on self-reports of memory ability in ecstasy users: A web-based study. *J. Psychopharmacol.* **2003**, *17*, 389–396. [CrossRef]
39. Scholey, A.B.; Parrott, A.C.; Buchanan, T.; Heffernan, T.M.; Ling, J.; Rodgers, J. Increased intensity of Ecstasy and polydrug usage in the more experienced recreational Ecstasy/MDMA users: A WWW study. *Addict. Behav.* **2004**, *29*, 743–752. [CrossRef] [PubMed]
40. Verster, J.C.; Benson, S.; Scholey, A. Motives for mixing alcohol with energy drinks and other nonalcoholic beverages, and consequences for overall alcohol consumption. *Int. J. Gen. Med.* **2014**, *7*, 285. [CrossRef] [PubMed]
41. Scholey, A.B.; Fowles, K.A. Retrograde enhancement of kinesthetic memory by alcohol and by glucose. *Neurobiol. Learn. Mem.* **2002**, *78*, 477–483. [CrossRef] [PubMed]
42. Scholey, A.; Benson, S.; Stough, C.; Stockley, C. Effects of resveratrol and alcohol on mood and cognitive function in older individuals. *Nutr. Aging* **2014**, *2*, 133–138.
43. Devilly, G.J.; Allen, C.; Brown, K. SmartStart: Results of a large point of entry study into preloading alcohol and associated behaviours. *Int. J. Drug Policy* **2017**, *43*, 130–139. [PubMed]
44. Devilly, G.J.; Hides, L.; Kavanagh, D.J. A Big Night out Getting Bigger: Alcohol Consumption, Arrests and Crowd Numbers, before and after Legislative Change. *PLoS ONE* **2019**, submitted for publication.
45. Tombaugh, T.N. Trail Making Test A and B: Normative data stratified by age and education. *Arch. Clin. Neuropsychol.* **2004**, *19*, 203–214. [CrossRef]
46. Bowie, C.R.; Harvey, P.D. Administration and interpretation of the Trail Making Test. *Nat. Protoc.* **2006**, *1*, 2277–2281. [PubMed]
47. Sorbello, J.G.; Devilly, G.J.; Allen, C.; Hughes, L.R.J.; Brown, K. Fuel-cell breathalyser use for field research on alcohol intoxication: An independent psychometric evaluation. *PeerJ* **2018**, *6*, e4418. [CrossRef] [PubMed]
48. Fernandez, A.L.; Marcopulos, B.A. A comparison of normative data for the Trail Making Test from several countries: Equivalence of norms and considerations for interpretation. *Scand. J. Psychol.* **2008**, *49*, 239–246. [CrossRef]
49. Rohsenow, D.J.; Howland, J. The role of beverage congeners in hangover and other residual effects of alcohol intoxication: A review. *Curr. Drug Abuse Rev.* **2010**, *3*, 76–79. [PubMed]
50. Köchling, J.; Geis, B.; Wirth, S.; Hensel, K.O. Grape or grain but never the twain? A randomized controlled multiarm matched-triplet crossover trial of beer and wine. *Am. J. Clin. Nutr.* **2019**, *109*, 345–352.
51. Benson, S.; Tiplady, B.; Scholey, A. Attentional and working memory performance following alcohol and energy drink: A randomised, double-blind, placebo-controlled, factorial design laboratory study. *PLoS ONE* **2019**. [CrossRef]
52. Stephens, R.; Ling, J.; Heffernan, T.M.; Heather, N.; Jones, K. Review A review of the literature on the cognitive effects of alcohol hangover. *Alcohol Alcohol.* **2008**, *43*, 163–170. [CrossRef]
53. Grange, J.A.; Stephens, R.; Jones, K.; Owen, L. The effect of alcohol hangover on choice response time. *J. Psychopharmacol.* **2016**, *30*, 654–661. [CrossRef] [PubMed]

© 2019 by the authors. Licensee MDPI, Basel, Switzerland. This article is an open access article distributed under the terms and conditions of the Creative Commons Attribution (CC BY) license (http://creativecommons.org/licenses/by/4.0/).

Article

The Impact of Alcohol Hangover on Simulated Driving Performance during a 'Commute to Work'—Zero and Residual Alcohol Effects Compared

Chris Alford [1,*], Callum Broom [1], Harriet Carver [1], Sean J. Johnson [2], Sam Lands [1], Rebecca Reece [1] and Joris C. Verster [3,4]

1. HAS - Health and Social Sciences, University of the West of England, Bristol BS16 1QY, UK; Callum_Broom@hotmail.com (C.B.); Hatty_Carver@hotmail.com (H.C.); Sam.Lands@alliancepharma.co.uk (S.L.); Rebecca.Reece@uwe.ac.uk (R.R.)
2. Centre for Trials Research, Cardiff University, Cardiff CF14 4YS, UK; JohnsonS11@cardiff.ac.uk
3. Division of Pharmacology, Utrecht Institute for Pharmaceutical Sciences (UIPS), Utrecht University, 3584CG Utrecht, The Netherlands; j.c.verster@uu.nl
4. Centre for Human Psychopharmacology, Swinburne University, Melbourne, VIC 3122, Australia
* Correspondence: chris.alford@uwe.ac.uk

Received: 31 March 2020; Accepted: 5 May 2020; Published: 12 May 2020

Abstract: Driving is increasing across the world and road traffic accidents are a major cause of serious injuries and fatalities. The link between alcohol consumption and impaired driving has long been established and has led to legislation in many countries, with enforcement of legal limits based on blood alcohol concentration levels. Alcohol hangover research is an emerging field with a range of laboratory and naturalistic studies now clearly demonstrating the significant impairments that can result from hangover, even when alcohol levels are measured at or close to zero the day following a social drinking occasion. Driving is a commonplace activity but requires competency with a range of complex and potentially demanding tasks. Driving impaired can have serious consequences, including death and serious injury. There have been only limited alcohol hangover driving studies. The studies presented examined the consequences of alcohol hangover with a driving simulator contrasting a group with zero residual alcohol (N = 26) next day and another with residual alcohol (N = 26) assessed with breathalyzer in the morning before undertaking a 20 min commute to work. All participants completed a morning drive after a night without alcohol consumption and another after a night of social drinking. The driving scenarios were relatively demanding including traffic and pedestrians, traffic lights and other potential hazards in a mixed rural and urban journey. Subjective hangover and workload were assessed in addition to a range of driving performance variables, including divided attention, steering control and driving violations. Analyses contrasted driving in the no alcohol condition with the residual alcohol condition. The combined groups data (N = 52) was contrasted with the zero and residual alcohol groups. Significant contrasts were found for a range of driving measures, including divided attention, vehicle control, and driving violations as well as perceived workload. The pattern of impairment was broadly similar across both groups, indicating that whether or not residual alcohol was present, consistent driving impairment was seen. The relatively high number of significant variables may reflect the increased cognitive demand of the 20 min commute drive including busy and complex urban environments. This was also reflected in the significant increase in perceived workload recorded across the 6 dimensions of the National Aeronautics and Space Administration Task Load Index (NASA-TLX). Associations between subjective measures and driving performance with hangover suggested a potential lack of awareness of impairment, though were limited in number. The overall findings indicate that the levels of impairment seen reflect those seen with alcohol impaired driving, even when breath alcohol is zero.

Keywords: alcohol; awareness of impairment; hangover; driving; residual alcohol

1. Introduction

Road traffic accidents are a leading cause of mortality, with around 1.35 million deaths recorded worldwide in 2016, ranked in the top 10 causes of mortality, the leading cause of death for 5–29 year olds, and with up to 35% of road accident deaths estimated as being alcohol related [1]. Whilst there has been much research into the impairing effects of alcohol intoxication on driving performance [2–5], there has been only limited research into the effects of alcohol hangover on driving [6–11]. Alcohol hangover is the most commonly experienced consequence of alcohol consumption [12] and is defined as the combination of negative mental and physical symptoms which can be experienced after a single episode of alcohol consumption, starting when blood alcohol concentration (BAC) approaches zero [13,14]. The annual costs of alcohol hangover to the economy in terms of absenteeism and presenteeism have been estimated at 169 billion dollars for the US and 4 billion pounds for the UK [15,16], corresponding with a significant loss of productivity [17]. The neurocognitive and psychomotor effects of alcohol hangover have been demonstrated in both laboratory and field studies [18,19], showing significant levels of impaired function on a range of tasks and impacting mood and subjective state in some but not all studies.

The one hour (100 km) on-road highway driving test is considered the 'gold standard' of driving assessments and has been the basis of a large body of research in the Netherlands examining the impact of acute dosing as well as residual effects of a range of psychoactive compounds including alcohol [20]. The test has also been successfully implemented in driving research demonstrating the acute and after effects of alcohol [9,21]. The principal measure is the standard deviation of lateral position (SDLP) that captures the weaving behavior of the car as it travels along the highway or motorway environment. The more impaired a driver is then the greater the weaving and increased measure of SDLP. The highway driving test was developed to measure monotonous driving performance with low external stimuli, in order to capture basic driving performance. However, other driving situations such as city driving comprise more skills than primarily steering control and vigilance performance. Different driving environments impose different levels of cognitive demand on the driver. The one hour driving test is undertaken in a relatively undemanding driving environment for a competent driver, such as a highway or motorway setting with an identified and clearly marked driving lane and without oncoming traffic. Driving assessments made under more challenging conditions have found significant impairment in driving control at blood alcohol concentration as low as 0.02%–0.03% BAC [3,22], and it has been shown that driving distractions can produce a two-fold increase in alcohol induced driving impairment [4].

A wide range of cognitive and psychomotor functions including planning, working memory, attention and perceptual motor tasks become more important in more cognitively demanding tasks such as a 'commute to work'. Modern simulators include a range of measures which complement SDLP and may provide a more comprehensive assessment of driving abilities needed in these circumstances. Whilst the one hour test has a significant bibliography to support it and provides comparators for assessing any new psychoactive compounds and medications, this does not represent the typical morning 'commute to work', which can include a range of rural as well as busy urban environments and is of shorter duration [23]. There is a need to assess shorter driving durations as the UK 2018 National Travel Survey [24] found that the average car journey was found to be just over 20 min. Typical commutes to work have been reported as averaging around 20–30 min in the UK in 2018 [23]. Given that a significant number of the working population will use a car for their morning commute, then the impact of alcohol hangover on a typical commute to work requires further investigation.

To date, there have been relatively few hangover driving studies [6–11]. Verster et al. [10] surveyed 343 Dutch truck drivers who reported impaired driving after drinking the previous night. Of concern, the drivers that participated in this study acknowledged continued driving, both professional and private, despite being aware that being hungover impaired their driving. Early driving studies reported impairment during hangover when performing swerve maneuvers around cones on a closed road, and performance on a high speed driving test in a simple driving simulator [7,8]. More recently, Verster et al. [9] reported significant driving impairment (increased SDLP) on a standardized 100-km highway driving test in the STISIM driving simulator. Significant increases in SDLP occurred mostly in the last half of the simulated drive. Robbins et al. [11] examined driving during hangover employing a 15 min test of simulated motorway driving, including hazards. They found that in the hangover condition participants, who mostly had residual alcohol though below the legal limit, drove at higher speeds and for longer periods over the speed limit, with greater speed variation interpreted as increased likelihood of traffic violations, although attention was not impaired. Together, these studies suggest that driving ability during hangover is impaired at various performance levels, which may also be present during a 'commute to work' drive.

Awareness of impairment plays an important role in determining if someone drives as well as affecting their driving behavior. Self estimates of acute alcohol intoxication vary with the level of consumption such that lower levels below the UK driving limit 0.08% BAC may be accurate or overestimates, transitioning to underestimates for higher levels [25,26]). There has been relatively little published research into the effects of alcohol hangover on self estimates of induced impairment, and yet this is an important aspect given that alcohol hangover cannot easily be quantified in the same way as a breathalyzer might assess alcohol induced impairment.

The current study was set out to assess the next-day consequences of a social drinking session on driving simulator performance on a representative commute to work drive. The study employed the commonly accepted naturalistic design [27]. The primary research question was to examine whether 'commute to work' driving is impaired during alcohol hangover. Previous driving studies investigating both acute alcohol intake and alcohol hangover have found significant decrements in performance despite increased mental effort [9]. This is therefore an important measure, particularly where cognitive demand is increased, and was therefore included in the current study. Awareness of impairment is also an important driving safety related factor and so the association between subjective state and driving performance was assessed.

Finally, the current definition of alcohol hangover states that the alcohol hangover starts when BAC approaches zero [13,14]. The few currently published driving studies did not differentiate between subjects that have a BAC of zero and subjects with residual alcohol still present [9,11] and included all subjects in the statistical analysis where they were treated as a single sample. However, given the established driving impairments found with acute alcohol administration at low dosages [3,22], it may be important to distinguish residual alcohol effects from hangover effects. Therefore, the current study directly contrasted hungover subjects with residual BAC with those with a BAC of zero.

2. Methods

2.1. Design and Participants

A within-subjects naturalistic design was used to investigate driving performance and subjective state contrasting two conditions: alcohol hangover (the morning after a social drinking occasion) and an alcohol-free control day (the morning after no alcohol consumption). Participants were free to arrange or postpone assessment days as they chose, depending on whether they had consumed alcohol on their social drinking occasion. The order of conditions was counterbalanced across participants, as was the order for the driving scenarios used to assess driving. The data represents the combination of three studies which were approved by the University of the West of England Faculty Ethics Committee (Approval numbers: ASP12-276, HAS.14.12.74, HAS.16.12.069). Informed consent was obtained from all participants, and the studies were undertaken in accordance with the British Psychological Society Code of Ethics and Conduct (2009). No financial payments were made for participation.

Participants included both students as well as others recruited through contacts of the experimenters. A total of N = 52 participants (21 men and 31 women) age range of 18–25 years, with a mean age of 20.9 were included. They were driving license holders with 1–6 years driving experience who self-reported as being in good health, not pregnant or using illicit drugs at the time of the investigation, non-smokers, with a daily consumption of less than 5 cups of coffee or other caffeinated drinks. Subjects were social drinkers and self-reported to experience hangovers regularly. There were 26 participants in each of the BAC zero and residual alcohol groups.

2.2. Procedure

Prospective participants completed screening questionnaires collecting demographic information and assessing health and pregnancy status, alcohol, caffeine, drug use and alcohol hangover experience. Screening also included a short 5 min simulated drive in the STISIM to check for simulator sickness propensity, promote adaptation, and familiarize participants with the driving simulator test [28,29].

During the study, participants attended the driving simulator twice in the morning for full driving assessment, once after an evening without alcohol consumption (no alcohol) and once after an alcohol consumption occasion (alcohol/hangover). They did not ride or drive to the university on post alcohol assessment days and their health status was checked verbally and by observation on attendance, followed by breathalyzer assessment. They were asked to follow their regular procedures on assessment days, although excluding caffeine consumption on the test days until after participation was completed. They completed subjective assessments rating hangover severity and subjective workload after completing the driving simulator test. Following completion of assessments participants health status was again checked before they were permitted to leave the University. On study completion they were provided with a debrief sheet including contacts for the experimenters, as well as advice on support for excessive alcohol use.

2.3. Breathalyzer Assessments

Breath alcohol concentration was measured with a Lion Alcometer 400/500 (Lion Laboratories, UK) by the experimenters who kept their hand over the visual display so that participants were blind to readings. Assessments were conducted directly before the driving simulator test.

2.4. Driving Simulator Test

A STISIM DriveTM driving fixed base simulator was used to measure driving performance (Systems Technology Inc, Hawthorne, CA, USA). The simulator comprises a driving frame housing the adjustable seat, standard foot controls (clutch, brake, accelerator), and the gear lever. The Driving scenario (see Figure 1) is displayed on a 40" LG monitor. The speed of the vehicle and current gear selection are displayed within the virtual dashboard at the bottom of the screen. The balanced scenarios comprised 60,000 feet of mixed rural and urban driving environments with speed limit set to 50 mph. This provided a driving assessment period of around 20 min in a virtual driving environment with UK signage and driving on the left hand carriageway. Potential hazards included crossings, stop signs, traffic lights, parked vehicles in the road, other vehicles as well as pedestrians and dogs crossing the road. A divided attention task was included with symbols displayed in the top left and right of the screen. The default symbol is a diamond and when either symbol changes to a triangle the participant is required to press the corresponding response button on the steering wheel. A range of buildings dominated the urban landscape, whilst trees and fields were prominent in the rural sections of the driving environment.

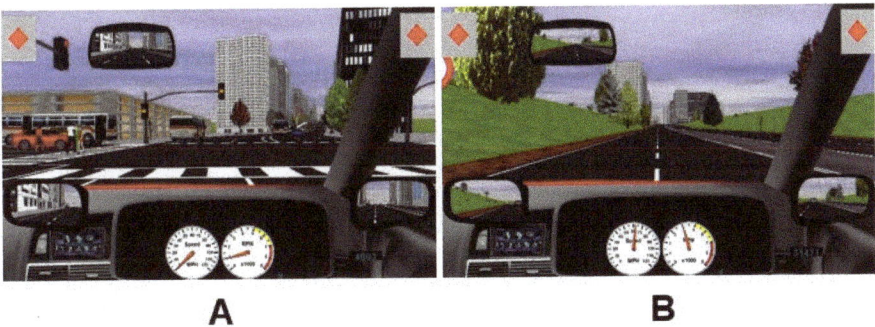

Figure 1. Examples from the STISIM driving test scenario. The STISIM driving test scenario included both city driving scenes (**A**) with a relatively high number of external stimuli (e.g., traffic lights, pedestrians crossing the road and rural roads (**B**) with less external stimuli (e.g., no traffic lights, lower traffic density). On the top left and right of the screen, symbols for the divided attention task are depicted to which subjects had to respond by pressing a button on the steering wheel.

The STISIM driving simulator records a range of parameters, including variables related to divided attention (mean response time, and the number of correct, incorrect and missed responses), basic vehicle control (mean speed and deviation, mean lane position and standard deviation of lateral position (SDLP)), driving errors (percentage of center line crossings (offside), percentage off road (nearside), excursions from lane) and driving violations (number of accidents and collisions, speed limit exceedances and tickets (e.g., failure to stop at stop sign or jumping red lights)).

2.5. Subjective Assessments

The Alcohol Hangover Severity Scale (AHSS) with an 11 point scale of 0–10 for a variety of symptoms, including fatigue, nausea and thirst with a total of 12 descriptors was used to assess subjective hangover [30]. Fatigue was selected as a single dimension to provide a measure of fatigue and sleepiness which are important consequences of hangover and associated with driving impairment [31,32]. Participants were also required to recall their alcohol consumption on their social nights out with the aid of a retrospective diary. The National Aeronautics and Space Administration Task Load Index (NASA-TLX) was used to assess perceived workload with an 101 point scale 0–100 with 5 unit divisions, for 6 workload dimensions including effort and performance, which are both important for assessing subjective impairment as well as the overall average or Raw TLX [33,34].

2.6. Statistical Analyses

Statistical analyses were undertaken using IBM SPSS Version 26 (IBM SPSS Statistics 2019 Armonk, NY, USA). In line with the research questions statistical analyses were based on contrasting driving performance data from the post alcohol/alcohol hangover with the no alcohol (control) condition, for both the residual and zero alcohol groups as A Priori factors, and gender and study as exploratory factors. Possible relationships between driving performance and subjective measures were also investigated. Separate MANOVAs were first performed for the driving performance and subjective data to limit type I errors, including independent variables within groups: measure (e.g., STISIM driving variable, subjective variable); alcohol (post-alcohol/hangover, no alcohol). The between groups factors were: residual alcohol (zero alcohol, residual alcohol); study (1,2,3); gender (female, male). Separate ANOVAS were run for each performance variable, and each subjective variable including alcohol and residual alcohol as factors. Significant findings with ANOVA allowed paired comparisons aligned with the research questions to be run for performance and subjective workload variables. These contrasted no alcohol (control) with the hangover condition for the zero alcohol (N = 26), and the residual alcohol (N = 26) groups. Nonparametric paired comparisons were run as distribution free confirmatory analyses. *p*-Values, two-tailed, were considered significant if $p < 0.05$. Effect size calculations for paired comparisons were undertaken using pooled variance estimates [35]. Measures of association (Pearson's *r*) were included where P-values were at least significant at $p < 0.01$ and supported by nonparametric equivalents, in order to limit type 1 errors given the number of variables included.

3. Results

All N = 52 participants had zero BAC on the morning after their no alcohol consumption night. Following a night of social drinking some participants were unable to recall their exact previous night's amount of alcohol consumption, though all participants did report they had a night of alcohol consumption. This was reflected in the results for the AHSS where group symptom means (4.7, SD 1.8) indicated they had a substantial hangover comparable with other studies. Breath analysis identified N = 26 participants who recorded 0% BAC, providing the zero alcohol group. A further N = 26 participants recorded residual alcohol with the breathalyzer, averaging 0.047% BAC (range 0.01–0.08% BAC) and made up the residual alcohol group. The AHSS hangover score was 5.1 for the zero BAC group and 4.3 for the residual alcohol group.

Multivariate analyses failed to reveal a significant effect for the exploratory factors gender and study for both the driving performance data and subjective data, and so these factors were removed whilst *A Priori* factors were retained, and data reanalyzed with the reduced model. MANOVA indicated a significant effect for alcohol for driving performance, as well as a significant effect for alcohol and an interaction between alcohol and residual alcohol for the subjective data.

3.1. Driving Performance

Multivariate analysis revealed an overall significant effect for alcohol ($F_{(1,50)} = 19.4$, $p < 0.0001$, $\eta^2 p = 0.30$). *A Priori* factors were included in the separate ANOVAs for the driving measures. Driving test results are presented in Table 1. Effect sizes fall within the small (0.2–0.4) to large (0.8–1+) effect size range [35].

Analysis of separate variables revealed significant differences between the alcohol hangover and no alcohol (control) conditions for components of the divided attention task including mean response time ($F_{(1,50)} = 15.6$, $p < 0.0001$, $\eta^2 p = 0.24$), incorrect responses ($F_{(1,50)} = 4.85$, $p = 0.03$, $\eta^2 p = 0.09$) and missed responses ($F_{(1,50)} = 4.48$, $p = 0.04$, $\eta^2 p = 0.08$) reflecting impairment with hangover. Driving control measures produced significant contrasts for mean speed ($F_{(1,50)} = 9.00$, $p = 0.004$, $\eta^2 p = 0.15$), center line crossing ($F_{(1,50)} = 6.49$, $p = 0.01$, $\eta^2 p = 0.12$), time off road ($F_{(1,50)} = 19.9$, $p < 0.0001$, $\eta^2 p = 0.29$) and excursions from lane ($F_{(1,50)} = 14.8$, $p < 0.0001$, $\eta^2 p = 0.23$), again reflecting poorer driving with hangover. Driving violations provided significant contrasts for accidents and collisions ($F_{(1,50)} = 4.59$, $p = 0.04$, $\eta^2 p = 0.08$), and tickets ($F_{(1,50)} = 5.99$, $p = 0.02$, $\eta^2 p = 0.11$). All these measures revealed impaired performance including slower responses, poorer steering control and more errors as well as increased traffic violations in the hangover condition compared to driving after no alcohol. Significant contrasts between the zero alcohol and residual alcohol group included mean speed ($F_{(1,50)} = 4.28$, $p = 0.04$, $\eta^2 p = 0.08$), speed limit exceedances ($F_{(1,50)} = 4.59$, $p = 0.04$, $\eta^2 p = 0.10$) and tickets ($F_{(1,50)} = 4.17$, $p = 0.046$, $\eta^2 p = 0.08$). Residual alcohol resulted in a greater mean speed, more speed limit exceedances and more tickets than the zero residual alcohol group.

A single interaction between alcohol and residual alcohol was found with accidents and collisions where the residual alcohol group experienced a greater number on hangover days compared to the zero alcohol group who showed a decrease, although with paired comparisons hangover driving failed to contrast with the no alcohol control for either the combined or zero alcohol groups, whilst the residual alcohol group had shown an increase with hangover.

Table 1. Driving Performance and Subjective Measures: Means (SDs).

	Combined Groups		Zero BAC Group		Residual Alcohol Group	
	Control	Hangover	Control	Hangover	Control	Hangover
Divided Attention						
Mean Response Time	2.4 (1.3)	3.5 (1.8) * [0.55]	2.3 (1.0)	3.4 (1.6) * [0.69]	2.4 (1.5)	3.7 (2.1) * [0.50]
Response Time Deviation	1.9 (1.1)	2.3 (1.0)	1.9 (1.1)	2.4 (1.2)	1.8 (1.2)	2.2 (0.8)
Correct Responses	15.2 (2.4)	14.2 (3.3)	15.5 (2.1)	14.4 (2.8)	14.9 (2.6)	14.0 (3.8)
Incorrect Responses	0.0 (0.0)	0.2 (0.5) * [0.21]	0.0 (0.0)	0.1 (0.3)	0.0 (0.0)	0.2 (0.6)
Missed Responses	1.1 (1.9)	2.1 (2.9) * [0.30]	0.9 (1.3)	2.0 (2.5)	1.3 (2.4)	2.2 (3.3)
Driving Control						
Mean Speed (mph)	31.7 (9.6)	35.9 (10.3) * [0.42]	29.1 (10.3)	33.7 (10.1) * [0.40]	34.3 (8.2)	38.1 (10.2) ++ [0.45]
Speed Deviation	10.5 (3.3)	10.6 (3.1)	10.3 (3.7)	10.5 (3.0)	10.7 (3.0)	10.7 (3.2)
Mean Lane Position	−6.4 (3.1)	−5.9 (3.2)	−7.0 (3.6)	−5.7 (3.2)	−5.8 (2.3)	−6.2 (3.2)
Lane Deviation	2.6 (1.8)	2.9 (1.7)	2.9 (2.0)	2.8 (1.6)	2.3 (1.5)	3.0 (1.8)
Center Line Crossing %	4.0 (3.1)	7.8 (10.6) * [0.36]	3.8 (2.8)	6.0 (4.7) * [0.48]	4.1 (3.4)	9.6 (14.2)
Off Road %	0.6 (0.7)	1.6 (1.7) * [0.68]	0.5 (0.6)	1.3 (1.6) * [0.61]	0.8 (0.8)	1.9 (1.8) * [0.72]
Excursions from Lane	10.6 (11.7)	18.9 (19.3) * [0.58]	9.7 (9.3)	20.2 (18.9) * [0.64]	11.5 (13.9)	17.6 (20.1) * [0.51]
Driving Violations						
Accidents and Collisions	1.3 (1.3)	1.8 (1.9)	1.6 (1.2)	1.1 (1.3)	0.9 (1.3)	2.5 (2.1) * [0.74]
Speed Limit Exceedances	8.7 (5.8)	9.8 (6.0)	7.0 (5.6)	9.2 (5.3)	10.4 (5.5)	10.5 (6.7)+
Tickets	0.8 (0.8)	1.2 (1.0) * [0.34]	0.6 (0.9)	1.0 (0.8)	1.0 (0.8)	1.4 (1.1)+
Subjective Assessments						
NASA-TLX Combined	38.4 (12.8)	53.8 (13.6) * [0.99]	40.6 (11.6)	51.3 (13.7) * [0.99]	36.2 (13.8)	56.3 (13.3) * [1.15]
Effort	41.6 (22.2)	60.5 (20.2) * [0.80]	44.4 (22.9)	60.4 (18.9) * [0.73]	38.8 (21.7)	60.6 (21.9) * [0.87]
Performance	47.4 (22.8)	60.9 (24.7) * [0.39]	47.1 (23.1)	51.0 (20.6)	47.7 (22.9)	70.8 (24.8) ++ [0.66]

Key: $p < 0.05$ * Control versus Hangover, + Zero BAC versus Residual Alcohol. Effect size [Cohen's d] Control versus Hangover.

For the combined groups, the results revealed significant impairments for three of the divided attention measures, indicating slower responses with more errors comprising missed responses and incorrect responses (false positives) in the hangover condition. Driving measures indicated faster driving (mean speed) with more center line crossing (going out of lane to the right), and time off road (going out of lane to the left) and more lane excursions (to left and right) in the hangover condition compared to after no alcohol. There were also a higher number of tickets (reflecting failure to stop at stop signs, jumping red lights). The paired comparisons for the zero alcohol group paralleled the combined groups except for divided attention task incorrect and missed responses, and tickets where differences were not significant. Contrasts between no alcohol and alcohol hangover for the residual alcohol group were similar to the zero alcohol group for the divided attention task, with both missed and incorrect responses failing to achieve significance, as did center line crossings in the driving control category. In contrast, there were a significantly higher number of accidents and collisions, reflecting the number of vehicles, pedestrians or animals hit, for the residual alcohol group but not for the zero alcohol group. Significant contrasts comparing hangover driving in the zero alcohol group with the residual alcohol group revealed faster driving, a greater number of speed exceedances and more tickets for the residual alcohol group.

Nonparametric paired comparisons were run as distribution free confirmatory analyses for the combined, zero and residual alcohol groups contrasting the no alcohol with the alcohol hangover condition. All statistically significant results with parametric tests were reflected in significant results with nonparametric analyses (Wilcoxon), except for missed responses for the zero alcohol group where a trend ($p < 0.07$) was observed.

3.2. Subjective Measures

Multivariate analysis combining the six NASA-TLX dimensions revealed an overall significant effect for alcohol ($F_{(1,50)} = 54.8$, $p < 0.0001$, $\eta^2 p = 0.52$) and alcohol x residual alcohol ($F_{(1,50)} = 5.19$, $p = 0.03$, $\eta^2 p = 0.09$) with the combined measures. Based on the literature, the focus for subjective assessment was on perceived effort and performance. These are presented in Table 1 along with the overall mean (Raw TLX) for the 6 workload dimensions. The parametric analyses provided the same significant contrasts as the nonparametric confirmatory analyses.

Analysis of the six component dimensions for the NASA-TLX with ANOVA revealed significant differences between the no alcohol (control) and alcohol hangover conditions for mental demand ($F_{(1,50)} = 18.8$, $p < 0.0001$, $\eta^2 p = 0.27$ $\eta^2 p$), physical demand ($F_{(1,50)} = 17.5$, $p < 0.0001$, $\eta^2 p = 0.26$), temporal demand ($F_{(1,50)} = 37.8$, $p < 0.0001$, $\eta^2 p = 0.43$), effort ($F_{(1,50)} = 33.3$, $p < 0.0001$, $\eta^2 p = 0.40$), performance ($F_{(1,50)} = 8.2$, $p = 0.006$, $\eta^2 p = 0.14$), frustration ($F_{(1,50)} = 16.3$, $p < 0.0001$, $\eta^2 p = 0.25$) and the overall means for the 6 component dimensions ($F_{(1,50)} = 54.8$, $p < 0.0001$, $\eta^2 p = 0.52$). All these measures showed an increase in perceived workload in the hangover condition compared to no alcohol. A significant alcohol x residual alcohol interaction was recorded for the overall means ($F_{(1,50)} = 5.19$, $p = 0.027$, $\eta^2 p = 0.09$), and performance ($F_{(1,50)} = 4.17$, $p = 0.046$, $\eta^2 p = 0.08$). For performance there was also a significant contrast between the zero alcohol and residual alcohol groups ($F_{(1,50)} = 5.69$, $p = 0.02$, $\eta^2 p = 0.10$) reflecting the greater increase in scores for the residual alcohol group with hangover (see Table 1.). The comparison between the zero alcohol and the residual alcohol groups indicates that the increase in performance score, reflecting poorer perceived performance, showed a significant increase with hangover in the residual alcohol group that was not seen in the zero alcohol group.

3.3. Associations Between Measures

The association between subjective measures and driving performance, reflecting awareness of subjective impairment, was a secondary research question. The selected subjective variables were the overall hangover rating as well as the fatigue component of the AHSS, and the NASA-TLX overall means together with the effort and performance components. The AHSS and NASA-TLX were correlated with each other as well as the driving performance measures. Variables achieving a minimum significance level of $p < 0.01$ with Pearson's r, and supported by significant nonparametric equivalents for at least one group are included in Table 2. Significant results included subjective ratings of hangover symptoms (AHSS), including fatigue, as well as overall subjective workload (NASA-TLX), and both effort and performance component ratings. Effect sizes were close to or in the large effect size range (≥ 0.37) [35]. The AHSS and NASA-TLX were not associated at the required significance level ($p < 0.01$).

Table 2. Associations between Driving Performance and Subjective Assessments (Pearson's *r*).

Measure	Subjective Assessment	Combined Groups	Zero BAC Group	Residual BAC Group
		Driving Control		
Mean Speed (mph)	AHSS	−0.359 **	0.173	−0.573 **
	Fatigue	−0.460 **	−0.232	−0.561 **
	Performance	−0.247	−0.706 **	−0.104
Speed Deviation	AHSS	0.051	−0.553 **	0.317
	NASA-TLX	−0.254	−0.542 **	0.003
Mean Lane Position	Residual BAC	0.154	-	0.555 **
Lane Deviation	Performance	−0.466	−0.542 **	−0.544
		Driving Violations		
Speed Exceedances	Effort	−0.357 **	−0.348	−0.367
		Residual BAC Assessments		
Residual BAC	Performance	0.467 **	-	0.336

Key: ** *p* < 0.01 (minimum significance level), and nonparametric correlation also significant.

For residual BAC, positive correlation with subjective performance indicated higher residual alcohol levels were associated with higher performance scores which reflect poorer perceptions of performance with this component of the NASA-TLX. Higher residual BAC was also associated with greater (less negative) mean lane position scores for the residual alcohol group, possibly reflecting poorer driving by being closer to the center of the road (scored at '0').

For the driving performance measures, higher scores generally indicate poorer driving. Increased mean speed and speed deviation were associated with lower hangover (AHSS, Fatigue) and workload ratings (Performance, NASA-TLX). Increased lane deviation was associated with lower performance ratings, reflecting better perceived performance, and lower subjective effort was associated with an increase in speed exceedances. However, significant associations with driving measures differed between the combined, zero and residual alcohol groups.

4. Discussion

This is the first study to demonstrate significant driving impairment with hangover during a typical commute to work. For the combined groups a total of 8 of the 15 driving variables including those associated with the divided attention task, as well as driving control, and driving violations which included attentional measures (e.g., tickets) showed significant impairment in the hangover condition. The means for the combined groups showed slower responses, more errors including poorer driving control and loss of attention for all the variables, even though they did not all individually achieve statistical significance. The levels of impairment during alcohol hangover seen here suggest that these are equivalent to the magnitude of driving impairment seen under conditions of alcohol intoxication at or above the 0.05–0.08% BAC level [21]. The range of neurocognitive impairments are supported by those reviewed by Kruisselbrink [19] as sensitive to alcohol hangover including perceptual motor function, complex attention and executive functioning. These results support earlier simulator studies, including the one hour highway driving test [9] and shorter motorway driving assessment [11]. The observed significant impairments indicate the need to address this issue, and also substantiate the use of our driving model for assessing hangover.

The likely explanation for the range of significant performance decrements observed on our driving test of relatively short duration is the increased cognitive demand imposed by the driving scenarios. Data from the NASA-TLX revealed that significantly increased effort was needed by the participants in both the zero and residual alcohol groups to perform (well on) the driving test, but this increased effort could not counteract driving performance impairments due to a hangover. This was also observed in the simulated highway driving study by Verster et al. [9]. Under more demanding cognitive load such as the complex driving environment used here, cognitive demand can more easily exceed capacity resulting in driving errors and decrements in performance. The impairments were therefore also recorded across a shorter driving period, rather than reflecting vigilance decrements seen in longer duration drives including the one hour driving test, in which the effects on driving were most pronounced in the second half of the test [9].

An important further aim of the study was to compare hangover driving performance for participants who had zero BAC with those who had residual levels of alcohol when driving next day in a mixed urban and rural environment during a 20 min commute to work. Although there was a residual BAC group, it is important to note that none of these participants was over the UK alcohol limit for driving (BAC 0.08%), and the mean BAC in the residual alcohol group (BAC 0.047%) was also below the 0.05% BAC limit common to several countries and recommended by the World Health Organization [1]. This observation is important as with current UK legislation all participants would be considered 'street legal', i.e., allowed to drive a car. When contrasting hungover driving performance of those with zero BAC and those with residual alcohol, the profiles of impairment for both these groups were similar to each other, suggesting that having residual BAC during hangover did not have a marked additional impact on driving ability. However, the residual alcohol group did drive significantly faster, resulting in significantly more speed exceedances, and they received significantly more tickets for driving violations, possibly indicating poorer attention and awareness. Increased speed, disinhibition and greater risk taking have been reported previously in studies of both alcohol intoxication, as well as alcohol hangover [3,4,11].

The impact of alcohol hangover on self awareness is important for driving and may determine whether or not a driver commutes to work next day, even though their residual alcohol BAC is within legal limits. Alcohol research has shown limited awareness of impairment in relation to driving particularly with higher BAC levels [25,26], and currently little is known in relation to alcohol hangover. Results from the present study yielded strong effects with hangover indicating that poorer driving (increased mean speed and speed deviation) was associated with reduced perceptions of hangover and fatigue (AHSS) as well as overall workload (NASA-TLX), even though workload was increased with hangover compared to no alcohol. In the hangover condition, reduced effort was associated with more speed exceedances (driving violations). Lower subjective performance ratings from the NASA-TLX, indicating perceptions of better performance, were similarly associated with poorer driving performance (increased mean speed and lane deviation). However, the finding that increased residual BAC was associated with higher ratings of poorer performance suggests some awareness, but includes the impact of the remaining alcohol. Overall, there were relatively few significant correlations between subjective measures and driving performance, this could reflect a limitation of the AHSS, which is based on individual symptoms that may not capture the complete hangover experience [36].

Taken together, these results suggest a potential lack of awareness of subjective impairment due to hangover. This was apparent in terms of lower hangover symptom scores including fatigue, and lower perceived workload including effort and better performance, being linked to more impaired driving. This is a concerning factor given that we do not currently have a direct objective measure of hangover that is equivalent to the alcohol breathalyzer and the driving performance scores show consistent impairment in the hangover condition. However, significant correlations with individual driving performance measures were limited and varied between the hangover groups, and therefore requires replication with larger samples, although the significant positive correlation with higher residual BAC associated with perceptions of poorer performance adds validity. Previous alcohol hangover studies have not shown a consistent relationship between estimated BAC for consumption the previous night, or hangover severity and objective performance on the next day [37,38]. Similarly, for this study, subjective hangover ratings (AHSS) were not significantly associated with overall subjective workload or effort and performance components.

The current study had a naturalistic design, so that participants were not monitored during the drinking session. As a result, participants consumed the type and amount of alcohol of choice, which varied between participants. Although this variability may be viewed as a limitation, it can also be regarded as a strength, because the naturalistic study design much more closely mimics a real-life drinking session, which includes non-standardized alcohol consumption and various behaviors (e.g., dancing, changing pubs) that cannot be replicated in randomized controlled trials [27]. In this context, it should be noted that the actual amount of alcohol consumed was of little relevance to the current study. More important, and irrespective of the alcohol consumed, was the premise that this would result in a next-day hangover. These criteria was met by all participants with overall hangover ratings broadly in line with earlier findings [37]. Recent research confirmed that hangovers can be experienced at any BAC level [38].

Participants were allocated to either the BAC zero or residual alcohol group after the data was collected. This was done for practical reasons, as the presence and severity of hangovers vary both within and between drinkers. Even in case of standardized alcohol administration in controlled trials, significant variability was observed in the presence and severity of hangover symptoms [38,39]. Given this, it was not possible to randomize participants and allocate them to a group before data collection took place. A related issue is how BAC was established. It has been demonstrated that breathalyzer assessments do not always correspond well with blood alcohol assessments [40]. Taking this into account, the allocation of participants to the BAC zero or residual alcohol group based on breath alcohol assessments only may not have been accurate. Future studies for which it is essential to accurately determine the presence of residual alcohol, should confirm BAC readings of zero obtained with a breathalyzer with assessments of ethanol in blood. The BAC zero group made up half of the sample, limiting the validity of the residual BAC correlations as half the 52 participants therefore had a zero score, but noting the correlations were strong with both parametric and nonparametric analyses.

Sleep disruption is a common component of alcohol hangover and naturalistic studies show reduced sleep with alcohol hangover compared to control no alcohol nights, so that overall sleep loss may be a significant component of hangover effects on waking performance next day [32]. In the current study, sleep was not recorded. Future studies should incorporate these assessments.

In the current study, participants were relatively young and inexperienced drivers. Their inclusion is warranted given that road traffic collisions are the leading cause of death for this age group [1], contributed to by a lack of observation and anticipation skills [41], increased recklessness and thrill seeking, and feelings of invincibility and over-confidence [42,43]. These combine to create 'skill-risk optimism' [44] whereby young drivers believe they possess high level driving skills and are unlikely to have an accident in risky-driving scenarios such as driving following a night of alcohol consumption. The research evidence, supported by the results presented here, demonstrates the opposite and may in part explain the lack of awareness of impairment amongst this population. However, what is unclear is whether the observed results in this study translate to older and more experienced drivers. Therefore, future research should also investigate the effects of alcohol hangover on driving across other age groups.

Finally, no significant sex differences were observed in the current study. This is in line with previous driving studies [9] and corresponds with the general absence in sex differences in the presence and severity of hangover symptoms at various BAC levels [45,46].

Positive aspects of the study included the sensitivity of the driving model, which included finding significant decrements in a range of driving variables including divided attention, driving control and driving violations reflecting impaired attention or risk taking, where other studies have failed to find such a wide range of effects.

5. Conclusions

Overall, this evaluation of driving hangover performance has been successful in demonstrating the marked impact of alcohol hangover in impairing driving performance, even though participants overall were below the legal alcohol limit for driving in several countries. The key findings are that significant impairments were seen in a range of driving measures, including complex attention and driving control for those with residual alcohol as well as those with zero alcohol, with only limited differences between the groups. The level of impairment seen here, which was comparable to driving while intoxicated at or above a BAC of 0.05%, indicates the dangers of driving whilst hungover, even when breath alcohol is at zero.

Author Contributions: Conceptualization, C.A., J.C.V.; data collection, C.B., H.C., S.L.; data analysis, C.A., R.R.; writing—original draft preparation, C.A.; writing—review and editing, C.A., S.J.J., J.C.V. All authors have read and agreed to the published version of the manuscript.

Funding: This research received no external funding.

Acknowledgments: We like to thank Janet Watkins at UWE for creating the STISIM Scenarios, and Paul White for advice on statistical analysis.

Conflicts of Interest: C.A. has undertaken sponsored research, or provided consultancy, for a number of companies and organizations including Airbus Group Industries, Astra, British Aerospace/BAeSystems, UK Civil Aviation Authority, Duphar, Farm Italia, Carlo Erba, Ford Motor Company, ICI, Innovate UK, Janssen, LERS Synthélabo, Lilly, Lorex/Searle, UK Ministry of Defence, More Labs, Quest International, Red Bull GmbH, Rhone-Poulenc Rorer, Sanofi Aventis, Vital Beverages. S.J. has undertaken sponsored research for Pfizer, AstraZeneca, Merck, Gilead, Novartis, Roche, Red Bull GmbH, the Department for Transport, and Road Safety Trust. Over the past three years, J.C.V. has received grants/research support from the Dutch Ministry of Infrastructure and the Environment, Janssen Research and Development, and Sequential, and has acted as a consultant/advisor for Clinilabs, More Labs, Red Bull GmbH, Sen-Jam Pharmaceutical, Toast!, and ZBiotics. The other authors have no potential conflicts of interest to disclose.

References

1. World Health Organization. *Global Status Report on Road Safety*; World Health Organization: Geneva, Switzerland, 2018.
2. Martin, T.L.; Solbeck, P.A.; Mayers, D.J.; Langille, R.M.; Buczek, Y.; Pelletier, M.R. A review of alcohol impaired driving: The role of blood alcohol concentration and complexity of the driving task. *J. Forensic Sci.* **2013**, *58*, 1238–1250. [CrossRef] [PubMed]
3. Ogden, E.J.D.; Moskowitz, H. Effects of alcohol and other drugs on driver performance. *Traffic Inj. Prev.* **2004**, *5*, 185–198. [CrossRef] [PubMed]
4. Van Dyke, N.A.; Fillmore, M.T. Distraction produces over-additive increases in the degree to which alcohol impairs driving performance. *Psychopharmacology* **2015**, *232*, 4277–4284. [CrossRef] [PubMed]
5. Irwin, C.; Iudakhina, E.; Desbrow, B.; McCartney, D. Effects of acute alcohol consumption on measures of simulated driving: A systematic review and meta-analysis. *Accid. Anal. Prev.* **2017**, *102*, 248–266. [CrossRef]
6. Verster, J.C. Alcohol hangover effects on driving and flying. *Int. J. Disabil. Hum. Dev.* **2007**, *6*, 361–367. [CrossRef]
7. Laurell, H.; Törnros, J. Investigation of alcoholic hang-over effects on driving performance. *Blutalkohol* **1983**, *20*, 489–499.
8. Törnros, J.; Laurell, H. Acute and hangover effects of alcohol on simulated driving performance. *Blut. Alcohol.* **1991**, *28*, 24–30.
9. Verster, J.C.; Bervoets, A.C.; de Klerk, S.; Vreman, R.A.; Olivier, B.; Roth, T.; Brookhuis, K.A. Effects of alcohol hangover on simulated highway driving performance. *Psychopharmacology* **2014**, *231*, 2999–3008. [CrossRef]
10. Verster, J.C.; Van Der Maarel, M.A.; McKinney, A.; Olivier, B.; De Haan, L. Driving during alcohol hangover among Dutch professional truck drivers. *Traffic Inj. Prev.* **2014**, *15*, 434–438. [CrossRef]
11. Robbins, C.J.; Russell, S.; Chapman, P. Student drivers the morning after drinking: A willingness to violate road rules despite typical visual attention. *Transp. Res. Part F Traffic Psychol. Behav.* **2019**, *62*, 376–389. [CrossRef]
12. Verster, J.C.; van Herwijnen, J.; Olivier, B.; Kahler, C.W. Validation of the Dutch version of the brief young adult alcohol consequences questionnaire (B-YAACQ). *Addict. Behav.* **2009**, *34*, 411–414. [CrossRef] [PubMed]
13. Van Schrojenstein Lantman, M.; van de Loo, A.J.; Mackus, M.; Verster, J.C. Development of a definition for the alcohol hangover: Consumer descriptions and expert consensus. *Curr. Drug Abuse Rev.* **2016**, *9*, 148–154. [CrossRef] [PubMed]

14. Verster, J.C.; Scholey, A.; van de Loo, A.J.A.E.; Benson, S.; Stock, A.-K. Updating the definition of the alcohol hangover. *J. Clin. Med.* **2020**, *9*, 823. [CrossRef]
15. Sacks, J.J.; Gonzales, K.R.; Bouchery, E.E.; Tomedi, L.E.; Brewer, R.D. 2010 National and State Costs of Excessive Alcohol Consumption. *Am. J. Prev. Med.* **2015**, *49*, e73–e79. [CrossRef] [PubMed]
16. Bhattacharya, A. Financial Headache. In *The Cost of Workplace Hangovers and Intoxication to the UK Economy*; Institute of Alcohol Studies: London, UK, 2019.
17. Frone, M.R. Employee psychoactive substance involvement: Historical context, key findings, and future directions. *Annu. Rev. Organ. Psychol. Organ. Behav.* **2019**, *6*, 273–297. [CrossRef]
18. Gunn, C.; Mackus, M.; Griffin, C.; Munafò, M.R.; Adams, S. A systematic review of the next-day effects of heavy alcohol consumption on cognitive performance. *Addiction* **2018**, *113*, 2182–2193. [CrossRef]
19. Kruisselbrink, L.D. The neurocognitive effects of alcohol hangover: Patterns of impairment/nonimpairment within the neurocognitive domain of the Diagnostic and Statistical Manual of Mental Disorders. In *Neuroscience of Alcohol: Mechanisms and Treatment*, 5th ed.; Preedy, V.R., Ed.; Academic Press: Cambridge, MA, USA, 2019; pp. 391–402.
20. Verster, J.C.; Roth, T. Standard operation procedures for conducting the on-the-road driving test, and measurement of the standard deviation of lateral position (SDLP). *Int. J. Gen. Med.* **2011**, *4*, 359–371. [CrossRef]
21. Mets, M.A.J.; Kuipers, E.; Senerpont Domis, L.M.; Leenders, M.; Olivier, B.; Verster, J.C. Effects of alcohol on highway driving in the STISIM driving simulator. *Hum. Psychopharmacol.* **2011**, *26*, 434–439. [CrossRef]
22. Laurell, H. Effects of small doses of alcohol on driver performance in emergency traffic situations. *Accid. Anal. Prev.* **1977**, *9*, 191–201. [CrossRef]
23. Statista. Average Commuting Time in Great Britain in October to December 2018*, by Mode of Transportation. Available online: https://www.statista.com/statistics/300712/average-time-taken-to-travel-to-work-in-the-united-kingdom/ (accessed on 31 March 2020).
24. National Travel Survey, UK. National Travel Survey Data Tables. 2018. Available online: https://www.gov.uk/government/collections/national-travel-survey-statistics (accessed on 31 March 2020).
25. Aston, E.R.; Liguori, A. Self-estimation of blood alcohol concentration: A review. *Addict. Behav.* **2013**, *38*, 1944–1951. [CrossRef]
26. Grant, S.; LaBrie, J.W.; Hummer, J.F.; Lac, A. How drunk am I? Misperceiving one's level of intoxication in the college drinking environment. *Psycho. Addic. Behav.* **2012**, *26*, 51–58. [CrossRef] [PubMed]
27. Verster, J.C.; van de Loo, A.J.A.E.; Adams, S.; Stock, A.-K.; Benson, S.; Alford, C.; Scholey, A.; Bruce, G. Advantages and limitations of naturalistic study designs and their implementation in alcohol hangover research. *J. Clin. Med.* **2019**, *8*, 2160. [CrossRef] [PubMed]
28. Domeyer, J.E.; Cassavaugh, N.D.; Backs, R.W. The use of adaptation to reduce simulator sickness in driving assessment and research. *Accid. Anal. Prev.* **2013**, *53*, 127–132. [CrossRef] [PubMed]
29. Keshavarz, B.; Hecht, H.; Lawson, B.D. Visually-induced motion sickness: Causes, characteristics, and countermeasures. In *Handbook of Virtual Environments, Design, Implementation, and Applications New York*, 2nd ed.; Hale, K.S., Stanney, K.M., Eds.; CRC Press Taylor: New York, NY, USA, 2015; pp. 647–698.
30. Penning, R.; McKinney, A.; Bus, L.D.; Olivier, B.; Slot, K.; Verster, J.C. Measurement of alcohol hangover severity: Development of the Alcohol Hangover Severity Scale (AHSS). *Psychopharmacology* **2013**, *225*, 803–810. [CrossRef]
31. Alford, C. Sleepiness, countermeasures and the risk of motor vehicle accidents. In *Drugs, Driving and Traffic Safety*; Verster, J.C., Pandi-Perumal, S.R., Ramaekers, J.G., De Gier, J.J., Eds.; Birkhauser: Boston, MA, USA, 2009; pp. 207–232.
32. Devenney, L.E.; Coyle, K.B.; Roth, T.; Verster, J.C. Sleep after Heavy Alcohol Consumption and Physical Activity Levels during Alcohol Hangover. *J. Clin. Med.* **2019**, *8*, 752. [CrossRef]
33. Hart, S.G.; Staveland, L.E. Development of NASA-TLX (Task Load Index): Results of Empirical and Theoretical Research. *Adv. Psychol.* **1988**, *52*, 139–183.
34. Hart, S.G. NASA-task load index (NASA-TLX): 20 years later. *Proc. Hum. Factors Ergon. Soc. Ann. Meet.* **2006**, *50*, 904–908. [CrossRef]
35. Lenhard, W.; Lenhard, A. Calculation of Effect Sizes. Psychometrica 2016. Available online: https://www.psychometrica.de/effectsize.html (accessed on 18 April 2020).

36. Verster, J.C.; van de Loo, A.J.A.E.; Benson, S.; Scholey, A.; Stock, A.-K. The assessment of overall hangover severity. *J. Clin. Med.* **2020**, *9*, 786. [CrossRef]
37. Alford, C.; Martinkova, Z.; Tiplady, B.; Reece, R.; Verster, J.C. The effects of alcohol hangover on mood and performance assessed at home. *J. Clin. Med.* **2020**, *9*, 1068. [CrossRef]
38. Verster, J.C.; Kruisselbrink, L.D.; Slot, K.A.; Anogeianaki, A.; Adams, S.; Alford, C.; Arnoldy, L.; Ayre, E.; Balikji, S.; Benson, S.; et al. Sensitivity to experiencing alcohol hangovers: Reconsideration of the 0.11% blood alcohol concentration (BAC) threshold for having a hangover. *J. Clin. Med.* **2020**, *9*, 179. [CrossRef]
39. Hensel, K.O.; Longmire, M.R.; Köchling, J. Should population-based research steer individual health decisions? *Aging* **2019**, *11*, 9231–9233. [CrossRef] [PubMed]
40. Verster, J.C.; Mackus, M.; van de Loo, A.J.A.E.; Garssen, J.; Scholey, A. The breathtaking truth about breath alcohol readings of zero. *Addict. Behav.* **2017**, *70*, 23–26. [CrossRef] [PubMed]
41. Cavallo, A.; Triggs, T.J. *Directions for Improving Young Driver Safety within Victoria: A Discussion Paper*; Monash University Accident Research Centre: Melbourne, Australia, 1996.
42. Clarke, S.; Robertson, I.T. A meta-analytic review of the Big Five personality factors and accident involvement in occupational and nonoccupational settings. *J. Occup. Organ. Psychol.* **2005**, *78*, 355–376. [CrossRef]
43. Falk, B.; Montgomery, H. Promoting traffic safety among young male driver by means of elaboration-based interventions. *Transp. Res. Part F Traffic Psychol. Behav.* **2009**, *12*, 1–11. [CrossRef]
44. White, M.J.; Cunningham, L.C.; Titchener, K. Young drivers' optimism bias for accident risk and driving skill: Accountability and insight experience manipulations. *Accid. Anal. Prev.* **2011**, *43*, 1309–1315. [CrossRef] [PubMed]
45. Van Lawick van Pabst, A.E.; Devenney, L.E.; Verster, J.C. Sex differences in the presence and severity of alcohol hangover symptoms. *J. Clin. Med.* **2019**, *8*, E867. [CrossRef]
46. Van Lawick van Pabst, A.E.; Devenney, L.E.; Verster, J.C. Correction: Van Lawick van Pabst. Sex differences in the presence and severity of alcohol hangover symptoms. *J. Clin. Med.* **2019**, *8*, 867. [CrossRef]

© 2020 by the authors. Licensee MDPI, Basel, Switzerland. This article is an open access article distributed under the terms and conditions of the Creative Commons Attribution (CC BY) license (http://creativecommons.org/licenses/by/4.0/).

Article

Alcohol Hangover and Multitasking: Effects on Mood, Cognitive Performance, Stress Reactivity, and Perceived Effort

Sarah Benson [1], Elizabeth Ayre [1], Harriet Garrisson [1], Mark A Wetherell [1,2], Joris C Verster [1,3] and Andrew Scholey [1,*]

1. Centre for Human Psychopharmacology, Swinburne University, Melbourne VIC 3122, Australia; sarahbenson@swin.edu.au (S.B.); eayre@swin.edu.au (E.A.); hgarrison@swin.edu.au (H.G.); mark.wetherell@northumbria.ac.uk (M.A.W.); J.C.Verster@uu.nl (J.C.V.)
2. Stress Research Group, Department of Psychology, Northumbria University, Northumberland Building, Newcastle upon Tyne NE1 8ST, UK
3. Division of Pharmacology, Utrecht Institute for Pharmaceutical Sciences (UIPS), Utrecht University, 3584CG Utrecht, The Netherlands
* Correspondence: andrew@scholeylab.com; Tel.: +44-(3)921-489-32

Received: 31 March 2020; Accepted: 14 April 2020; Published: 17 April 2020

Abstract: The aim of this study was to examine the effects of hangover on mood, multitasking ability, and psychological stress reactivity to cognitive demand. Using a crossover design and semi-naturalistic methodology, 25 participants attended the laboratory in the morning following a night of (i) alcohol abstinence and (ii) alcohol self-administration during a typical night out (with order counterbalanced across participants). They completed a four-module multitasking framework (MTF, a widely used laboratory stressor) and a battery of questionnaires assessing mood, hangover symptom severity, and previous night's sleep. The effects of the MTF on mood and perceived workload were also assessed. Participants in the hangover condition reported significantly lower alertness and contentment coupled with a higher mental fatigue and anxiety. Multitasking ability was also significantly impaired in the hangover condition. Completion of the cognitive stressor increased reported levels of mental demand, effort, and frustration, and decreased perceived level of performance. MTF completion did not differentially affect mood. Lastly, participants rated their sleep as significantly worse during the night prior to the hangover compared with the control condition. These findings confirm the negative cognitive and mood effects of hangover on mood. They also demonstrate that hangover is associated with greater perceived effort during task performance.

Keywords: hangover; alcohol; internet; attention; executive function; working memory

1. Introduction

The effects of binge drinking on cognitive functioning beyond intoxication, to the subsequent alcohol hangover, has received increasing attention over the past few years. The alcohol hangover is generally obtained following alcohol intake equivalent to a blood alcohol concentration (BAC) of at least 0.10%–0.12% [1], although it can occur following lower BACs [2], and commences once BAC approaches 0.00% [3]. It is characterized by a cluster of physical symptoms, negative mood, and impaired cognitive functioning. Commonly reported symptoms include headache [1,4], nausea [1,5], anxiety [5–7], fatigue [1,6,8], reduced alertness [1,6,9,10], and concentration difficulties [1,11,12].

Hangover-induced cognitive impairment has been described across several domains including spatial and visual abilities [13], attention [14–17], memory, information processing [18], and reaction time [18–20]. Conversely, some studies have failed to find convincing evidence for hangover-induced cognitive impairment [10,11,16]. A recent meta-analysis of 11 articles reported hangover-induced impairment to short and long-term memory, psychomotor performance, and sustained attention, but not divided attention [21]. As suggested by the authors, accumulating mental fatigue caused by prolonged attentional demand is more evident in sustained as opposed to divided attention tasks that may drive the differential attention effects [22].

Importantly, each of the studies [15,17,23,24] included in the meta-analysis [21] assessing the effects of hangover on divided attention used tasks with relatively few stimuli and limited task-switching. Furthermore, three of the four studies were published decades ago and appear under-powered [15,23,24]. More recent studies have utilized assessments of compound behaviors requiring more complex divided attention across several stimuli, such as simulated driving [25,26] and flying [27,28] ability. These have consistently demonstrated hangover-induced performance impairment. Indeed, driving and flying are complex skills which require several cognitive processes and as such, are not strict measures of divided attention per se. Nevertheless, complex tasks are a realistic representation of real-world activities where attention is often divided across several stimuli streams, causing cognitive demand [29] and a stress reactivity response [30], characterized by increased negative mood, cortisol, and self-reported stress [30,31].

Stress reactivity is increased in certain populations such as recreational drug users [32]. Given that the alcohol hangover is characterized by worsened cognition and mood, individuals with a hangover may display an exaggerated response to stress. The effects of hangover on stress reactivity have not been previously assessed but warrant investigation. Particularly in the context of potential negative implications of performing everyday multitasking and potentially stress-inducing activities, such as driving, studying, or working, with a hangover.

Real-life stressors typically involve simultaneous exposure to multiple stressors. The Purple multitasking framework (MTF) requires simultaneous attention and response to several stimuli and, therefore, may have better ecological validity than previously used divided attention tasks [33]. The MTF has previously been shown to elicit a response typical of workload stress (where mental resources cannot meet ongoing demands). That is reduced self-rated calmness, elevated stress, and state anxiety coupled with increased perceived demand, effort, and frustration [32–34].

The aim of this investigation was to examine the effects of hangover on mood, stress reactivity, multitasking performance, and perceived effort of performing the tasks. Specifically, we hypothesized that compared with no hangover, hangover would be associated with more negative mood, higher stress reactivity, poorer performance, and greater perceived effort during multitasking.

2. Experimental Section

2.1. Method

The present study was approved by the Swinburne University Human Research Ethics Committee (SUHREC, approval number 2016/061) and was conducted in accordance with the Declaration of Helsinki.

2.2. Design

This study utilized a semi-naturalistic, crossover design whereby, in the hangover condition, participants consumed alcohol during a 'typical' night out and attended the laboratory the following morning. The testing visit of the no-hangover condition was held following a night of alcohol abstinence.

2.3. Participants

Thirty-six participants enrolled in the study. However, seven participants failed to complete the two testing visits and one participant reported alcohol intake the evening prior to the hangover-free visit and thus, was excluded. The final sample consisted of 25 participants (76% female) with a mean age of 25.32 years (range = 18–35 years).

All participants were healthy, not taking any medications that could potentially interact with alcohol and experienced an alcohol hangover frequently. Exclusion criteria were current or past alcohol abuse and current or past psychiatric disorders.

2.4. Measures

2.4.1. Blood Alcohol Concentration (BAC)

To ensure a BAC of 0.00% prior to the commencement of each testing visit, BAC readings were recorded using a regularly calibrated Lion Alcolmeter SD400PA.

2.4.2. Consensus Sleep Diary (CSD)

All sleep parameters were collected using the 9-item CSD [35].

2.4.3. Alcohol Consumption Questions

Participants were questioned on their alcohol intake on the evening prior to the testing visits. Specifically, participants were asked the number of standard drinks of various drink types (i.e., beer, cider, wine, spirits, and alcohol mixed with energy drink) that were consumed and the number of hours that they had spent drinking.

2.4.4. Estimated BAC (eBAC)

Responses to the alcohol consumption questions were used to calculate eBAC. eBAC obtained on the night of drinking was calculated by averaging the total body water (TBW) estimates of Forrest [36], Watson [37], Seidl [38], Widmark [39], and Ulrich [40] (males only) calculations. The mean TBW was then used in the following eBAC formula:

$$BAC = (G/(TBW) - \beta \times t \tag{1}$$

where G is the amount of alcohol consumed in grams; β is the metabolic rate in gram per hour; and t is time in hours.

2.4.5. Multitasking Framework

The Purple multitasking framework (Purple MTF: Purple Research Solutions, UK) has been shown to induce cognitive demand, stress, negative mood, and anxiety [32,33,41–43]. The task requires attention to be given to four tasks that are presented simultaneously, each on one of four quadrants on a computer screen (see Figure 1).

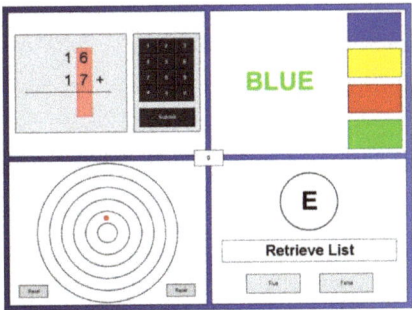

Figure 1. Layout of the Purple multitasking framework (MTF). The task requires participants to simultaneously perform four cognitive tasks. These were (clockwise from top left) Mental Arithmetic, Stroop, Letter Search, and Visual Tracking.

In the centre of the screen, a counter displaying the overall performance score, based on accuracy and reaction time, is presented. The tasks used in the current study were Mental Arithmetic, Stroop, Letter Search, and Visual Tracking. Each task was set to the 'difficult' level and the battery was completed over 20 min (for a detailed description of the tasks and scoring, please see [44]). As well as an overall score, each task was scored separately.

2.4.6. NASA-Task Load Index (NASA-TLX)

Participants were asked to provide their subjective assessment of the workload presented by the Purple MTF using the NASA-TLX [45]. Workload was assessed on six dimensions: Mental Demand, Physical Demand, Temporal Demand, Own Performance, Effort, and Frustration.

2.4.7. Single Item Hangover Symptom Severity Score

Overall hangover severity was measured using a single visual analogue scale (VAS) asking participants to rate 'how severe is your hangover' on a scale from 0, being 'no hangover symptoms', to 10, being 'very severe hangover symptoms'.

2.4.8. Alcohol Hangover Severity Scale (AHSS)

The AHSS [46] measures hangover severity according to 12-items measured on a 10-point Likert scale. The AHSS is found to be reliable (Cronbach's $\alpha = 0.85$) and valid ($r = 0.92$ correlation with the Acute Hangover Scale (AHS; [4])) [46].

2.4.9. Stress and Fatigue Visual Analogue Mood Scale (Stress and Fatigue VAMS)

Self-reported levels of stress and fatigue were each measured using a single-item visual analogue scale labeled stress and fatigue with the endpoints labeled 'not at all' and 'extremely'.

2.4.10. Spielberger State-Trait Anxiety Inventory, State Portion (STAI-S)

The STAI consists of two parts evaluating trait and state anxiety, respectively. This study used the state (STAI-S; [47]) portion only. Participants responded to 20 statements (e.g., "I feel calm", "I am relaxed") on a 4-point scale ranging from 'not at all' to 'very much so'. Scores are combined to give a measure of current anxiety.

2.4.11. Bond-Lader Visual Analogue Mood Scales (Bond-Lader VAMS)

The Bond-Lader VAMS [48] have been utilized in numerous pharmacological, psychopharmacological, and medical trials. These scales comprise a total of 16 lines measuring 100 mm long and anchored at either end by antonyms (e.g., alert-drowsy, calm-excited). Participants indicate their current subjective position between the antonyms on the line. Individual item scores were calculated as the distance along the line. Outcomes are three factor analysis-derived scores: Alertness, calmness, and contentment.

2.5. Procedure

Participants were required to attend a screening visit and two testing visits; one while experiencing a hangover and one following at least 24-h of alcohol abstinence. During the screening visit, participants provided a written informed consent before being assessed for eligibility. Participants then underwent training and practice in completing the Bond-Lader VAMS and Stress and Fatigue VAMS, a cognitive dual attention task (to be reported elsewhere), the Purple MTF, and the NASA-TLX.

Participants were instructed not to consume any caffeine the morning of their testing visits or any alcohol within 24-h prior to the no-hangover visits. Participants were also asked for the food consumed prior to the testing visits to be held consistent. Lastly, participants booked in their first testing visit, which could be either the hangover or no-hangover condition, depending on the participants planned drinking activities.

At the commencement of each testing visit, participants were breathalyzed to ensure a BAC of 0.00%. Participants then completed the pre-stressor Bond-Lader VAMS, the individual Stress and Fatigue VAMS, and the STAI-S. They then underwent the Purple MTF, followed by the post-stressor Bond-Lader VAMS, Stress and Fatigue VAMS, and STAI-S. Finally, they completed the NASA-TLX, alcohol consumption questions, the single-item measure of overall hangover severity, the AHSS, and lastly, the CSD.

2.6. Statistical Analysis

All analyses were performed using SPSS version 25 (IBM Corp, Armonk, NY, USA). The effects of the Purple MTF on the three mood dimensions from the Bond-Lader VAMS ('alert', 'calm', 'content'), anxiety measured using the STAI-S, and the individual VAMS items ('stress', 'mental fatigue') were assessed using two-way (Hangover; present, absent × Time; pre, post) repeated measures ANOVA. All other variables were analyzed using paired sample *t*-tests comparing hangover with control conditions.

Effect sizes (Cohen's d) were calculated for all significant *t*-test findings using 'Equation 8' [49] to account for dependence in the data. All testing was two-tailed, and comparisons were planned prior to testing.

3. Results

3.1. Consensus Sleep Diary (CSD)

In the hangover condition, participants rated their sleep quality as significantly worse ($t(24) = 2.70, p = 0.012, d = 0.54$) and reported significantly more awakenings ($t(24) = 2.47, p = 0.021, d = 1.80$) than in the control condition. However, the length of awakenings and time taken to fall asleep did not differ between the conditions.

3.2. eBAC

The mean eBAC level on the night prior to the hangover visit was 0.135% (±0.001%).

3.3. Hangover Symptom Severity

As shown in Table 1, at the hangover visit, participants were significantly negatively affected on each hangover symptom included in the AHSS apart from 'sweating'.

3.4. Multitasking Performance

The total MTF score was significantly lower in the hangover compared with the no-hangover condition ($t(24) = 2.26, p = 0.033, d = 0.46$). Performance on the four individual tasks did not differ with condition.

3.5. Mood

As displayed in Figure 2, completion of the Purple MTF did not differentially affect mood ratings in either hangover condition. However, there were significant main effects of hangover on alertness ($F(1,24) = 54.63, p < 0.001$), contentedness ($F(1,24) = 16.34, p < 0.001$), anxiety ($F(1,24) = 10.97, p = 0.003$), and mental fatigue ($F(1,24) = 40.70, p < 0.001$). There were also significant main effects of time (pre-post) on calmness ($F(1,24) = 8.42, p = 0.008$, stress ($F(1,24) = 5.86, p = 0.023$), and mental fatigue ($F(1,24) = 5.90, p = 0.023$). Prior to completing the Purple MTF, participants reported significantly worse levels of alertness ($t(24) = 7.33, p < .001, d = 1.47$), contentedness ($t(24) = 3.34, p = 0.002, d = 0.68$), mental fatigue ($t(24) = 4.90, p < 0.001, d = 0.99$), and anxiety ($t(24) = 2.58, p = 0.016, d = 0.21$) in the hangover, compared to the no-hangover condition.

Table 1. Effects of alcohol hangover on the single item severity score and items from the Alcohol Hangover Severity Scale (AHSS). Means and standard deviations are presented with t-statistic, associated p-value, and Cohen's d effect size.

Item	Condition		t-Value	p-Value	Cohen's d
	Control	Hangover			
Single-Item Severity Scale					
'How severe is your hangover?'	0.05(0.16)	4.67(2.36)	10.07	<0.001	3.07
Alcohol Hangover Severity Scale					
Fatigue	2.40(2.36)	7.36(2.18)	9.08	<0.001	1.82
Apathy	0.56(0.96)	3.64(2.80)	5.69	<0.001	1.35
Concentration problems	1.00(1.41)	5.88(2.22)	10.54	<0.001	2.20
Clumsiness	0.68(0.90)	4.72(3.14)	6.46	<0.001	1.54
Confusion	0.48(1.12)	3.48(3.20)	5.30	<0.001	1.38
Thirst	1.44(2.20)	6.24(2.51)	8.86	<0.001	1.78
Sweating	0.40(1.04)	1.20(1.94)	2.02	0.055	-
Shivering	0.16(0.62)	1.84(3.00)	2.80	0.010	0.69
Stomach pain	0.08(0.28)	2.36(2.84)	4.04	<0.001	1.10
Nausea	0.40(1.00)	3.12(2.76)	4.95	<0.001	1.14
Dizziness	0.36(0.91)	4.28(2.64)	7.50	<0.001	1.75
Heart pounding	0.40(1.23)	2.24(2.52)	3.71	0.001	0.82
Total score	0.70(0.74)	3.86(1.83)	8.90	<0.001	2.03

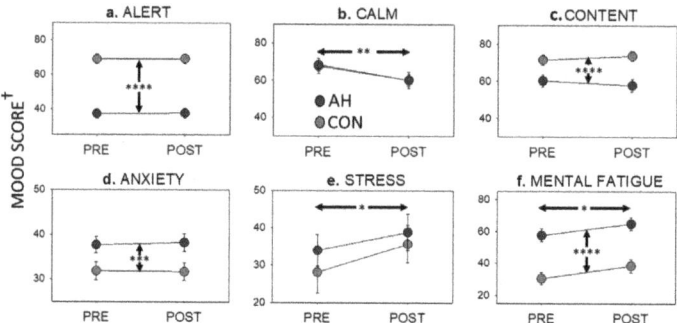

Figure 2. Effects of hangover and completing the Purple MTF on dimensions of mood measured pre- and post-Purple MTF. (**a**) Alert, (**b**) calm, and (**c**) content are derived from the Bond-Lader scales; (**d**) anxiety is derived from the Spielberger State-Trait Anxiety Inventory (STAI state portion); (**e**) stress and (**f**) mental fatigue are single visual analogue scales. Graphs depict means with standard errors of the mean (SEM). Vertical and horizontal arrows indicate a significant main effect of hangover and time, respectively (*, $p < 0.05$; **, $p < 0.01$; ***, $p < 0.005$; ***, $p < 0.001$). † Scales range from 0–100 except (**d**) anxiety which ranges from 20–80.

3.6. NASA-TLX

Participants rated the workload demand of completing the Purple MTF as more mentally demanding ($t(24) = 2.19$, $p = 0.039$, $d = 0.46$) and effortful ($t(24) = 2.29$, $p = 0.031$, $d = 0.47$) during the hangover condition. Additionally, participants rated their performance as worse ($t(24) = 2.75$, $p = 0.011$, $d = 0.55$) and the task as more frustrating ($t(24) = 2.26$, $p = 0.033$, $d = 0.46$) in the hangover condition, see Figure 3.

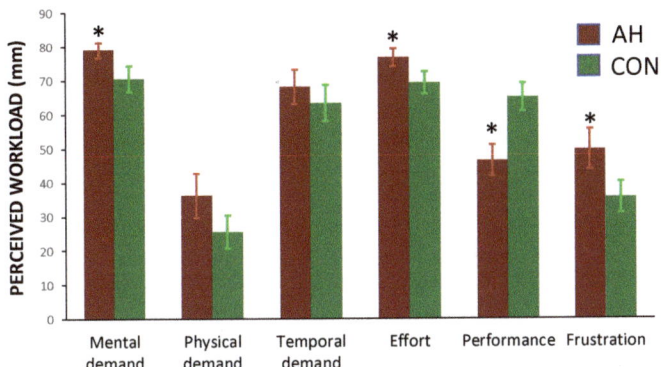

Figure 3. Mean (standard error) perceived levels of workload demand in the hangover (AH) and control (CON) conditions. * $p < 0.05$ between conditions.

4. Discussion

The current study assessed the effects of alcohol hangover on mood, multitasking performance and stress reactivity, and perceived demand. Several of our hypotheses were supported. Compared with no hangover, hangover was associated with significantly more negative mood (lower alertness and contentment, higher anxiety, and mental fatigue). Additionally, hangover was associated with poorer multitasking performance and greater perceived effort during multitasking. Counter to our hypothesis, stress reactivity was not differentially affected by hangover.

Hangover was reliably induced by using a semi-naturalistic study design and allowing participants to self-administer alcohol, according to the estimated BAC calculations, participants obtained a mean eBAC of 0.135%, a level beyond that deemed required to cause a hangover [1,50]. Self-reports also provided support for the link between hangover and poor sleep quality [16,51,52]. However, our findings were inconsistent with previous sleep studies that show a faster sleep onset after alcohol [53–55].

The negative mood associated with hangover was manifested as significantly lower alertness and contentment, and higher anxiety and mental fatigue. These findings are consistent with symptoms commonly reported in the hangover literature [1,5–10]. They are also supported by significant effects, here and elsewhere, on individual items on the AHSS which gauge elements of mood—namely fatigue, apathy, concentration problems, and confusion.

Participant self-reports also confirmed the presence of a hangover and hangover symptoms, demonstrating that the manipulation was successful. Every item of the AHSS differed significantly between visits with the exception of 'sweating' (p-value for 'sweating' = 0.055 two-tailed). Inspection of the effect sizes shows that the largest effect by far is for the single hangover item (Cohen's $d = 3.07$). This adds further support to the recent argument that the single item measure can capture hangover better than more granular items [56].

The second most prominent effect ($d = 2.20$) was for 'concentration problems'. This supports the documented cognitive impairment seen with hangover. In the current study, this was manifested as significantly poorer overall performance on the MTF but not on any individual task thereof. These results, along with previous evidence for hangover-induced impairment to more complex [25–28] but not simple divided attention [15,17,23,24] suggest that hangover differentially impairs tasks which draw on multiple domains. This result may help explain certain inconsistencies in the literature on hangover and cognitive performance (as outlined in the introduction). Unlike previous studies that have utilized assessments of divided attention across relatively few stimuli with limited task-switching, the current study used a cognitive stressor with multitasking across four tasks. Unlike many other laboratory stressors, this model has better correspondence with the more usual day-to-day stress that typically requires attention and response to several tasks concurrently [57]. Additionally, it may be that behavioral outcomes which are aggregates of several cognitive domains may be most sensitive to the negative effects of hangover. This has clear implications for behaviors such as driving which draw on several cognitive domains simultaneously. Further research is needed utilizing cognitive measures which mirror the processes involved in driving.

There are additional layers of cognitive control and function that are required for complex multitasking. For example, multitasking places a considerable load on working memory and executive processes, in addition to the resources needed for task performance, thus, likely taxing cognitive capacities [58,59]. These data suggest that hangover depletes cognitive resources making the effects of cognitive demand more profound than in a no hangover state.

Contrary to our hypothesis, completion of the cognitive stressor failed to elicit further changes in mood. The fact that two of the mood measures (calmness and stress) were significantly worse due to completing the MTF (independent of hangover condition) strongly suggests that the measure is sensitive to change in this cohort. Inspection of the pattern of results suggests that, for the other measures, the mood state of participants prior to completion of the MTF was already low and unlikely to change further.

Supporting our hypothesis, NASA-TLX scores revealed that being exposed to the cognitive stressor resulted in greater subjective demand in the hangover group. Compared with the control condition, hangover significantly increased ratings of perceived effort, mental demand, and frustration. Additionally, participants were aware of their poorer performance in the hangover condition. This is unlike alcohol intoxication, where individuals become 'uncalibrated' and, thus, overestimate their performance [60]. Despite awareness of poorer performance, it appears that in the hangover state individuals are unable to draw on additional cognitive resources to compensate and meet ongoing task demands. This has clear implications for behaviors requiring complex multitasking—including driving.

The physiological mechanisms of hangover are not well understood, but likely involve multiple processes [61], including actual and perceived immune status [62,63].

Limitations of this study included relying on self-reported alcohol intake and eBAC calculations to estimate BAC obtained to induce the hangover, and the lack of objective stress (i.e., cortisol) and sleep (i.e., actigraphy) measurement. Since excessive drinking commonly results in memory impairment [64–66], participants' accuracy of recalling their drinking behaviors may be questioned, a limitation inherent in hangover semi-naturalistic and naturalistic study designs (see [2]). Given the lack of empirical investigation into the effects of hangover on "real-life" multitasking ability and stress reactivity, this study fills an important research gap.

5. Conclusions

The findings presented here show that hangover negatively affects performance on an ecologically valid, complex multitasking tasks and mood. This is the first report of the effects of hangover on psychological stress reactivity and results indicated that, despite a greater perceived workload in the hangover condition, mood, and/or stress were not differentially affected following the completion of the cognitively demanding task. Future research should include a variety of complex 'real-life' measures of cognition and be directed towards determining the physiological changes that occur with a hangover.

Author Contributions: Conceptualization, A.S. and S.B.; methodology, M.A.W., S.B., A.S. and J.C.V.; software, M.A.W.; formal analysis, S.B.; data curation, S.B., E.A., H.G.; writing—original draft preparation, S.B.; writing—review and editing, S.B., E.A., H.G., M.A.W., J.C.V., A.S.; visualization, A.S.; supervision, S.B. and A.S.; project administration, S.B. All authors have read and agreed to the published version of the manuscript.

Acknowledgments: E.A. and H.G. are recipients of Swinburne University Postgraduate Research Awards.

Conflicts of Interest: Over the past 36 months, A.S. has held research grants from Abbott Nutrition, Arla Foods, Bayer, BioRevive, DuPont, Fonterra, Kemin Foods, Nestlé, Nutricia-Danone, and Verdure Sciences. He has acted as a consultant/expert advisor to Bayer, Danone, Naturex, Nestlé, Pfizer, Sanofi, Sen-Jam Pharmaceutical, and has received travel/hospitality/speaker fees from Bayer, Sanofi, and Verdure Sciences. Over the past 36 months, J.C.V. has held grants from the Dutch Ministry of Infrastructure and the Environment, Janssen, Nutricia, and Sequential, and acted as a consultant/expert advisor to Clinilabs, More Labs, Red Bull, Sen-Jam Pharmaceutical, Toast!, and ZBiotics.

References

1. Verster, J.C.; Stephens, R.; Penning, R.; Rohsenow, D.; McGeary, J.; Levy, D.; McKinney, A.; Finnigan, F.; Piasecki, T.M.; Adan, A. The alcohol hangover research group consensus statement on best practice in alcohol hangover research. *Curr. Drug Abuse Rev.* **2010**, *3*, 116. [CrossRef] [PubMed]
2. Verster, J.C.; van de Loo, A.J.; Adams, S.; Stock, A.K.; Benson, S.; Scholey, A.; Alford, C.; Bruce, G. Advantages and Limitations of Naturalistic Study Designs and their Implementation in Alcohol Hangover Research. *J. Clin. Med.* **2019**, *8*, 2160. [CrossRef] [PubMed]
3. Verster, J.C.; Scholey, A.; van de Loo, A.J.; Benson, S.; Stock, A.K. Updating the Definition of the Alcohol Hangover. *J. Clin. Med.* **2020**, *9*, 823. [CrossRef] [PubMed]
4. Rohsenow, D.J.; Howland, J.; Minsky, S.J.; Greece, J.; Almeida, A.; Roehrs, T.A. The Acute Hangover Scale: A new measure of immediate hangover symptoms. *Addict. Behav.* **2007**, *32*, 1314–1320. [CrossRef] [PubMed]
5. Verster, J.C. The alcohol hangover–A puzzling phenomenon. *Alcohol Alcohol.* **2008**, *43*, 124–126. [CrossRef] [PubMed]
6. Mc Kinney, A.; Coyle, K. Alcohol hangover effects on measures of affect the morning after a normal night's drinking. *Alcohol Alcohol.* **2006**, *41*, 54–60. [CrossRef]
7. Prat, G.; Adan, A.; Sanchez-Turet, M. Alcohol hangover: A critical review of explanatory factors. *Hum. Psychopharmacol.* **2009**, *24*, 259–267. [CrossRef]
8. Wiese, J.G.; Shlipak, M.G.; Browner, W.S. The alcohol hangover. *Ann. Intern Med.* **2000**, *132*, 897–902. [CrossRef]
9. Takahashi, M.; Li, W.; Koike, K.; Sadamoto, K. Clinical effectiveness of KSS formula, a traditional folk remedy for alcohol hangover symptoms. *J. Nat. Med.* **2010**, *64*, 487–491. [CrossRef]
10. Verster, J.C.; van Duin, D.; Volkerts, E.R.; Schreuder, A.H.; Verbaten, M.N. Alcohol hangover effects on memory functioning and vigilance performance after an evening of binge drinking. *Neuropsychopharmacology* **2003**, *28*, 740–746. [CrossRef]
11. Finnigan, F.; Schulze, D.; Smallwood, J.; Helander, A. The effects of self-administered alcohol-induced 'hangover' in a naturalistic setting on psychomotor and cognitive performance and subjective state. *Addiction* **2005**, *100*, 1680–1689. [CrossRef]
12. Slutske, W.S.; Piasecki, T.M.; Hunt-Carter, E.E. Development and initial validation of the Hangover Symptoms Scale: Prevalence and correlates of hangover symptoms in college students. *Alcohol. Clin. Exp. Res.* **2003**, *27*, 1442–1450. [CrossRef]
13. Kim, D.-J.; Kim, W.; Yoon, S.-J.; Choi, B.-M.; Kim, J.-S.; Go, H.J.; Kim, Y.-K.; Jeong, J. Effects of alcohol hangover on cytokine production in healthy subjects. *Alcohol* **2003**, *31*, 167–170. [CrossRef]
14. Anderson, S.; Dawson, J. Neuropsychological correlates of alcoholic hangover. *S. Afr. J. Sci.* **1999**, *95*, 145.
15. Roehrs, T.; Yoon, J.; Roth, T. Nocturnal and next-day effects of ethanol and basal level of sleepiness. *Hum. Psychopharmacol. Clin. Exp.* **1991**, *6*, 307–311. [CrossRef]
16. Howland, J.; Rohsenow, D.J.; Littlefield, C.A.; Almeida, A.; Heeren, T.; Winter, M.; Bliss, C.A.; Hunt, S.; Hermos, J. The effects of binge drinking on college students' next-day academic test-taking performance and mood state. *Addiction* **2010**, *105*, 655–665. [CrossRef] [PubMed]
17. McKinney, A.; Coyle, K.; Penning, R.; Verster, J.C. Next day effects of naturalistic alcohol consumption on tasks of attention. *Hum. Psychopharmacol. Clin. Exp.* **2012**, *27*, 587–594. [CrossRef]
18. Grange, J.A.; Stephens, R.; Jones, K.; Owen, L. The effect of alcohol hangover on choice response time. *J. Psychopharmacol.* **2016**, *30*, 654–661. [CrossRef]

19. McKinney, A.; Coyle, K. Next-day effects of alcohol and an additional stressor on memory and psychomotor performance. *J. Stud. Alcohol Drugs* **2007**, *68*, 446–454. [CrossRef]
20. McKinney, A.; Coyle, K. Next day effects of a normal night's drinking on memory and psychomotor performance. *Alcohol Alcohol.* **2004**, *39*, 509–513. [CrossRef]
21. Gunn, C.; Mackus, M.; Griffin, C.; Munafò, M.R.; Adams, S. A systematic review of the next-day effects of heavy alcohol consumption on cognitive performance. *Addiction* **2018**, *113*, 2182–2193. [CrossRef]
22. Langner, R.; Willmes, K.; Chatterjee, A.; Eickhoff, S.B.; Sturm, W. Energetic effects of stimulus intensity on prolonged simple reaction-time performance. *Psychol. Res.* **2010**, *74*, 499–512. [CrossRef]
23. Collins, W.E.; Chiles, W.D. Laboratory performance during acute alcohol intoxication and hangover. *Hum. Factors J. Hum. Factors Ergon. Soc.* **1980**, *22*, 445–462. [CrossRef]
24. Collins, W.E. *Performance Effects of Alcohol Intoxication and Hangover at Ground Level and at Simulated Altitude*; Civil Aerospace Medical Institute: Oklahoma City, OK, USA, October 1979.
25. Verster, J.C.; Bervoets, A.C.; de Klerk, S.; Vreman, R.A.; Olivier, B.; Roth, T.; Brookhuis, K.A. Effects of alcohol hangover on simulated highway driving performance. *Psychopharmacology* **2014**, *231*, 2999–3008. [CrossRef]
26. Törnros, J.; Laurell, H. *Acute and Hang-over Effects of Alcohol on Simulated Driving Performance*; VTI Särtryck 169; Statens Väg-och Trafikinstitut: Linköping, Sweden, 1991.
27. Yesavage, J.A.; Leirer, V.O. Hangover effects on aircraft pilots 14 hours after alcohol ingestion: A preliminary report. *Am. J. Psychiatry* **1986**, *143*, 1546–1550.
28. Yesavage, J.A.; Dolhert, N.; Taylor, J.L. Flight simulator performance of younger and older aircraft pilots: Effects of age and alcohol. *J. Am. Geriatr. Soc.* **1994**, *42*, 577–582. [CrossRef]
29. Dzubak, C.M. Multitasking: The good, the bad, and the unknown. *J. Assoc. Tutoring Prof.* **2008**, *1*, 1–12.
30. Wetherell, M.A.; Carter, K. The multitasking framework: The effects of increasing workload on acute psychobiological stress reactivity. *Stress Health* **2014**, *30*, 103–109. [CrossRef]
31. Wetherell, M.A.; Hyland, M.E.; Harris, J.E. Secretory immunoglobulin A reactivity to acute and cumulative acute multi-tasking stress: Relationships between reactivity and perceived workload. *Biol. Psychol.* **2004**, *66*, 257–270. [CrossRef]
32. Wetherell, M.A.; Atherton, K.; Grainger, J.; Brosnan, R.; Scholey, A.B. The effects of multitasking on psychological stress reactivity in recreational users of cannabis and MDMA. *Hum. Psychopharmacol. Clin. Exp.* **2012**, *27*, 167–176. [CrossRef]
33. Wetherell, M.A.; Sidgreaves, M.C. Secretory immunoglobulin-A reactivity following increases in workload intensity using the Defined Intensity Stressor Simulation (DISS). *Stress Health J. Int. Soc. Investig. Stress* **2005**, *21*, 99–106. [CrossRef]
34. Scholey, A.B.; Owen, L.; Gates, J.; Rodgers, J.; Buchanan, T.; Ling, J.; Heffernan, T.; Swan, P.; Stough, C.; Parrott, A.C. Hair MDMA samples are consistent with reported ecstasy use: Findings from a study investigating effects of ecstasy on mood and memory. *Neuropsychobiology* **2011**, *63*, 15–21. [CrossRef]
35. Carney, C.E.; Buysse, D.J.; Ancoli-Israel, S.; Edinger, J.D.; Krystal, A.D.; Lichstein, K.L.; Morin, C.M. The Consensus Sleep Diary: Standardizing Prospective Sleep Self-Monitoring. *Sleep* **2012**, *35*, 287–302. [CrossRef]
36. Forrest, A. The estimation of Widmark's factor. *J. Forensic Sci. Soc.* **1986**, *26*, 249–252. [CrossRef]
37. Watson, P.E.; Watson, I.D.; Batt, R.D. Prediction of blood alcohol concentrations in human subjects. Updating the Widmark Equation. *J. Stud. Alcohol.* **1981**, *42*, 547–556. [CrossRef]
38. Seidl, S.; Jensen, U.; Alt, A. The calculation of blood ethanol concentrations in males and females. *Int. J. Leg. Med.* **2000**, *114*, 71–77. [CrossRef]
39. Widmark, E.M.P. *Principles and Applications of Medicolegal Alcohol Determination*; Biomedical Publications: Davis, CA, USA, 1981.
40. Ulrich, L.; Cramer, Y.; Zink, P. Relevance of individual parameters in the calculation of blood alcohol levels in relation to the volume of intake. *Blutalkohol* **1987**, *24*, 192.
41. Scholey, A.; Haskell, C.; Robertson, B.; Kennedy, D.; Milne, A.; Wetherell, M. Chewing gum alleviates negative mood and reduces cortisol during acute laboratory psychological stress. *Physiol. Behav.* **2009**, *97*, 304–312. [CrossRef]
42. Kennedy, D.O.; Little, W.; Scholey, A.B. Attenuation of laboratory-induced stress in humans after acute administration of Melissa officinalis (Lemon Balm). *Psychosom. Med.* **2004**, *66*, 607–613. [CrossRef]

43. Kennedy, D.O.; Pace, S.; Haskell, C.; Okello, E.J.; Milne, A.; Scholey, A.B. Effects of cholinesterase inhibiting sage (Salvia officinalis) on mood, anxiety and performance on a psychological stressor battery. *Neuropsychopharmacology* **2006**, *31*, 845. [CrossRef]
44. Benson, S.; Downey, L.A.; Stough, C.; Wetherell, M.; Zangara, A.; Scholey, A. An acute, double-blind, placebo-controlled cross-over study of 320 mg and 640 mg doses of Bacopa monnieri (CDRI 08) on multitasking stress reactivity and mood. *Phytother. Res.* **2014**, *28*, 551–559. [CrossRef]
45. Hart, S. Development of a multi-dimensional workload rating scale: Results of empirical and theoretical research. *Hum. Ment. Workload.* **1988**, *52*, 139–183.
46. Penning, R.; McKinney, A.; Bus, L.D.; Olivier, B.; Slot, K.; Verster, J.C. Measurement of alcohol hangover severity: Development of the Alcohol Hangover Severity Scale (AHSS). *Psychopharmacology* **2013**, *225*, 803–810. [CrossRef]
47. Spielberger, C.; Gorsuch, R.; Lushene, R. *The State Trait Anxiety Inventory Manual*; Consulting Psychologists Press: Palo Alto, CA, USA, 1969.
48. Bond, A.; Lader, M. The use of analogue scales in rating subjective feelings. *Br. J. Med. Psychol.* **1974**, *47*, 211–218. [CrossRef]
49. Morris, S.B.; DeShon, R.P. Combining effect size estimates in meta-analysis with repeated measures and independent-groups designs. *Psychol. Methods* **2002**, *7*, 105. [CrossRef]
50. Verster, J.C.; Kruisselbrink, L.D.; Slot, K.A.; Anogeianaki, A.; Adams, S.; Alford, C.; Arnoldy, L.; Ayre, E.; Balikji, S.; Benson, S. Sensitivity to experiencing alcohol hangovers: Reconsideration of the 0.11% Blood Alcohol Concentration (BAC) threshold for having a hangover. *J. Clin. Med.* **2020**, *9*, 179. [CrossRef] [PubMed]
51. Rohsenow, D.J.; Howland, J.; Minsky, S.J.; Arnedt, J.T. Effects of heavy drinking by maritime academy cadets on hangover, perceived sleep, and next-day ship power plant operation. *J. Stud. Alcohol.* **2006**, *67*, 406–415. [CrossRef]
52. van Schrojenstein Lantman, M.; Roth, T.; Roehrs, T.; Verster, J.C. Alcohol Hangover, Sleep Quality, and Daytime Sleepiness. *Sleep Vigil.* **2017**, *1*, 37–41. [CrossRef]
53. Roehrs, T.; Roth, T. Sleep, sleepiness, and alcohol use. *Alcohol Res. Health* **2001**, *25*, 101–109.
54. Ebrahim, I.O.; Shapiro, C.M.; Williams, A.J.; Fenwick, P.B. Alcohol and sleep I: Effects on normal sleep. *Alcohol. Clin. Exp. Res.* **2013**, *37*, 539–549. [CrossRef]
55. Thakkar, M.M.; Sharma, R.; Sahota, P. Alcohol disrupts sleep homeostasis. *Alcohol* **2015**, *49*, 299–310. [CrossRef]
56. Verster, J.C.; van de Loo, A.J.; Benson, S.; Scholey, A.; Stock, A.K. The assessment of overall hangover severity. *J. Clin. Med.* **2020**, *9*, 786. [CrossRef]
57. Wetherell, M.A.; Craw, O.; Smith, K.; Smith, M.A. Psychobiological responses to critically evaluated multitasking. *Neurobiol. Stress* **2017**, *7*, 68–73. [CrossRef]
58. Wolff, N.; Gussek, P.; Stock, A.K.; Beste, C. Effects of high-dose ethanol intoxication and hangover on cognitive flexibility. *Addict. Biol.* **2018**, *23*, 503–514. [CrossRef]
59. Monsell, S.; Driver, J. *Control of Cognitive Processes: Attention and Performance XVIII*; MIT Press: Cambrigde, MA, USA, 2000; Volume 18.
60. Tipple, C.; Benson, S.; Scholey, A. A review of the physiological factors associated with alcohol hangover. *Curr. Drug Abuse Rev.* **2016**, *9*, 93–98. [CrossRef]
61. Tiplady, B.; Franklin, N.; Scholey, A. Effect of ethanol on judgments of performance. *Br. J. Psychol.* **2004**, *95*, 105–118. [CrossRef]
62. van de Loo, A.; Mackus, M.; van Schrojenstein Lantman, M.; Kraneveld, A.; Brookhuis, K.; Garssen, J.; Scholey, A.; Verster, J. Susceptibility to alcohol hangovers: The association with self-reported immune status. *Int. J. Environ. Res. Public Health* **2018**, *15*, 1286. [CrossRef]
63. van de Loo, A.J.; van Schrojenstein Lantman, M.; Mackus, M.; Scholey, A.; Verster, J.C. Impact of mental resilience and perceived immune functioning on the severity of alcohol hangover. *BMC Res. Notes* **2018**, *11*, 526. [CrossRef]
64. White, A.; Hingson, R. The burden of alcohol use: Excessive alcohol consumption and related consequences among college students. *Alcohol Res.* **2013**, *35*, 201–218.

65. Leitz, J.R.; Morgan, C.J.; Bisby, J.A.; Rendell, P.G.; Curran, H.V. Global impairment of prospective memory following acute alcohol. *Psychopharmacology* **2009**, *205*, 379–387. [CrossRef]
66. Crego, A.; Holguín, S.R.; Parada, M.; Mota, N.; Corral, M.; Cadaveira, F. Binge drinking affects attentional and visual working memory processing in young university students. *Alcohol. Clin. Exp. Res.* **2009**, *33*, 1870–1879. [CrossRef] [PubMed]

© 2020 by the authors. Licensee MDPI, Basel, Switzerland. This article is an open access article distributed under the terms and conditions of the Creative Commons Attribution (CC BY) license (http://creativecommons.org/licenses/by/4.0/).

Article
The Effects of Alcohol Hangover on Executive Functions

Craig Gunn [1,*], Graeme Fairchild [2], Joris C. Verster [3,4,5] and Sally Adams [1,*]

1 Addiction and Mental Health Group, Department of Psychology, University of Bath, Bath BA2 7AY, UK
2 Department of Psychology, University of Bath, Bath BA2 7AY, UK; G.Fairchild@bath.ac.uk
3 Division of Pharmacology, Utrecht University, 3584CG Utrecht, The Netherlands; J.C.Verster@uu.nl
4 Institute for Risk Assessment Sciences (IRAS), Utrecht University, 3584CM Utrecht, The Netherlands
5 Centre for Human Psychopharmacology, Swinburne University, Melbourne, VIC 3122, Australia
* Correspondence: craig.a.gunn@gmail.com (C.G.); S.Adams@bath.ac.uk (S.A.)

Received: 3 April 2020; Accepted: 14 April 2020; Published: 17 April 2020

Abstract: Recent research has suggested that processes reliant on executive functions are impaired by an alcohol hangover, yet few studies have investigated the effect of hangovers on core executive function processes. Therefore, the current study investigated the effect of hangovers on the three core components of the unity/diversity model of executive functions: the ability to switch attention, update information in working memory, and maintain goals. Thirty-five 18-to-30-year-old non-smoking individuals who reported experiencing a hangover at least once in the previous month participated in this study. They completed tasks measuring switching (number-switching task), updating (n-back task), and goal maintenance (AX Continuous Performance Test, AX-CPT) whilst experiencing a hangover and without a hangover in a 'naturalistic' within-subjects crossover design. Participants made more errors in the switching task ($p = 0.019$), more errors in both the 1- ($p < 0.001$) and 2-back ($p < 0.001$) versions of the n-back, and more errors in the AX-CPT ($p = 0.007$) tasks when experiencing a hangover, compared to the no-hangover condition. These results suggest that an alcohol hangover impairs core executive function processes that are important for everyday behaviours, such as decision-making, planning, and mental flexibility.

Keywords: alcohol; hangover; executive functions; working memory; cognition

1. Introduction

An alcohol hangover is a combination of mental and physical symptoms, experienced the day after a single episode of heavy drinking, when blood alcohol concentration (BAC) approaches zero [1]. It is the most common negative consequence of heavy drinking and can impair cognitive processes, such as sustained attention, memory, and psychomotor skills [2,3]. However, relatively few studies have investigated the effect of alcohol hangovers on core components of executive functions.

Executive functions are higher-order cognitive processes used in everyday behaviours, such as decision-making, mental flexibility, and planning. Recent studies have indicated that executive functions may be negatively influenced by alcohol hangovers. Studies have suggested that performance on tasks of interference control [4,5] and response inhibition [6] is impaired when subjects are experiencing a hangover, suggesting poorer inhibitory control, which may negatively influence decisions around subsequent alcohol use [7] and emotion regulation [8]. Furthermore, findings showing poorer spatial working memory [4], reward learning [9], prospective memory [10,11], semantic verbal fluency [10], and performance on backward visual span tasks [12] indicate that executive functions are impaired whilst experiencing a hangover. A recent report by the Institute of Alcohol Studies suggested that the cost of hangover-related reductions in work productivity could be as high as £1.4 billion per annum in the UK [13]. As effective workplace performance relies on an individual's ability to make decisions, organise tasks, and plan, detrimental effects of hangovers on executive functions may contribute toward these costs. Therefore, it is important to understand how these processes may be influenced the morning after a night of heavy alcohol consumption, i.e., during a hangover.

Executive functions are utilised when behaviours need to be controlled (rather than when they are 'automatic'), when cognitive processes are combined, or when individuals need to switch attention between tasks [14]. The unity/diversity model conceptualises executive functions as being composed of two core components, alongside a single common factor that is utilised in all executive function tasks [15]. The two components represent the ability to switch attention from one task/mental set to another (switching) and the ability to update information within working memory (updating). The common factor of the unity/diversity model represents the ability to maintain and manage goals, in order to effectively complete tasks (goal maintenance). All executive function tasks utilise aspects of these core components. As hangover-related impairments have been observed in higher-order cognitive processes, such as prospective memory [10], it is possible that hangovers influence these core components of executive function.

Attentional switching requires allocation of attentional resources to effectively switch from one task or mental set to another [16]. Recent studies have indicated that a hangover may be a state in which individuals experience high cognitive load [17] and thus have fewer available resources to switch attention [18,19]. When available cognitive resources are low, completion of executive function tasks becomes ineffective or inefficient [20–22]. Factors associated with heavy alcohol consumption, such as a reduction in glutamatergic and an increase in GABAergic, dopaminergic, and serotonergic neurotransmission, may also influence attentional switching [23,24]. During hangover, dopaminergic neurotransmission may be reduced, and noradrenaline may be elevated [9,25], suggesting that switching could become impaired. Furthermore, studies have highlighted that fatigue (which is one of the most commonly reported symptoms of a hangover [26]) can lead to impairments in switching [27].

Thus far, studies investigating attentional switching in individuals experiencing a hangover have yielded mixed results. One study induced hangovers experimentally [19] and reported no effect on switch costs, reflecting the additional time needed to switch attention to the new rule set. However, experimental hangover manipulations involve administering lower doses of alcohol than are typically consumed when drinking in everyday life [28], and this practice could influence the effects of a hangover [29]. Two naturalistic studies, which involve assessing the impact of hangovers experienced following real-life drinking, investigated the effects of a hangover on perseveration errors, which are erroneous responses made according to the previously correct rule or set, reflecting a switching failure. One reported that a hangover did not influence switching performance in a non-student sample [4], whereas another study using a student sample indicated poorer switching accuracy when experiencing a hangover, as compared to a control condition [30]. It is possible that hungover individuals attempt to maintain performance on switching tasks by either sacrificing accuracy to maintain the speed of their responses or by sacrificing speed to maintain accuracy (i.e., a 'speed–accuracy trade-off').

To our knowledge, no studies have investigated the effects of an alcohol hangover on updating or goal maintenance; however, there are indications that both processes could be negatively affected by a hangover. Goal maintenance is an important process utilised to complete all executive function tasks [15]. For example, an individual completing a task at work (e.g., writing a report) would need to keep his or her overall goal in mind whilst planning, organising, and making decisions about the individual task subcomponents. If goal maintenance is impaired, individuals may be less effective or efficient at completing complex tasks with multiple subcomponents. As previously mentioned, studies have indicated impairments in working memory performance, prospective memory, and semantic verbal fluency—all tasks requiring executive functions—during a hangover [4,10–12]. Therefore, it is possible that a common factor underlying hangover-related impairments in each of these tasks is a deficit in the ability to maintain goals. Inhibitory control is also impaired when experiencing a hangover [4–6] and is a key part of goal maintenance [15], further suggesting goal maintenance could be influenced by a hangover. In addition, reduced cognitive resources during a hangover may influence goal maintenance by biasing individuals toward reacting to external events (i.e., bottom-up stimulus-driven processing) rather than proactive control of one's actions (i.e., actively sustaining goal representations through top-down processing) [18,31]. The AX Continuous Performance Task (AX-CPT) can be used to assess goal maintenance and can differentiate between proactive and reactive control [32].

The process of updating information in working memory can become impaired by high cognitive load and when there is a reduction in available cognitive resources [33]. As previously mentioned, cognitive resources may be reduced during a hangover [18,19], negatively affecting the ability to update information in working memory. Furthermore, a study of the cognitive effects of pain indicate that a headache can impair performances on tasks measuring updating [34]. By using an n-back task with conditions that vary in difficulty, studies have also demonstrated that cognitive load selectively influenced the disrupting effect of pain on updating [35]. In addition, studies have indicated that updating can be impaired following sleep deprivation [36]. As a headache is a 'core' hangover symptom [1], and individuals experience sleep disruptions after heavy alcohol consumption (e.g., decreased efficiency and REM sleep, increased night-time awakenings [37], and decreased sleep duration [38]), updating ability may also be compromised by a hangover. To assess this possibility, we used an n-back working memory task with two conditions that vary in difficulty (1-back and 2-back).

In summary, the current study aimed to investigate the effects of an alcohol hangover on all three core components of the unity/diversity model of executive functions: switching, updating, and goal maintenance. Specifically, we hypothesised that participants experiencing a hangover would show impairments in: (1) switching, (2) updating, and (3) goal maintenance, as compared to the no-hangover control condition. We also hypothesised that participants would adopt a more reactive control style on the AX-CPT task in the hangover condition, as compared to the no-hangover condition, and that the magnitude of impairments in goal maintenance, updating, and switching abilities would be positively associated with hangover severity. As performance on executive function tasks may be related to an individual's confidence in his or her ability to complete tasks (self-efficacy: the belief we have in our ability to execute the actions required for specifically designated performances, usually assessed as our degree of confidence that we can perform specific tasks [39]), and self-efficacy to complete tasks is lower when individuals are experiencing a hangover [12,40], we also explored the relationship between self-efficacy and task performance. We hypothesised that performance in goal maintenance, updating, and switching tasks would be positively associated with self-efficacy to complete these tasks.

2. Materials and Methods

2.1. Participants

Thirty-eight participants were recruited from a student population by poster/flyer and digital advertisements, the University of Bath's research participation scheme, word of mouth, and direct approach by the researcher. Inclusion criteria required participants to consume at least 6 (female) or 8 (male) units of alcohol in a typical heavy drinking session, to be aged between 18 and 30 years old, to be non-smokers, and to be in general good mental and physical health. To exclude the potential confound of hangover-resistance, only participants who reported experiencing a hangover in the past month were recruited. Participants who were pregnant/breast-feeding, taking medication or recreational drugs, consuming > 400 mg of caffeine per day, had a current or past personal or family history of drug dependency, or had a diagnosed sleep disorder were excluded. Three participants withdrew before completing both conditions; thus, 35 participants (14 males; 21 females) completed the study. The University of Bath Psychology research ethics committee approved this research, ethics code: 18-328.

2.2. Design

An experimental 'naturalistic' design, with one within-subjects factor of condition (hangover and no-hangover) was used. The naturalistic design is a valid method when one is interested in examining the real-life cognitive effects of alcohol hangover, and it has been successfully implemented in many hangover studies [41]. The hangover condition took place on a morning following an evening of heavy alcohol consumption, and the no-hangover condition on a morning following no alcohol consumption for at least 24 h prior to testing. Order of testing was counterbalanced across subjects, whereby 53% of participants completed the hangover condition first.

2.3. Measures

Participants completed three cognitive tasks assessing different components of executive function: switching, updating, and goal maintenance.

2.3.1. Number-Switching Task

A cued-switching task was used to measure switching [42]. In this task, participants were presented with a string of numbers (1, 2, 3, 4, 6, 7, 8, and 9) appearing within a shape (square or diamond). A cue (square/diamond without number) appeared for 650 ms before the number stimulus. Participants were instructed to respond depending on the 'rule', which was indicated by the colour of the shape. Participants responded with 'z' if the number was odd or 'x' if the number was even, when presented within a blue shape, and responded with 'n' if the number was lower than 5 or 'm' if the number was higher than 5, when presented within an orange shape. The rule was switched every 4 trials, in a sequential manner. The primary outcome measures were switch costs, which were calculated by subtracting RT for the second trial following a rule change (P2) from the first trial following a rule change (P1) and perseveration errors, i.e., erroneous responses made according to the prior rule set. Schematic representations of each task are presented in Figure 1.

2.3.2. The N-Back Task

The letter version of the n-back task was used to measure updating [43]. In this task, participants viewed a string of letters (random presentation) and were asked to indicate whether the letter was the same as the letter presented in a previous trial (i.e., n-back). Letters were presented for 500 ms, with an inter-trial interval (blank screen) for 1500 ms. Participants were asked to respond with 'm' when the letter was the same as n-back (target trials), and 'z' when it was not the same (non-target trials). The task consisted of two 1-back (letter same as the previous trial) and two 2-back (letter same as the one presented before the last trial) blocks presented in alternating blocks (i.e., 1-back, 2-back, 1-back, and 2-back). There were 45 trials in each block, with target stimuli (those that are valid n-back trials) presented 33% of the time. The primary outcome measure for this task was errors to target stimuli.

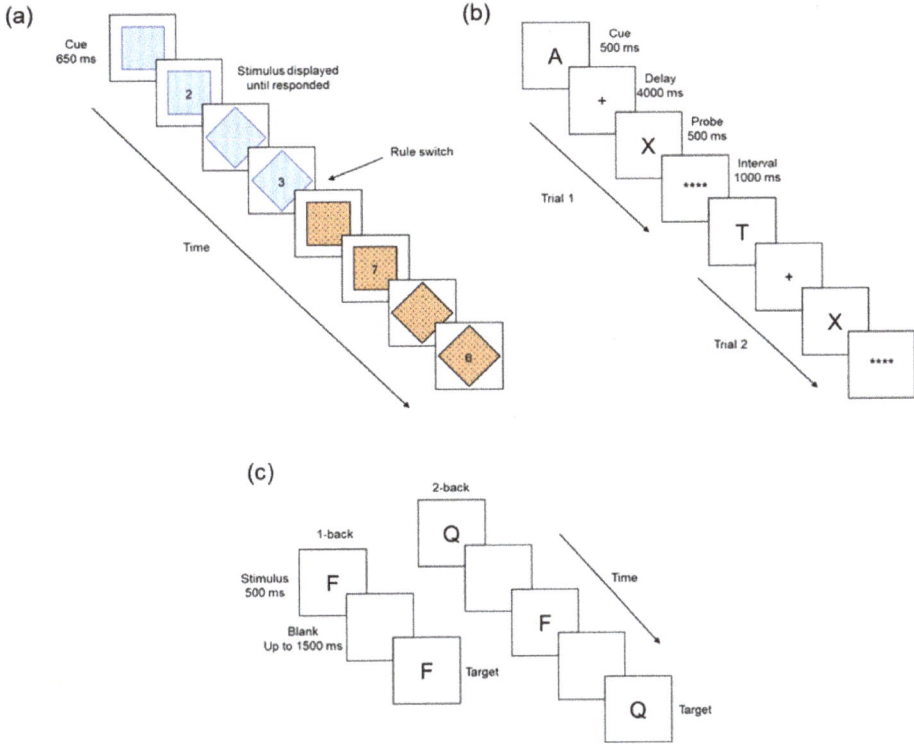

Figure 1. Schematic representations of each cognitive task. (**a**) In the switching task, participants are presented with a cue (empty blue/orange shape), followed by a number stimulus. Participants respond according to the rule (determine odd/even or higher/lower than 5), indicated by the colour of the shape (blue or orange). (**b**) In the AX Continuous Performance Task (AX-CPT) task, participants are presented with a cue-probe pair separated by a long delay (+). When **** appeared on the screen, participants respond by pressing the 'm' key when the cue is 'A' and probe is 'X'; otherwise, participants respond with the 'z' key. The first trial is an example of a target trial (AX) and the second trial is an example of a BX non-target trial type (the cue "T" is incorrect in this case). (**c**) In the n-back task, participants respond with the 'm' key when the target is the same as the stimulus presented either 1 or 2 trials earlier (e.g., if the target is the same as the previous letter in the 1-back version); otherwise, participants respond with the 'z' key.

2.3.3. The AX Continuous Performance Task

The AX Continuous Performance Task (AX-CPT) can be used to assess goal maintenance and can differentiate between proactive and reactive aspects of cognitive control [32,44,45]. Participants respond to a probe on the basis of a preceding cue. A letter cue was presented on screen for 500 ms, followed by a long delay of 4000 ms (displayed as '+') [32]. Participants were then presented with a letter probe for 500 ms, followed by an inter-trial interval of 1000 ms (displayed as '****'). Participants responded to probes by pressing 'm' on the keyboard for cue-probe targets or 'z' for non-targets. Target responses are when an 'A' cue is followed by an 'X' probe (AX-type trial), and non-target trials are responses to all other letter sequences. 'AY-type' trials are when an 'A' cue is followed by any probe other than 'X'; 'BX-type' trials are those when any cue other than 'A' are followed by an 'X' probe; and 'BY-type' trials occur when any cue other than 'A' is followed by any probe other than 'X'. Target trials (AX) were presented with 70% frequency, and non-targets with 30% frequency; non-target trial frequency was equally distributed so that non-cue–probe (e.g., BX-type), cue–non-probe (e.g., AY-type), and non-cue–non-probe (e.g., BY-type) trials each occurred 10% of the time. A total of 120 trials were presented in a single block, and the primary outcome measure was the number of errors for each trial type. Participants utilising reactive control selectively retrieve contextual information when stimuli are presented, and they are less likely to actively maintain contextual information. In the AX-CPT task, reactive control can be observed with increased errors in 'BX-type' trials as participants react to a valid stimulus (the 'X'), but without actively maintaining the preceding invalid cue (not an 'A'). Thus, if individuals with a hangover are biased toward reactive control processes, we would expect to observe an increase in erroneous responses to 'BX-type' trials relative to the no-hangover control condition.

2.3.4. Subjective Measures

Self-reported alcohol consumption on the previous night was used to calculate estimated peak BAC (eBAC), using the Widmark formula [46]. Hangover severity was measured by using a 1-item hangover severity scale and modified Alcohol Hangover Severity Scale (mAHSS; [47]). Participants were also asked to rate how confident they felt about completing the tasks effectively (self-efficacy) on an 11-point scale (0 = cannot do at all; 10 = certainly can do; [39]), following practice trials on each cognitive task. Following each task, participants were asked to complete the Rating Scale of Mental Effort (RSME) which assessed the degree of effort involved in performing the respective task [48].

2.4. Procedure

Participants were given information about the study and were booked in for two sessions (hangover and no-hangover), according to when they next expected to experience a hangover or have a no-hangover day. Time of day of testing was as similar as possible for both sessions. Participants were screened to ensure they met inclusion criteria and gave written informed consent before the study started. Participants self-reported their previous night's alcohol consumption by using pictorial prompts labelled with alcohol unit content and caffeine consumption on the day of testing. Participants were breathalysed and completed the 1-item hangover severity scale and mAHSS, to verify their condition (hangover and no-hangover) before completing the three cognitive tasks in a randomised counterbalanced order. Following practice trials, participants rated their self-efficacy before completing each task. Following completion of each task, participants completed the RSME. Participants then arranged the second testing session at least 36 h later, to prevent crossover effects. Upon completion of both conditions, participants were paid £10 and received a full debrief.

2.5. Statistical Analysis

Statistical analysis was conducted in accordance with our preregistered protocol [49]. Outliers were removed if they were > 1.5 * Inter-Quartile Range and > 2 SD from the mean. Analysis was also conducted by winsorizing the outliers, which did not impact the results presented below. For the switching task, trials following an error and trials with RT > 2500 ms were omitted from analysis. Participants for whom < 50% trials were available were removed from analysis (n = 5). Error trials were omitted from RT analysis [50]. Repeated measures ANOVAs were conducted with order and sex as between-subject factors, using SPSS (version 25). Effect sizes are reported as Cohen's d. Due to the possible effects of acute intoxication at BAC > 0.02% [51], a sensitivity analysis was conducted to see if residual alcohol concentrations during a hangover influenced cognitive performance. A sensitivity analysis, excluding one participant with a BAC > 0.02%, yielded similar results; therefore, this participant is included in the analyses presented below.

3. Results

3.1. Participant Characteristics

The average age of participants was 20.23 years (SD = 2.81; range = 18–30), and they consumed an average of 13.28 alcohol units the evening before the hangover condition (SD = 5.13; range = 5–28.5). The mean eBAC calculated for the evening before the hangover condition was 0.16% (SD = 0.08; range = 0.01%–0.37%). None of the participants consumed alcohol before the no-hangover control condition or reported experiencing a hangover (i.e., all participants scored zero on the hangover-severity scale). Although eBAC calculations for some participants were low (e.g., 0.01%), all participants in the hangover condition reported having a hangover (severity scale score > 0). A sensitivity analysis indicated that excluding participants with an eBAC < 0.05% the night before the hangover condition did not alter results, and these participants were therefore included in the analyses reported below. A visual representation of the range of eBAC values in the sample is provided in Figure 2. There was no difference in caffeine consumption between the hangover and no-hangover conditions ($p = 0.781$).

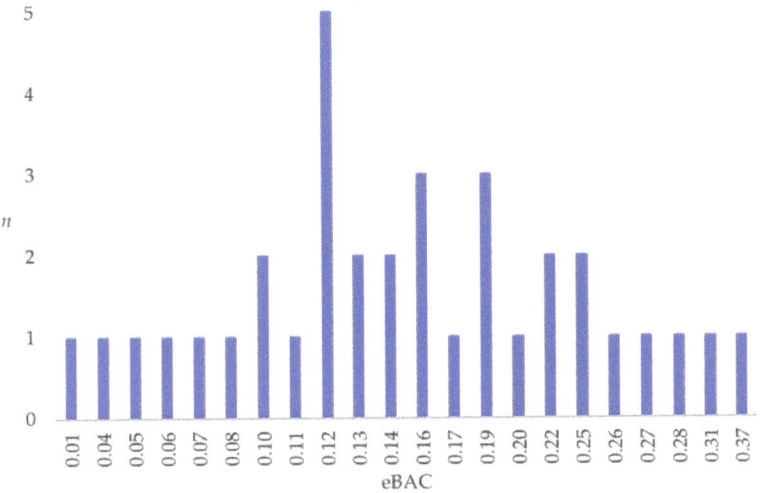

Figure 2. A visual representation of the range of eBAC values during the drinking episode preceding the hangover condition and number of participants experiencing each eBAC value.

3.2. Effects of Hangover on Switching

For reaction times, the analysis of switch costs indicated a trend-level main effect for condition ($F (1, 26) = 3.359$, $p = 0.078$, $d = 0.72$), whereby switch costs were marginally greater in the hangover relative to the no-hangover condition. There was also a condition *order interaction ($F (1, 26) = 9.850$, $p = 0.004$, $d = 1.23$) indicating performance improved (lower switch costs) across testing days when the first testing session was the hangover condition ($F (1, 26) = 13.748$, $p = 0.001$, $d = 1.45$). However, there were no significant differences between testing days for those who completed the task for the second, time when hungover ($p = 0.387$). There were no other significant effects or interactions. Results for main effects on each task are presented graphically in Figure 3, condition * order interactions are presented in Figure 4, and means and SDs are presented in Table 1.

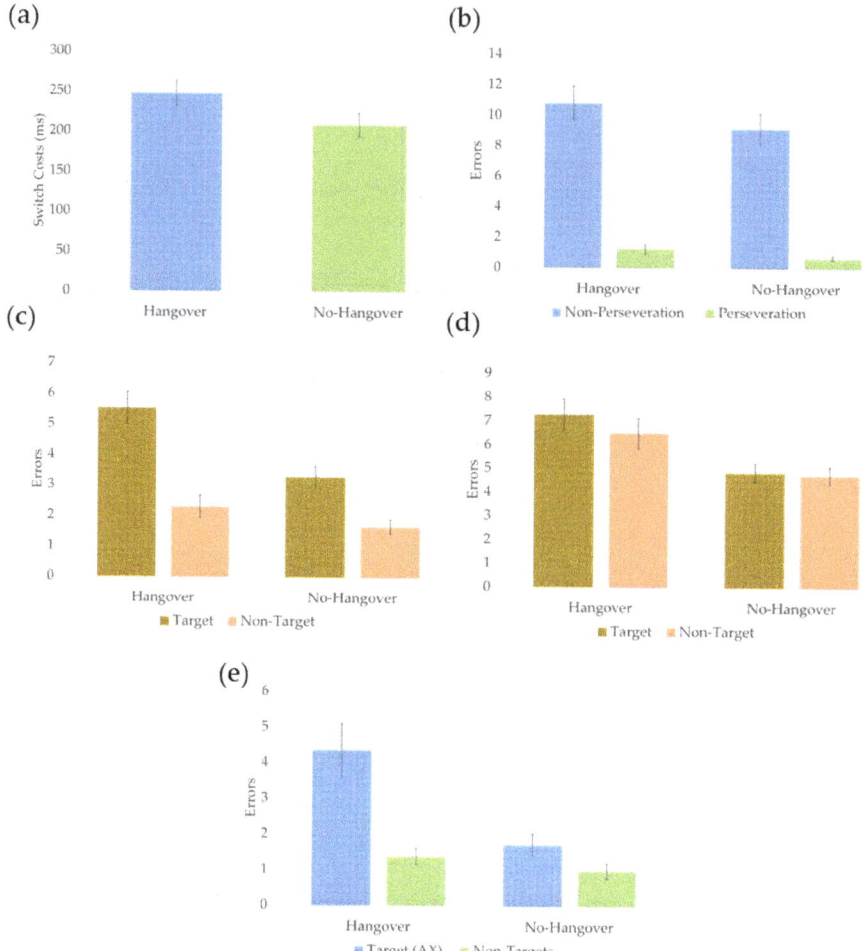

Figure 3. Graphical representations of the main effects from the three cognitive tasks. (**a**) Relative to the no-hangover condition, mean switch costs on the switching task tended to be greater when individuals were experiencing a hangover. (**b**) Relative to the no-hangover condition, mean errors on the switching task were higher when individuals were experiencing a hangover. (**c**) Relative to the no-hangover condition, errors for non-target and target stimuli in the 1-back version of the n-back task were greater in the hangover condition. (**d**) Relative to the no-hangover condition, errors in the 2-back task were greater overall in the hangover condition. (**e**) Relative to the no-hangover condition, errors on AX trials of the AX-CPT task were greater in the hangover condition. The error bars represent ±1 standard error of the mean.

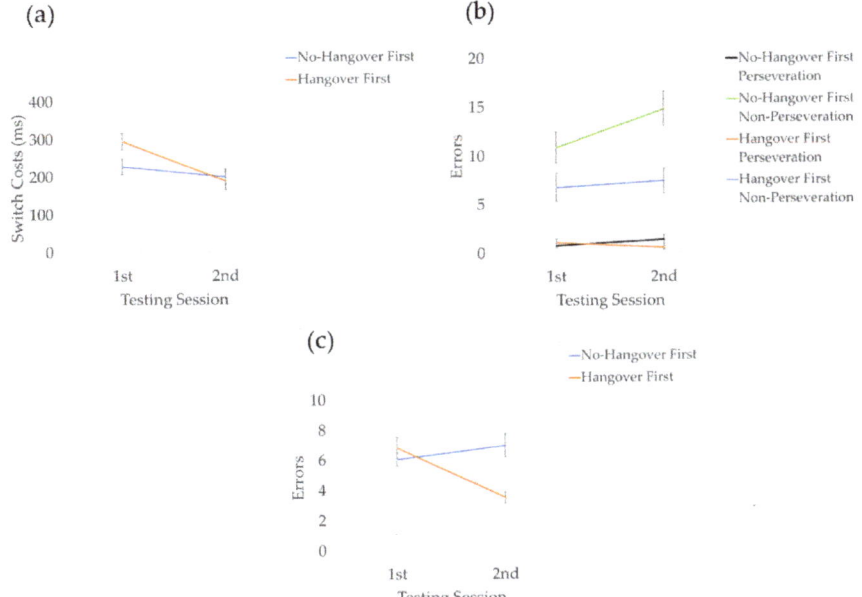

Figure 4. Graphical representations of the condition * order interactions. (**a**) Switching speed decreased (lower switch costs) across testing days when the first testing session was the hangover condition, but not when the first testing session was the no-hangover condition. (**b**) Switching accuracy declined (greater number of non-perseveration errors) across testing days when the first testing session was the no-hangover condition, but not when the first testing session was the hangover condition. (**c**) Updating performance improved (fewer errors) across the testing days for those completing the hangover condition first, but not for those completing the no-hangover condition first. The error bars represent ±1 standard error of the mean.

Table 1. Means, standard deviations, and group comparisons for each variable.

Variable	Hangover		No-Hangover		p	Effect Size
	M	SD	M	SD		
Switching Task						
Switch Cost (ms)	247.83	87.09	208.95	82.27	0.078	$d = 0.72$
Switch Errors	6.01	2.97	4.92	2.68	0.019 *	$d = 1.08$
n-back Working Memory Task						
1-back errors	3.92	1.78	2.47	1.31	<0.001 *	$d = 1.64$
2-back errors	6.89	3.12	4.78	1.66	<0.001 *	$d = 1.63$
AX-CPT Task						
Target Errors (AX-type trials)	4.48	4.33	1.79	163	<0.001 *	$d = 1.52$
Non-Target Errors	1.39	1.24	1.00	1.23	0.081	$d = 0.69$
Hangover Severity						
1-Item Hangover Severity	3.83	1.84	0	0	<0.001 *	$d = 2.08$
mAHSS	2.40	1.31	0.24	0.26	<0.001 *	$d = 1.72$
Alcohol Consumption						
Alcohol Units (night before testing)	13.28	5.13	0	0	<0.001 *	$d = 5.27$
eBAC	0.16%	0.08	0	0	<0.001 *	$d = 4.04$

Table 1. Cont.

Variable	Hangover		No-Hangover		p	Effect Size
	M	SD	M	SD		
Subjective Measures						
RSME Switching	77.27	23.7	58.72	22.78	0.001 *	d = 0.72
RSME n-back	76.41	24.22	58.79	20.81	0.001 *	d = 0.65
RSME AX-CPT	59.69	23.70	47.41	28.67	0.008 *	d = 0.55
Self-efficacy Switching	6.88	2.14	8.76	1.28	< 0.001 *	d = 1.00
Self-efficacy n-back	6.31	2.18	6.74	2.5	0.384	d = 0.14
Self-efficacy AX-CPT	8.53	1.38	9.06	1.41	0.051	d = 0.35

Notes: M, mean; SD, standard deviation; mAHSS, modified Alcohol Hangover Severity Scale; RSME, Rating Scale of Mental Effort; eBAC, estimated Blood Alcohol Concentration; AX-CPT, AX-Continuous Performance Task. The asterisk indicates that the difference between the hangover and no-hangover conditions was significant.

For errors, analysis indicated a main effect of condition ($F_{(1, 22)} = 6.392$, $p = 0.019$, $d = 1.08$), whereby errors were greater overall in the hangover relative to no-hangover condition. There was also a main effect of error type ($F_{(1, 26)} = 77.544$, $p < 0.001$, $d = 3.75$), whereby there was a greater number of non-perseveration than perseveration errors. An order * error-type interaction indicated non-perseveration errors were greater for those who completed the no-hangover condition first than those who completed the hangover condition first ($F_{(1, 22)} = 8.301$, $p = 0.009$, $d = 1.23$). There was also a greater number of non-perseveration errors than perseveration errors in both orders of condition ($ps < 0.001$). A condition * order interaction ($F_{(1, 26)} = 7.483$, $p = 0.012$, $d = 1.17$) indicated that performance significantly declined across testing days when the first testing session was the no-hangover condition ($F_{(1, 22)} = 11.650$, $p = 0.002$, $d = 1.45$), whereas there were no significant differences between testing days for those who completed the task for the second time, when sober ($p = 0.872$). The analysis also indicated that participants who were hungover during their second session made greater errors in the hangover condition than those who were hungover during their first session ($F_{(1, 22)} = 12.958$, $p = 0.002$, $d = 1.54$). A condition * order * error-type interaction indicated that order effects were restricted to non-perseveration errors ($F_{(1, 26)} = 6.428$, $p = 0.019$, $d = 1.08$) (see Figure 4b). There were no other significant effects or interactions.

3.3. Effects of Hangover on Updating

To investigate the effect of a hangover on updating, each version of the n-back task was analysed separately. For the 1-back version, there was a main effect of condition ($F_{(1, 31)} = 20.734$, $p < 0.001$, $d = 1.64$), whereby errors were greater in the hangover than the no-hangover condition. There was also a main effect of trial type ($F_{(1, 31)} = 25.399$, $p < 0.001$, $d = 1.81$), whereby there as a greater number of errors for target than non-target trials. Furthermore, there was a condition*trial type interaction ($F_{(1, 31)} = 7.444$, $p = 0.01$, $d = 0.98$). Pairwise comparisons indicated errors were greater in the hangover condition than the no-hangover condition for both target ($F_{(1, 33)} = 21.700$, $p < 0.001$, $d = 1.62$) and non-target trials ($F_{(1, 33)} = 4.454$, $p = 0.042$, $d = 0.74$). Furthermore, errors for target trials were greater than errors for non-target trials within both the hangover ($F_{(1, 33)} = 24.087$, $p < 0.001$, $d = 1.71$) and no-hangover conditions ($F_{(1, 33)} = 19.080$, $p < 0.001$, $d = 1.52$). There were no other significant effects or interactions.

For the more difficult 2-back version of the task, there was a main effect of condition ($F_{(1, 31)} = 20.708$, $p < 0.01$, $d = 1.63$), whereby errors were greater in the hangover than the no-hangover condition. There was also a condition*order interaction ($F_{(1, 31)} = 6.732$, $p = 0.014$, $d = 0.93$) that indicated performance significantly improved across testing days for those completing the hangover condition first ($F_{(1, 31)} = 28.528$, $p < 0.001$, $d = 1.92$), whereas there were no significant differences between testing days for those who completed the task for a second time, whilst hungover ($p = 0.198$) (see Figure 4c). Our analysis also indicated that participants who were sober during their first session made a greater number of errors in the no-hangover condition than those who were sober during their second session (F (1, 22) = 12.958, $p = 0.002$, $d = 1.54$). There were no other significant effects or interactions.

3.4. Effects of Hangover on Goal Maintenance

Target and non-target trials were analysed separately, to avoid comparing stimuli presented 70% of the time to non-target stimuli, which were presented 10% of the time each [45]. A 2 (condition) * 2 (order) repeated measures ANOVA indicated a main effect of condition only (F (1, 29) = 16.643, $p < 0.001$, $d = 1.52$), whereby AX-type trial errors were greater in the hangover than the no-hangover condition.

Errors for non-target trials (BX-, BY-, and AY-type trials) were analysed separately. Increased errors on BX-type trials in the hangover relative to the no-hangover condition are indicative of a shift toward a reactive control style. There was a trend-level main effect of condition (F (1, 28) = 3.279, $p = 0.081$, $d = 0.69$) whereby non-target errors tended to be greater in the hangover relative to the no-hangover condition. In addition, there was a main effect of trial type (F (1, 31) = 28.829, $p < 0.001$, $d = 2.84$), whereby there were more errors on AY-type relative to BY-type and BX-type trials and more errors on BX-type relative to BY-type trials. There were no other significant effects or interactions.

3.5. Subjective Measures

A series of paired-samples t-tests was used to analyse RSME scores for each task. For the switching task, perceived mental effort was greater (t (28) = 3.899, $p = 0.001$, $d = 0.72$) in the hangover condition than the no-hangover condition. For the n-back task, perceived mental effort was also greater (t (33) = 3.767, $p = 0.001$, $d = 0.65$) in the hangover condition than the no-hangover condition. Furthermore, perceived mental effort for the AX-CPT task was greater (t (31) = 2.818, $p = 0.008$, $d = 0.50$) in the hangover condition than the no-hangover condition. There were lower self-efficacy scores in the hangover relative to the no-hangover condition for the switching task (t (33) = 5.816, $p < 0.001$, $d = 1.00$). This difference was marginally significant for the AX-CPT task ($p = 051$), but not the n-back task ($p = 0.384$). There were no sex differences in hangover severity ($p = 0.790$) or eBAC ($p = 0.195$).

3.6. Correlational Analysis

Bivariate correlational analysis provided no evidence that hangover-severity scores (as measured by the mAHSS; $ps > 0.178$) and self-efficacy scores ($ps > 0.098$) were associated with performance on the switching, n-back, or AX-CPT tasks. Bivariate correlational analysis also provided no evidence that eBAC was related to hangover severity ($p = 0.229$) or task performance ($ps \geq 0.161$).

4. Discussion

This study demonstrated that switching, updating, and goal maintenance are all impaired during an alcohol hangover. Thus, in terms of the unity/diversity model of executive functions [15], all of the core components of executive function appear to be negatively influenced by a hangover. Errors for non-target trial types on the AX-CPT task (i.e., AY-type, BX-type, or BY-type trials) showed a trend toward being greater in the hangover than the no-hangover condition. Moreover, contrary to our hypothesis, there was no evidence that performance on switching, updating, and goal-maintenance tasks was related to hangover severity. There was also no evidence that hangover-related impairments in task performance were related to self-efficacy during switching, updating, and goal-maintenance task performance. However, the participants felt that they needed to expend greater mental effort to complete each task when experiencing a hangover than when not hungover. Furthermore, there was no influence of sex on cognitive performance when hungover for any of the tasks.

In line with a previous naturalistic study of hangovers [30], our results from the switching task indicate that individuals make a greater number of errors, reflective of deficits in task switching, when they are experiencing a hangover, as opposed to when they are not hungover. This suggests that a hangover impairs an individual's ability to switch attention from one task or mental set to another effectively. Although studies that experimentally induce hangovers often administer lower doses of alcohol than observed in real-life drinking [2], our null results for an effect of a hangover on switch costs are in line with previous experimental research [19]. Therefore, it appears as though individuals may maintain speed of switching, but become less accurate, when experiencing a hangover, as compared to not being hungover. For switching, our results also tentatively indicated an interaction of condition with order, further suggesting a speed-accuracy trade-off. Those completing the hangover condition first appear to sacrifice time (switch costs) to maintain accuracy during the hangover condition, whereas those completing the hangover condition second appear to sacrifice accuracy to maintain speed.

Our results indicate poorer performance on both the 1-back and 2-back versions of the n-back task in the hangover compared to no-hangover condition. This suggests that an individual's ability to update information in working memory is impaired during a hangover. As the 1-back version of the task is relatively easy and places a comparatively low load on working memory, the current results suggest that participants with a hangover experienced an increased cognitive load, relative to during a non-hungover state. This is in line with previous research suggesting that a hangover reduces the amount of cognitive resources available [18,19], and it is consistent with our results indicating greater mental effort to complete tasks. Although hangover symptoms, such as headache and fatigue, are known to impair an individual's ability to update information via increased cognitive load [34], our results indicate no evidence of an association between performance on any task and overall hangover-severity scores. This suggests that hangover-related impairments in executive functions are likely due to factors other than simple cognitive interference due to the presence of negative symptoms. For example, it is possible that physical alterations in hangovers, such as dopaminergic or noradrenergic transmission [9,25], or immune effects (indexed via cytokine levels) [52,53], influence cognition [54]. The observed interaction of condition and order tentatively suggests that those completing the hangover condition first appear to have greater improvement in their second session than those completing the no-hangover condition first. This could indicate an expectancy effect, whereby, when the first condition is during a hangover, participants expect their second performance on the task (i.e., when sober) to be greatly improved.

Results from the current study indicate poorer goal maintenance during hangovers, as reflected by a greater number of errors on the core AX trials of the AX-CPT task in the hangover compared to the no-hangover condition. This suggests that an individual's ability to maintain and manage goals is impaired whilst experiencing a hangover. Goal maintenance is thought to represent the 'common factor' of the unity/diversity model, and an important aspect of maintaining goals is inhibitory control [15]. Therefore, impaired goal maintenance during a hangover may contribute toward findings of previous studies of executive functions that have reported impaired prospective memory, semantic verbal fluency [10], working memory, [12], and inhibitory control [5,6,30] during a hangover relative to a no-hangover condition. Contrary to our hypothesis, there was no evidence that participants were biased toward reactive control during a hangover, suggesting participants engaged in proactive control during this task, but were ineffective in doing so (as evidenced by increased errors on the core AX-type trials). However, it is possible that the current study did not have sufficient power to observe effects on reactive control, due to the low number of non-target trials on this task. As goal maintenance is important for many everyday behaviours that rely on executive functions, such as planning, decision making, organising, and other 'higher-order' skills, future studies should investigate the influence of hangovers on these processes.

The current results should be viewed in light of the following strengths and limitations. The crossover, within-subjects design could be considered a strength of the current study, because each subject serves as his or her own control. Furthermore, the naturalistic design, although the naturalistic design is limited in its control over alcohol consumption, it can be considered a strength as it involves investigating the impact of real-life drinking, rather than an experimentally induced hangover, which might involve consuming lower levels of alcohol [41]. However, the study is limited in its ability to generalise beyond the narrow demographics of this student population (i.e., to other age groups, education levels, etc.). Another limitation is the use of the Widmark formula, which should be viewed as a rough estimate of alcohol consumption. Future studies should explore directly measuring BAC during the heavy drinking occasion, possibly via wearable technology. Although each task used in this study was chosen to reflect switching, updating, or goal maintenance, these tasks are cognitively complex (i.e., they measure multiple executive and non-executive functions). One technique that could be utilised in future studies, to overcome variability attributable to task stimuli, rather than the respective executive function component, is the adoption of a latent variable approach, which is a statistical technique that can capture common variance across multiple measures (e.g., [55]).

5. Conclusions

Results from the current study indicate that all domains of the unity/diversity model of executive functions are negatively affected by alcohol hangover. Executive functions are important cognitive processes which are utilised in everyday behaviours, such as planning, decision-making, and emotion regulation. Thus, impairments in a range of executive functions could have broad implications for a wide variety of everyday activities, including in the workplace. For example, employees who go to work when experiencing a hangover may negatively influence the productivity and working environment of others [13]. Future studies should aim to investigate the impact of hangover-induced executive dysfunction on the performance of everyday tasks.

Author Contributions: C.G., G.F. and S.A. made substantial contributions to conception and design; C.G. and S.A. analysed the data; C.G. and S.A. drafted the manuscript; and all authors revised it critically, for important intellectual content, and approved the final version of the manuscript. All authors have read and agreed to the published version of the manuscript.

Funding: This research was funded by a University Research Studentship Award from the University of Bath.

Acknowledgments: We would like to thank all of those who took time to participate in this study; without them, this research would not have been possible.

Conflicts of Interest: Over the past three years, Joris Verster has received grants/research support from the Dutch Ministry of Infrastructure and the Environment, Janssen, Nutricia, and Sequential and has acted as a consultant/advisor for Clinilabs, More Labs, Red Bull, Sen-Jam Pharmaceutical, Toast! and ZBiotics. The other authors have no conflicts of interest to declare.

References

1. Van Schrojenstein Lantman, M.; van de Loo, A.; Mackus, M.; Verster, J. Development of a definition for the alcohol hangover: Consumer descriptions and expert consensus. *Curr. Drug Abus. Rev.* **2017**, *9*, 148–154. [CrossRef] [PubMed]
2. Gunn, C.; Mackus, M.; Griffin, C.; Munafò, M.R.; Adams, S. A systematic review of the next-day effects of heavy alcohol consumption on cognitive performance. *Addiction* **2018**, *113*, 2182–2193. [CrossRef] [PubMed]
3. McGee, R.; Kypri, K. Alcohol-related problems experienced by university students in New Zealand. *Aust. N. Z. J. Public Health* **2004**, *28*, 321–323. [CrossRef] [PubMed]
4. Devenney, L.E.; Coyle, K.B.; Verster, J.C. Cognitive performance and mood after a normal night of drinking: A naturalistic alcohol hangover study in a non-student sample. *Addict. Behav. Rep.* **2019**, *10*, 100197. [CrossRef]
5. McKinney, A.; Coyle, K.; Penning, R.; Verster, J.C. Next day effects of naturalistic alcohol consumption on tasks of attention. *Hum. Psychopharmacol.* **2012**, *27*, 587–594. [CrossRef]
6. Gunn, C.; Verster, J.C.; Adams, S. The effects of alcohol hangover on response inhibition and attentional bias towards alcohol-related stimuli. *Alcohol. Clin. Exp. Res.* **2019**, *43*, 185. [CrossRef]
7. Noël, X.; Bechara, A.; Dan, B.; Hanak, C.; Verbanck, P. Response inhibition deficit is involved in poor decision making under risk in nonamnesic individuals with alcoholism. *Neuropsychology* **2007**, *21*, 778–786. [CrossRef]
8. Schmeichel, B.J.; Volokhov, R.N.; Demaree, H.A. Working memory capacity and the self-regulation of emotional expression and experience. *J. Pers. Soc. Psychol.* **2008**, *95*, 1526–1540. [CrossRef]
9. Howse, A.D.; Hassall, C.D.; Williams, C.C.; Hajcak, G.; Krigolson, O.E. Alcohol hangover impacts learning and reward processing within the medial-frontal cortex. *Psychophysiology* **2018**, *2017*, e13081. [CrossRef]
10. Heffernan, T.; Samuels, A.; Hamilton, C.; McGrath-Brookes, M. Alcohol hangover has detrimental impact upon both executive function and prospective memory. *Front. Psychiatry* **2019**, *10*, 1–6. [CrossRef]

11. Heffernan, T. A state of alcohol hangover impedes everyday prospective memory. *Front. Hum. Neurosci.* **2018**, *12*, 348. [CrossRef] [PubMed]
12. Howland, J.; Rohsenow, D.J.; Greece, J.A.; Littlefield, C.A.; Almeida, A.; Heeren, T.; Winter, M.; Bliss, C.A.; Hunt, S.; Hermos, J. The effects of binge drinking on college students' next-day academic test-taking performance and mood state. *Addiction* **2010**, *105*, 655–665. [CrossRef] [PubMed]
13. Bhattacharya, A. *Financial Headache: The Cost of Workplace Hangovers and Intoxication to the UK Economy*; Institute for Alcohol Studies: London, UK, 2019.
14. Husain, M. Executive function and behaviour. In Proceedings of the 3rd Congress of the European Academy of Neurology, Amsterdam, The Netherlands, 24–27 June 2017.
15. Friedman, N.P.; Miyake, A. Unity and diversity of executive functions: Individual differences as a window on cognitive structure. *Cortex* **2017**, *86*, 186–204. [CrossRef] [PubMed]
16. Lépine, R.; Bernardin, S.; Barrouillet, P. Attention switching and working memory spans. *Eur. J. Cogn. Psychol.* **2005**, *17*, 329–345. [CrossRef]
17. Zink, N.; Bensmann, W.; Beste, C.; Stock, A.-K. Alcohol hangover increases conflict load via faster processing of subliminal information. *Front. Hum. Neurosci.* **2018**, *12*, 316. [CrossRef]
18. Scholey, A.; Ayre, B.; Terpstra, C.; Benson, S. Alcohol hangover results in reduced attentional resources. *Alcohol. Clin. Exp. Res.* **2019**, *43*, 188A. [CrossRef]
19. Wolff, N.; Gussek, P.; Stock, A.-K.; Beste, C. Effects of high-dose ethanol intoxication and hangover on cognitive flexibility. *Addict. Biol.* **2016**, *23*, 503–514. [CrossRef]
20. Eysenck, M.W.; Derakshan, N.; Santos, R.; Calvo, M.G. Anxiety and cognitive performance: Attentional control theory. *Emotion* **2007**, *7*, 336–353. [CrossRef]
21. Lavie, N.; Dalton, P. *Load Theory of Attention and Cognitive Control*; Oxford University Press: Oxford, UK, 2014; pp. 56–75.
22. Lavie, N.; Hirst, A.; de Fockert, J.W.; Viding, E. Load theory of selective attention and cognitive control. *J. Exp. Psychol. Gen.* **2004**, *133*, 339–354. [CrossRef]
23. Stock, A.-K.; Beste, C. Binge drinking and the differential influence of ethanol on cognitive control subprocesses: A novel field of neurotoxicology. *Arch. Toxicol.* **2014**, *88*, 9–10. [CrossRef]
24. Goldstein, R.Z.; Volkow, N.D. Dysfunction of the prefrontal cortex in addiction: Neuroimaging findings and clinical implications. *Nat. Rev. Neurosci.* **2011**, *12*, 652. [CrossRef] [PubMed]
25. Maki, T.; Toivonen, L.; Koskinen, P.; Naveri, H.; Harkonen, M.; Leinonen, H. Effect of ethanol drinking, hangover, and exercise on adrenergic activity and heart rate variability in patients with a history of alcohol-induced atrial fibrillation. *Am. J. Cardiol.* **1998**, *82*, 317–322. [CrossRef]
26. Penning, R.; McKinney, A.; Verster, J.C. Alcohol hangover symptoms and their contribution to the overall hangover severity. *Alcohol Alcohol.* **2012**, *47*, 248–252. [CrossRef]
27. Van Der Linden, D.; Frese, M.; Meijman, T.F. Mental fatigue and the control of cognitive processes: Effects on perseveration and planning. *Acta Psychol. (Amst)* **2003**, *113*, 45–65. [CrossRef]
28. Verster, J.C.; de Klerk, S.; Bervoets, A.C.; Kruisselbrink, L.D. Editorial: Can hangover immunity be really claimed? *Curr. Drug Abus. Rev.* **2014**, *6*, 253–254. [CrossRef]
29. Scholey, A.; Benson, S.; Kaufman, J.; Terpstra, C.; Ayre, E.; Verster, J.C.; Allen, C.; Devilly, G. Effects of alcohol hangover on cognitive performance: Findings from a field/internet mixed methodology study. *J. Clin. Med.* **2019**, *8*, 440. [CrossRef] [PubMed]
30. Devenney, L.E.; Coyle, K.B.; Verster, J.C. Memory and attention during an alcohol hangover. *Hum. Psychopharmacol.* **2019**, *34*, 1–7. [CrossRef] [PubMed]
31. Speer, N.K.; Jacoby, L.L.; Braver, T.S. Strategy-dependent changes in memory: Effects on behavior and brain activity. *Cogn. Affect. Behav. Neurosci.* **2003**, *3*, 155–167. [CrossRef] [PubMed]
32. Gonthier, C.; Macnamara, B.N.; Chow, M.; Conway, A.R.A.; Braver, T.S. Inducing proactive control shifts in the AX-CPT. *Front. Psychol.* **2016**, *7*, 1–14. [CrossRef] [PubMed]
33. Botto, M.; Basso, D.; Ferrari, M.; Palladino, P. When working memory updating requires updating: Analysis of serial position in a running memory task. *Acta Psychol. (Amst)* **2014**, *148*, 123–129. [CrossRef]
34. Moore, D.J.; Keogh, E.; Eccleston, C. Headache impairs attentional performance. *Pain* **2013**, *154*, 1840–1845. [CrossRef] [PubMed]
35. Moore, D.J.; Eccleston, C.; Keogh, E. Cognitive load selectively influences the interruptive effect of pain on attention. *Pain* **2017**, *158*. [CrossRef] [PubMed]

36. Martínez-Cancino, D.P.; Azpiroz-Leehan, J.; Jiménez-Angeles, L. *The Effects of Sleep Deprivation in Working Memory Using the N-Back Task*; IFMBE: Rome, Italy, 2015; Volume 49, pp. 421–424. [CrossRef]
37. Rohsenow, D.J.; Howland, J.; Arnedt, J.T.; Almeida, A.B.; Greece, J.; Minsky, S.; Kempler, C.S.; Sales, S. Intoxication with Bourbon versus Vodka: Effects on hangover sleep and next-day neurocognitive performance in young adults. *Alcohol. Clin. Exp. Res.* **2010**, *34*, 509–518. [CrossRef] [PubMed]
38. Verster, J.C. Sleep after an evening of heavy drinking and its impact on daytime sleepiness and alcohol hangover severity. *Sleep Biol. Rhythm.* **2007**, *5* (Suppl. 1), A18.
39. Chow, J.T.; Hui, C.M.; Lau, S. A depleted mind feels inefficacious: Ego-depletion reduces self-efficacy to exert further self-control. *Eur. J. Soc. Psychol.* **2015**, *45*, 754–768. [CrossRef]
40. Finnigan, F.; Schulze, D.; Smallwood, J.; Helander, A. The effects of self-administered alcohol-induced "hangover" in a naturalistic setting on psychomotor and cognitive performance and subjective state. *Addiction* **2005**, *100*, 1680–1689. [CrossRef]
41. Verster, J.C.; Van De Loo, A.J.; Adams, S.; Stock, A.; Benson, S.; Scholey, A.; Alford, C.; Bruce, G. Advantages and limitations of naturalistic study designs and their implementation in alcohol hangover research. *J. Clin. Med.* **2019**, *8*, 2160. [CrossRef]
42. Monsell, S.; Sumner, P.; Waters, H. Task-set reconfiguration with predictable and unpredictable task switches. *Mem. Cogn.* **2003**, *31*, 327–342. [CrossRef] [PubMed]
43. Attridge, N.; Eccleston, C.; Noonan, D.; Wainwright, E.; Keogh, E. Headache impairs attentional performance: A conceptual replication and extension. *J. Pain* **2017**, *18*, 29–41. [CrossRef]
44. Braver, T.S.; Rush, B.K.; Satpute, A.B.; Racine, C.A.; Barch, D.M. Context processing and context maintenance in healthy aging and early stage dementia of the Alzheimer's type. *Psychol. Aging* **2005**, *20*, 33–46. [CrossRef]
45. Paxton, J.L.; Barch, D.M.; Racine, C.A.; Braver, T.S. Cognitive control, goal maintenance, and prefrontal function in healthy aging. *Cereb. Cortex* **2008**, *18*, 1010–1028. [CrossRef] [PubMed]
46. Kypri, K.; Langley, J.; Stephenson, S. Episode-centred analysis of drinking to intoxication in university students. *Alcohol Alcohol.* **2005**, *40*, 447–452. [CrossRef] [PubMed]
47. Hogewoning, A.; Van de Loo, A.; Mackus, M.; Raasveld, S.J.; De Zeeuw, R.; Bosma, E.; Bouwmeester, N.; Brrokhuis, K.A.; Garssen, J.; Verster, J.C. Characteristics of social drinkers with and without a hangover after heavy alcohol consumption. *Subst. Abus. Rehabil.* **2016**, *7*, 161–167. [CrossRef]
48. Zijlstra, F.R.; Van Doorn, L. *The Construction of a Scale to Measure Perceived Effort*; Department of Philosophy and Social Sciences, Delft University of Technology: Delft, The Netherlands, 1985.
49. Gunn, C.; Fairchild, G.; Verster, J.; Adams, S. The Effects of Alcohol Hangover on Executive Functions osf.io/n3ydu. Available online: https://osf.io/n3ydu (accessed on 28 January 2019).
50. Longman, C.S.; Lavric, A.; Monsell, S. The coupling between spatial attention and other components of task-set: A task-switching investigation. *Q. J. Exp. Psychol.* **2016**, *69*, 2248–2275. [CrossRef] [PubMed]
51. Holloway, F.A. *Low-Dose Alcohol Effects on Human Behavior and Performance: A Review of Post-1984 Research*; Oklahoma Univ Health Sciences Center Oklahoma City: Oklahoma City, OK, USA, 1994.
52. Van de Loo, A.; Hogewoning, A.; Raasveld, S.J.; De Zeeuw, R.; Bosma, E.R.; Bouwmeester, N.H.; Lukkes, M.; Brookhuis, K.A.; Knipping, K.; Garssen, J.; et al. Saliva cytokine concentrations the day after heavy alcohol consumption in drinkers suffering from a hangover versus those who claim to be hangover resistant. *Alcohol Alcohol.* **2015**, *50* (Suppl. 1). [CrossRef]
53. Kim, D.-J.; Kim, W.; Yoon, S.-J.; Choi, B.-M.; Kim, J.-S.; Go, H.J.; Kim, Y.-K.; Jeong, J. Effects of alcohol hangover on cytokine production in healthy subjects. *Alcohol* **2003**, *31*, 167–170. [CrossRef]
54. Tipple, C.T.; Benson, S.; Scholey, A. A review of the physiological factors associated with alcohol hangover. *Curr. Drug Abus. Rev.* **2017**, *9*, 93–98. [CrossRef]
55. Korucuoglu, O.; Sher, K.J.; Wood, P.K.; Saults, J.S.; Altamirano, L.; Miyake, A.; Bartholow, B.D. Acute alcohol effects on set-shifting and its moderation by baseline individual differences: A latent variable analysis. *Addiction* **2017**, *112*, 442–453. [CrossRef]

© 2020 by the authors. Licensee MDPI, Basel, Switzerland. This article is an open access article distributed under the terms and conditions of the Creative Commons Attribution (CC BY) license (http://creativecommons.org/licenses/by/4.0/).

Article

Alcohol Hangover Differentially Modulates the Processing of Relevant and Irrelevant Information

Antje Opitz *, Christian Beste and Ann-Kathrin Stock

Cognitive Neurophysiology, Department of Child and Adolescent Psychiatry, Faculty of Medicine, TU Dresden, Fetscherstr. 74, 01307 Dresden, Germany; christian.beste@ukdd.de (C.B.); ann-kathrin.stock@ukdd.de (A.-K.S.)
* Correspondence: antje.opitz@ukdd.de

Received: 6 February 2020; Accepted: 9 March 2020; Published: 12 March 2020

Abstract: Elevated distractibility is one of the major contributors to alcohol hangover-induced behavioral deficits. Yet, the basic mechanisms driving increased distractibility during hangovers are still not very well understood. Aside from impairments in attention and psychomotor functions, changes in stimulus-response bindings may also increase responding to distracting information, as suggested by the theory of event coding (TEC). Yet, this has never been investigated in the context of alcohol hangover. Therefore, we investigated whether alcohol hangover has different effects on target-response bindings and distractor-response bindings using a task that allows to differentiate these two phenomena. A total of $n = 35$ healthy males aged 19 to 28 were tested once sober and once hungover after being intoxicated in a standardized experimental drinking setting the night before (2.64 gr of alcohol per estimated liter of body water). We found that alcohol hangover reduced distractor-response bindings, while no such impairment was found for target-response bindings, which appeared to be unaffected. Our findings imply that the processing of distracting information is most likely not increased, but in fact decreased by hangover. This suggests that increased distractibility during alcohol hangover is most likely not caused by modulations in distractor-response bindings.

Keywords: alcohol; hangover; distraction; stimulus-response binding; distractor-response binding; theory of event coding

1. Introduction

Driving under the influence of alcohol is prohibited and legally sanctioned in most countries, as alcohol drastically impairs the required cognitive and motor skills, thus putting the driving individual as well as co-drivers and bystanders at substantial risk of accidents and injury [1–5]. Even after the acute intoxication has worn off and blood alcohol levels (BAC) have returned to 0.00‰, cognitive, attentional, and psychomotor functions likely remain impaired to a certain degree [6–11]. While there is currently no neurobiochemical marker that could help to reliably detect the presence of alcohol hangover [12], being hungover may not only be disastrous for driving a vehicle, but also for various daily activities including safety-sensitive tasks at work and at home. Surprisingly, the mechanisms underlying such hangover-associated behavioral deficits are still not very well understood, as compared to the effects of acute intoxication.

Alcohol hangover is commonly defined as a "combination of mental and physical symptoms, experienced the day after a single episode of heavy drinking, starting when BAC approaches zero" [13]. The findings on cognitive and behavioral effects of alcohol-induced hangover in different functional domains are not always consistent (for review, please see Mackus et al. [11]), but there is consensus that attentional deficits are one of the factors that likely underlie hangover-associated driving impairments [7,11,14]. Another important factor for impaired driving skills are slowing and impairments of motor responses and coordination during hangover [6,7,11,14,15]. Even though distraction is recognized as a major risk factor for impaired driving [16], we still know rather little about whether and how alcohol hangover changes the way we process and respond to distracting information, other than that hangover may influence response selection and conflict monitoring, thus impairing goal-directed behavior [17–19]. Specifically, only very little is known about whether alcohol hangover differently modulates the integration and processing of irrelevant vs. relevant information, even though this may give rise to conflicts and distraction at later information processing stages. Against this background, obtaining a better understanding of hangover-induced changes in distractor processing is important. From a mechanistic point of view, driving performance, as well as some forms of occupational performance (for review, please see Mackus et al. [11]), may not only be compromised by hangover-induced impairments in attention and psychomotor functions, but also by the phenomenon of stimulus-response (S-R) binding. In this context, the theory of event coding (TEC) [20] has provided compelling evidence for why and how we respond to distracting information (Figure 1). While it has not been developed to explain alcohol effects per se, it is a widely known theory which may be used to elucidate mechanisms potentially underlying hangover-induced changes. In short, it states that sensory information on all kinds of different stimuli will be "bound" (i.e., associated) in a so-called object file, while all response-related information is bound/associated in the so-called "response file". Additionally, stimulus and response features become associated with each other in the "event file", thus constituting the S-R link [21]. Importantly, the event file binds stimulus and response features irrespective of whether a given stimulus feature/sensory information is functionally relevant for the response we intend or need to carry out [21]. Instead, temporal co-occurrence is the main determinant of whether sensory input becomes associated with a given response [21]. As a result, any distractor that will occur at the same time as a response-relevant (target) stimulus feature, will also find entry into the event file. Following a pattern-completion logic, the entire event file, including the previously associated response, will become activated once any part of a pre-existing event file is re-encountered [20]. Importantly, this also holds true for distractor information so that task-irrelevant information may lead to a similar reactivation of an event file as task-relevant information [20]. This way, a distractor, especially if previously encountered, may prime certain responses - even when they are no longer needed or correct.

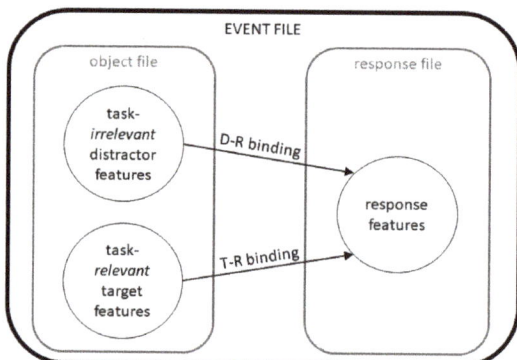

Figure 1. Schematic illustrations of the bindings postulated by the theory of event coding (TEC). Abbreviations: D-R = distractor-response; T-R = target-response.

While previous studies have suggested that acute alcohol intoxication does not seem to strongly impair S-R binding [22], this has never been investigated in the hungover state, which may functionally differ from the intoxicated state [17,18]. Of note, there are several mechanisms which could drive such differences, including differential modulation of *gamma-aminobutyric acid* (GABAergic) signaling by ethanol and its metabolite acetaldehyde, which is likely responsible for feeling hungover [23–25], and immunological parameters, including those related to inflammation [26,27]. To the best of our knowledge, it has furthermore never been investigated whether potential alcohol effects on this phenomenon could be different for the S-R binding of response-relevant target information vs. that of response-irrelevant distractor information. Stock et al. [22] investigated acute intoxication effects using a paradigm, which does not separately account for target- and distractor-response binding. We therefore chose a paradigm that allows to make this dissociation in order to investigate our research question in detail. Since the sensory processing of stimulus information might be altered during hangover due to increased sensory information processing speed [17], it could be conceivable that S-R binding, including distractor-response binding, is in fact enhanced during alcohol hangover. Such an enhancement could constitute a risk factor for distractibility and resulting safety issues for activities like driving (via the automatic reactivation of previously associated and thus potentially incorrect responses [28]).

To investigate this research question, we subjected $n = 35$ healthy young males to an experimental within-subject study design that was comparable to the procedure used in previously published hangover studies from our group [18,19]. Each participant was once tested sober and once after having consumed 2.64 gr of alcohol per estimated liter of body water on the night before testing. A paradigm introduced by Moeller et al. [29] was used to separately investigate the effects of target-response and distractor-response binding. In short, the paradigm combines target and distractor stimuli in a shared visual array that requires different manual responses (button presses) from the participants. By systematically and independently varying the repetition vs. change of target information, distractor information, and required response, the task allows to dissociate the different binding effects: The effect of distractor-response (D-R) binding should become evident in the statistical main effects of the distractor (i.e., be reflected by the effect size of distractor repetition vs. distractor change). Likewise, the effects of target-response (T-R) binding should become evident in statistical main effects of the target (i.e., be reflected by the effect size of whether the target and/or required response changed). Lastly, distractor-response binding may also show in the statistical interaction of distractor and response effects, as the automatic response tendency induced by the distractor should vary across response repetition vs. response change [30]: Distractor repetition should be beneficial whenever the repetition of a given response is required because the repetition of the distractor then reactivates the correct response. Opposing this, distractor repetition should be detrimental whenever a given response needs to be changed because the repetition of the distractor then reactivates an incorrect response.

In summary, we investigated whether S-R binding is altered during alcohol hangover against the background of potentially different sensory processing of distracting stimulus information [17,18]. We further investigated whether there are potentially different hangover effects on the binding of task-/response-relevant vs. irrelevant information, as this might explain why and how distractibility may be increased during alcohol hangover.

2. Experimental Section

2.1. Participants

The original study sample comprised $n = 37$ healthy males aged between 19 and 28 years, who were recruited from the local university (TU Dresden) and local nightlife district. Inclusion criteria were right-handedness, normal or corrected-to-normal vision, absence of color blindness, no acute or chronic somatic psychological, or neurological illnesses, and no intake of medication affecting kidney or liver function, or the central nervous system. Participants' drinking habits were assessed during a telephone screening using the alcohol use disorders identification test (AUDIT) [31,32]. An AUDIT score below 16 specifies low-to-moderate risk of alcohol use disorder (AUD) [31]. Almost all participants scored below this cut off and therefore reported no drinking habits at high-risk. In total, $n = 4$ participants scored between 16 and 19 indicating harmful use of alcohol [31] but did not meet the criteria for the diagnosis of an AUD according to the International Classification of Diseases (ICD-10). None of the participants scored above 19, which would have indicated a high risk/presence of an AUD [31]. Although harmful drinking may potentially have detrimental effects on cognitive functions [33], previous findings from our lab suggested that young frequent binge drinkers (who engaged in equal, and/or more binge-drinking than our current sample) seem to have no significant behavioral impairments in executive functioning [34].

Exclusion criteria used to minimize the number of individuals who might not cope well with the amount of experimentally administered alcohol were an overall AUDIT score below 2 points, less than 13 self-reported binge-drinking nights over the past 12 months, and not being markedly drunk at least once in the past 12 months. Exclusion criteria used to minimize the number of individuals with high alcohol tolerance and/or high risk of AUD were an overall AUDIT score above 19 points [31], and self-reported (nearly) daily occurrences of either binge drinking behavior, and/or alcohol-induced memory problems, and/or failure to fulfill daily routine tasks due to alcohol consumption. All participants provided written informed consent and received a reimbursement of 80€. The study was conducted according to the Declaration of Helsinki. Ethical approval was given by the ethics committee of the Faculty of Medicine of the TU Dresden, Germany (EK293082014). Eventually, $n = 35$ participants (23.1 ± 2.7 years old) were included in the statistical analyses because one participant was excluded due to performance accuracy close to chance level (i.e., below 55%) in at least one experimental task condition and another participant was excluded because of his high residual breath alcohol concentration (BrAC = 0.45‰) at the hangover session, which would have taken a waiting time of 4 to 5 additional hours until sobriety and he was not able to invest the required time on that day. Please also note that the sample in this publication largely overlapped with that of a previous publication investigating the influence of alcohol hangover on meta-control/the interplay between controlled and automated processes [19].

2.2. Experimental Design

The experimental design was identical to that described in previous publications by our group [18,19], and is illustrated in Figure 2.

Each participant was tested twice (once sober and once hungover) with a delay of a minimum of 48 h to a maximum of 7 days between the sober and hangover sessions. The order of both sessions was counterbalanced across the sample. Participants were asked to neither use nicotine, nor caffeine or guarana within four hours before the start of each session. At the start of each session, BrAC was measured using the breathalyzer "Alcotest 3000" following the instructions by the manufacturer (Drägerwerk, Lübeck, Germany). The experimental procedure in the sober and hangover sessions only proceeded in case of a BrAC of 0.00‰. On the night before the hangover session, participants were experimentally intoxicated at our laboratory in order to induce hangover symptoms. For this, we asked groups of 4 to 8 participants to join a drinking session on a Friday or Saturday evening (starting at 20:00 and ending at approximately 01:30/02:00). The subsequent hangover session was planned for the following morning (that is either on Saturday or on Sunday, starting between 09:00 and 11:00) to slightly reduce sleep duration, as the amount of sleep time is negatively related to hangover severity [35].

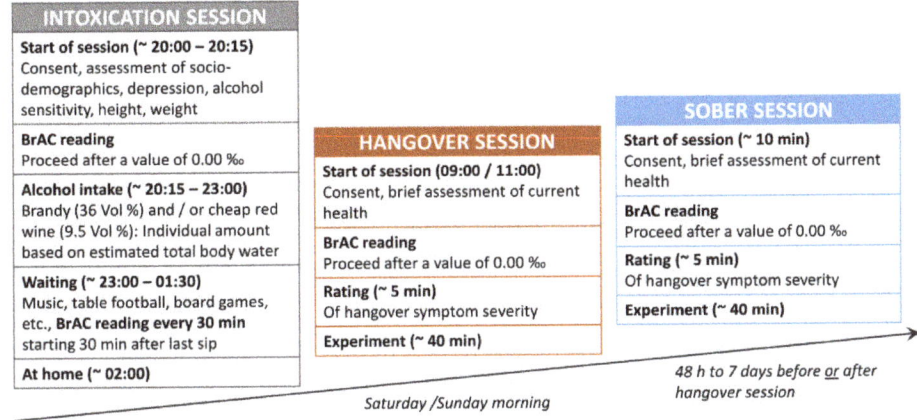

Figure 2. Outline of the intoxication, hangover, and sober sessions. Each participant was tested twice, once sober and once hungover, with a delay of a min. 48 h to max. 7 days between both sessions. The order of both sessions was counterbalanced across all participants. Written informed consent and a brief evaluation of the current health status was provided at the start of each session. Rating of hangover symptoms and recording of task performance was only started in case of a breath alcohol concentration (BrAC) of zero on both sessions. To induce alcohol hangover symptoms, participants were experimentally intoxicated at our laboratory on a Friday or Saturday evening starting from 20:00 (8 pm). Participants provided their written consent and completed questionnaires on sociodemographic data, height, weight, alcohol sensitivity, and depression. BrAC was measured before they started drinking to ensure sobriety. An individually-measured alcohol amount was consumed from approximately 20:15 to 23:00. BrAC was measured 30, 60, 90, and 120 min after the last alcoholic drink was finished. At around 01:30, participants were brought home via cab, given that their BrAC values were already declining, they had no impairments of consciousness, and no major walking/coordination impairments. Participants came back to the laboratory in the next morning between 09:00 and 11:00 to complete their hangover session.

2.3. Experimental Intoxication Procedure

The experimental intoxication procedure applied to cause hangover symptoms followed the same protocol as described in previous publications [18,19] and is briefly outlined in Figure 2. By using a version of the equation by Widmark [36] and Watson et al. [37], an individual amount of alcohol was calculated for each participant at the beginning of the intoxication session. It estimates the total body water (TBW) in liters based on an individual's sex, age, weight, and height and specifies the alcohol amount in grams which is required to be dissolved in the body water to attain a given intoxication level (2.64 gr of alcohol per estimated liter of body water). This limited the provided alcohol amount so that it was physiologically impossible to reach a BrAC of more than 2.0‰ (which would only have been possible to reach if all of the alcohol had been dissolved in TBW at once). Realistically, the amount of provided alcohol translates to a mean BrAC of 1.6‰ in case it is consumed at once and on an empty stomach (i.e., with an expected resorption deficit of about 20%). However, participants were encouraged to eat a full dinner before the intoxication session, as a full stomach typically increases the resorption deficit (approx. 30%–40%). Additionally, the experimenters ensured a minimum consumption duration of 2 h in order to prevent the participants from drinking the entire amount at once and potentially experiencing "overshooting" (i.e., a very rapid increase in blood alcohol levels that might initially lead to otherwise unexpectedly high blood alcohol concentrations and associated complications like vomiting). As a consequence of these precautions, the participants were expected to achieve a mean BrAC of approximately 1.2‰ with only a small chance of reaching BrAC values beyond

1.6‰. The equation used to calculate the individual amount of alcohol and the protocol used to record the alcohol consumption of each participant is available online at https://osf.io/ktgyr/. As alcoholic beverages with high congener content are more likely to induce a (more severe) hangover [7], we only offered cheap red wine (9.5 vol %) and/or cheap brandy (36 vol %). To keep the quantity and speed of alcohol consumption similar across beverages, the experimenters served standardized portions of 50 mL brandy (14 g alcohol) or 200 mL red wine (15 g alcohol). For each standardized portion of alcoholic beverage, participants could choose whether they wanted to drink it pure, chilled on ice, or mixed with orange lemonade, caffeine-free coke, or ginger ale. Participants were supplied with tap water and snacks (wine gums and chips), the use of which was not recorded. Participants were additionally allowed to smoke cigarettes while drinking, as this has been suggested to enhance hangover severity [38,39]. In total, $n = 9$ participants appreciated this offer, $n = 8$ of whom reported to smoke regularly. BrAC was assessed 30, 60, 90, and 120 min after the last alcoholic beverage was consumed.

2.4. Questionnaires

At both the sober and hangover session, participants rated the severity of their hangover symptoms. For this purpose, we used the 11-point Likert-scale by van Schrojenstein Lantman and colleagues [13]. This rating scale enlists 23 hangover symptoms, and rating options range from 0 points (no symptoms) to 10 points (extreme symptoms). We additionally asked participants to rate their sleep quality/sleep problems of the previous night in the same way. Furthermore, participants indicated the number of sleeping hours they had received the night prior to each testing session. Of note, they were explicitly asked to rate each symptom irrespective of whether or not they had been drinking the night before and of whether or not they attributed their symptoms to alcohol.

At the intoxication session, participants provided sociodemographic information and filled in Beck's depression inventory (BDI) [40] to identify depressive symptoms, as depression could have potentially affected cognitive performance. Finally, the alcohol sensitivity questionnaire (ASQ) was used to assess susceptibility to known alcohol effects, like experiencing a hangover or feeling relaxed [41]. The ASQ further distinguishes between alcohol-related experiences associated with lower doses (ASQ_{light}) and with heavier doses (ASQ_{heavy}). While we ran detailed analyses on the role of alcohol sensitivity in a previous publication [19], we refrained from detailed alcohol sensitivity analyses in the current study, as this did not seem to contribute to hangover-associated behavioral changes in the previous study [19].

2.5. Task

In order to be able to dissociate the effects of alcohol hangover on the processing and binding of response-relevant (target) information and response-irrelevant (distractor) information, we used a paradigm introduced by Moeller et al. [29]. Figure 3 illustrates the paradigm.

The stimuli were presented on a 17″ high quality flat screen monitor and keyboard responses were recorded using Presentation® software (Version 16.5, Neurobehavioral Systems, Inc., Berkeley, CA, USA). The time course of a single trial was as follows: A white fixation cross was centrally presented on a black screen for 500 ms. This was followed by a centrally presented prime stimulus consisting of five horizontally aligned letters. The presentation of the prime array was either ended by a button press, or after 1500 ms had elapsed (in case of a missing response). Next, a central white fixation cross was presented for 500 ms. In case of an incorrect or missed response to the prime, a 500 ms error feedback was additionally displayed on the screen (i.e., the word "Fehler", translating to "error"). Then, another five letters were centrally displayed as the probe array. Like for the prime array, the presentation of the probe array was either terminated by a button press, or after 1500 ms had elapsed (in case of a missing response). The probe array was also followed by the presentation of a central white fixation cross. In the case of an incorrect or missed response to the probe, an additional 500 ms error feedback was presented on the screen (i.e., the word "Fehler", translating to "error").

The inter-trial interval (ITI) randomly varied between 700 and 1100 ms. The paradigm comprised a total of 576 trials, which were divided into six equally-sized blocks. Participants were offered to take breaks in-between these blocks.

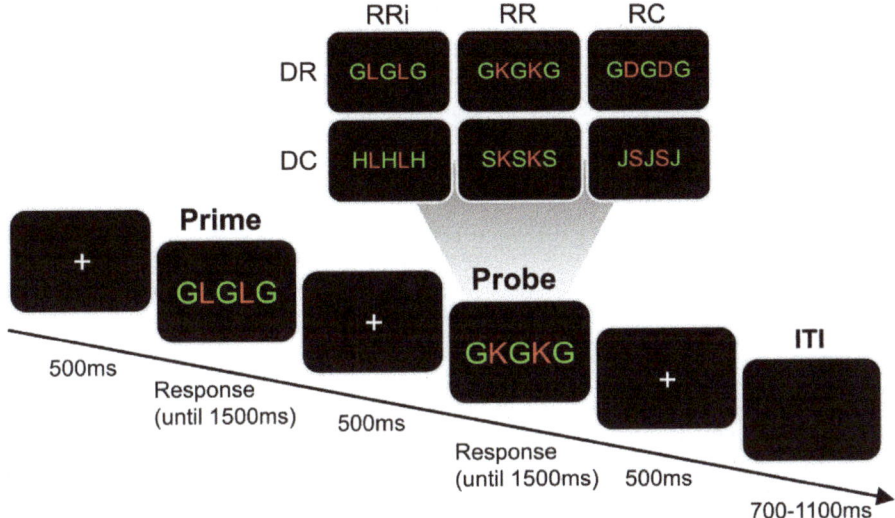

Figure 3. Illustration of the time course in milliseconds (ms) of a single trial in the distractor-response paradigm [29]. In each trial, a white fixation cross was centrally shown on a black screen for 500 ms. This was followed by a centrally presented prime array consisting of five letters. The prime array presentation was terminated either by a button press in response to the red target letters, or after 1500 ms (in case of a missing response). After another presentation of a central white fixation cross for 500 ms, a probe array was centrally presented. The probe array also consisted of five letters and its presentation was again terminated by a button press in response to the red target letters or ended after 1500 ms. The inter-trial interval (ITI) was jittered between 700 and 1100 ms. The upper half of the figure illustrates exemplary probe arrays for each combination of conditions in relation to the prime array used for illustration in this figure ("G L G L G"). Depending on whether the target and/or distractor stimulus were repeated or changed between prime and probe array, six different combinations of conditions were distinguished. In terms of the distractor stimuli, distractor repetition (DR) and distractor change (DC) were distinguished. In terms of the required response, response repetition when the target stimulus was identical (RRi), response repetition when the target stimulus differed (RR), and response change (RC) were distinguished.

With respect to S-R mapping, two out of eight letter stimuli (S, D, F, G, H, J, K, L) each formed a group. This resulted in four stimulus groups, each associated with a different response button on a standard QWERTZ keyboard. In detail, S and D required pressing the left Ctrl button with the left middle finger; F and G required pressing the left Alt button with the left index finger; H and J required pressing the right Win button with the right index finger; and K and L required pressing the right Ctrl button with the right middle finger. In each array, there were always two target letter stimuli which were shown in red font color. They were horizontally flanked by three distractor letter stimuli, which were shown in green font color. Please note that the target and distractor stimuli were never identical and always associated with different response buttons, even though only the targets required a single button press response. Participants were asked to respond to the red target letters while ignoring the green distractor letters. They were instructed to respond by pressing the appropriate button as quickly and as accurately as possible. Each trial asked for a first response to the prime array and a second response to the subsequent probe array.

The task allows to distinguish six combinations of conditions resulting from the modification of the distractor stimulus and/or response between prime and probe. All conditions are exemplarily illustrated in Figure 3. Regarding the distractor stimuli, there could be a distractor repetition (DR) or a distractor change (DC) between the prime and probe. Regarding the required response, there could be a repetition of the correct response button when the target stimulus was identical for both prime and probe (RRi), a repetition of the correct response button when the target stimulus differed between prime and probe (RR), and a change of the correct response button (RC). Each combination of conditions was presented equally often. In each of the conditions, all possible combinations of probe target and probe distractor were equally likely (considering that target and distractor stimuli could never require the same response). Prime and probe distractors were identical in all DR trials, while the prime distractor was randomly picked from one of the other response groups in all DC trials (e.g., when the probe distractor was the letter "G", the prime distractor could not be the letter "G" or "F"). Additionally, the prime target was randomly picked from one of the other response groups in the RC condition. A total of 30 practice trials was performed before the experiment started. The task itself took around 40 min to finish.

2.6. Statistical Analyses

Statistical analyses included only trials in which participants responded correctly to both the prime and probe arrays. For each participant, trials with prime and/or probe response times (RTs) beyond ±2 standard deviations from the individual's mean RT in the respective task condition were rejected in order to reduce outlier effects. The obtained correct probe RTs and probe accuracy were analyzed with SPSS Statistics 25 (IBM Corp. Released 2017. IBM SPSS Statistics for Windows, Version 25.0. Armonk, NY, USA: IBM Corp.) using separate repeated-measures ANOVAs. Status (hangover vs. sober), response (response repetition to identical target (RRi) vs. response repetition to other target (RR) vs. response change (RC)), and distractor (distractor repetition (DR) vs. distractor change (DC)) were used as within-subject factors. Greenhouse–Geisser correction was applied whenever necessary. Post hoc multiple comparisons were Bonferroni-corrected. All behavioral variables were analyzed for normal distribution indicated by Kolmogorov–Smirnov tests. As this assumption was not met for RTs in a few conditions and for most accuracy conditions, all significant main effects and post hoc tests were additionally tested for significance using non-parametric Wilcoxon signed-rank tests. All descriptive statistics provide the mean value and the standard error of the mean (SEM) as a measurement of variability. Descriptive and behavioral data, as well as all statistical analyses, are available online at https://osf.io/ktgyr/.

3. Results

3.1. Sample

Out of $n = 35$ participants, $n = 17$ had their hangover session before their sober session and $n = 18$ had their sober session before their hangover session. Sociodemographic characteristics, questionnaire scores, and alcohol-related data are detailed in Table 1. Subjective sleep and hangover ratings on both sessions are provided in Table 2. A total of $n = 2$ participants reported to have never experienced a hangover over their entire lifetime as assessed by the first ASQ item. Only one of them belonged to a total of $n = 2$ participants who indicated to not have experienced any overall hangover at the hangover session, meaning they rated the first item of the hangover severity rating by van Schrojenstein Lantman et al. [13] with 0 points. Nevertheless, both of these participants reported light hangover complaints for some of the symptoms in this rating, suggesting that none of them were entirely free of any hangover symptoms (please note that add-on analyses on hangover severity and symptoms can be found in the Supplementary Materials). Furthermore, a paired t-test showed that the number of standard drinks consumed at the intoxication session (16.74 drinks ± 0.25) was significantly larger than the maximal number of drinks participants indicated to be able to drink before experiencing a hangover (8.94 drinks ± 0.69; $t_{(32)}$ = |10.696|; $p < 0.001$). Thus, the amount of alcohol administered in this study exceeded the self-reported hangover threshold.

Table 1. Sociodemographic, questionnaire, and alcohol-related data of the included sample. All values are reported as means ± standard error of the mean and (range). Please note that all relevant items were averaged for the ASQ_{light} score (alcohol-related experiences associated with lower doses). For the ASQ_{heavy} score (alcohol-related experiences associated with heavier doses), all relevant items except for "passing out" were averaged, as only a single participant had indicated to have ever passed out due to alcohol drinking.

Characteristic	Included Sample (n = 35)
Age in years	23.14 ± 0.47 (19–28)
Height in cm	181.89 ± 0.98 (170–195)
Weight in kg	76.87 ± 1.69 (63–105)
Cigarettes smoked per day	0.78 ± 0.37 (0–10)
Hours of sport per week	4.63 ± 0.56 (0–16)
BDI score	3.37 ± 0.69 (0–19)
AUDIT score	10.26 ± 0.58 (5–19)
ASQ total score	8.23 ± 0.40 (3.25–13.29)
ASQ_{light} score	5.16 ± 0.30 (1.63–9.25)
ASQ_{heavy} score	13.17 ± 0.78 (5.20–24.0)
Individual alcohol amount indicated in mL of brandy (36 vol %)	418.00 ± 5.81 (369–516)
Alcohol consumption duration in minutes	181.43 ± 4.29 (111–243)
BrAC 30 min after end of consumption	1.32 ± 0.03 (1.05–1.69)
BrAC 60 min after end of consumption	1.24 ± 0.02 (1.01–1.56)
BrAC 90 min after end of consumption	1.15 ± 0.02 (0.91–1.40)
BrAC 120 min after end of consumption	1.08 ± 0.03 (0.83–1.43)

BDI = Beck depression inventory; AUDIT = alcohol use disorders identification test; ASQ = alcohol sensitivity questionnaire; BrAC = breath alcohol concentration.

Table 2. Subjective ratings of sleep and hangover symptoms on both sessions. Hangover symptoms were rated on an 11-point Likert scale ranging from 0 points (no symptoms) to 10 points (extreme symptoms). Of note, participants were asked to truthfully rate the severity of each symptom on both sessions, irrespective of whether they had been drinking the night before the sober session. The mild symptom severity for the sober session and the resulting minimal variance of this rating may have contributed to the fact that nearly all hangover symptoms differed significantly between the sober and the hangover session (as was intended by the study). These comparisons were performed with uncorrected paired t-tests. p-Values are reported in the column "Difference." All values are given as means ± standard error of the mean and (range).

Symptom	Sober	Hangover	Presence at Hangover Session	Difference
Hours of sleep in previous night	7.40 ± 0.15 (5.5–9)	5.54 ± 0.19 (4–8)	-	$p < 0.001$ **
Overall hangover severity	0 ± 0 (0–0)	3.63 ± 0.39 (0–10)	94.3%	$p < 0.001$ **
Thirst	0.91 ± 0.28 (0–7)	3.46 ± 0.37 (0–8)	91.4%	$p < 0.001$ **
Concentration problems	0.61 ± 0.18 (0–4)	3.11 ± 0.37 (0–8)	88.6%	$p < 0.001$ **
Tired	1.12 ± 0.21 (0–4)	4.57 ± 0.39 (1–10)	88.6%	$p < 0.001$ **
Sleepiness	1.00 ± 0.23 (0–4)	3.71 ± 0.42 (0–9)	85.7%	$p < 0.001$ **
Weakness	0.30 ± 0.11 (0–2)	2.31 ± 0.34 (0–10)	85.7%	$p < 0.001$ **
Headache	0.06 ± 0.04 (0–1)	2.69 ± 0.38 (0–8)	82.9%	$p < 0.001$ **
Clumsy	0.39 ± 0.15 (0–3)	1.97 ± 0.32 (0–6)	71.4%	$p < 0.001$ **
Dizziness	0.03 ± 0.03 (0–1)	1.69 ± 0.27 (0–6)	71.4%	$p < 0.001$ **

Table 2. *Cont.*

Symptom	Sober	Hangover	Presence at Hangover Session	Difference
Sensitivity to light	0.24 ± 0.12 (0–3)	1.57 ± 0.32 (0–8)	60%	$p < 0.001$ **
Reduced appetite	0.21 ± 0.16 (0–5)	1.97 ± 0.41 (0–9)	57.1%	$p = 0.001$ **
Sweating	1.18 ± 0.33 (0–7)	1.09 ± 0.21 (0–4)	51.4%	$p = 0.784$
Nausea	0.03 ± 0.03 (0–1)	1.46 ± 0.35 (0–7)	48.6%	$p < 0.001$ **
Apathy	0 ± 0 (0–0)	0.89 ± 0.20 (0–4)	48.6%	$p < 0.001$ **
Shivering	0.36 ± 0.13 (0–3)	1.20 ± 0.29 (0–6)	45.7%	$p = 0.010$ *
Confusion	0.15 ± 0.08 (0–2)	0.91 ± 0.22 (0–5)	45.7%	$p < 0.001$ **
Heart pounding	0.30 ± 0.10 (0–2)	0.94 ± 0.23 (0–5)	42.9%	$p = 0.012$ *
Stomach pain	0.24 ± 0.16 (0–5)	1.06 ± 0.32 (0–7)	40%	$p = 0.012$ *
Anxiety	0.39 ± 0.14 (0–3)	0.94 ± 0.25 (0–5)	40%	$p = 0.028$ *
Regret	0.12 ± 0.12 (0–4)	0.83 ± 0.31 (0–10)	37.1%	$p = 0.003$ **
Sleeping problems	0.21 ± 0.14 (0–4)	0.91 ± 0.32 (0–8)	31.4%	$p = 0.027$ *
Depression	0.21 ± 0.16 (0–5)	0.66 ± 0.28 (0–9)	28.6%	$p = 0.021$ *
Vomiting	0.03 ± 0.03 (0–1)	0.60 ± 0.28 (0–9)	22.9%	$p = 0.055$
Heart racing	0.15 ± 0.08 (0–2)	0.57 ± 0.25 (0–6)	22.9%	$p = 0.138$

* $p < 0.05$, ** $p < 0.01$.

3.2. Behavioral Data

The correct probe RTs and probe accuracy are illustrated in Figure 4.

The repeated-measures ANOVA for correct probe RTs (in case of correct prime response) revealed task-specific effects: There was a main effect of response ($F_{(2,68)} = 374.42$; $p < 0.001$; $\eta_p^2 = 0.917$). Post hoc comparisons revealed significant differences between all conditions, with the fastest responses in RRi trials (493 ms ± 9), followed by RR trials (613 ms ± 14), and RC trials (770 ms ± 13) (all $p < 0.001$). The main effect of distractor ($F_{(1,34)} = 54.56$; $p < 0.001$; $\eta_p^2 = 0.616$) showed faster responses in DR trials (617 ms ± 11) than in DC trials (634 ms ± 11). Furthermore, there was an interaction of response and distractor ($F_{(2,68)} = 9.21$; $p = 0.001$; $\eta_p^2 = 0.213$). Post hoc comparisons revealed that participants responded significantly faster in DR than in DC trials on RRi condition ($p < 0.001$; RRi-DR: 479 ms ± 9; RRi-DC: 508 ms ± 9), and on RR condition ($p = 0.006$; RR-DR: 605 ms ± 14; RR-DC: 621 ms ± 14), but not on RC condition ($p = 0.254$). We additionally computed the size of the distractor effect (DC minus DR) for each response condition. Post hoc comparisons revealed that the distractor effect in RRi trials (29 ms ± 3) was significantly greater than in RR trials (16 ms ± 5) ($p = 0.023$) and in RC trials (7 ms ± 4) ($p \leq 0.001$). RR and RC trials did not significantly differ in the size of their distractor effect ($p = 0.471$). All main and interaction effects of status were non-significant (all $F \leq 1.16$; all $p \geq 0.295$).

Regarding probe accuracy, the repeated-measures ANOVA showed a main effect of response ($F_{(2,68)} = 79.07$; $p < 0.001$; $\eta_p^2 = 0.699$). Post hoc comparisons revealed significant differences between all conditions, with the highest accuracy in RRi trials (93.5 % ± 0.7), followed by RR trials (89.4 % ± 1.0) and then by RC trials (83.5 % ± 1.2) (all $p < 0.001$). Importantly, an interaction of status and distractor was obtained ($F_{(1,34)} = 5.02$; $p = 0.032$; $\eta_p^2 = 0.129$), which is illustrated in Figure 5. Post hoc comparisons showed differences between distractor repetition and distractor change in the sober and hangover assessment. In the sober assessment, accuracy was significantly higher for DR trials (89.8 % ± 1.1) than for DC trials (88.4 % ± 1.1) ($p = 0.027$), while there was no significant distractor difference during the hangover assessment ($p = 0.487$). We additionally computed the size of the distractor effect (DC minus DR) for each status. A post hoc paired t-test showed that the size of the distractor effect was significantly smaller in the hangover assessment (|1.0| % ± 0.8) than in the sober assessment (|1.4| % ± 0.5) ($t_{(34)} = |2.24|$; $p = 0.032$). To investigate whether alcohol sensitivity or hangover severity modulated this effect, we correlated the overall ASQ score, the ASQ heavy score, and the overall hangover severity item by van Schrojenstein Lantman [13] with an average of all hungover DR trials, an average of all hungover DC trials, and the distractor effect size at hangover assessment as well as with the difference in distractor effect sizes between the sober and hangover assessment, and with the size of the hangover effect (SOB minus HANG) in DR and DC trials. None of these uncorrected correlations between ASQ scores, hangover severity, and performance parameters was significant (all $p \geq 0.064$). All other main and interaction effects of the ANOVA were non-significant (all $F \leq 2.26$; all $p \geq 0.118$).

To summarize, we found that alcohol hangover affects the processing of response-irrelevant (distractor) information, but not the processing of response-relevant (target) information. The former is reflected by the interaction of status and distractor in probe accuracy, while the latter is reflected by non-significant hangover effects on response conditions in both correct probe RTs and probe accuracy.

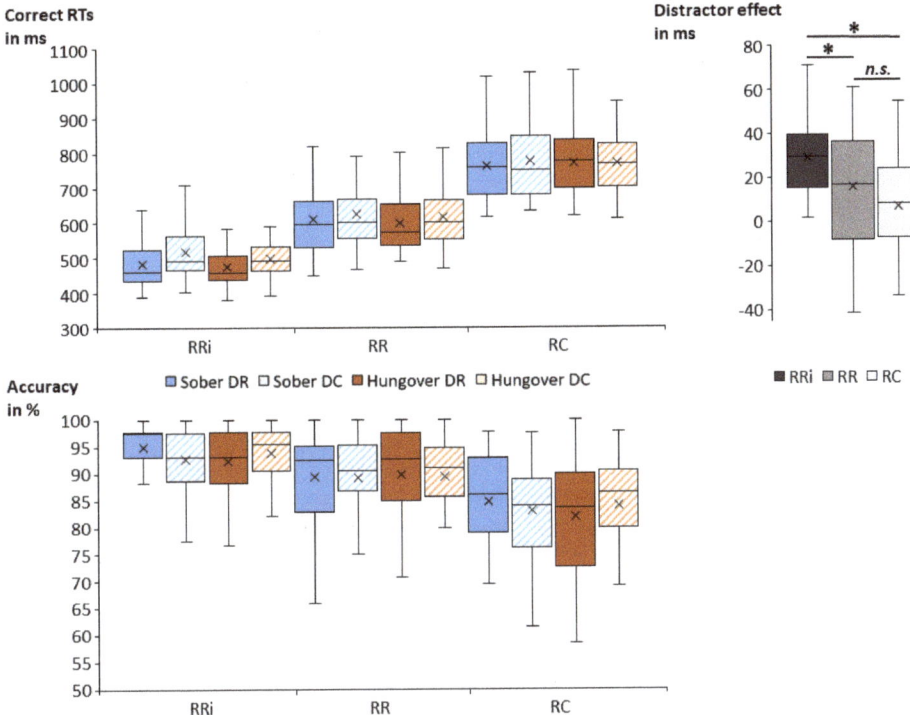

Figure 4. Box plots of the obtained mean correct probe response times (RTs in milliseconds, upper left graph) and mean probe accuracy (in percent, lower graph) for each combination of conditions. We observed significantly faster and more accurate responses in RRi than in RR and RC. We further observed significantly faster responses in DR than in DC. For probe response times, the size of the distractor effect (i.e., the difference between DC and DR, upper right graph) was significantly greater in RRi than in RR and RC, while it was not significantly different between RR and RC. While we found no significant main effects/overall differences between the sober and hangover session (i.e., irrespective of all other conditions), we found a significant interaction between status and distractor, which is depicted in Figure 5. Abbreviations: DR = distractor repetition; DC = distractor change; RRi = response repetition when the target stimulus was identical; RR = response repetition when the target stimulus differed; RC = response change; ms = milliseconds; * = $p < 0.05$; n.s. = non-significant.

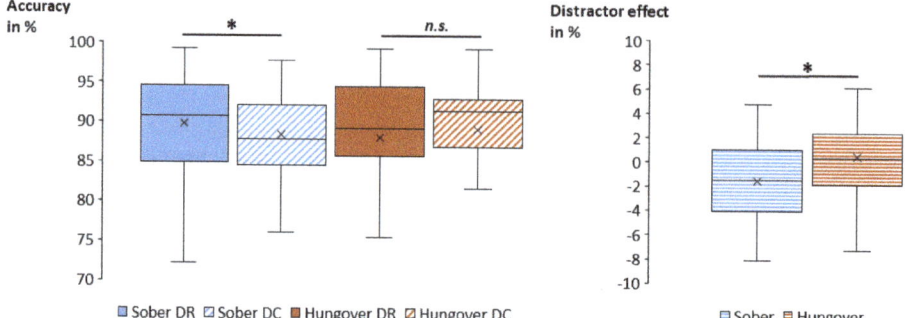

Figure 5. Box plots of the obtained mean probe accuracy (in percent) for the significant interaction of status and distractor are illustrated in the left graph. While the accuracy was significantly higher in DR trials (89.8 % ± 1.1) than in DC trials (88.4 % ± 1.1) when being sober, there was no such significant difference when being hungover. Consequently, the size of the distractor effect (the difference between DC and DR) was significantly smaller in the hangover assessment than in the sober assessment (right graph). Abbreviations: DR = distractor repetition; DC = distractor change; * = $p < 0.05$; n.s. = non-significant.

4. Discussion

Distractibility is one of the major contributors of hangover-induced behavioral deficits in daily activities, like when driving a car [6–11,16]. As the basic mechanisms causing hangover-induced distractibility are still not very well understood, our study aimed to investigate the underlying mechanisms of distraction, as proposed by the TEC [20]. This theory postulates that stimulus-response (S-R) bindings are established in so-called event files. Given previous findings that sensory information seems to be accumulated more quickly when hungover [17], we hypothesized that S-R bindings should consequently be enhanced in the hungover state. We further hypothesized that alcohol hangover has different effects on the binding of response-relevant (target) information and the binding of response-irrelevant (distracting) information. To test these hypotheses, we used a paradigm by Moeller et al. [29] that allows to differentiate the effects of target-response binding and distractor-response binding. Each participant was tested once sober and once hungover after being intoxicated in a standardized experimental drinking setting the night before. Behavioral testing did not start until a BrAC of 0.00‰ was measured on both appointments.

Interestingly, alcohol hangover modulated distractor-response binding in the accuracy data, while target-response binding was not similarly affected: During the sober assessment, distractor repetition was advantageous in responding to the relevant target stimuli, and this effect was found regardless of target and response alterations (repetition vs. change). During the hungover assessment, this advantage of distractor repetition was no longer evident, as the distractor effect was significantly smaller. Of note, the sober distractor repetition effect can be explained by the TEC [20]: Whenever response-irrelevant distracting information is presented at the same time as response-relevant target information, both irrelevant and relevant stimulus features, as well as the response, are bound in event files [20,42]. Once the distracting information of the pre-existing event file reappears, the entire event file becomes (re)activated following a pattern-completion logic [20]. Changes in the configuration of presented stimuli and required response do however require readjustments of the current event file. As compared to distractor changes, distractor repetition should therefore require less extensive changes or adaptations in the reactivated event file. Hence, less cognitive resources should be needed in case of distractor repetition. This ultimately results in faster and more accurate responses. Even though the TEC may explain why and how distracting information modulates behavior in a sober state, this theory does not provide a straightforward explanation for the resulting pattern observed during alcohol hangover. Given that target-response binding was seemingly unaffected by hangover (as substantiated by the lack of interactions between response and status), it can be assumed that event files were still established during the hungover state. Within the framework of the TEC, the fact that the distractor effect became substantially smaller during hangover may be explained by reduced distractor processing and/or by enhanced target processing: Alcohol hangover may have affected the reactivation of the event file by recurrent distractor features. This would imply that the sensory processing of distracting information may be altered during hangover. Given that the accumulation of sensory information is likely more efficient in hangover state [17], we had initially expected distractor-response bindings to be strengthened. Based on Stock et al. [17], it was however not clear, whether all or only selected sensory information could be processed in a more efficient way. It is therefore also possible that relevant and irrelevant sensory information may be better discriminated during hangover. As a consequence, response-irrelevant distracting information might be processed to a lesser degree than during the sober state. By comparison, target information would then be processed more efficiently. In line with this, the typically higher activation of target information is assumed to have a greater impact on S-R bindings than the typically lower distractor information [43]. It could hence be speculated that the activation of the more dominant target information might be even higher during hangover due to its more efficient information accumulation [17]. Given that response and target repetition effects were however not enhanced during hangover, it seems more likely that the observed changes were primarily driven by decreased distractor processing, rather than by absolute increases in target processing. Interestingly, this may theoretically help to neutralize some of the cognitive side effects of alcohol hangover, rather than aggravate them. Further adding to this, the lack of target-associated effects matches the findings of a previous publication demonstrating that target-based S-R associations were not systematically modulated by acute alcohol intoxication [22].

Irrespective of hangover state, we found task-specific distractor-response binding effects in response times, which are in line with previous findings on the paradigm used in this study [30,44]: The event file-associated automatic response tendency induced by the distractor was advantageous in case of response repetition while the benefit of distractor repetition was no longer evident in case of response change. Additionally, the size of the distractor repetition effect was most pronounced in case of response and target repetition, while the effect was smaller in case of target change, regardless of the required response (repeated vs. changed). This result pattern may again be explained by the TEC: Whenever distractor repetition and response repetition coincide, the distractor reactivates an event file containing a response that is still correct in the current setting, thus resulting in faster responding. Whenever there is a response change in the presence of distractor repetition, the reactivated response no longer holds true for the current task requirement, resulting in a time-taking "unbinding" process of the event file [42,45]. Additionally, the reactivation of the event file is suggested to be stronger, the more stimulus and response features are repeated/activated [42,45]. That is, target and distractor repetition lead to a stronger reactivation and thus faster responding than distractor repetition alone, even when the response is the same in both cases (i.e., target change and target repetition). These sober findings are thus well in line with various studies investigating S-R binding [22,29,30,46].

Putting our results in a broader context, it should be noted that previous studies investigating distractor-induced cognitive and behavioral conflicts reported hangover-related impairments of selective attention when applying classical Stroop and Erikson flanker tasks [47,48]. Further detailing this, a previous study of our group showed that hangover enhanced the flanker effect only in case of (subliminally) increased conflict load, suggesting that the overall strain on cognitive control resources may play an important role for hangover effects on distractibility [18]. Yet, all these studies have in common that the applied tasks did not allow differentiation between response-relevant target and response-irrelevant distractor processing to the same degree as the current study. Furthermore, they put a stronger focus on conflict processing than on binding phenomena. While we only found selective impairments in distractor-response bindings, this does not exclude the possibility that other cognitive and attentional mechanisms, which were not assessed by our paradigm, may also be impaired and thus contribute to the hangover-related enhanced distractibility reported in other studies [14,47,48]. Therefore, further research is needed on other attentional mechanisms that could help explain increased distractibility in the hangover state. Furthermore, it would be conceivable to investigate these mechanisms in a naturalistic environment [49] and to investigate performance in real-life challenges like an actual exam or quantifiable work performance. This would help to gain insights into whether the consequences of alcohol hangover could be more severe in these situations and could help to derive better preventive measures.

Limitations

The study sample was limited to males, as the ethics committee did not approve inducing such an intoxication in females. This is rather unfortunate, as other studies have reported females to experience more severe hangover symptoms than males [24,50], potentially because females tend to metabolite ethanol more slowly than males. In this light, females might experience greater hangover effects, which could translate to more severely impaired behavioral performance. Additionally, it has been suggested that the quality and incidences of experiencing alcohol hangover might change with age [51,52]. Based on these aspects, further studies including both sexes and adults of different age groups would provide a more complete picture. Furthermore, we did not collect data on drinking and detailed hangover history beyond the scope of the ASQ and AUDIT (which only covers the last 12 months). In doing so, we might have missed details on hangover and binge-drinking history, which could potentially have explained additional variance. In this context, it should however also be noted that as a result of the recruitment process, our study sample was quite homogenous with respect to drinking patterns. As our sample was likely to have shown comparatively minor differences in their hangover and/or binge-drinking history, we did not systematically investigate this. Studies that do not apply such inclusion criteria might however benefit from assessing more details on their sample's drinking and hangover history.

5. Conclusions

In summary, we investigated whether and how alcohol hangover modulates S-R bindings, as this may explain why distractibility might be increased during hangover, resulting in various behavioral and performance difficulties. Specifically, we investigated whether hangover has different effects on target-response bindings vs. distractor-response bindings. While hangover did not affect target-response bindings, it decreased distractor-response bindings, resulting in the absence of distractor effects. These findings implicate that heightened distractibility during hangover is likely not due to increased distractor-response binding. Hence, other cognitive mechanisms relating to selective attention are more likely to underlie the adverse effects of hangover on driving skills and other daily activities. Therefore, more studies investigating the different aspects of distractibility and distractor processing are needed to conclusively understand how alcohol hangover may increase distractibility.

Supplementary Materials: The following is available online at http://www.mdpi.com/2077-0383/9/3/778/s1.

Author Contributions: Conceptualization, methodology, validation, all authors; formal analysis, A.O. and A.-K.S.; investigation, A.O. and A.-K.S.; resources, C.B. and A.-K.S.; data curation, A.O. and A.-K.S.; writing—original draft preparation, A.O. and A.-K.S.; writing—review and editing, all authors; visualization, A.O.; supervision, C.B. and A.-K.S.; project administration, A.-K.S.; funding acquisition, C.B. and A.-K.S. All authors have read and agreed to the published version of the manuscript.

Funding: This research was funded by the Deutsche Forschungsgemeinschaft (DFG), grant numbers TRR 265 B07 and FOR 2698.

Acknowledgments: The authors thank all individuals who took part in this study. Open Access Funding by the Publication Funds of the TU Dresden.

Conflicts of Interest: The authors declare no conflict of interest. The funders had no role in the design of the study; in the collection, analyses or interpretation of data; in the writing of the manuscript or in the decision to publish the results.

References

1. Fell, J.C. Approaches for reducing alcohol-impaired driving: Evidence-based legislation, law enforcement strategies, sanctions, and alcohol-control policies. *Forensic Sci. Rev.* **2019**, *31*, 161–184. [PubMed]
2. Ogden, E.J.D.; Moskowitz, H. Effects of alcohol and other drugs on driver performance. *Traffic Inj. Prev.* **2004**, *5*, 185–198. [CrossRef] [PubMed]
3. Lefio, Á.; Bachelet, V.C.; Jiménez-Paneque, R.; Gomolán, P.; Rivas, K. A systematic review of the effectiveness of interventions to reduce motor vehicle crashes and their injuries among the general and working populations. *Rev. Panam. Salud Publica Pan Am. J. Public Health* **2018**, *42*, e60. [CrossRef]
4. Rezaee-Zavareh, M.S.; Salamati, P.; Ramezani-Binabaj, M.; Saeidnejad, M.; Rousta, M.; Shokraneh, F.; Rahimi-Movaghar, V. Alcohol consumption for simulated driving performance: A systematic review. *Chin. J. Traumatol. Zhonghua Chuang Shang Za Zhi* **2017**, *20*, 166–172. [CrossRef] [PubMed]
5. Verster, J.C.; van de Loo, A.J.A.E.; Downey, L.A. Driving while hungover: The necessity of biomarkers of the alcohol hangover state. *Curr. Drug Abuse Rev.* **2014**, *7*, 1–2. [CrossRef]
6. Seppälä, T.; Leino, T.; Linnoila, M.; Huttunen, M.; Ylikahri, R. Effects of hangover on psychomotor skills related to driving: Modification by fructose and glucose. *Acta Pharmacol. Toxicol. (Copenh.)* **1976**, *38*, 209–218. [CrossRef]
7. Rohsenow, D.J.; Howland, J.; Arnedt, J.T.; Almeida, A.B.; Greece, J.; Minsky, S.; Kempler, C.S.; Sales, S. Intoxication with Bourbon Versus Vodka: Effects on Hangover, Sleep, and Next-Day Neurocognitive Performance in Young Adults. *Alcohol. Clin. Exp. Res.* **2010**, *34*, 509–518. [CrossRef]
8. Verster, J.C.; Maarel, M.A.V.D.; McKinney, A.; Olivier, B.; Haan, L.D. Driving During Alcohol Hangover Among Dutch Professional Truck Drivers. *Traffic Inj. Prev.* **2014**, *15*, 434–438. [CrossRef]
9. Verster, J.C.; Bervoets, A.C.; de Klerk, S.; Vreman, R.A.; Olivier, B.; Roth, T.; Brookhuis, K.A. Effects of alcohol hangover on simulated highway driving performance. *Psychopharmacology (Berl.)* **2014**, *231*, 2999–3008. [CrossRef]
10. Høiseth, G.; Fosen, J.T.; Liane, V.; Bogstrand, S.T.; Mørland, J. Alcohol hangover as a cause of impairment in apprehended drivers. *Traffic Inj. Prev.* **2015**, *16*, 323–328. [CrossRef]
11. Mackus, M.; Griffin, C.; Munafò, M.R.; Adams, S. A systematic review of the next-day effects of heavy alcohol consumption on cognitive performance. *Addiction* **2018**, *113*, 2182–2193.
12. Mackus, M.; van de Loo, A.J.A.E.; Raasveld, S.J.; Hogewoning, A.; Sastre Toraño, J.; Flesch, F.M.; Korte-Bouws, G.A.H.; van Neer, R.H.P.; Wang, X.; Nguyen, T.T.; et al. Biomarkers of the alcohol hangover state: Ethyl glucuronide (EtG) and ethyl sulfate (EtS). *Hum. Psychopharmacol. Clin. Exp.* **2017**, *32*, e2624. [CrossRef]
13. van Schrojenstein Lantman, M.; van de Loo, A.; Mackus, M.; Verster, J. Development of a Definition for the Alcohol Hangover: Consumer Descriptions and Expert Consensus. *Curr. Drug Abuse Rev.* **2017**, *9*, 148–154. [CrossRef]
14. McKinney, A.; Coyle, K.; Verster, J. Direct comparison of the cognitive effects of acute alcohol with the morning after a normal night's drinking. *Hum. Psychopharmacol.* **2012**, *27*, 295–304. [CrossRef] [PubMed]
15. Karadayian, A.G.; Cutrera, R.A. Alcohol hangover: Type and time-extension of motor function impairments. *Behav. Brain Res.* **2013**, *247*, 165–173. [CrossRef]
16. Shiferaw, B.; Stough, C.; Downey, L. Drivers' visual scanning impairment under the influences of alcohol and distraction: A literature review. *Curr. Drug Abuse Rev.* **2014**, *7*, 174–182. [CrossRef]
17. Stock, A.-K.; Hoffmann, S.; Beste, C. Effects of binge drinking and hangover on response selection sub-processes-a study using EEG and drift diffusion modeling: Binge-drinking and hangover. *Addict. Biol.* **2017**, *22*, 1355–1365. [CrossRef]
18. Zink, N.; Bensmann, W.; Beste, C.; Stock, A.-K. Alcohol Hangover Increases Conflict Load via Faster Processing of Subliminal Information. *Front. Hum. Neurosci.* **2018**, *12*, 316. [CrossRef]
19. Opitz, A.; Hubert, J.; Beste, C.; Stock, A.-K. Alcohol Hangover Slightly Impairs Response Selection but not Response Inhibition. *J. Clin. Med.* **2019**, *8*, 1317. [CrossRef]
20. Hommel, B. Action control according to TEC (theory of event coding). *Psychol. Res. Psychol. Forsch.* **2009**, *73*, 512–526. [CrossRef]
21. Hommel, B. How Much Attention Does an Event File Need? *J. Exp. Psychol. Hum. Percept. Perform.* **2005**, *31*, 1067–1082. [CrossRef]

22. Stock, A.-K.; Bensmann, W.; Zink, N.; Münchau, A.; Beste, C. Automatic aspects of response selection remain unchanged during high-dose alcohol intoxication. *Addict. Biol.* **2019**, e12852. [CrossRef]
23. Martí-Prats, L.; Sánchez-Catalán, M.J.; Orrico, A.; Zornoza, T.; Polache, A.; Granero, L. Opposite motor responses elicited by ethanol in the posterior VTA: The role of acetaldehyde and the non-metabolized fraction of ethanol. *Neuropharmacology* **2013**, *72*, 204–214. [CrossRef] [PubMed]
24. Prat, G.; Adan, A.; Sánchez-Turet, M. Alcohol hangover: A critical review of explanatory factors. *Hum. Psychopharmacol. Clin. Exp.* **2009**, *24*, 259–267. [CrossRef]
25. Wiese, J.G.; Shlipak, M.G.; Browner, W.S. The alcohol hangover. *Ann. Intern. Med.* **2000**, *132*, 897–902. [CrossRef]
26. Le Daré, B.; Lagente, V.; Gicquel, T. Ethanol and its metabolites: Update on toxicity, benefits, and focus on immunomodulatory effects. *Drug Metab. Rev.* **2019**, *51*, 545–561. [CrossRef]
27. Tipple, C.T.; Benson, S.; Scholey, A. A Review of the Physiological Factors Associated with Alcohol Hangover. *Curr. Drug Abuse Rev.* **2017**, *9*, 93–98. [CrossRef]
28. Klapp, S.T. One version of direct response priming requires automatization of the relevant associations but not awareness of the prime. *Conscious. Cogn.* **2015**, *34*, 163–175. [CrossRef]
29. Moeller, B.; Schächinger, H.; Frings, C. Irrelevant Stimuli and Action Control: Analyzing the Influence of Ignored Stimuli via the Distractor-Response Binding Paradigm. *J. Vis. Exp.* **2014**, 51571. [CrossRef]
30. Opitz, A.; Beste, C.; Stock, A.-K. Using temporal EEG signal decomposition to identify specific neurophysiological correlates of distractor-response bindings proposed by the theory of event coding. *NeuroImage* **2020**, *209*, 116524. [CrossRef]
31. World Health Organization. AUDIT: The Alcohol Use Disorders Identification Test: Guidelines for use in primary care/Thomas F. Babor... [et al.], 2nd ed. Available online: https://apps.who.int/iris/handle/10665/67205 (accessed on 5 March 2020).
32. Dybek, I.; Bischof, G.; Grothues, J.; Reinhardt, S.; Meyer, C.; Hapke, U.; John, U.; Broocks, A.; Hohagen, F.; Rumpf, H.-J. The Reliability and Validity of the Alcohol Use Disorders Identification Test (AUDIT) in a German General Practice Population Sample. *J. Stud. Alcohol* **2006**, *67*, 473–481. [CrossRef]
33. Montgomery, C.; Fisk, J.E.; Murphy, P.N.; Ryland, I.; Hilton, J. The effects of heavy social drinking on executive function: A systematic review and meta-analytic study of existing literature and new empirical findings: EXECUTIVE FUNCTION AND ALCOHOL USE. *Hum. Psychopharmacol. Clin. Exp.* **2012**, *27*, 187–199. [CrossRef] [PubMed]
34. Bensmann, W.; Kayali, Ö.F.; Beste, C.; Stock, A.-K. Young frequent binge drinkers show no behavioral deficits in inhibitory control and cognitive flexibility. *Prog. Neuropsychopharmacol. Biol. Psychiatry* **2019**, *93*, 93–101. [CrossRef]
35. van Schrojenstein Lantman, M.; Mackus, M.; Roth, T.; Verster, J. Total sleep time, alcohol consumption, and the duration and severity of alcohol hangover. *Nat. Sci. Sleep* **2017**, *9*, 181–186. [CrossRef]
36. Widmark, E.M.P. *Die Theoretischen Grundlagen Und Die Praktische Verwendbarkeit Der Gerichtlich-Medizinischen Alkoholbestimmung*; Urban und Schwarzenberg: Berlin, Germany, 1932.
37. Watson, P.E.; Watson, I.D.; Batt, R.D. Total body water volumes for adult males and females estimated from simple anthropometric measurements. *Am. J. Clin. Nutr.* **1980**, *33*, 27–39. [CrossRef] [PubMed]
38. Epler, A.J.; Tomko, R.L.; Piasecki, T.M.; Wood, P.K.; Sher, K.J.; Shiffman, S.; Heath, A.C. Does Hangover Influence the Time to Next Drink? An Investigation Using Ecological Momentary Assessment. *Alcohol. Clin. Exp. Res.* **2014**, *38*, 1461–1469. [CrossRef] [PubMed]
39. Jackson, K.M.; Rohsenow, D.J.; Piasecki, T.M.; Howland, J.; Richardson, A.E. Role of Tobacco Smoking in Hangover Symptoms Among University Students. *J. Stud. Alcohol Drugs* **2013**, *74*, 41–49. [CrossRef]
40. Beck, A.T. An Inventory for Measuring Depression. *Arch. Gen. Psychiatry* **1961**, *4*, 561. [CrossRef]
41. Fleming, K.A.; Bartholow, B.D.; Hilgard, J.; McCarthy, D.M.; O'Neill, S.E.; Steinley, D.; Sher, K.J. The Alcohol Sensitivity Questionnaire: Evidence for Construct Validity. *Alcohol. Clin. Exp. Res.* **2016**, *40*, 880–888. [CrossRef]
42. Colzato, L.S.; Warrens, M.J.; Hommel, B. Priming and binding in and across perception and action: A correlational analysis of the internal structure of event files. *Q. J. Exp. Psychol.* **2006**, *59*, 1785–1804. [CrossRef]
43. Hommel, B.; Müsseler, J.; Aschersleben, G.; Prinz, W. The Theory of Event Coding (TEC): A framework for perception and action planning. *Behav. Brain Sci.* **2001**, *24*, 849–937. [CrossRef] [PubMed]

44. Frings, C.; Rothermund, K.; Wentura, D. Distractor repetitions retrieve previous responses to targets. *Q. J. Exp. Psychol.* **2007**, *60*, 1367–1377. [CrossRef] [PubMed]
45. Hommel, B.; Colzato, L. Visual attention and the temporal dynamics of feature integration. *Vis. Cogn.* **2004**, *11*, 483–521. [CrossRef]
46. Petruo, V.A.; Stock, A.-K.; Münchau, A.; Beste, C. A systems neurophysiology approach to voluntary event coding. *NeuroImage* **2016**, *135*, 324–332. [CrossRef]
47. Devenney, L.E.; Coyle, K.B.; Verster, J.C. Memory and attention during an alcohol hangover. *Hum. Psychopharmacol. Clin. Exp.* **2019**, *34*, e2701. [CrossRef]
48. McKinney, A.; Coyle, K.; Penning, R.; Verster, J.C. Next day effects of naturalistic alcohol consumption on tasks of attention. *Hum. Psychopharmacol. Clin. Exp.* **2012**, *27*, 587–594. [CrossRef]
49. Verster, J.C.; van de Loo, A.J.A.E.; Adams, S.; Stock, A.-K.; Benson, S.; Scholey, A.; Alford, C.; Bruce, G. Advantages and Limitations of Naturalistic Study Designs and Their Implementation in Alcohol Hangover Research. *J. Clin. Med.* **2019**, *8*, 2160. [CrossRef]
50. van Lawick van Pabst, A.E.; Devenney, L.E.; Verster, J.C. Sex Differences in the Presence and Severity of Alcohol Hangover Symptoms. *J. Clin. Med.* **2019**, *8*, 867. [CrossRef]
51. Tolstrup, J.S.; Stephens, R.; Grønbaek, M. Does the Severity of Hangovers Decline with Age? Survey of the Incidence of Hangover in Different Age Groups. *Alcohol. Clin. Exp. Res.* **2014**, *38*, 466–470. [CrossRef]
52. Thumin, F.; Wims, E. The Perception of the Common Cold, and other Ailments and Discomforts, as Related to Age. *Int. J. Aging Hum. Dev.* **1975**, *6*, 43–49. [CrossRef]

© 2020 by the authors. Licensee MDPI, Basel, Switzerland. This article is an open access article distributed under the terms and conditions of the Creative Commons Attribution (CC BY) license (http://creativecommons.org/licenses/by/4.0/).

Article

Alcohol Hangover Slightly Impairs Response Selection but not Response Inhibition

Antje Opitz, Jan Hubert, Christian Beste and Ann-Kathrin Stock *

Cognitive Neurophysiology, Department of Child and Adolescent Psychiatry, Faculty of Medicine, TU Dresden, Fetscherstr. 74, 01307 Dresden, Germany
* Correspondence: ann-kathrin.stock@ukdd.de

Received: 31 July 2019; Accepted: 23 August 2019; Published: 27 August 2019

Abstract: Alcohol hangover commonly occurs after an episode of heavy drinking. It has previously been demonstrated that acute high-dose alcohol intoxication reduces cognitive control, while automatic processes remain comparatively unaffected. However, it has remained unclear whether alcohol hangover, as a consequence of binge drinking, modulates the interplay between cognitive control and automaticity in a comparable way. Therefore, the purpose of this study was to investigate the effects of alcohol hangover on controlled versus automatic response selection and inhibition. $N = 34$ healthy young men completed a Simon Nogo task, once sober and once hungover. Hangover symptoms were experimentally induced by a standardized administration of alcoholic drinks (with high congener content) on the night before the hangover appointment. We found no significant hangover effects, which suggests that alcohol hangover did not produce the same functional deficits as an acute high-dose intoxication. Yet still, add-on Bayesian analyses revealed that hangover slightly impaired response selection, but not response inhibition. This pattern of effects cannot be explained with the current knowledge on how ethanol and its metabolite acetaldehyde may modulate response selection and inhibition via the dopaminergic or GABAergic system.

Keywords: alcohol; hangover; cognitive control; automatism; Simon Nogo task; response selection; response inhibition

1. Introduction

Alcohol hangover is an unpleasant state that may occur after an episode of heavy drinking, that is, once the breath/blood alcohol concentration (BAC) returns to 0.0‰. It subsumes several aversive mental and physical symptoms like headaches, vomiting, tiredness, sweating, circulatory problems or depressed mood [1]. It has furthermore been shown to slow psychomotor speed as well as information processing and has been suggested to impair several cognitive functions, including attention, memory, and executive functioning [2–7]. As a consequence, alcohol hangover is associated with impaired workplace productivity and safety [8], as well as reduced driving abilities [4,9]. Given that binge drinking is quite prevalent [10–12], alcohol hangover has been estimated to also be very prevalent and result in huge economic and societal costs due to its debilitating effects [2].

Importantly, regular binge-drinking does not only increase the likelihood and frequency of hangover, it also strongly increases the risk of developing alcohol use disorder (AUD) [13–15]. While it is still unclear whether or how these two consequences are functionally linked, it has been shown that the acute cognitive effects of a high-dose alcohol intoxication resemble the pattern of cognitive deficits observed in AUD patients—both produce pronounced impairments in cognitive control/executive functioning, while behavioural automaticity is comparatively preserved [16–20]. This imbalance between behavioural control and automaticity has repeatedly been shown to play an important role in the development and maintenance of AUD [21] but it has never been investigated whether such specific effects can also be found during alcohol hangover (which should be a regular occurrence in frequent binge drinkers). Demonstrating the persistence of such detrimental effects during hangover (i.e., beyond acute intoxication) would provide an important functional link, which may help to explain why and how regular binge drinking increases the risk of developing AUD [13–15]: If impairments of cognitive control functions persisted beyond acute intoxication, the poor cognitive functioning on the day following alcohol use might promote continued aberrant drinking as well as other decisions that may be detrimental to the overall health of affected individuals. This is especially relevant as acute alcohol intoxication, alcohol hangover and AUD are all characterized by changes in the dopaminergic and GABAergic neurotransmitter systems [22–27]. Specifically, ethanol and its major metabolite acetaldehyde, which has been suggested to strongly contribute to hangover symptoms [28,29], show similar effects on dopaminergic neural transmission by increasing dopaminergic signalling [23,26]. In contrast, ethanol enhances GABAergic signalling [24,25], while acetaldehyde has been suggested to decrease GABAergic signalling [27,30], even though this effect is still debated [31,32]. Overall, both the dopaminergic and GABAergic neurotransmitter systems play a strong modulatory role for cognitive control including response selection [33–37] and might therefore provide a functional link between all three phenomena.

We therefore set out to investigate the effects of alcohol hangover on controlled versus automatic behavior in young healthy males. They were subjected to a previously established, counter-balanced within-subject study design where hangover symptoms are experimentally induced via standardized administration of alcoholic beverages with high congener content [38]. In order to assess differences in controlled versus automatic response selection, we used the Simon Nogo task [39], which is an extended Simon task [40,41], where 30% of the trials require to inhibit all motor responses. The typical Simon effect reflects a stimulus-response (S-R) conflict [41,42]. This conflict is thought to arise between automatic "direct route" processes that promote "unconditionally automatic" responding on the task-irrelevant stimulus side and controlled "indirect route" processing of the correct response based on the "conditional" processing of task-relevant stimulus features [43]. When stimulus and responding hand have the same laterality (S-R congruency), these two kinds of processing indicate the same response. This overlap allows to rely on automatic response selection so that very little control is needed. When stimulus and responding hand have an incongruent laterality, the two processes interfere with each other. As a consequence, increased control is required to overcome the incorrect automatic response tendencies, which ultimately impairs response selection [40,41,43–45]. Aside from this typical Simon effect, opposing effects are observed in case of response inhibition which is typically better in case of S-R incongruence, than in case of S-R congruency [18,39,46]. The reason for this inversion of effects is that, other than response selection, response inhibition is typically worse in case of fast and automatic response tendencies (as compared to slower, more controlled responding) [18,39,47,48]. As a consequence, correct responding requires more top-down control/is more error-prone in incongruent Go trials and congruent Nogo trials (as compared to congruent Go trials and incongruent Nogo trials). Importantly, the detrimental effects of an acute high-dose alcohol intoxication have recently been found to be most pronounced in these conditions, suggesting that top-down control is more severely impaired by alcohol intoxication than automatic processing [18]. As we wanted to test whether the same effects can be found during alcohol hangover, we hypothesized that hangover induces a comparable pattern (i.e., stronger impairments in incongruent Go trials and congruent Nogo trials, as compared to congruent Go trials and incongruent Nogo trials). Given that alcohol hangover might however not necessarily replicate the same data pattern as acute intoxication [38], we decided to conduct Bayesian analyses on non-significant hangover effects in order to substantiate whether the null or alternative hypothesis is more likely, given the obtained data.

2. Experimental Section

2.1. Participants and Sample Size Estimation

In a previous within-subject hangover study conducted by our group [38], the detected hangover effects yielded effect sizes between $\eta_p^2 = 0.26/f = 0.59$ and $\eta_p^2 = 0.39/f = 0.80$. When using those, as well as the correlation of $r = 0.356$ among repeated (sober accuracy) measures obtained in a previous study using the same paradigm [18], for an a priori estimation of required sample size with G*power software [49], we obtained a sample size between $n = 7$ and $n = 4$, when yielding for an alpha error probability of 5% and a power of 95%. The effect sizes for acute high-dose intoxication on this task were between $\eta_p^2 = 0.09/f = 0.31$ and $\eta_p^2 = 0.16/f = 0.44$ [18], which yields samples sizes between $n = 19$ and $n = 11$ in the same a priori estimation. When being a little more conservative and assuming medium effect sizes of $f = 0.25$ [50], the a priori power analysis yielded a required sample size of $n = 30$. In order to compensate for drop-outs and minor issues, we therefore initially recruited $n = 37$ healthy male participants. These were aged 19–28 years and recruited via online advertising, flyers and postings at the local University (TU Dresden) and in the local nightlife district (Dresden Neustadt). All participants were right-handed and had normal or corrected to normal vision. Inclusion criteria and eligibility was assessed during a telephone screening. All included participants reported to have no chronic, somatic, neurological or psychiatric diseases and to not take any medication affecting normal central nervous system (CNS), liver or kidney function. We further assessed the alcohol use disorders identification test (AUDIT) [51]. In order to exclude individuals with a high likelihood of AUD and/or high alcohol tolerance, we excluded all applicants with an overall AUDIT score above 19 [51]. We further excluded all individuals reporting to binge-drink (i.e., consume 8 or more units of alcohol on a single occasion), or have alcohol-induced memory issues, or fail to do things that were normally expected from them "daily or almost daily." In order to exclude light drinkers (who might not be able to cope well with the alcohol amounts administered in this study), we excluded individuals who indicated to binge-drink only once a month or less in the last 12 months. As a consequence, the minimum AUDIT cutoff for inclusion was 2 points. Lastly, individuals who stated to not having been noticeably drunk on at least one occasion in the last 12 months were also excluded. All participants provided written informed consent and were reimbursed with 80 €. The study was approved by the Ethics Committee of the Faculty of Medicine of the TU Dresden and was conducted in accordance of the Declaration of Helsinki.

2.2. Experimental Design and Hangover Provoking Procedure

The study design and experimental intoxication to provoke hangover symptoms followed the protocol used in a previous study of our lab [38], which is illustrated in Figure 1.

In short, each participant was tested twice (once sober and once hungover) with a delay of no less than 48 h and no more than 7 days between sober and hangover appointment. The order of both appointments was counterbalanced between all participants (that is, half of the participants had their sober appointment before their hangover appointment; the other half had their hangover appointment before their sober appointment). Breath alcohol concentration (BAC) was assessed at the start of each appointment. The experiment was not started until participants reached a BAC of 0.00‰. We used the breathalyser "Alcotest 3000" to measure BAC as instructed by the manufacturer (Drägerwerk, Lübeck, Germany). In order to provoke hangover symptoms, we invited 4 to 8 participants to our laboratory on the night before their hangover appointment. These drinking appointments were always scheduled for Friday or Saturday evening (20:00 starting time; ending usually 01:30 to 02:00), while the subsequent hangover appointment was always scheduled for the morning of the following day (i.e., either Saturday or Sunday, starting time between 09:00 and 11:00). This resulted in a slight reduction of sleeping time, which seems to be inversely associated with hangover severity [52].

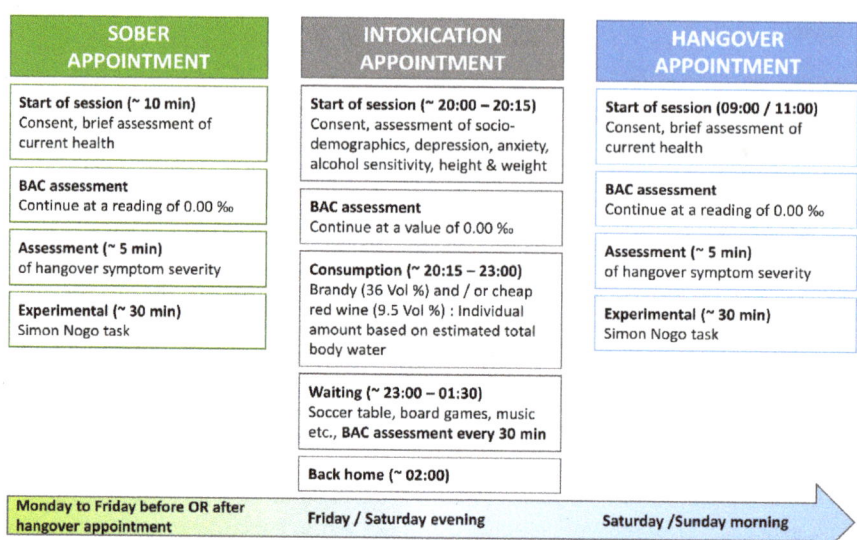

Figure 1. Illustration of sober, intoxication and hangover appointments, as adapted from Zink et al. [38]. Each participant was tested on a sober and a hangover appointment, which were at least 48 h and no more than 7 days apart. The order of both appointments was counterbalanced across the sample. Each session started with providing written consent and a brief assessment of the current health. Data collection (assessment of hangover symptoms and paradigm performance) was not started before participants had reached a blood alcohol concentration (BAC) of 0.00‰ on both appointments. In order to provoke alcohol hangover symptoms, participants were invited to the laboratory on a Friday or Saturday evening at 20:00. After providing their written consent, participants filled out questionnaires assessing sociodemographic data, depression, anxiety, alcohol sensitivity, height and weight. To assure sobriety at the start of drinking, BAC was assessed before alcohol consumption. An individually calculated amount of alcohol was consumed from around 20:15 to 23:00. Participants were asked to stay in the laboratory until 01:30. BAC readings were taken every 30 min starting half an hour after the last sip of alcohol. Eventually, participants were taken home by taxi and returned the next morning between 09:00 and 11:00 for their hangover appointment.

In the night of experimental intoxication, an individual amount of alcohol was determined for each participant using a version of the equation by Widmark [53] and Watson [54]. We aimed to reach BAC values of no more than 1.6‰, by assessing how much alcohol needed to be added to the total body water in order to reach a concentration of 2.0‰. Given that the expected resorption deficit is about 20% on an empty stomach in case all alcohol is consumed at once, this ensures that a BAC of 1.6‰ is very unlikely to be exceeded. Furthermore, the consumption duration was stretched over at least 2 h by the experimenters and all participants were instructed to partake on a full stomach, where the resorptions deficit is usually higher (about 30–40%). Hence, participants were expected to reach a mean BAC of approximately 1.2‰ with a small probability of achieving a BAC beyond 1.6‰. The tool used to determine individual amounts of alcohol and to document alcohol consumption can be found online at https://osf.io/9ykpg/.

Due to the greater likelihood of causing a severe hangover, only alcoholic drinks with a high congener content were offered [55,56]. Therefore, each participant could choose between drinking cheap brandy (36 Vol %) and/or cheap red wine (9.5 Vol %). Both drinks were served by the experimenters in standardized portions of 200 mL red wine (15 g alcohol) or 50 mL brandy (14 g alcohol), so that the speed and amount of alcohol consumption were similar across drinks. Participants could furthermore choose whether they wanted to consume each drink pure, chilled on ice or mixed with caffeine-free coke, ginger ale, or orange lemonade. Tap water and snacks (chips, wine gums) were available at all times and their intake was not controlled or documented. Moreover, participants were permitted to smoke while drinking, because this is assumed to increase hangover symptoms [57,58]. This opportunity was taken by $n = 8$ participants, of which $n = 7$ claimed to be regular smokers. BAC was measured 30, 60, 90 and 120 min after the end of alcohol consumption. Lastly, participants were encouraged to neither use caffeine, guarana nor nicotine within four hours before the start of each appointment.

2.3. Questionnaires

At the beginning of the intoxication appointment (before alcohol administration), participants provided sociodemographic details and filled in the anxiety sensitivity index (ASI) [59] to assess fears of physical symptoms of anxiety itself. As hangover symptoms like nausea, heart racing, shivering or concentration problems are similar to the anxiety symptoms assessed by the ASI, this allowed us to obtain an estimation of how unpleasant physical symptoms of hangover might be for the participant (please note that we did so as there is currently no reliable questionnaire to assess the affective rating of hangover symptoms). Moreover, Beck's depression inventory (BDI) [60] was used to determine depressive symptoms, as these may have a potential influence on cognitive (task) performance. To evaluate sensitivity to common alcohol effects, participants were asked to fill in the alcohol sensitivity questionnaire (ASQ) [61]. The ASQ enables to distinguish between alcohol related experiences referring to lighter drinking (like being more talkative) and alcohol related experiences referring to heavier drinking (like experiencing a hangover). High ASQ scores point to low alcohol sensitivity. At the beginning of each sober and hangover appointment, participants were asked to rate their hangover symptom severity on a Likert-scale ranging from 0 (no symptoms) to 10 (extreme symptoms) consisting of 23 items as used by van Schrojenstein Lantman and colleagues [1]. Of note, we added another item to assess sleeping problems/sleep quality. Finally, participants stated how many hours of sleep they obtained the previous nights, respectively.

2.4. Task

We used a so-called Simon Nogo paradigm in order to assess the effects of alcohol hangover on automatic versus top-down response selection and inhibition [39]. Importantly, we previously employed this paradigm in a study demonstrating that an acute alcohol intoxication of ~1.1‰ had stronger detrimental effects when high levels of top-down control were required for correct responding, as compared to less controlled/more automatic response processing [18]. The paradigm is schematically illustrated in Figure 2.

The task was presented on a 17" CRT monitor, which displayed a central white fixation cross and two lateralized white frame boxes on black background throughout the entire duration of the task. Every trial started with the synchronous 200 ms presentation of a target stimulus (single yellow letter; either "A" or "B") in one of the two boxes and a distractor stimulus (three white horizontal lines) in the respective other box. Participants were instructed to press the left Ctrl button on a regular QWERTZ keyboard with the left index finger in response to the target letter "A" and to press the right Ctrl button with the right index finger in response to the target letter "B." Importantly, participants were further instructed to only respond when the target letters were printed in a regular font (i.e., "A"/"B"; Go condition) but to refrain from all responses whenever the target letters were printed italic and bold (i.e., "*A*"/"*B*"; Nogo condition). The first given response ended the trial. In Nogo trials, any response was coded as a false alarm, while it was either coded as "correct" or "incorrect" in Go trials. If no response was given, the trial was terminated after 1700 ms and either coded as a "miss" (Go condition) or as a "correct omission" (Nogo condition). The inter-trial interval (ITI) was jittered between 1300 and 1700 ms. Overall, the experiment comprised 360 trials, which were subdivided into three equally large blocks. The participants were offered breaks in between the blocks and took approximately 15 min to complete the task.

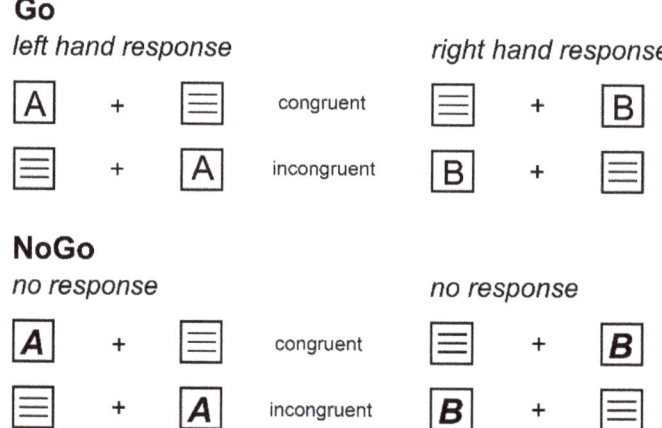

Figure 2. Illustration of the experimental paradigm showing all possible stimulus combinations. On the upper half, the Go condition (70% of all trials) is illustrated. It was indicated by regular letter stimuli. Regardless of stimulus location, stimulus "A" always required a left button press while stimulus "B" always required a right button press. Whenever stimulus location and required response button were located on the same side, the trial was labelled as congruent. Whenever stimulus location and required response button were located on opposite sides, the trial was labelled as incongruent. On the lower half, the Nogo condition (30% of all trials) is illustrated. It was indicated by bold and italic letter stimuli ("*A*" and "*B*"), which required to withhold all motor responses and thus display no response at all.

Of note, each of the two target stimuli occurred equally often on each side and in each condition. Whenever the target stimulus appeared on the side of the correct response (i.e., left "A" and right "B"), we coded the trial as congruent (50% of Go and Nogo trials). When the target stimulus appeared on the opposing side (i.e., right "A" and left "B"), we coded the trial as incongruent (50% of Go and Nogo trials). In summary, each experimental block contained 70% Go trials and 30% Nogo trials and in each of those two conditions, each target letter and congruency rating occurred equally often. The order of trials was separately randomized in each block.

2.5. Statistical Analyses

The behavioural data (accuracy) was analysed using a repeated-measures ANOVA. Hangover status (hangover vs. sober), condition (Go vs. Nogo), and congruency (congruent vs. incongruent) were used as within-subject factors. Reported values of the ANOVA underwent Greenhouse-Geisser correction and parametric post hoc tests were Bonferroni-corrected, whenever necessary. All descriptive statistics were reported using the mean value and the standard error of the mean (SEM) as a measurement of variability. All variables were tested for normal distribution using Kolmogorov-Smirnov tests. As this prerequisite was not given for most accuracy variables, all significant main effects and post hoc tests were additionally tested for significance using non-parametric Wilcoxon signed-rank tests. Lastly, in order to investigate whether the null or alternative hypothesis was more likely in case of non-significant main or interaction effects of hangover status, Bayesian analyses were computed as suggested by Masson [62]. In short, this method returns the likelihood of the H_0 and H_1, given the obtained data. According to Raftery [63], $P(H_i|D)$ values of 50–75% can be regarded as weak evidence for a given hypothesis, values of 75–95% can be regarded as positive evidence for a given hypothesis, values of 95–99% can be regarded as strong evidence for a given hypothesis, and values above 99% can be regarded as very strong evidence for a given hypothesis. Behavioural data and all statistical analyses can be found online at https://osf.io/9ykpg/.

3. Results

3.1. Sample Characteristics

Out of the initially recruited $n = 37$ participants, one participant was excluded from the sample due to a high residual BAC at the start of his hangover appointment (0.45‰), as this would have required 4 to 5 h of waiting time which he could not afford to spend on that day. Another participant was excluded due to technical problems on the sober appointment. A last participant was excluded due to performance accuracy below 60% in Nogo trials on both appointments (his performance deviated from the mean performance of the respective conditions by ≥ 3.57 standard deviations). As a result, $n = 34$ participants entered statistical analyses. Of those, $n = 17$ had their sober appointment before their hangover appointment and $n = 17$ had their hangover appointment before their sober appointment. Sociodemographic characteristics as well as hangover-related data are provided in Table 1. Subjective ratings of sleep and hangover symptoms on both appointments are given in Table 2.

Table 1. Sociodemographic and alcohol-related data of all included participants. All values are given as means ± standard error of the mean (range). Of note, missing alcohol sensitivity questionnaire (ASQ) values (in case a participant indicated never having experienced a given alcohol-associated phenomenon) were not interpolated. All ASQ lighter drinking items as well as almost all ASQ heavier drinking items except for the item "passing out" were averaged, as only one participant reported ever having passed out after heavy drinking.

Characteristic	Included Sample ($n = 34$)
Age in years	23.21 ± 0.48 (19–28)
Height in cm	181.59 ± 0.99 (170–195)
Weight in kg	77.37 ± 1.73 (63–105)
Cigarettes smoked per day	0.80 ± 0.38 (0–10)
Hours of sport per week	4.65 ± 0.58 (0–16)
BDI Score	3.32 ± 0.70 (0–19)
ASI Score	13.34 ± 1.45 (1–33)
AUDIT Score	9.94 ± 0.54 (5–18)
ASQ Score total	8.40 ± 0.41 (3.25–13.29)
ASQ Score of light-drinking	5.27 ± 0.31 (1.63–9.25)
ASQ Score of heavy-drinking	13.43 ± 0.79 (5.20–24.00)
Individual measured alcohol amount of brandy (36 Vol %) in mL	419.71 ± 5.95 (369–516)
Alcohol consumption duration in minutes	182.50 ± 4.33 (111–243)
BAC 30 min after end of consumption	1.31 ± 0.03 (1.05–1.69)
BAC 60 min after end of consumption	1.24 ± 0.02 (1.01–1.56)
BAC 90 min after end of consumption	1.15 ± 0.02 (0.91–1.40)
BAC 120 min after end of consumption	1.07 ± 0.03 (0.83–1.43)

BDI = Beck Depression Inventory, ASI = Anxiety Sensitivity Index, AUDIT = Alcohol Use Disorders Identification Test, ASQ = Alcohol Sensitivity Questionnaire, BAC = Breath Alcohol Concentration.

Table 2. Subjective sleep and hangover symptoms on both appointments. Hangover symptoms were rated on an 11-point Likert scale ranging from 0 (no symptoms) to 10 (extreme symptoms). Of note, participants were asked to truthfully rate the severity of each symptom on both testing days, irrespective of whether they had been drinking the night before the sober appointment. The mild symptom severity for the sober appointment and the resulting minimal variance of this rating may have contributed to the fact that nearly all hangover symptoms differed significantly between the sober and the hangover testing (as was intended by the study). These comparisons were run with uncorrected paired t-tests. p-values are given in the column "Difference." All values are reported as means ± standard error of the mean (range).

Symptom	Sober	Hangover	Difference
Hours of sleep in previous night	7.35 ± 0.15 (5.50–9)	5.58 ± 0.19 (4–8)	$p < 0.001$ **
Overall hangover severity	0 ± 0 (0–0)	3.82 ± 0.42 (0–10)	$p < 0.001$ **
Headache	0.06 ± 0.04 (0–1)	2.68 ± 0.40 (0–8)	$p < 0.001$ **

Table 2. Cont.

Symptom	Sober	Hangover	Difference
Nausea	0.03 ± 0.03 (0–1)	1.71 ± 0.39 (0–7)	$p < 0.001$ **
Concentration problems	0.61 ± 0.18 (0–4)	3.21 ± 0.40 (0–8)	$p < 0.001$ **
Regret	0.12 ± 0.12 (0–4)	1.00 ± 0.36 (0–10)	$p = 0.003$ **
Sleepiness	0.97 ± 0.22 (0–4)	3.91 ± 0.44 (0–9)	$p < 0.001$ **
Heart pounding	0.30 ± 0.10 (0–2)	0.94 ± 0.24 (0–5)	$p = 0.010$ *
Vomiting	0.03 ± 0.03 (0–1)	0.82 ± 0.34 (0–9)	$p < 0.030$ *
Tired	1.09 ± 0.21 (0–4)	4.71 ± 0.40 (1–10)	$p < 0.001$ **
Shivering	0.36 ± 0.13 (0–3)	1.24 ± 0.29 (0–6)	$p = 0.006$ **
Clumsy	0.42 ± 0.15 (0–3)	2.12 ± 0.33 (0–6)	$p < 0.001$ **
Weakness	0.30 ± 0.11 (0–2)	2.50 ± 0.37 (0–10)	$p < 0.001$ **
Dizziness	0.03 ± 0.03 (0–1)	1.88 ± 0.33 (0–8)	$p < 0.001$ **
Apathy	0 ± 0 (0–0)	1.03 ± 0.23 (0–5)	$p < 0.001$ **
Sweating	1.06 ± 0.28 (0–6)	1.24 ± 0.32 (0–9)	$p = 0.555$
Stomach pain	0.24 ± 0.16 (0–5)	1.29 ± 0.38 (0–8)	$p < 0.001$ **
Confusion	0.15 ± 0.08 (0–2)	1.15 ± 0.28 (0–7)	$p = 0.001$ **
Sensitivity to light	0.30 ± 0.13 (0–3)	1.68 ± 0.33 (0–8)	$p < 0.001$ **
Thirst	0.79 ± 0.22 (0–4)	3.44 ± 0.37 (0–8)	$p < 0.001$ **
Heart racing	0.15 ± 0.08 (0–2)	0.65 ± 0.26 (0–6)	$p = 0.084$
Anxiety	0.48 ± 0.16 (0–3)	1.24 ± 0.35 (0–9)	$p = 0.016$ *
Depression	0.21 ± 0.16 (0–5)	0.76 ± 0.29 (0–9)	$p = 0.011$ *
Reduced appetite	0.21 ± 0.16 (0–5)	2.12 ± 0.44 (0–9)	$p < 0.001$ **
Sleeping problems	0.21 ± 0.14 (0–4)	1.00 ± 0.35 (0–8)	$p = 0.018$ *

* $p < 0.05$, ** $p < 0.01$.

3.2. Test-Retest Reliability of the Experimental Paradigm

As we tested the participants on two consecutive appointments, we decided to assess the test-retest reliability of the Simon Nogo paradigm. For this, we compared the first and second appointment of the entire group, while disregarding the experimental manipulation. As appointment order had been balanced across the group, half of the performance data in the first appointment was "sober," while the other half was "hungover." The same was true for the second appointment. We separately determined the test-retest reliability of Go and Nogo trials, as they assess different cognitive domains. Doing so, we found that the test-retest reliability was good in both measures, as all $r \geq 0.803$ and all $p < 0.001$.

3.3. Behavioral Data

The accuracy data are shown in Figure 3. The repeated-measures ANOVA for accuracy revealed a main effect of condition ($F_{(1,33)} = 12.284$; $p = 0.001$; $\eta_p^2 = 0.271$), with higher accuracy in Go (97.0% ± 0.40) than in Nogo trials (93.3% ± 1.03). There was also a main effect of congruency ($F_{(1,33)} = 6.944$; $p = 0.013$; $\eta_p^2 = 0.174$), showing higher accuracy in incongruent (95.7% ± 0.54) than in congruent trials (94.7% ± 0.68). Yet, it should be noted that this effect does not contradict the Simon effect, as these numbers average Go trials (where congruent trials are typically performed better than incongruent trials) and Nogo trials (where congruent trials are typically performed worse than incongruent trials). In addition, there was an interaction of condition x congruency ($F_{(1,33)} = 16.03$; $p < 0.001$; $\eta_p^2 = 0.327$). In line with previous studies of this task [18,39], we found opposing effects of stimulus congruency on Go versus Nogo trials. Specifically, there was a typical Simon effect (congruent > incongruent) in Go trials ($t_{(33)} = 2.134$; $p = 0.040$; cong = 97.4% ± 0.33; incong = 96.6% ± 0.54), which was inverted (congruent < incongruent) in Nogo trials ($t_{(33)} = -3.851$; $p = 0.001$; cong = 91.9% ± 1.30; incong = 94.7% ± 0.84). In line with this, a post-hoc paired t-test showed a positive Simon effect (congruent minus incongruent) in Go trials (0.9% ± 0.40) and negative Simon effect in Nogo trials (−2.8% ± 0.74), which significantly differed from each other ($t_{(33)} = 4.004$; $p < 0.001$). The interaction of hangover status x condition was non-significant ($F_{(1,33)} = 3.478$; $p = 0.071$; $\eta_p^2 = 0.095$) but add-on Bayesian analysis provided positive evidence for the alternative hypothesis, given the obtained data ($P_{BIC}(H_1|D) = 94.3\%$) (please see Table 3 for details) [63]. We therefore decided to conduct post-hoc analyses. Post-hoc paired t-tests showed significantly higher accuracy in sober Go trials (97.4% ± 0.39) than in hangover Go trials (96.6% ± 0.51) ($t_{(33)}$ = 2.053; p = 0.048). No such effect was obtained for Nogo trials ($t_{(33)}$ = −1.125; p = 0.269). Additional correlation analyses showed that hangover Go accuracy was significantly correlated with the hangover effect (sober minus hangover) (r = −0.664, p < 0.001), while sober Go accuracy was not (p = 0.291). This suggests that the size of the hangover effect was mainly determined by changes induced by the hangover status. In order to explore whether these measures showed a performance difference between participants who reported light versus heavy hangover symptoms, we performed a median split of the group. All participants with an overall hangover severity rating of 0–3 on the 11-point Likert scale were classified as "low hangover severity" (n = 19), while all participants with an overall hangover severity rating of 4–10 were classified as "high hangover severity" (n = 15). Mann-Whitney-U tests for independent samples showed that Go accuracy on the hangover day was lower in individuals with high hangover severity (95.50% ± 0.89) than in individuals with low hangover severity (97.43% ± 0.51) (p = 0.040). No such effect was found for the Go accuracy performance difference between sober and hangover day (p = 0.891). All other main and interaction effects of the hangover status were non-significant in the repeated-measures ANOVA (all $F \leq 0.364$; $p \geq 0.550$).

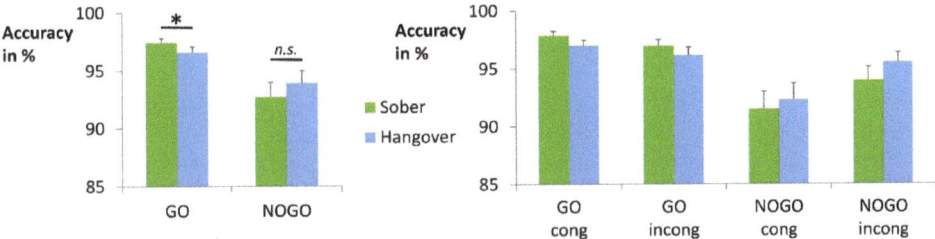

Figure 3. Both graphs visualize the obtained sober vs. hungover accuracy (percentage of correct trials) with error bars depicting the standard error of the mean. Left graph: While we obtained no significant difference between the sober and hangover appointments, Bayesian statistics strongly suggested that there was an interaction between hangover status and condition. As depicted, post hoc paired t-tests detailing this effect showed significantly decreased accuracy in the Go condition, but not in the Nogo condition. Right graph: All factor combinations are depicted. There was a significant interaction between condition and congruency, as Go accuracy was higher in the congruent condition than in the incongruent condition, while Nogo accuracy was higher in the incongruent condition than in the congruent condition. This effect did, however, not vary between sober and hangover appointments.

In order to assess whether there was truly no other effect of hangover status (i.e., whether the null hypothesis H_0 was more likely than the alternative hypothesis H_1), we furthermore ran add-on Bayesian analyses [62] for all main and interaction effects of the hangover status. With the exception of hangover status x condition (on which we already elaborated above), all Bayesian analyses provided positive evidence for the null hypothesis, that is the assumption that hangover status did indeed not modulate the data (please see Table 3 for details).

Table 3. Bayesian analyses for each effect of hangover status.

Effect	BF	$P_{BIC}(H_0|D)$	$P_{BIC}(H_1|D)$
Main effect hangover status	5.56	0.848	0.152
Interaction hangover status x condition	0.06	0.057	0.943
Interaction hangover status x congruency	5.77	0.852	0.148
Interaction hangover status x condition x congruency	5.92	0.855	0.145

BF = Bayes factor, $P_{BIC}(H_0|D)$ = posterior probability for the null hypothesis (i.e., the probability of the null hypothesis, given the obtained data); $P_{BIC}(H_1|D)$ = posterior probability for the alternative hypothesis.

In summary, we found no significant hangover effects on task performance. Yet, Bayesian analyses suggested a slight detrimental effect of hangover status on response selection, as Go accuracy was slightly worse at the hangover appointment than at the sober appointment. This effect was slightly larger for individuals with high hangover severity on the day of hangover testing. Given that no such effect was observed for the Nogo condition, there was no such corresponding effect on response inhibition.

3.4. Add-On Analyses—Alcohol Sensitivity

Previous studies suggested that individuals with lower sensitivity to the effects of alcohol are more hangover-resistant [64] and might show altered cognitive control processes in alcohol-related contexts [65,66]. It could hence be possible that alcohol sensitivity modulated hangover effects in our study. Based on this, we had a closer look at alcohol sensitivity in add-on analyses, as described below.

In total, $n = 2$ participants stated in the ASQ that they had never experienced a hangover before (first item). Yet, only one of these two participants reported not having overall hangover symptoms on the day of the hangover appointment (rating of "0" on the hangover severity scale by van Schrojenstein Lantman et al. [1]). Despite this, he still indicated mild complaints in some of the symptoms that were separately assessed in the list by van Schrojenstein Lantman et al. [1]. Further given that none of the participants rated all of the assessed hangover symptoms as "0" on the hangover day, we concluded that none of them were truly hangover-resistant, as defined by never experiencing any kind of hangover symptoms on the day after heavy drinking. Lastly, we correlated the total ASQ score, the ASQ heavy score and the ASQ hangover item (maximal number of drinks which could be consumed before experiencing hangover) with the overall hangover severity Likert scale rating on the hangover test day. This showed no significant correlations (all uncorrected $p \geq 0.150$).

In order to compare the self-reported number of drinks in the ASQ hangover item with the individually calculated amount of brandy (36 Vol %) administered during the intoxication appointment, we converted the latter into standard drink units (25 mL brandy or 100 mL red wine, which equal 7–7.5 gr of alcohol, were defined as one standard drink). A paired t-test revealed that the administered amount of alcohol exceeded the self-reported hangover threshold, as the number of experimentally administered standard drinks (16.81 drinks ± 0.25) was significantly larger than the maximal number of drinks which could be consumed before developing a hangover (9.09 drinks ± 0.69) ($t_{(31)} = -10.388$; $p < 0.001$).

In order to investigate whether alcohol sensitivity or hangover sensitivity correlated with the achieved intoxication, we correlated the overall ASQ score, the ASQ heavy score and ASQ hangover item with the BAC measured 30/60/90/120 min after the end of consumption on the evening of experimental drinking. None of the uncorrected correlations between ASQ and BAC were significant (all $p \geq 0.192$).

Lastly, we investigated whether alcohol sensitivity or hangover sensitivity correlated with the relevant behavioural parameters. For this, we correlated the overall ASQ score, the ASQ heavy score and ASQ hangover item with the Go accuracy at the sober appointment, the Go accuracy at the hangover appointment and the Go accuracy difference between both appointments. None of the uncorrected correlations between ASQ and performance were significant (all $p \geq 0.068$).

In summary, we found that none of the included participants were entirely resistant to hangover symptoms after high-dose alcohol administration. We further found that alcohol and hangover sensitivity did not significantly correlate with intoxication levels, subjective hangover ratings or performance in hangover-affected cognitive domains. Yet, it should be noted that the lack of such effects could potentially have been caused by the standardized alcohol administration (i.e., participants were not allowed to drink ad libitum) and/or the extremely homogenous study sample.

4. Discussion

Alcohol hangover is an unpleasant state that is reached after an episode of heavy drinking and starts once breath and blood alcohol levels have returned to zero [1]. Aside from symptoms like headaches and gastrointestinal problems [1], a wide range of mild cognitive impairments have been reported, including attention, memory, task-switching, and psychomotor performance deficits [2–5,7,67,68]. Importantly, these domains are also known to be impaired during acute alcohol intoxication [11,17,38,69,70]. It has furthermore been shown that both intoxication and hangover are at least partly characterized by similar neurobiochemical changes, like increased dopaminergic signalling [23,26]. Against this background, cognitive functions that are impaired by acute intoxication might also be impaired during alcohol hangover. Yet, the reports on this are still somewhat inconsistent, with some studies finding comparable effects, while others fail to do so [6,11,20,38,67,71,72]. Given that alcohol has repeatedly been reported to impair controlled behaviour to a larger degree than automatic behaviour [21], we decided to focus on how alcohol hangover impairs controlled versus automatic response selection processes. We experimentally induced a standardized intoxication and investigated whether the subsequent alcohol hangover impaired controlled response selection more severely than automatic response selection, as previously reported for acute alcohol intoxication using a Simon Nogo task [18]. Participants were tested once sober and once hungover (always at a BAC of 0.00‰).

Replicating the typical task effect of the Simon Nogo paradigm and thus further underpinning its validity, we found inverse effects of S-R congruency on response selection and response inhibition during both appointments [18,39,73]: Whenever automatic S-R mapping yielded the correct response, it was beneficial for response selection, but detrimental for response inhibition. Whenever automatic S-R mapping yielded the wrong response, the compensatory controlled S-R mapping was detrimental for response selection, but beneficial for response inhibition. These findings are in line with the assumptions of the dual-process theory [43]—In the context of lateralized stimuli, it distinguishes between "direct route" processes and "indirect route" processes. The theory postulates that the task-irrelevant stimulus lateralization gives rise to direct route processes that result in rather automatic response tendencies on the side of the stimulus. In contrast to that, indirect route processing refers to the controlled selection of the correct response based on task-relevant stimulus features (in our case letter identification) [43]. As direct route processing is faster and more expedient than indirect route processing, it improves response selection whenever there is an overlap between the two processes (i.e., in congruent trials, where both indicate the same response). Hence, response selection is improved by mostly relying on direct route processes in case of congruent stimuli. In case of incongruent stimuli, responses are typically slower and more error-prone as indirect route processing is required to overcome the wrong automatic response tendency. In contrast to response selection (Go trials), response inhibition (Nogo trials) is typically most impaired by fast and automatic response tendencies [18,39,47,48]. As a consequence, inhibition worsens whenever response selection is mainly driven by direct route processing (i.e., in case of congruent trials). In case of stimulus incongruence, the more controlled indirect route processing reduces automatic response tendencies, which supports response inhibition and ultimately improves performance.

In this context, the adverse effects of acute high-dose alcohol intoxication on response selection and inhibition have recently been shown to be most pronounced when high levels of control and effort were needed for correct task performance (i.e., in incongruent Go trials and in congruent Nogo trials) [18]. This suggests that top-down control is more strongly impaired by alcohol intoxication than automatic response selection processing. In terms of our main research question, we expected alcohol hangover to induce a comparable pattern of adverse effects. However, our data showed that the interaction between condition and congruency was not significantly modulated by hangover status. Furthermore, add-on Bayesian analysis provided positive evidence for rejecting the assumption that alcohol hangover alters the interplay between controlled and automatic processes in a similar fashion as intoxication. As both acute alcohol intoxication and AUD show similar impairment patterns (i.e., strongly impaired control functions and relatively preserved behavioural automaticity) [16–20], we had initially hypothesized that alcohol hangover might be characterized by similar changes, thus providing a functional link that helps to explain why and how regular binge drinking increases the risk of developing AUD [13–15]. Specifically, we had reasoned that cognitive control impairments which persist beyond intoxication, might contribute to poor behavioural choices, thus (partly) explaining the increased AUD risk of regular binge-drinkers. Given that we obtained positive evidence for the absence of such effects for congruency on its own, as well as the interaction of congruency and condition, the hypothesis of comparable effects between intoxication and hangover needs to be rejected. With respect to our initial hypothesis, this suggests that premorbid deficits in cognitive control, especially in the domain of inhibition, might contribute more to the observed link between binge-drinking and AUD, than the effects of repeated alcohol hangover. While our data does not provide concise information on why we did not find this effect, it seems conceivable that our participants were either not impaired, or healthy enough to compensate detrimental effects of hangover onto cognition (e.g., by increasing their effort during task performance).

Even though top-down cognitive control does hence not seem to be impaired by alcohol hangover, we found evidence suggesting that response selection might be slightly impaired. While we only observed a non-significant trend for this effect, Bayesian analyses strongly suggested that alcohol hangover seems to impair response selection in Go trials, but not response inhibition in Nogo trials. Importantly, this was observed regardless of S-R mapping. While these findings are not in line with our initial hypotheses, they match previous reports that alcohol hangover reduces information processing efficiency during response selection and increases the threshold for response execution in a number categorization task [5]. Importantly, the observed detrimental effect of hangover on response selection was slightly larger in individuals with high hangover severity, as compared to individuals with low hangover severity. Taken together, these findings might indicate a compensatory strategy for perceived deficits during hangover, where participants are more cautious and hesitant to respond. This also matches our descriptive data (please see Figure 3): Although we did not find significant hangover effects in the Nogo condition, alcohol hangover appears to have slightly improved response inhibition, rather than impairing it. The neurobiological basis of this, however, remains rather unclear, as we did not directly measure correlates of neurotransmission or neurotransmitter levels. It has been suggested that acetaldehyde, which is likely present during alcohol hangover, increases dopaminergic signalling but decreases GABAergic signalling [27]. But as dopamine and GABA have beneficial effects on both response selection and response inhibition [35–37,74,75], this does however not help to explain the finding of slightly impaired selection and relatively preserved inhibition.

There are two limitations pertaining to sex and age. We refrained from recruiting females as the ethical approval for this was denied. On average, females tend to have a slower metabolism of alcohol, which might lead to greater overall impairments [11]. Furthermore, heavy drinking has been shown to impair response inhibition more strongly in females than in males [76]. For these reasons, females could be likely to show greater deficits in cognitive performance during alcohol hangover. We also did not recruit adults over the age of 30 in order to have a homogeneous young and healthy study sample. Alcohol hangover frequency and the quality of hangover symptoms seem to change with age [77,78]. Hence, further studies considering both sexes as well as middle-aged to older adults are needed to investigate whether the hangover effects we observed may be generalized. Furthermore, we did not explicitly record participants' history of binge drinking beyond the last 12 months, which were assessed with the AUDIT. We also did not assess details on hangover history beyond the ASQ. Assessing this data in more detail might have helped to explain differences in hangover severity and cognitive performance during hangover. As we pre-selected individuals according to their drinking habits, we however had a very homogenous sample, which likely showed only minimal variation with respect to this factor. Lastly, we omitted placebo administration, as it would have been rather easy to distinguish between placebo and alcoholic drinks, given the large amounts we served. In line with this, a recent study demonstrated no expectancy effects on cognitive performance during alcohol hangover by briefing one group of participants with the correct study purpose [79].

5. Conclusions

In summary, we investigated whether and how alcohol hangover modulates controlled versus automatic response selection and behaviour. For this, we used a Simon Nogo task that has previously been used to demonstrate stronger intoxication-related impairments in controlled behaviour than in automatic behaviour. The lack of significant hangover effects shows that alcohol hangover did not produce the same functional impairments as an acute high-dose intoxication. Yet, add-on Bayesian analyses suggested that alcohol hangover slightly impairs response selection but not response inhibition. These analyses also showed that alcohol hangover does most likely not alter controlled versus automatic response selection processes. As the pattern of effects is thus not comparable with intoxication or AUD, cognitive alcohol hangover symptoms cannot help to explain why or how regular binge drinking increases the risk of developing AUD.

Author Contributions: Conceptualization, A.O., C.B. and A.-K.S.; methodology, A.O., C.B. and A.-K.S.; validation, A.O., J.H., C.B. and A.-K.S.; formal analysis, A.O., J.H. and A.-K.S.; investigation, A.O., J.H. and A.-K.S.; resources, C.B. and A.-K.S.; data curation, A.O., J.H. and A.-K.S.; writing—original draft preparation, A.O., J.H. and A.-K.S.; writing—review and editing, all authors; visualization, A.O. and A.-K.S.; supervision, C.B. and A.-K.S.; project administration, A.-K.S.; funding acquisition, C.B. and A.-K.S.

Funding: This research was funded by the Deutsche Forschungsgemeinschaft (DFG), grant number TRR 265 B07.

Acknowledgments: The authors thank all individuals who took part in this study. Open Access Funding by the Publication Funds of the TU Dresden.

Conflicts of Interest: The authors declare no conflict of interest. The funders had no role in the design of the study; in the collection, analyses or interpretation of data; in the writing of the manuscript or in the decision to publish the results.

References

1. Van Schrojenstein Lantman, M.; JAE van de Loo, A.; Mackus, M.; Verster, J.C. Development of a Definition for the Alcohol Hangover: Consumer Descriptions and Expert Consensus. *Curr. Drug Abuse Rev.* **2017**, *9*, 148–154. [CrossRef]
2. Stephens, R.; Ling, J.; Heffernan, T.M.; Heather, N.; Jones, K. A review of the literature on the cognitive effects of alcohol hangover. *Alcohol Alcohol.* **2008**, *43*, 163–170. [CrossRef] [PubMed]
3. Ling, J.; Stephens, R.; Heffernan, T.M. Cognitive and psychomotor performance during alcohol hangover. *Curr. Drug Abuse Rev.* **2010**, *3*, 80–87. [CrossRef] [PubMed]
4. Gunn, C.; Mackus, M.; Griffin, C.; Munafò, M.R.; Adams, S. A systematic review of the next-day effects of heavy alcohol consumption on cognitive performance. *Addiction* **2018**, *113*, 2182–2193. [CrossRef] [PubMed]
5. Grange, J.A.; Stephens, R.; Jones, K.; Owen, L. The effect of alcohol hangover on choice response time. *J. Psychopharmacol.* **2016**, *30*, 654–661. [CrossRef] [PubMed]
6. Wolff, N.; Gussek, P.; Stock, A.-K.; Beste, C. Effects of high-dose ethanol intoxication and hangover on cognitive flexibility. *Addict. Biol.* **2018**, *23*, 503–514. [CrossRef] [PubMed]
7. Stock, A.-K.; Hoffmann, S.; Beste, C. Effects of binge drinking and hangover on response selection sub-processes-a study using EEG and drift diffusion modeling: Binge-drinking and hangover. *Addict. Biol.* **2017**, *22*, 1355–1365. [CrossRef] [PubMed]
8. Bush, D.M.; Lipari, R.N. Workplace Policies and Programs Concerning Alcohol and Drug Use. In *The CBHSQ Report*; Substance Abuse and Mental Health Services Administration: Rockville, MD, USA, 2013.
9. Verster, J.C.; Bervoets, A.C.; de Klerk, S.; Vreman, R.A.; Olivier, B.; Roth, T.; Brookhuis, K.A. Effects of alcohol hangover on simulated highway driving performance. *Psychopharmacology* **2014**, *231*, 2999–3008. [CrossRef]
10. Jernigan, D.H. *Global Status Report: Alcohol and Young People*; World Health Organization: Geneva, Switzerland, 2001.
11. Montgomery, C.; Fisk, J.E.; Murphy, P.N.; Ryland, I.; Hilton, J. The effects of heavy social drinking on executive function: A systematic review and meta-analytic study of existing literature and new empirical findings. *Hum. Psychopharmacol.* **2012**, *27*, 187–199. [CrossRef]
12. WHO. Global Status Report on Alcohol and Health 2018. Available online: http://www.who.int/substance_abuse/publications/global_alcohol_report/en/ (accessed on 27 August 2019).

13. Courtney, K.E.; Polich, J. Binge drinking in young adults: Data, definitions and determinants. *Psychol. Bull.* **2009**, *135*, 142–156. [CrossRef]
14. Knight, J.R.; Wechsler, H.; Kuo, M.; Seibring, M.; Weitzman, E.R.; Schuckit, M.A. Alcohol abuse and dependence among US college students. *J. Stud. Alcohol* **2002**, *63*, 263–270. [CrossRef] [PubMed]
15. Shnitko, T.A.; Gonzales, S.W.; Grant, K.A. Low cognitive flexibility as a risk for heavy alcohol drinking in non-human primates. *Alcohol* **2018**, *74*, 95–104. [CrossRef] [PubMed]
16. Bjork, J.M.; Gilman, J.M. The effects of acute alcohol administration on the human brain: Insights from neuroimaging. *Neuropharmacology* **2014**, *84*, 101–110. [CrossRef] [PubMed]
17. Field, M.; Schoenmakers, T.; Wiers, R.W. Cognitive processes in alcohol binges: A review and research agenda. *Curr. Drug Abuse Rev.* **2008**, *1*, 263–279. [CrossRef] [PubMed]
18. Chmielewski, W.X.; Zink, N.; Chmielewski, K.Y.; Beste, C.; Stock, A.-K. How high-dose alcohol intoxication affects the interplay of automatic and controlled processes. *Addict. Biol.* **2018**. [CrossRef] [PubMed]
19. Stock, A.-K.; Schulz, T.; Lenhardt, M.; Blaszkewicz, M.; Beste, C. High-dose alcohol intoxication differentially modulates cognitive subprocesses involved in response inhibition. *Addict. Biol.* **2014**, *21*, 136–145. [CrossRef] [PubMed]
20. Stock, A.-K.; Riegler, L.; Chmielewski, W.X.; Beste, C. Paradox effects of binge drinking on response inhibition processes depending on mental workload. *Arch. Toxicol.* **2015**, *90*, 1429–1436. [CrossRef] [PubMed]
21. Stock, A.-K. Barking up the Wrong Tree: Why and How We May Need to Revise Alcohol Addiction Therapy. *Front. Psychol.* **2017**, *8*, 884. [CrossRef]
22. Chastain, G. Alcohol, neurotransmitter systems and behavior. *J. Gen. Psychol.* **2006**, *133*, 329–335. [CrossRef]
23. Di Chiara, G. Alcohol and dopamine. *Alcohol Health Res. World* **1997**, *21*, 108–114.
24. Iversen, L.L.; Iversen, S.D.; Bloom, F.E.; Roth, R.H. *Introduction to Neuropsychopharmacology*; Oxford University Press: New York, NY, USA, 2009; ISBN 978-0-19-538053-8.
25. Kumar, S.; Porcu, P.; Werner, D.F.; Matthews, D.B.; Diaz-Granados, J.L.; Helfand, R.S.; Morrow, A.L. The role of GABA(A) receptors in the acute and chronic effects of ethanol: A decade of progress. *Psychopharmacology* **2009**, *205*, 529–564. [CrossRef] [PubMed]
26. Melis, M.; Diana, M.; Enrico, P.; Marinelli, M.; Brodie, M.S. Ethanol and acetaldehyde action on central dopamine systems: Mechanisms, modulation and relationship to stress. *Alcohol* **2009**, *43*, 531–539. [CrossRef] [PubMed]
27. Martí-Prats, L.; Sánchez-Catalán, M.J.; Orrico, A.; Zornoza, T.; Polache, A.; Granero, L. Opposite motor responses elicited by ethanol in the posterior VTA: The role of acetaldehyde and the non-metabolized fraction of ethanol. *Neuropharmacology* **2013**, *72*, 204–214. [CrossRef] [PubMed]
28. Prat, G.; Adan, A.; Sánchez-Turet, M. Alcohol hangover: A critical review of explanatory factors. *Hum. Psychopharmacol. Clin. Exp.* **2009**, *24*, 259–267. [CrossRef] [PubMed]
29. Wiese, J.G.; Shlipak, M.G.; Browner, W.S. The alcohol hangover. *Ann. Intern. Med.* **2000**, *132*, 897–902. [CrossRef] [PubMed]
30. Kuriyama, K.; Ohkuma, S.; Taguchi, J.; Hashimoto, T. Alcohol, acetaldehyde and salsolinol-induced alterations in functions of cerebral GABA/benzodiazepine receptor complex. *Physiol. Behav.* **1987**, *40*, 393–399. [CrossRef]
31. Correa, M.; Salamone, J.D.; Segovia, K.N.; Pardo, M.; Longoni, R.; Spina, L.; Peana, A.T.; Vinci, S.; Acquas, E. Piecing together the puzzle of acetaldehyde as a neuroactive agent. *Neurosci. Biobehav. Rev.* **2012**, *36*, 404–430. [CrossRef]
32. Quertemont, E.; Grant, K.A.; Correa, M.; Arizzi, M.N.; Salamone, J.D.; Tambour, S.; Aragon, C.M.G.; McBride, W.J.; Rodd, Z.A.; Goldstein, A.; et al. The role of acetaldehyde in the central effects of ethanol. *Alcohol. Clin. Exp. Res.* **2005**, *29*, 221–234. [CrossRef]
33. Bar-Gad, I.; Morris, G.; Bergman, H. Information processing, dimensionality reduction and reinforcement learning in the basal ganglia. *Prog. Neurobiol.* **2003**, *71*, 439–473. [CrossRef]
34. Plenz, D. When inhibition goes incognito: Feedback interaction between spiny projection neurons in striatal function. *Trends Neurosci.* **2003**, *26*, 436–443. [CrossRef]
35. Willemssen, R.; Falkenstein, M.; Schwarz, M.; Müller, T.; Beste, C. Effects of aging, Parkinson's disease and dopaminergic medication on response selection and control. *Neurobiol. Aging* **2011**, *32*, 327–335. [CrossRef] [PubMed]

36. Plessow, F.; Fischer, R.; Volkmann, J.; Schubert, T. Subthalamic deep brain stimulation restores automatic response activation and increases susceptibility to impulsive behavior in patients with Parkinson's disease. *Brain Cogn.* **2014**, *87*, 16–21. [CrossRef] [PubMed]
37. De la Vega, A.; Brown, M.S.; Snyder, H.R.; Singel, D.; Munakata, Y.; Banich, M.T. Individual differences in the balance of GABA to glutamate in pFC predict the ability to select among competing options. *J. Cogn. Neurosci.* **2014**, *26*, 2490–2502. [CrossRef] [PubMed]
38. Zink, N.; Bensmann, W.; Beste, C.; Stock, A.-K. Alcohol Hangover Increases Conflict Load via Faster Processing of Subliminal Information. *Front. Hum. Neurosci.* **2018**, *12*, 316. [CrossRef] [PubMed]
39. Chmielewski, W.X.; Beste, C. Testing interactive effects of automatic and conflict control processes during response inhibition—A system neurophysiological study. *Neuroimage* **2017**, *146*, 1149–1156. [CrossRef] [PubMed]
40. Simon, J.R. The effects of an irrelevant directional cue on human information processing. In *Stimulus-Response Compatibility: An Integrated Perspective*; Proctor, R.W., Reeve, T.G., Eds.; Elsevier: Amsterdam, The Netherlands, 1990; pp. 31–86.
41. Leuthold, H. The Simon effect in cognitive electrophysiology: A short review. *Acta Psychol.* **2011**, *136*, 203–211. [CrossRef] [PubMed]
42. Hommel, B. The Simon effect as tool and heuristic. *Acta Psychol.* **2011**, *136*, 189–202. [CrossRef] [PubMed]
43. De Jong, R.; Liang, C.C.; Lauber, E. Conditional and unconditional automaticity: A dual-process model of effects of spatial stimulus-response correspondence. *J. Exp. Psychol. Hum. Percept. Perform.* **1994**, *20*, 731–750. [CrossRef]
44. Kornblum, S.; Hasbroucq, T.; Osman, A. Dimensional overlap: Cognitive basis for stimulus-response compatibility—A model and taxonomy. *Psychol. Rev.* **1990**, *97*, 253–270. [CrossRef] [PubMed]
45. Keye, D.; Wilhelm, O.; Oberauer, K.; Stürmer, B. Individual differences in response conflict adaptations. *Front. Psychol.* **2013**, *4*, 947. [CrossRef]
46. Chmielewski, W.X.; Mückschel, M.; Beste, C. Response selection codes in neurophysiological data predict conjoint effects of controlled and automatic processes during response inhibition. *Hum. Brain Mapp.* **2018**, *39*, 1839–1849. [CrossRef] [PubMed]
47. Donkers, F.C.L.; van Boxtel, G.J.M. The N2 in go/no-go tasks reflects conflict monitoring not response inhibition. *Brain Cogn.* **2004**, *56*, 165–176. [CrossRef] [PubMed]
48. Dippel, G.; Chmielewski, W.; Mückschel, M.; Beste, C. Response mode-dependent differences in neurofunctional networks during response inhibition: An EEG-beamforming study. *Brain Struct. Funct.* **2016**, *221*, 4091–4101. [CrossRef] [PubMed]
49. Faul, F.; Erdfelder, E.; Buchner, A.; Lang, A.-G. Statistical power analyses using G*Power 3.1: Tests for correlation and regression analyses. *Behav. Res. Methods* **2009**, *41*, 1149–1160. [CrossRef] [PubMed]
50. Cohen, J. *Statistical Power Analysis for the Behavioral Sciences*, 2nd ed.; reprint; Psychology Press: New York, NY, USA, 2009; ISBN 978-0-8058-0283-2.
51. Babor, T.F.; Higgins-Biddle, J.C.; Saunders, J.B.; Monteiro, M.G. *The Alcohol Use Disorders Identification Test Guidelines for Use in Primary Care*, 2nd ed.; World Health Organization: Geneva, Switzerland, 2001.
52. Van Schrojenstein Lantman, M.; Mackus, M.; Roth, T.; Verster, J. Total sleep time, alcohol consumption and the duration and severity of alcohol hangover. *Nat. Sci. Sleep* **2017**, *9*, 181–186. [CrossRef] [PubMed]
53. Widmark, E.M.P. *Die Theoretischen Grundlagen Und Die Praktische Verwendbarkeit Der Gerichtlich-Medizinischen Alkoholbestimmung*; Urban & Schwarzenberg: Berlin, Germany, 1932.
54. Watson, P.E.; Watson, I.D.; Batt, R.D. Total body water volumes for adult males and females estimated from simple anthropometric measurements. *Am. J. Clin. Nutr.* **1980**, *33*, 27–39. [CrossRef] [PubMed]
55. Rohsenow, D.J.; Howland, J.; Arnedt, J.T.; Almeida, A.B.; Greece, J.; Minsky, S.; Kempler, C.S.; Sales, S. Intoxication With Bourbon Versus Vodka: Effects on Hangover, Sleep and Next-Day Neurocognitive Performance in Young Adults. *Alcohol. Clin. Exp. Res.* **2010**, *34*, 509–518. [CrossRef]
56. Verster, J.C. The alcohol hangover-a puzzling phenomenon. *Alcohol Alcohol.* **2008**, *43*, 124–126. [CrossRef]
57. Epler, A.J.; Tomko, R.L.; Piasecki, T.M.; Wood, P.K.; Sher, K.J.; Shiffman, S.; Heath, A.C. Does Hangover Influence the Time to Next Drink? An Investigation Using Ecological Momentary Assessment. *Alcohol. Clin. Exp. Res.* **2014**, *38*, 1461–1469. [CrossRef]
58. Jackson, K.M.; Rohsenow, D.J.; Piasecki, T.M.; Howland, J.; Richardson, A.E. Role of Tobacco Smoking in Hangover Symptoms Among University Students. *J. Stud. Alcohol Drugs* **2013**, *74*, 41–49. [CrossRef]

59. Reiss, S.; Peterson, R.A.; Gursky, D.M.; McNally, R.J. Anxiety sensitivity, anxiety frequency and the prediction of fearfulness. *Behav. Res. Ther.* **1986**, *24*, 1–8. [CrossRef]
60. Beck, A.T. An Inventory for Measuring Depression. *Arch. Gen. Psychiatry* **1961**, *4*, 561. [CrossRef] [PubMed]
61. Fleming, K.A.; Bartholow, B.D.; Hilgard, J.; McCarthy, D.M.; O'Neill, S.E.; Steinley, D.; Sher, K.J. The Alcohol Sensitivity Questionnaire: Evidence for Construct Validity. *Alcohol. Clin. Exp. Res.* **2016**, *40*, 880–888. [CrossRef] [PubMed]
62. Masson, M.E.J. A tutorial on a practical Bayesian alternative to null-hypothesis significance testing. *Behav. Res.* **2011**, *43*, 679–690. [CrossRef] [PubMed]
63. Raftery, A.E. Bayesian Model Selection in Social Research. *Sociol. Methodol.* **1995**, *25*, 111. [CrossRef]
64. Piasecki, T.M.; Alley, K.J.; Slutske, W.S.; Wood, P.K.; Sher, K.J.; Shiffman, S.; Heath, A.C. Low sensitivity to alcohol: Relations with hangover occurrence and susceptibility in an ecological momentary assessment investigation. *J. Stud. Alcohol Drugs* **2012**, *73*, 925–932. [CrossRef] [PubMed]
65. Bailey, K.; Bartholow, B.D. Alcohol words elicit reactive cognitive control in low-sensitivity drinkers. *Psychophysiology* **2016**, *53*, 1751–1759. [CrossRef]
66. Fleming, K.A.; Bartholow, B.D. Alcohol cues, approach bias and inhibitory control: Applying a dual process model of addiction to alcohol sensitivity. *Psychol. Addict. Behav.* **2014**, *28*, 85–96. [CrossRef]
67. Devenney, L.E.; Coyle, K.B.; Verster, J.C. Cognitive performance and mood after a normal night of drinking: A naturalistic alcohol hangover study in a non-student sample. *Addict. Behav. Rep.* **2019**, *10*, 100197. [CrossRef]
68. Scholey, A.; Benson, S.; Kaufman, J.; Terpstra, C.; Ayre, E.; Verster, J.; Allen, C.; Devilly, G. Effects of Alcohol Hangover on Cognitive Performance: Findings from a Field/Internet Mixed Methodology Study. *J. Clin. Med.* **2019**, *8*, 440. [CrossRef]
69. Harrison, E.L.R.; Fillmore, M.T. Alcohol and distraction interact to impair driving performance. *Drug Alcohol Depend.* **2011**, *117*, 31–37. [CrossRef] [PubMed]
70. Brumback, T.; Cao, D.; McNamara, P.; King, A. Alcohol-induced performance impairment: A 5-year re-examination study in heavy and light drinkers. *Psychopharmacology* **2017**, *234*, 1749–1759. [CrossRef] [PubMed]
71. Devenney, L.E.; Coyle, K.B.; Verster, J.C. Memory and attention during an alcohol hangover. *Hum. Psychopharmacol. Clin. Exp.* **2019**. [CrossRef] [PubMed]
72. Finnigan, F.; Hammersley, R.; Cooper, T. An examination of next-day hangover effects after a 100 mg/100 mL dose of alcohol in heavy social drinkers. *Addiction* **1998**, *93*, 1829–1838. [CrossRef] [PubMed]
73. Wolff, N.; Chmielewski, W.; Buse, J.; Roessner, V.; Beste, C. Paradoxical response inhibition advantages in adolescent obsessive compulsive disorder result from the interplay of automatic and controlled processes. *Neuroimage Clin.* **2019**, *23*, 101893. [CrossRef] [PubMed]
74. Quetscher, C.; Yildiz, A.; Dharmadhikari, S.; Glaubitz, B.; Schmidt-Wilcke, T.; Dydak, U.; Beste, C. Striatal GABA-MRS predicts response inhibition performance and its cortical electrophysiological correlates. *Brain Struct. Funct.* **2015**, *220*, 3555–3564. [CrossRef]
75. Ramdani, C.; Carbonnell, L.; Vidal, F.; Béranger, C.; Dagher, A.; Hasbroucq, T. Dopamine precursors depletion impairs impulse control in healthy volunteers. *Psychopharmacology* **2015**, *232*, 477–487. [CrossRef]
76. Nederkoorn, C.; Baltus, M.; Guerrieri, R.; Wiers, R.W. Heavy drinking is associated with deficient response inhibition in women but not in men. *Pharmacol. Biochem. Behav.* **2009**, *93*, 331–336. [CrossRef] [PubMed]
77. Tolstrup, J.S.; Stephens, R.; Grønbaek, M. Does the severity of hangovers decline with age? Survey of the incidence of hangover in different age groups. *Alcohol. Clin. Exp. Res.* **2014**, *38*, 466–470. [CrossRef]
78. Thumin, F.; Wims, E. The Perception of the Common Cold and other Ailments and Discomforts, as Related to Age. *Int. J. Aging Hum. Dev.* **1975**, *6*, 43–49. [CrossRef]
79. Devenney, L.E.; Coyle, K.B.; Verster, J.C. The impact of expectancy on cognitive performance during alcohol hangover. *BMC Res. Notes* **2018**, *11*, 730. [CrossRef] [PubMed]

© 2019 by the authors. Licensee MDPI, Basel, Switzerland. This article is an open access article distributed under the terms and conditions of the Creative Commons Attribution (CC BY) license (http://creativecommons.org/licenses/by/4.0/).

Article

Alcohol Hangover Does Not Alter the Application of Model-Based and Model-Free Learning Strategies

Julia Berghäuser [1], Wiebke Bensmann [2], Nicolas Zink [2], Tanja Endrass [1], Christian Beste [2] and Ann-Kathrin Stock [2,*]

1. Chair of Addiction Research, Institute for Clinical Psychology and Psychotherapy, Faculty of Psychology TU Dresden, Chemnitzer Str. 46, 01062 Dresden, Germany; Julia.berghaeuser@tu-dresden.de (J.B.); tanja.endrass@tu-dresden.de (T.E.)
2. Cognitive Neurophysiology, Department of Child and Adolescent Psychiatry, Faculty of Medicine, TU Dresden, Fetscherstr. 74, 01307 Dresden, Germany; Wiebke.Bensmann@ukdd.de (W.B.); Nicolas.Zink@ukdd.de (N.Z.); Christian.Beste@ukdd.de (C.B.)
* Correspondence: Ann-Kathrin.Stock@ukdd.de

Received: 2 April 2020; Accepted: 7 May 2020; Published: 13 May 2020

Abstract: Frequent alcohol binges shift behavior from goal-directed to habitual processing modes. This shift in reward-associated learning strategies plays a key role in the development and maintenance of alcohol use disorders and seems to persist during (early stages of) sobriety in at-risk drinkers. Yet still, it has remained unclear whether this phenomenon might be associated with alcohol hangover and thus also be found in social drinkers. In an experimental crossover design, $n = 25$ healthy young male participants performed a two-step decision-making task once sober and once hungover (i.e., when reaching sobriety after consuming 2.6 g of alcohol per estimated liter of total body water). This task allows the separation of effortful model-based and computationally less demanding model-free learning strategies. The experimental induction of alcohol hangover was successful, but we found no significant hangover effects on model-based and model-free learning scores, the balance between model-free and model-based valuation (ω), or perseveration tendencies (π). Bayesian analyses provided positive evidence for the null hypothesis for all measures except π (anecdotal evidence for the null hypothesis). Taken together, alcohol hangover, which results from a single binge drinking episode, does not impair the application of effortful and computationally costly model-based learning strategies and/or increase model-free learning strategies. This supports the notion that the behavioral deficits observed in at-risk drinkers are most likely not caused by the immediate aftereffects of individual binge drinking events.

Keywords: alcohol; cognitive effort; decision making; hangover; model-based; model-free

1. Introduction

Alcohol is a widely used, and often abused, substance that may cause a number of different adverse effects during acute intoxication, but also thereafter [1]. Especially after the consumption of larger-than-usual doses, there is a high risk of developing alcohol hangover [2], which is defined as the "the combination of negative mental and physical symptoms which can be experienced after a single episode of alcohol consumption, starting when blood alcohol concentration (BAC) approaches zero" [3]. Symptoms that are commonly reported during hangover include nausea and vomiting, headaches and stomach pains, clumsiness and weakness, tiredness and sleepiness, depressive symptoms and apathy, dizziness and confusion, as well as concentration problems [4]. Based on such recurring subjective reports and in line with studies postulating reduced workplace productivity and safety during alcohol hangover [5], it is often readily assumed that various physiological and cognitive functions are impaired during hangover. Yet, this seemingly apparent conclusion has become challenged by several studies showing that not all functional domains appear to be (equally) impaired [6]. On the physiological level, for example, it has been reported that hangover reduces performance in athletics [7] and in military contexts [8], but there are also contradictory findings suggesting that holiday activities like hiking performance do not seem to be objectively impaired in hungover individuals (even though study participants reported greater subjective exhaustion) [9]. Likewise, there are repeated reports of impaired cognition in the domains of attention and memory [6,10], which can however not be found in all studies and tasks investigating these phenomena [6,11,12]. Despite such heterogeneous findings and despite the fact that most of the tested functional domains require investing voluntary effort, which is considered to be (potentially) straining, it has never been systematically investigated whether alcohol hangover might actually reduce the ability and/or willingness to invest cognitive effort, rather than the general ability to perform a given task. Yet, this could help to explain the observed heterogeneity of effects, like why physical impairments have been reported in working contexts [7,8], but not necessarily in recreational activities [9]. Beyond this, improved knowledge about alcohol effects on engagement of effortful cognitive processes could also help to better understand phenomena like hangover-related increases in workplace absenteeism [13], or under which circumstances hungover individuals might still be able to compensate deficits by means of increased effort [14,15].

When investigating alcohol effects on the investment of cognitive effort, one can make use of the fact that behavior may be generated by relying on different strategies that vary in how much voluntary effort and control they require. This is all the more important, as both acute and chronic effects of aberrant alcohol consumption seem to strongly impair performance in tasks that require effortful cognitive top-down control, while performance is rather unaltered in tasks that require substantially less effortful automatic processes [16–20]. However, many of these studies typically make the participants perform both hard and easy tasks, thus confounding the findings with the factor of task difficulty (which should not be confused with effort). More importantly, the tasks used in these studies did typically not provide the participants with the possibility to choose a strategy for themselves, or to arbitrate between more and less straining strategies. Investigating the arbitration between effortful top-down controlled "model-based" behavior and less demanding "model-free" behavior is not only of scientific interest, but also of clinical relevance: A better understanding of the mechanisms underlying intra- and inter-individual differences can elucidate the behavioral and psychological changes that have been associated with problematic drinking patterns like binge drinking [21,22] and shown to drive and maintain alcohol use disorders (AUDs) [16,23–25].

In the framework of reinforcement learning, model-based and model-free learning can be distinguished from each other as two classes of methods. Model-based learning uses an internal model of the environment and enables us to take appropriate actions through planning, which is based on that model and on the expected outcomes of the available choice options. While this is computationally demanding, model-based learning can quickly incorporate and adjust to changes in environmental structures or in outcomes and is thus associated with adaptive and flexible (goal-directed) behavior [26]. In contrast to this, a model-free strategy does not use a model of the world. Instead, model-free behavior uses prediction errors to learn the (outcome) values of the available choice options. Those values are stored in scalar quantities and can be easily accessed so that model-free learning is computationally cheap. The downside of this strategy is that changes in the environment or in outcomes can only slowly be incorporated in the values of the choice options through trial-end-error learning, which makes model-free learning less adaptive [26,27].

The arbitration between more or less demanding cognitive-behavioral strategies can be assessed with the Markov decision task (also called two-step task), which was specifically designed to disentangle model-based and model-free learning strategies [28]. This task requires participants to make two successive decisions, which lead to an outcome (differently sized gains or losses) in the end of each trial. The outcomes change throughout the course of the task, which necessitates constant updating (learning). Crucially, the first-level decision leads to one of two second-level states, and therefore to different associated choice options, with certain probabilities. This transition structure of the task can be used for the model-based learning strategy or can be neglected in case of model-free learning, which allows to computationally distinguish the two strategies [27,28].

We applied the two-step task and subsequent parameter modeling to $n = 25$ healthy young men, who participated in a within-subject experimental design [11] where they were tested once sober and once hungover (i.e., after a night of experimentally induced drinking). Our hypotheses were based on the findings that AUD patients and heavy binge drinkers (BD) have previously been shown to demonstrate significant reductions in effortful controlled model-based cognitive strategies, thus inducing an imbalance between model-free and model-based behavior (as compared to healthy controls) [21,25]. While there is broad consensus that AUD patients shift from model-based to model-free behavior [23], it should however be noted that this pattern could not be completely observed in all studies investigating the phenomenon. For example, Voon et al. did not find differences between abstinent AUD patients and healthy matched controls [29], while Sebold et al. could not reproduce their initial findings of selectively impaired goal-directed functions [25] in a larger AUD sample [30]. Furthermore supporting our hypotheses that alcohol might shift the balance between model-based and model-free behavior, it has been demonstrated that the BD-associated imbalance in favor of rather effortless model-free behavior seems to normalize as the time since the last binging event increases [21]. Lastly, increased perseveration tendencies (i.e., reduced cognitive flexibility) have been observed in BD, but not in case of abstinent AUD patients [21,29,31]. Thus, increased perseveration tendencies might also potentially be found during alcohol hangover. Therefore, we hypothesized that alcohol hangover could induce qualitatively similar effects, albeit probably to a lesser degree. The investigation of AUD patients and BDs alone does not allow for any conclusions about whether the shift from model-based to model-free behavior observed in these groups reflects premorbid deficits. Yet, the lack of such "premorbid" changes in control participants with a positive family history of AUD [32] as well as in otherwise healthy, young BDs [33] suggests that this might not be the case and that this behavioral shift is rather a consequence of excessive alcohol consumption. Against the background that habitual binge drinking might induce a shift from model-based to model-free behavior that can still be observed after the end of an acute binge-like intoxication [21], we hence hypothesized that this was also the case during the hangover following a single binge drinking episode.

2. Materials and Methods

2.1. Participants

Healthy young men aged 18–30 were recruited via flyers and online ads at the local university (TU Dresden). In order to be included in the sample, all participants underwent an extended telephone screening, during which their somatic, neurological, and psychiatric well-being, as well as their alcohol consumption were assessed with the help of a semi-structured interview by experienced neuropsychologists. They had to report to have normal or corrected-to-normal vision, be free of psychiatric and neurologic disorders, as well as somatic diseases (especially those affecting the gastrointestinal tract, liver, and kidneys). Likewise, they had to report not taking any medication or illicit drugs either regularly, or during their participation in the study (including a sufficient number of preceding days in case the metabolism of a given substance took more than 18 h). With respect to alcohol consumption habits, we required all included participants to have scores between 2 and 19 points in the Alcohol Use Disorder Identification Test (AUDIT) [34]. Additionally, they were required to have voluntarily engaged in binge drinking (defined as consuming 8 or more standard units of alcohol on a single evening) between 13 and 150 times in the past year and to recall at least one event within the past year when they were markedly drunk (defined as experiencing alcohol-induced gait, motor, or speech impediments). Individuals who had less than 2 points in the AUDIT or drank less than these lower limits were excluded in order to minimize the risk of including participants who might become unwell after drinking the alcohol dose we experimentally administered to induce intoxication and subsequent hangover. We further excluded individuals who had more than 19 points in the AUDIT (as scores of 20 points or more "clearly warrant further diagnostic evaluation for alcohol dependence" [34]), drank more than our pre-defined upper limits (as binge drinking on 3 or more days a week shows that binge-like alcohol consumption is no longer limited to social drinking on weekends), and/or reported having at least weekly alcohol-induced memory problems and/or at least near-daily failures to fulfill routine tasks that were expected of them (as this would have indicated a high and likely clinically relevant degree of alcohol-related cognitive dysfunction). In sum, these upper thresholds were implemented to minimize the likelihood of including individuals with strong alcohol tolerance and a high risk for AUD. The study was approved for males only by the ethics committee of the Faculty of Medicine of the TU Dresden, Germany (EK293082014). All participants provided written informed consent at the start of each study appointment while (still) sober. They received a compensation of 80€ for study participation.

There were no previous studies investigating the size of hangover effects in Markov decision tasks, but studies on other cognitive control domains reported effect sizes between $f = 0.32$ and $f = 0.6$ for their reported hangover effects in comparable within-subject study designs [11,17,35]. Based on this, we estimated the required sample size for two repeated measures sessions (sober vs. hangover) and five relevant measures (MF-score, MB-score, final score, ω, π) at an alpha error probability of 5% and a power of 95% for an estimated medium effect size of $f = 0.30$ (assuming a default inter-correlation of 0.5). This yielded a required sample size of $n = 23$. Based on this initial sample size estimation, $n = 25$ subjects matching all of the criteria detailed above were eventually included in the sample and underwent experimental testing as well as statistical analyses. Please note that the sample used in this publication strongly overlaps with that of a previous publication, which investigated alcohol hangover effects on attentional processes during varying conflict loads in a prime and flanker context [11].

2.2. Experimental Design

Importantly, we used the same study design as already reported in our previous publication [11]. In short, each participant was invited to the lab for three different appointments the order of which was balanced across the sample so that half of the participants first performed the paradigm sober and then hungover, while the other half first performed the paradigm hungover and then sober. Participants could not start with any of their appointments unless they were entirely sober at the start of each appointment. The required breath alcohol concentration (BrAC) of 0.00‰ was controlled using the breathalyzer "Alcotest 3000" following the instructions by the manufacturer (Drägerwerk, Lübeck, Germany). Participants were further required to refrain from using legal stimulants like coffee, taurine, or guarana in the three hours preceding each appointment and to eat a full dinner before participating in the intoxication appointment.

On both the sober and the hangover appointment, the participants rated their subjective hangover symptoms on a Likert scale (see Section 2.3 for details) and then performed a total of four conceptually unrelated behavioral tasks. The results of two of these tasks have been previously been published [11], and the results of the third task, which assessed mental rotation and response inhibition, have not been published or submitted anywhere, as of yet. The task reported in this study was always conducted last (i.e., approximately 60–75 min after the start of the appointment). The sober appointments were conducted on weekdays and between 2 and 7 days apart from the hangover appointments, which were always conducted on Saturday or Sunday (starting time between 09:00 and 11:00) after a previous night of experimentally induced alcohol intoxication. These intoxication appointments took place on Friday or Saturday, starting at 20:00. For each intoxication appointment, we invited between 2 and 6 subjects to the lab. They were asked to fill in a sociodemographic questionnaire and then consume an individually determined amount of 2.6375 g of alcohol per estimated liter of total body water (TBW), which was determined with an equation by Widmark [36] and Watson et al. [37]. The details of equation as well as the protocols used to document drinking can be found in our previous publication [11] and in the data sheet provided in the Supplementary Materials. In line with recommendations from previous experimental studies [2,38–41], the administered amount of alcohol was expected to result in a mean peak intoxication of ~1.2 ‰ on the full stomach we asked participants to have (i.e., at a resorption deficit of ~40%), and no more than 1.6‰ on an empty stomach, which we asked participants to avoid (i.e., at a resorption deficit of ~20%). Due to the ratio of TBW and administered alcohol, it was physically impossible to exceed a peak intoxication of 2.0‰ (i.e., at a hypothetical resorption deficit of 0%). Additionally, the experimenters did not issue more than half of a participant's drinks within the first hour of drinking so that participants were kept from consuming the entire amount at once (consumption typically took 2–3 h). Participants got their drinks from the experimenters and could choose whether they wanted 200 mL red wine (9.5 Vol % equaling 15 g of alcohol) or 50 mL brandy (36 Vol % equaling 14 g of alcohol) with each refill. These two drinks were chosen for their comparatively high congener content, which is thought to increase hangover severity (as compared to beverages with lower congener content, like vodka or white wine) [42–44]. Drinks could be mixed with caffeine-free softdrinks (coke, orange lemonade, ginger ale) and ice cubes. Participants were further provided with unlimited access to snacks (chips and wine gum) and tap water, the consumption of which were not monitored. They were furthermore allowed to smoke. Participants were free to socially interact, listen to music, play board and card games, or table soccer during the intoxication appointment. 30, 60, 90, and 120 min after the individual end of their consumption, participants were asked to provide BrAC measurements. They were then sent home via taxi around 1:30 to 02:00 in the morning (given decreasing BrAC values and no clouded awareness and/or major motor impairments). They were invited to come back the following day at either 09:00 or 10:30 for their hangover appointment. This was done for two reasons: Firstly, we wanted to test the participants as soon as possible after reaching the sobriety criterion of 0.00‰ because hangover-associated cognitive deficits of social drinkers might be most pronounced at this time point [45] (if they failed to reach this criterion at the originally scheduled time, they were asked to wait until BrAC had returned to 0.00‰). Secondly, it has been recommended to standardize sleep time in experimental hangover induction [41,46] as reduced sleeping time could be associated with more severe hangover symptoms (although reduced sleep time and quality are of course also directly associated with alcohol intoxication itself) [42,44,47–49]. Yet still, alcohol effects on sleep do not seem to necessarily mediate hangover effects on cognitive performance [40,44]. Lastly, it should be noted that while we experimentally standardized sleeping times across the sample for optimal comparability across participants, both the time at which participants could go to bed and the estimated average sleeping time were oriented towards normal behavior in young healthy social drinkers, as previously reported in a study with a naturalistic study design (in that study, average drinking started between 20:06 and 21:06, average bedtimes were between 02:49 and 03:18 am, and the average sleep duration was between 05:36 and 05:58 h) [47].

2.3. Questionnaires

At the start of the intoxication session and before alcohol administration, subjects provided sociodemographic information. At the beginning of both the sober and hangover session, participants were asked to rate the subjective severity of 22 hangover symptoms suggested by van Schrojenstein Lantman et al. [4,50] on an 11-point Likert-scale ranging from 0 (no symptoms) to 10 (extreme symptoms). Importantly, participants were asked to truthfully rate the severity of each symptom irrespective of whether they had consumed alcohol the night before or attributed their symptoms to alcohol consumption. Furthermore, subjects reported the hours of sleep during the previous night.

2.4. Two-Step Decision-Making Task

In order to investigate whether alcohol hangover reduced cognitively effortful model-based behavior and/or increased the less costly model-free behavior, we used a modified two-step decision-making task based on Daw et al. [28] and Kool et al. [27], which was embedded in a space game. Each trial consisted of two sequential decisions that led to a final outcome. As can be seen in Figure 1, two different spaceships were presented in the beginning of each trial to represent the choice options at first stage. The spaceships were associated with a transition probability of 80% (common transition) to reach one of two planets, and a transition probability of 20% (rare transition) to reach the other planet. These planets indicated the second stage options. At the second stage, two new choice options were presented in the form of different aliens. The subjects were told that the aliens mine in "space mines" where they could find either treasures (representing positive outcomes), or antimatter (negative outcomes), or nothing. The outcomes for each of the four second-level choice options slowly changed throughout the task. Therefore, the value of each option had to be constantly updated. The outcomes ranged from −4 to +5 points and magnitudes were slowly drifting according to a Gaussian random walk: The outcomes for each of the four options at second stage were calculated independently so that they ranged from 0 to 1 and slowly changed with a drift rate of 0.2. The resulting scores were then transformed into points. The transition distribution and reward distribution were the same for all subjects. Those distributions were simulated beforehand in order to ensure that model-based engagement would lead to higher final scores (for details, please see section "Simulation of Transition and Reward Distribution" in the Supplementary Materials). In other words, this made sure that the more costly model-based strategy always paid off more than the model-free strategy. At the end of each trial, a bar was presented to indicate the current total score. We modified the original two-step task by Daw et al. [28] in several ways based on simulation results of Kool et al. [27] in order to allow for a stronger relationship between model-based learning strategy and reward payoff: Firstly, we used a simpler, more distinguishable transition probability of 80:20 (instead of the original 70:30) to reduce rare transition trials, which was intended to reduce the trade-off between pay-off and cognitive costs. Secondly, we increased the drift rates of second stage outcomes and used a broader range of reward probabilities (Gaussian random walk: M = 0, SD = 0.20, reflecting boundaries = [0 1] vs. originally: Gaussian random walk: M = 0, SD = 0.025, reflecting boundaries = [0.25 0.75]) to induce faster changes of rewards, which was intended to reduce the possibility of easy adaptations of model-free learning and thereby increase the relative advantage of model-based learning. Thirdly, we used points instead of binary probabilistic outcomes to increase the information gain of each trial and thus reduce the necessity to integrate information over several outcomes per choice option. Taken together, these modifications should have resulted in a higher pay-off for the more cognitive costly model-based strategy. The task consisted of 250 trials, which were divided into two equally sized blocks. The main goal of the subjects was to collect as much treasure (points) as possible. The screen position (left or right) of the two choice options was randomized across trials for stimuli at both stages. If no response was made via button press on a standard keyboard within the 2 s response limit, participants received a penalty loss of 5 points and the trial was repeated. Prior to the main task, subjects received detailed instructions and tutorials, including 25 practice trials to familiarize them with the task. The task was presented with Presentation software (Neurobehavioral Systems Inc., Berkeley, CA, USA). Trial timing is illustrated in Figure 1a. We used different planet and alien stimulus sets at the two appointments in order to minimize carry-over effects between the sober and the hungover appointment.

Importantly, the task allows the detection and dissociation of model-free vs. model-based decision-making behavior. This becomes especially apparent after rare transitions that ultimately lead to high rewards: In those cases, an entirely model-free agent would repeat the choices that resulted in this reward (i.e., the agent would choose the same action/spaceship again) in accordance with basic reinforcement principles, which state that the probability to choose an option again is higher when this option was previously rewarded. In contrast, a model-based agent would take into account the model of the task, i.e., the knowledge of the transition probabilities between stages. In that case, the probability to choose the same action again would be much lower, because the valuation system would take into account that the other option has a much higher probability to lead to the promising second stage option. Therefore, a model-based agent would likely switch the first stage choice under circumstances of high rewards after rare transitions.

Following this logic, first stage choice behavior can be utilized to determine and distinguish signatures of model-based and model-free learning. For that purpose, stay probabilities can be computed, i.e., the probability to choose the same first stage option again, as a function of previous outcome (win or loss) and transition type (common or rare). Those probabilities can be used to calculate a model-free score (MF-score) and a model-based score (MB-score) for each subject in order to analyze the reliance on the respective learning systems [25]. The MF-score indicates the pure influence of previous reward on the first stage choice pattern:

$$\text{MF-score} = (\text{Stay}_{\text{win common}} + \text{Stay}_{\text{win rare}}) - (\text{Stay}_{\text{loss common}} + \text{Stay}_{\text{loss rare}}).$$

In contrast, the MB-score reflects the interaction effect of previous reward and transition type on stay probability, and thus the consideration of the model of the task for first stage choices:

$$\text{MB-score} = (\text{Stay}_{\text{win common}} + \text{Stay}_{\text{loss rare}}) - (\text{Stay}_{\text{win rare}} + \text{Stay}_{\text{loss common}}).$$

Both scores consider choice behavior with regard to the previous trial, but ignore performance throughout the whole task, which can be provided by computational modeling accounts.

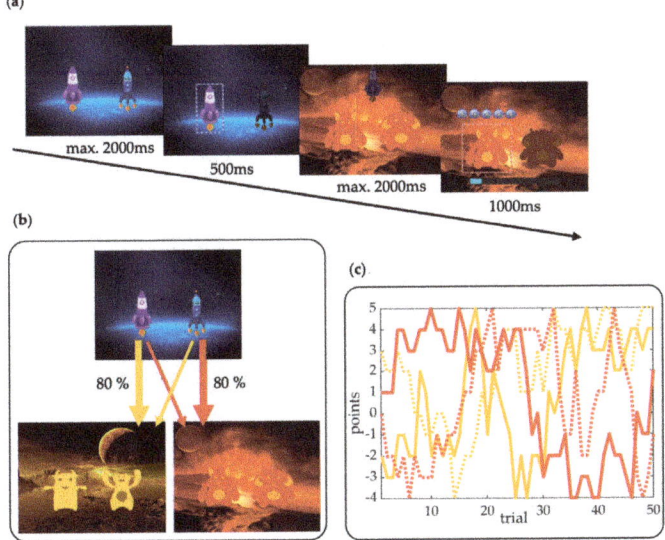

Figure 1. Two-step decision-making task. (**a**) An exemplary trial sequence as well as the trial timing are depicted: At the first stage, two spaceships were presented. Participants indicated their response choice via a button press, followed by a transition two the second stage. Two aliens represented second-stage choice options and participants made their second response choice via another button press. Response choices were indicated by boxes around the respective spaceship/alien and trial outcomes are indicated by blue spheres (space treasure) representing the number of gained points (+5 shown) or pink spheres (antimatter) representing the number of lost points (not shown). The response time limit was 2 s for each of the two choices. According to the transition structure (**b**), a transition could either be common (80% probability) or rare (20% probability). After the second stage response, the outcome was presented. (**c**) The outcomes (+5 to −4 points) of the four choice options are presented for the first 50 trials. Each line represents a second-stage choice option (alien) for the yellow and red planet, respectively.

The dual-system reinforcement-learning model is an established computational model for the task we used. It assumes a mixture of model-based and model-free learning strategies [27,28]: During the course of the task, expected values (Q-values) will be learned for each choice option (*a*) in each state (*s*) at the two stages (*i*). The model-free value (Q_{MF}) is updated at each trial (*t*) according to a state-action-reward-state-action, or SARSA(λ) temporal difference learning algorithm [27,51]. After each action, an update takes place to calculate a new estimate of the value of the chosen option based on the agent's experience. The general updating rule is:

$$Q_{MF}(s,a) = Q_{MF}(s,a) + \alpha \delta_{i,t} \tag{1}$$

where α denotes the learning rate and δ is the reward prediction error:

$$\delta_{i,t} = r_{i,t} + Q_{MF}(s_{i+1,t}, a_{i+1,t}) - Q_{MF}(s_{i,t}, a_{i,t}) \tag{2}$$

with r denoting the received reward. The learning rate determines to which extent the new information provided by the reward prediction error is incorporated in the value estimate. At first stage, the reward prediction error is solely driven by the Q-value of the option that is chosen at second stage, since no reward is delivered at the first stage:

$$\delta_{1,t} = Q_{MF}(s_{2,t}, a_{2,t}) - Q_{MF}(s_{1,t}, a_{1,t}). \tag{3}$$

At the second stage, the reward prediction error is driven by the received reward, since no third stage is available:

$$\delta_{2,t} = r_{2,t} - Q_{MF}(s_{2,t}, a_{2,t}). \tag{4}$$

The Q-values for both stages are updated at the end of each trial. For the update of the first stage model-free Q-value, a decay-rate parameter for eligibility traces (λ) is used to additionally down-weight the second stage prediction error:

$$Q_{MF}(s_{1,t}, a_{1,t}) = Q_{MF}(s_{1,t}, a_{1,t}) + \alpha \lambda \delta_{2,t}. \tag{5}$$

In contrast to this, the model-based strategy for the first stage update considers the transition probability P (model of the environment) between stages and combines this knowledge with the values of second stage options. It is assumed that the transition probability is fixed and known to the agent:

$$Q_{MB}(s_{1,t}, a_j) = P(s_{2A}|s_1, a_j) \max Q_{MF}(s_{2A,t}, a) + P(s_{2B}|s_1, a_j) \max Q_{MF}(s_{2B,t}, a) \tag{6}$$

where j denotes the index of the first stage choice options and s_{2A} and s_{2B} the two different states at the second stage. At the second stage, the updating rule for values is the same as for the model-free strategy.

To select an action at the first stage, the model-free and model-based Q-values are combined and weighted by the parameter ω:

$$Q_{net}(s_1, a_j) = \omega Q_{MB}(s_1, a_j) + (1 - \omega) Q_{MF}(s_1, a_j). \tag{7}$$

A low weighting parameter ($\omega < 0.5$) indicates a stronger reliance on the model-free strategy, whereas high values ($\omega > 0.5$) indicate a stronger influence of the model-based strategy. At second stage, both learning strategies use the model-free Q-value for action selection.

The probability to choose an action at each stage is computed according to a sofmax rule:

$$P(a_{i,t} = a|s_{i,t}) = \frac{exp(\beta[Q_{net}(s_{i,t}, a) + \pi \cdot rep(a)])}{\sum_{a'} exp(\beta[Q_{net}(s_{i,t}, a') + \pi \cdot rep(a')])} \tag{8}$$

where the inverse temperature β determines the stochasticity of the choices. Higher β values indicate that the agent is more likely to choose the action with the highest Q-value (i.e., high expected outcome) and lower β values indicate a tendency towards random choice (i.e., that the agent's decisions are less determined by this learning strategy). Additionally, a choice "stickiness" parameter π was included, which was multiplied with an indicator variable rep(a) that indicates whether the same action was chosen again, or not. This parameter indicates perseveration ($\pi > 0$) or switching ($\pi < 0$) tendency. Lastly, optimal choice rates were separately calculated for each stage [52]. These rates reflect whether decisions were made in favor of the option with the higher Q-value or not, and thus reflect the probability of choosing the optimal option. The model fitting was conducted with Matlab 2018b

(The MathWorks, Inc., Natick, MA, USA), with empirical priors using Sam Gershman's mfit toolbox to find the maximum a posteriori parameter estimates [27,53].

In summary, the MF- and MB-scores represent different influences of simple reinforcement learning vs. effortful goal-directed computation for action selection which is based on the experience of the previous trial. Whereas the computational model considers choice behavior over the course of the whole task, the weighting parameter omega (ω) indicates the relative contribution of model-free and model-based strategies to decision-making and thus the extent of cognitive investment. The choice stickiness parameter π indicates the arbitration between behavioral perseveration and switching. Lastly, β values indicate to what degree the participant is likely to choose the response that is associated with the highest expected outcome. Together with the optimal choice rate, these variables reflect whether decisions were made in favor of the option with the higher Q-value. Finally, the sum of all collected outcomes (final score) and reaction times for choices at first and second stage may be used to compare the overall performance.

2.5. Statistical Analyses

To compare task performance between the sober and hangover session, we used the Bayesian procedure for related samples provided by SPSS Statistics 25 (IBM Corp., Armonk, NY, USA), which computes a traditional (paired samples) *t*-test and the Bayes Factor (BF). For this, we used default settings (Adaptive Gauss-Lobatto Quadrature approach, Tolerance = 0.000001, maximum iterations = 2000) with a noninformative prior (diffuse prior distribution). To check the normality assumption, we used the Shapiro-Wilk-Tests and conducted additional non-parametric tests, whenever necessary.

The BF indicates the ratio of the data likelihood given the null hypothesis versus the data likelihood given the alternative hypothesis: A value above one indicates (more) relative evidence for the null hypothesis whereas values below one indicate (more) relative evidence for the alternative hypothesis. Values above three are considered as positive evidence for the null hypothesis, i.e., no difference between the sober and hangover session [54].

Since we were mainly interested in hangover-associated differences in model-based and model-free learning, we analyzed MB-scores and MF-scores, which were calculated for each subject and session, as well as the weighting parameter ω. To analyze perseveration tendencies, we focused on the choice stickiness parameter π. All other parameters provided by the computation model were analyzed in an exploratory fashion. The Bayesian information criterion (BIC) was used to verify comparable model fit between sessions. To further examine whether the participants had based their decisions on hybrid Q-value estimation to a similar degree in both of their sessions, we analyzed their optimal choice rates for each stage [52]. Finally, we used the sum of all collected outcomes (final score) and reaction times for choosing at first and second stage to compare the overall performance.

Given that we balanced the order of the two appointments across participants, used two different task versions/stimuli on the first and second appointment, and further randomized stimulus positions on the screen for each trial, we did not anticipate any confounding effects of appointment order. For this reason, appointment order was not included as a factor in any of the analyses presented in the results section but add-on analyses of this factor can be found in the section "Investigation of Hypothetical Task Order Effects" of the Supplementary Materials.

The raw behavioral data as well as the analyzed data (including the syntax) can be accessed at https://osf.io/vzpn3/.

3. Results

3.1. Sample Characteristics and Intoxication Procedure

The included participants were on average 21.5 years old (SD = 2.3; range 18–27), 183.4 cm tall (SD = 7.0; range 167–198), and weighed 80.5 kg (SD = 11.6; range 56.5–96.5). This resulted in an average individual alcohol amount of 432.0 mL brandy (SD = 39.01; range 349–497) at 36 Vol. %. Participants took on average 174.4 min (SD = 28.3; range 115–230) to consume the alcohol. The mean BrAC was

1.17 ‰ (SD = 0.23; range 0.75–1.63) 30 min after the end of consumption, 1.09‰ (SD = 0.22; range 0.65–1.51) 60 min after the end of consumption, 1.04 ‰ (SD = 0.17; range 0.65–1.41) 90 min after the end of consumption, and 0.94 ‰ (SD = 0.14; range 0.67–1.16) 120 min after the end of consumption.

As would have been expected from the study design, participants reported a shorter average sleep duration in hangover session (mean = 6.05 h; SD = 0.83; range 4.50–8.00) than in the sober session (mean = 8.10 h; SD = 1.39; range 5.50–10.00). Hence, our participants slept approximately two hours less before the hungover appointment than before the sober appointment. Of note, this is very similar to the hangover-associated 1 h and 50 min sleep reduction reported in a previous, naturalistic study by Hogewoning et al. (where hungover participants had slept 7 h and 26 min on sober nights and 5 h and 36 min on hungover nights) [47]. Given that none of the task-relevant behavioral and estimated measures worsened during hangover (for details, please refer to the following text sections), there was however no need to control for the shorter sleeping time before the hangover session.

Based on the recruitment criterion that all participants had to have some degree of binge drinking experience in order to minimize the risk of severe adverse effects during alcohol administration, the mean AUDIT score of the sample was 10.1 points (SD = 2.8; range 4–16). Out of the $n = 25$ participants, $n = 19$ had scores between 8 and 15 points, which has been linked to hazardous alcohol use that does however not require clinical intervention [34]. Only $n = 1$ participant had a score of 16, which is the lower boundary for "brief counseling and continued monitoring" recommended by WHO guidelines [34]. Yet, none of the participants obtained a score of 20 or higher and none of the participants met the criteria for the diagnosis of an AUD according to the International Classification of Diseases (ICD-10). The subjective ratings for overall hangover severity and the severity of individual hangover symptoms are presented in Table 1.

Table 1. Symptom severity ratings on both appointments.

Item	Sober	Hungover	p
Overall hangover severity	0.167 ± 0.637	3.640 ± 2.119	<0.001
Regret	0.000 ± 0.000	0.440 ± 1.261	0.039
Headache	0.240 ± 0.831	2.600 ± 2.769	0.001
Sensitivity to light	0.040 ± 0.200	1.680 ± 2.076	0.001
Concentration problems	0.440 ± 0.961	3.640 ± 2.464	<0.001
Clumsy	0.080 ± 0.400	2.120 ± 1.716	<0.001
Confusion	0.000 ± 0.000	1.120 ± 1.166	0.001
Dizziness	0.040 ± 0.200	2.400 ± 2.380	<0.001
Anxiety	0.080 ± 0.277	0.560 ± 0.961	0.020
Depression	0.000 ± 0.000	0.640 ± 1.497	0.008
Apathy	0.120 ± 0.440	1.400 ± 1.780	0.004
Stomach pain	0.120 ± 0.440	0.480 ± 1.229	0.129
Nausea	0.160 ± 0.800	1.520 ± 1.828	0.001
Vomiting	0.040 ± 0.200	0.800 ± 1.384	0.011
Reduced appetite	0.240 ± 1012	1.440 ± 2.022	0.021
Thirst	0.440 ± 1083	3.840 ± 2.267	<0.001
Heart pounding	0.160 ± 0.554	1.280 ± 1.768	0.003
Heart racing	0.000 ± 0.000	0.400 ± 0.707	0.015
Shivering	0.080 ± 0.400	1.083 ± 1.176	0.001
Weakness	0.040 ± 0.200	2.480 ± 2.084	<0.001
Sweating	0.080 ± 0.277	0.920 ± 1.552	0.004
Tired	0.560 ± 1083	4.080 ± 2.448	<0.001
Sleepiness	0.440 ± 0.961	3.680 ± 2.410	<0.001
Sleeping problems	0.120 ± 0.332	0.720 ± 1.242	0.036

Average ± SD rating of each symptom on a Likert-scale ranging from 0 (no symptoms) to 10 (extreme symptoms), as suggested by van Schrojenstein Lantman et al. [4,50]. Participants had been asked to rate each item on both appointments, irrespective of whether or not they had consumed alcohol the night before the sober appointment and also irrespective of whether they attributed a given complaint to alcohol hangover. Whenever the average rating was greater than zero on both appointments, the appointments were compared using paired Wilcoxon signed-rank tests. Whenever all of the ratings in the sober session were zero, the hungover appointment was compared to zero using one sample Wilcoxon signed-rank tests. Uncorrected p-values of the conducted tests are given in the right column.

3.2. Two-Step Decision-Making Task

Descriptive statistics are shown in Table 2 and Figure 2 shows the first stage choice behavior for the sober and hangover session.

Table 2. Descriptive task statistics for the two-step decision-making task for the sober and hangover session.

	Mean	SEM	SD	Min	Max
sober					
MF-score	0.08	0.04	0.18	−0.25	0.39
MB-score	0.50	0.07	0.33	−0.10	1.10
Final score	357.12	24.90	124.50	66	564
First stage RT	491	28	141	136	728
Second stage RT	585	27	136	198	886
hangover					
MF-score	0.06	0.04	0.18	−0.28	0.35
MB-score	0.49	0.05	0.25	<−0.01 *	0.98
Final score	361.48	19.34	96.71	180	521
First stage RT	489	24	121	204	617
Second stage RT	597	12	62	505	762

MF-score: model-free score; MB-score: model-based score; final score: accumulated outcomes at the end of the task (in points); RT: reaction time in msec. * The true value lies between −0.01 and 0.00.

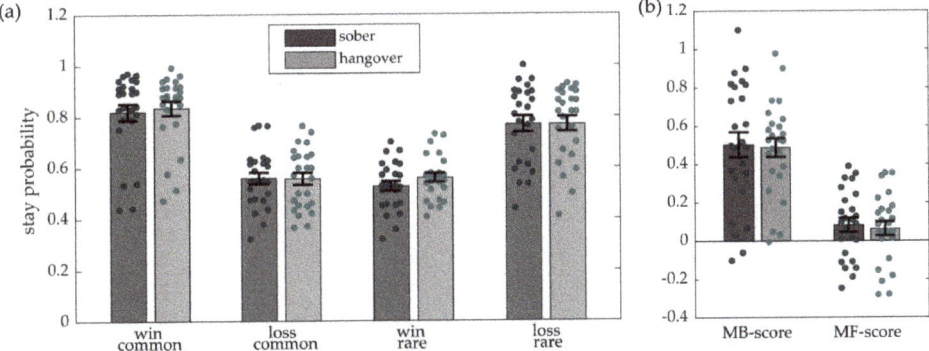

Figure 2. First stage choice behavior. Dots indicate values of individual participants and bars indicate group means with error bars depicting the standard error of the mean. (**a**) Stay probability (choosing the same options as in the previous trial) for win and loss trials as a function of transition (common vs. rare). The sober session is depicted in dark grey and the hangover session is depicted in light grey. (**b**) Model-based score (MB-score), reflecting the interaction between outcome and transition type, and model-free score (MF-score), reflecting the main effect of outcome, for the sober session (dark grey) and the hangover session (light grey).

There was no significant difference between the sober and hangover session with respect to either MB-score ($t_{(24)} = 0.38$, $p = 0.80$) or MF-score ($t_{(24)} = 0.38$, $p = 0.71$). Bayesian analyses indicated positive evidence in favor of the null hypothesis, i.e., the assumption that the MB-score (BF = 6.30) and the MF-score (BF = 6.05) did not differ between the sober and hangover session. This suggests that the degree of model-based and model-free learning was not changed by alcohol hangover.

With respect to the overall task performance, we observed that participants earned comparable cumulative points at the end of the task (final score). These outcomes did not significantly differ between sessions ($t_{(24)} = -0.11$, $p = 0.91$) and Bayesian analysis provided positive evidence in favor of the null hypothesis (BF = 6.46). The reaction times at the first stage and at the second stage did also not significantly differ between the sober and hangover session (first stage: $t_{(24)} = 0.11$, $p = 0.91$; $Z = -0.23$, $p = 0.82$; second stage: $t_{(24)} = -0.53$, $p = 0.60$). Again, Bayesian analyses provided positive evidence in favor of the null hypothesis, i.e., no difference between the sober and hangover session in response latency (first stage: BF = 6.46; second stage: BF = 5.69). These findings suggest that neither overall task performance, nor response speed are modulated by alcohol hangover.

Table 3 shows all estimated parameters based on the hybrid dual-system reinforcement-learning model. The model fit by means of BIC did not statistically differ between sessions ($t_{(24)} = -0.28$, $p = 0.79$, BF = 6.27). In the sober session, the average BIC was 504.07 (SEM = 21.48) and in the hangover session, the average BIC was 508.72 (SEM = 18.12). At the first stage, subjects reached optimal choice rates with an average choice rate of 0.66 (SEM = 0.03) in the sober session and with an average choice rate of 0.70 (SEM = 0.03) in the hangover session. Choice rates did not significantly differ between sessions ($t_{(24)} = -0.83$, $p = 0.42$; $Z = -0.72$, $p = 0.48$). At the second stage, the average optimal choice rate was 0.77 (SEM = 0.02) in the sober session and 0.76 (SEM = 0.03) in the hangover session. Like for the first stage, the second stage choice rates did not significantly differ between sessions ($t_{(24)} = 0.33$, $p = 0.75$; $Z = -0.69$, $p = 0.50$). Bayesian analyses provided positive evidence in favor of the null hypothesis (no difference between the sober and hangover session) for optimal choice rates at the first stage (BF = 4.68) and at the second stage (BF = 6.17). Therefore, both BIC and optimal choice rates indicate that participants similarly applied the hybrid Q-learning model in both sessions. In this context, please note that the decision process is assumed to include some randomness. With regard to the dynamic task environment (slowly changing rewards), it is reasonable that subject explored the other choice option from time to time, which is also reflected in the optimal choice rates. In such a dynamic environment, even perfectly adjusted behavior could not yield choice rates of (or close to) 1.

Table 3. Distribution of estimated parameters based on the hybrid dual-system reinforcement-learning model for the sober and hangover session.

Percentile	ω	α	β	λ	π
sober					
25	0.70	0.81	3.39	0.00	0.11
50	0.83	0.89	4.70	0.48	0.16
75	0.90	1.00	5.48	0.84	0.19
hangover					
25	0.68	0.75	3.16	0.28	0.09
50	0.88	0.86	4.04	0.51	0.20
75	0.95	0.98	5.52	0.79	0.23

The weighting parameter ω represents the balance between model-based ($\omega > 0.5$) and model-free learning ($\omega < 0.5$). The learning rate α indicates to what extent new information is incorporated in the Q-value update. The inverse temperature β determines the randomness of decision-making. The decay-rate parameter λ represents the degree to which experience in later stages influences first stage Q-value update. The choice stickiness parameter π indicates perseveration tendencies ($\pi > 0$).

Most importantly, we found no significant difference between the sober and hangover session in the weighting parameter ω ($t_{(24)} = -0.48$, $p = 0.63$; $Z = -1.39$, $p = 0.17$). Further supporting this, Bayesian analyses yielded positive evidence for the null hypothesis (no difference between sessions; BF = 5.81), indicating that the balance between model-based and model-free learning was not affected by hangover status. This null finding is in accordance with the results of the MB-score (no session effect) and provides evidence for unaffected goal-directed learning in the context of this task.

The choice "stickiness" parameter π, which indicates a perseveration tendency in case of values above zero, did also not differ between the sober and hangover session ($t_{(24)} = -1.43$, $p = 0.17$), but the obtained BF of 2.52 provided only weak evidence in favor of the null hypothesis.

An exploratory analysis of the learning rate α revealed no significant differences between the sober and hangover session ($t_{(24)} = 0.73$, $p = 0.47$; $Z = -1.39$, $p = 0.17$). We also found no significant difference between sessions with respect to the inverse temperature β ($t_{(24)} = 0.64$, $p = 0.53$), which represents the randomness of decisions, i.e., the reliance on Q-values in decision-making. Likewise, the decay-rate parameter λ did not statistically differ between the sober and hangover session ($t_{(24)} = -0.48$,

$p = 0.64$; $Z = -0.55$, $p = 0.58$). Bayesian analyses provided positive evidence for the null hypothesis (i.e., no difference between the sober and hangover session), for learning rate α (BF = 5.03), inverse temperature β (BF = 5.33), and decay-rate parameter λ (BF = 5.82). Thus, our exploratory analyses suggest that none of these parameters seems to be modulated by alcohol hangover.

3.3. Add-On Analyses of Alcohol Consumption Habits

Given that a recent study found acute alcohol intoxication effects on model-based behavior to be modulated by drinking problems (as assessed with the AUDIT) [55], we ran exploratory add-on analyses to investigate whether AUDIT scores correlated with any of the functionally relevant descriptive or estimated parameters in the sober and/or hungover session. As can be seen in Table 4, we did not find any significant correlation in either the sober or the hungover session. Bayesian analyses (default settings for Bayesian Pearson Correlation: Tolerance = 0.0001, maximum iterations = 2000; uniform prior; Jeffreys–Zellner–Siow Bayes Factor) provided positive evidence in favor of the null hypothesis (no relationship between AUDIT and task performance) for MF-score, MB-score, weighting parameter ω, and choice stickiness parameter π; and weak evidence in favor of the null hypothesis for the final score (earned cumulative points at the end of the task). We therefore refrained from using the AUDIT as a control variable/covariate in any of the main analyses.

Table 4. Correlation between AUDIT and two-step task performance in the sober and hangover session.

	r (p)	BF	τ (p)
sober			
MF-score	0.24 (0.24)	3.26	0.15 (0.32)
MB-score	0.21 (0.32)	3.94	0.16 (0.28)
Final score	0.30 (0.15)	2.35	0.21 (0.17)
ω	0.21 (0.31)	3.88	0.16 (0.28)
π	0.06 (0.76)	6.21	−0.05 (0.74)
hangover			
MF-score	<−0.01 * (0.98)	6.50	0.03 (0.85)
MB-score	0.16 (0.45)	4.90	0.13 (0.37)
Final score	−0.11 (0.60)	5.67	−0.14 (0.34)
ω	0.08 (0.72)	6.10	0.11 (0.45)
π	−0.01 (0.95)	6.49	0.03 (0.85)

BF: Bayes Factor; MF-score: model-free score; MB-score: model-based score; final score: accumulated outcomes at the end of the task (in points); weighting parameter ω: balance between model-based ($\omega > 0.5$) and model-free learning ($\omega < 0.5$); Choice stickiness parameter π: indicates perseveration tendencies ($\pi > 0$). * The true value lies between −0.01 and 0.00.

4. Discussion

Aberrant alcohol consumption has repeatedly been demonstrated to be associated with negative cognitive, affective, and behavioral consequences [1]. While the effects of acute intoxication and long-term abuse are comparatively well-researched, much less is known about the cognitive and behavioral effects of alcohol hangover. In this study, we used an experimental cross-over design to test the hypothesis that alcohol hangover decreases model-based and increases model-free behavior. A total of $n = 25$ healthy young men were tested with a two-step task. Each participant was tested once sober and once hungover, i.e., after having consumed a standardized amount of alcohol in an experimental setting. Several behavioral and computational modeling parameters were then compared across the two sessions. Our study motivation and hypotheses had been based on several studies showing that alcohol seems to have much stronger detrimental effects on goal-directed/model-based processes that require high levels of cognitive effort, than on model-free processes which typically require substantially lower levels of effort. This observation has repeatedly been made in the context of acute, binge-like intoxication levels [18,19,56–58], and in the context of AUD [16,20,23]. In social BDs, who do not fulfil enough criteria for an AUD diagnosis, results are generally more mixed, but there are also repeated reports of impairments in the domain of goal-directed (executive) functions [21,22]. With respect to the arbitration between goal-directed and habitual behavior, both AUD and BD have been linked to reductions in effortful controlled model-based cognitive strategies [21,25]. At least in BD, this imbalance seems to normalize as the time that has passed since the last binging episode increases [21]. Moreover, perseveration tendencies seem to be altered in BD [21,31]. Based on these findings, we had hypothesized that alcohol hangover might induce qualitatively similar effects, albeit probably to a lesser degree.

The employed two-step decision making task based on Daw et al. [28] and Kool et al. [27] allows for the quantification of model-based and model free behavior by contrasting first stage stay probabilities in case of all combinations of gain/loss and common/rare transitions on the one hand and by estimating individual parameters of task performance with a computational model on the other hand. The underlying logic is that model-free behavior is solely based on previous rewards/losses and does not consider transition probability, which makes it computationally cheap, but also rather inflexible. Following this strategy, first stage choices are repeated whenever that choice has been rewarded, and switched when that choice has been not rewarded or has been punished. In contrast to this, model-based choices should additionally account for transition probabilities, which makes it computationally more demanding and effortful, but also more flexible and adaptive. Following this strategy, first stage choices tend to be repeated whenever a choice has been rewarded on a common transition or punished on a rare transition, and switched when a choice has been rewarded on a rare transition or punished on a common transition.

Even though the experimental induction of hangover was effective (as demonstrated by significant increases in 21 out of 22 assessed hangover symptoms, as well as overall hangover severity [59]), we did not find evidence for any hangover effects in the MF- and MB-score, as well as in the computationally deduced weighting parameter ω (which represents the balance between the two strategies), or in the overall outcome (obtained score). Instead, Bayesian analyses provided positive evidence that there was likely no difference between the sober and the hungover session. The lack of response time effects further suggests that the application of goal-directed strategies was not maintained at the cost of a speed-accuracy tradeoff. The BIC parameter, which allows to compare the model fit across sessions, further suggested that the observed comparability across sessions was not caused by differences in the goodness of the model fit. Likewise, we found no evidence for increased perseveration tendencies (π) during hangover. Subsequent Bayesian analysis failed to provide conclusive evidence for either hypothesis, but still favored the null hypothesis over the alternative hypothesis at an anecdotal level. Hence, all of our findings are in favor of the assumption that alcohol hangover does not alter the balance between model-based and model-free learning strategies, or increase perseveration tendencies. Still, it would be commendable to also investigate other potential facets of alcohol hangover effects on automatic and/or habitual behavior with other promising new paradigms [60,61]. Add-on exploratory analyses further showed that there were also no hangover effects on the learning rate (α), the randomness of decision-making (β), or the down-weighing of previous experience (λ).

It should however be noted that the weighting parameter omega was numerically higher than in other studies with healthy young samples [32,62,63], which indicates a stronger preference for the model-based learning strategy in the investigated sample/applied task. This could be due to the manipulations in task administration (e.g., more prominent transition probabilities as well as usage of simulated distributions for outcomes and transitions), which facilitates the application of model-based, computationally demanding strategies [27]. A higher reliance on a model-based system is typically found when high incentives shift the cost-benefit-arbitration in favor of a computationally costly strategy [64,65]. Given this strong preference for model-based over model-free strategies in both sessions, we can assume a general willingness to exert cognitive effort and thereby exclude the possibility that the lack of hangover effects could be due to a lack of overall motivation to perform the task as instructed. It however remains an open question whether high intrinsic motivation or changes in task settings have led to the more pronounced dominance of the model-based learning system, as compared to other studies in the field. In addition, the task and its parameters do not allow to distinguish between the ability and the willingness/motivation to exert cognitive control: While we found no decline in the application of effortful model-based strategies, our data does not allow to exclude the theoretical possibility the participants' awareness of their hangover symptoms and/or associated expectations of decreased performance motivated them to exert more effort than during the sober session, thus masking small to medium detrimental effects of alcohol hangover on effortful model-based strategies.

To the best of our knowledge, this is the first publication that explicitly investigates the arbitration between effortful model-based and computationally less demanding model-free learning in alcohol hangover. Yet still, the finding that hangover does not reduce the ability and/or willingness to invest cognitive effort (despite the experimentally applied sleep restriction and the fact that binge drinking is known to decrease sleep quality [42,44,47–49]) adds to the general literature on cognitive hangover effects, where cognitive effects that can be reliably observed during alcohol intoxication or AUD cannot always be reproduced during alcohol hangover [6,10–12,17], and are not necessarily modulated or worsened by light hangover-associated sleep impediments [40,44]. Given that there is an ongoing debate on whether or not habitual binge drinking impairs cognitive control functions that require high levels of effort [22,66] and all of our participants had been recruited to engage in binge drinking at least one a month (in order to minimize the risk of severe adverse side effects during experimental intoxication), we ran add-on analyses to investigate the potential effects of alcohol use severity on the investigated measures at both sessions. Of note, none of these analyses provided evidence for an association between alcohol use severity (indicated by AUDIT scores) and changes in any of the measures relevant to the arbitration between model-based and model-free behavior. Of note, this finding is in line with a study by Doñamayor et al. [21], who compared both female and male binge drinkers (mean AUDIT score of 16) to healthy controls (mean AUDIT score of 5) of similar age as our sample. While they reported a shift from goal-directed behavior to habitual behavior in binge drinkers, they found no statistical relationship between AUDIT scores, the weighting parameter, model-free scores, or model-based scores across the entire sample, even though they had a similar overall mean and greater variance in AUDIT scores across the entire sample [21]. It also matches reports by Patzelt et al. [67], who found no correlation between alcohol use (as assessed with the AUDIT) and model-based scores in over 900 adult Amazon Mechanical Turk participants.

Given that we only investigated young healthy males, it should be critically discussed whether the null finding reported in this study would also have been found females, or in other age groups. Females tend to metabolize alcohol more slowly than males [68] and have been suggested to report greater subjective hangover symptoms than males [69–71]. Lastly, women have been suggested to show greater cognitive impairments than males in case of regular binge drinking [22] and alcohol abuse [72]. Given that women might hence be more vulnerable to the negative cognitive effects of alcohol, our results might unfortunately not be readily generalizable to female populations, thus necessitating further studies. Furthermore, it has been shown that general cognitive and executive functions (e.g., processing speed or working memory) interact with model-based learning [62,73]. It could hence be possible that the typically high functioning levels found in young healthy samples protected our participants from detrimental effects of intoxication. Given that old age has repeatedly been associated with decreased cognitive functions in various domains, including model-based decision making [74,75], and further given that hangover severity might also differ with age [76,77], it could be conceivable that the combination of reduced cognitive resources and altered alcohol hangover might render elderly individuals more vulnerable towards the potential detrimental effects of alcohol on model-based processing. It should therefore be investigated whether our null finding can be reproduced in older samples as well. Lastly, we did not control for factors such as reward sensitivity or the subjective cost of control, which may depend on both internal and external factors. Correcting for the expected value of control as suggested by Shenhav et al. [78] (e.g., controlling for aspects like reward sensitivity, task difficulty, or anterior cingulate cortex activation) might potentially provide new insights and/or help identify functional subgroups. In line with this, it would also have been interesting to assess whether subjective ratings of motivation and invested effort differed between the sober and hungover appointment and/or whether they correlated with any of the assessed parameters.

5. Conclusions

In summary, we investigated whether alcohol hangover shifts decision making strategies from a more model-based to a more model-free approach. We asked $n = 25$ young healthy male social drinkers to perform a two-step decision-making task once while sober and once while hungover. Behavioral and modeling parameters were compared across appointments. The lack of significant hangover effects and the positive Bayesian evidence for the null hypothesis in all but one investigated parameters suggest that alcohol hangover, which results from a single binge drinking episode, does not impair the application of effortful and computationally costly model-based learning strategies and/or increase model-free learning strategies. While this finding still awaits confirmation in females and other age groups, it adds to a growing body of literature suggesting that behavioral deficits observed in at-risk drinkers [20,21,25] might not be a mere consequence of alcohol consumption alone [20,23,24,30,55,67]. When applying this finding to a clinical context, it suggests that the behavioral and psychological changes that have been associated with problematic drinking patterns like binge drinking [21,22] and shown to drive and maintain alcohol use disorders (AUD) [16,23–25], are not likely to arise as a consequence of hangover (alone).

Supplementary Materials: The following are available online at http://www.mdpi.com/2077-0383/9/5/1453/s1: Simulation of Transition and Reward Distribution (including Figures S1 and S2, which display reward distributions), Investigation of Hypothetical Task Order Effects (including Tables S1–S4, which provide descriptive statistics and estimated parameters with respect to hypothetical appointment order effects), a data sheet containing the equation used to determine individually served alcohol amounts and document drinking during experimental intoxication.

Author Contributions: Conceptualization, C.B. and A.-K.S.; methodology, J.B., T.E., C.B., and A.-K.S.; validation, J.B. and A.-K.S.; formal analysis, J.B.; investigation, W.B., N.Z., and A.-K.S.; resources, T.E., C.B., and A.-K.S.; data curation, J.B., W.B., N.Z., and A.-K.S.; writing—original draft preparation, J.B. and A.-K.S.; writing—review and editing, all authors; visualization, J.B.; supervision, T.E., C.B., and A.-K.S.; project administration, A.-K.S.; funding acquisition, T.E., C.B., and A.-K.S. All authors have read and agreed to the published version of the manuscript.

Funding: This research was funded by the Deutsche Forschungsgemeinschaft (DFG), grant number TRR 265 B07 to C.B. and A.-K.S., and grant number TRR 265 B01 to T.E.

Acknowledgments: The authors thank all individuals who took part in this study. We thank Wouter Kool for sharing his computational modeling scripts on Github. Open Access Funding by the Publication Funds of the TU Dresden.

Conflicts of Interest: The authors declare no conflict of interest. The funders had no role in the design of the study; in the collection, analyses or interpretation of data; in the writing of the manuscript or in the decision to publish the results.

References

1. WHO. Global Status Report on Alcohol and Health 2018. Available online: http://www.who.int/substance_abuse/publications/global_alcohol_report/en/ (accessed on 30 September 2019).
2. Verster, J.C.; Kruisselbrink, L.D.; Slot, K.A.; Anogeianaki, A.; Adams, S.; Alford, C.; Arnoldy, L.; Ayre, E.; Balikji, S.; Benson, S.; et al. Sensitivity to Experiencing Alcohol Hangovers: Reconsideration of the 0.11% Blood Alcohol Concentration (BAC) Threshold for Having a Hangover. *J. Clin. Med.* **2020**, *9*, 179. [CrossRef] [PubMed]
3. Verster, J.C.; Scholey, A.; van de Loo, A.J.A.E.; Benson, S.; Stock, A.-K. Updating the Definition of the Alcohol Hangover. *J. Clin. Med.* **2020**, *9*, 823. [CrossRef]
4. Van Schrojenstein Lantman, M.; van de Loo, A.J.A.E.; Mackus, M.; Verster, J.C. Development of a Definition for the Alcohol Hangover: Consumer Descriptions and Expert Consensus. *Curr. Drug Abus. Rev.* **2017**, *9*, 148–154. [CrossRef] [PubMed]
5. Bush, D.M.; Lipari, R.N. Workplace Policies and Programs Concerning Alcohol and Drug Use. In *The CBHSQ Report*; Substance Abuse and Mental Health Services Administration (US): Rockville, MD, USA, 2013.

6. Gunn, C.; Mackus, M.; Griffin, C.; Munafò, M.R.; Adams, S. A systematic review of the next-day effects of heavy alcohol consumption on cognitive performance. *Addiction* **2018**, *113*, 2182–2193. [CrossRef] [PubMed]
7. O'Brien, C.P.; Lyons, F. Alcohol and the athlete. *Sports Med.* **2000**, *29*, 295–300. [CrossRef]
8. Barker, C.T. The alcohol hangover and its potential impact on the UK armed forces: A review of the literature on post-alcohol impairment. *J. R. Army Med. Corps* **2004**, *150*, 168–174. [CrossRef]
9. Verster, J.C.; Anogeianaki, A.; Kruisselbrink, D.; Alford, C.; Stock, A.-K. Relationship between Alcohol Hangover and Physical Endurance Performance: Walking the Samaria Gorge. *J. Clin. Med.* **2019**, *9*, 114. [CrossRef]
10. Ling, J.; Stephens, R.; Heffernan, T.M. Cognitive and psychomotor performance during alcohol hangover. *Curr. Drug Abus. Rev.* **2010**, *3*, 80–87. [CrossRef]
11. Zink, N.; Bensmann, W.; Beste, C.; Stock, A.-K. Alcohol Hangover Increases Conflict Load via Faster Processing of Subliminal Information. *Front. Hum. Neurosci.* **2018**, *12*, 316. [CrossRef]
12. Stock, A.-K.; Hoffmann, S.; Beste, C. Effects of binge drinking and hangover on response selection sub-processes-a study using EEG and drift diffusion modeling. *Addict. Biol.* **2017**, *22*, 1355–1365. [CrossRef]
13. Crofton, J. Extent and costs of alcohol problems in employment: A review of British data. *Alcohol Alcohol.* **1987**, *22*, 321–325. [PubMed]
14. Devenney, L.E.; Coyle, K.B.; Verster, J.C. Memory and attention during an alcohol hangover. *Hum. Psychopharmacol.* **2019**, *34*, e2701. [CrossRef] [PubMed]
15. Verster, J.C.; Bervoets, A.C.; de Klerk, S.; Vreman, R.A.; Olivier, B.; Roth, T.; Brookhuis, K.A. Effects of alcohol hangover on simulated highway driving performance. *Psychopharmacology* **2014**, *231*, 2999–3008. [CrossRef]
16. Stock, A.-K. Barking up the Wrong Tree: Why and How We May Need to Revise Alcohol Addiction Therapy. *Front. Psychol.* **2017**, *8*, 884. [CrossRef] [PubMed]
17. Opitz, A.; Hubert, J.; Beste, C.; Stock, A.-K. Alcohol Hangover Slightly Impairs Response Selection but not Response Inhibition. *J. Clin. Med.* **2019**, *8*, 1317. [CrossRef]
18. Chmielewski, W.X.; Zink, N.; Chmielewski, K.Y.; Beste, C.; Stock, A.-K. How high-dose alcohol intoxication affects the interplay of automatic and controlled processes. *Addict. Biol.* **2018**. [CrossRef]
19. Stock, A.-K.; Bensmann, W.; Zink, N.; Münchau, A.; Beste, C. Automatic aspects of response selection remain unchanged during high-dose alcohol intoxication. *Addict. Biol.* **2019**, e12852. [CrossRef]
20. Heinz, A.; Kiefer, F.; Smolka, M.N.; Endrass, T.; Beste, C.; Beck, A.; Liu, S.; Genauck, A.; Romund, L.; Banaschewski, T.; et al. Addiction Research Consortium: Losing and regaining control over drug intake (ReCoDe)-From trajectories to mechanisms and interventions. *Addict. Biol.* **2019**, e12866. [CrossRef]
21. Doñamayor, N.; Strelchuk, D.; Baek, K.; Banca, P.; Voon, V. The involuntary nature of binge drinking: Goal directedness and awareness of intention. *Addict. Biol.* **2018**, *23*, 515–526. [CrossRef]
22. Montgomery, C.; Fisk, J.E.; Murphy, P.N.; Ryland, I.; Hilton, J. The effects of heavy social drinking on executive function: A systematic review and meta-analytic study of existing literature and new empirical findings. *Hum. Psychopharmacol.* **2012**, *27*, 187–199. [CrossRef]
23. Heinz, A.; Beck, A.; Halil, M.G.; Pilhatsch, M.; Smolka, M.N.; Liu, S. Addiction as Learned Behavior Patterns. *J. Clin. Med.* **2019**, *8*, 1086. [CrossRef]
24. Heinz, A.; Deserno, L.; Zimmermann, U.S.; Smolka, M.N.; Beck, A.; Schlagenhauf, F. Targeted intervention: Computational approaches to elucidate and predict relapse in alcoholism. *Neuroimage* **2017**, *151*, 33–44. [CrossRef]
25. Sebold, M.; Deserno, L.; Nebe, S.; Schad, D.J.; Garbusow, M.; Hägele, C.; Keller, J.; Jünger, E.; Kathmann, N.; Smolka, M.; et al. Model-based and model-free decisions in alcohol dependence. *Neuropsychobiology* **2014**, *70*, 122–131. [CrossRef] [PubMed]
26. Dayan, P.; Niv, Y. Reinforcement learning: The good, the bad and the ugly. *Curr. Opin. Neurobiol.* **2008**, *18*, 185–196. [CrossRef] [PubMed]
27. Kool, W.; Cushman, F.A.; Gershman, S.J. When Does Model-Based Control Pay Off? *PLoS Comput. Biol.* **2016**, *12*, e1005090. [CrossRef] [PubMed]
28. Daw, N.D.; Gershman, S.J.; Seymour, B.; Dayan, P.; Dolan, R.J. Model-based influences on humans' choices and striatal prediction errors. *Neuron* **2011**, *69*, 1204–1215. [CrossRef] [PubMed]
29. Voon, V.; Derbyshire, K.; Rück, C.; Irvine, M.A.; Worbe, Y.; Enander, J.; Schreiber, L.R.N.; Gillan, C.; Fineberg, N.A.; Sahakian, B.J.; et al. Disorders of compulsivity: A common bias towards learning habits. *Mol. Psychiatry* **2015**, *20*, 345–352. [CrossRef] [PubMed]

30. Sebold, M.; Nebe, S.; Garbusow, M.; Guggenmos, M.; Schad, D.J.; Beck, A.; Kuitunen-Paul, S.; Sommer, C.; Frank, R.; Neu, P.; et al. When Habits Are Dangerous: Alcohol Expectancies and Habitual Decision Making Predict Relapse in Alcohol Dependence. *Biol. Psychiatry* **2017**, *82*, 847–856. [CrossRef]
31. Carbia, C.; Cadaveira, F.; López-Caneda, E.; Caamaño-Isorna, F.; Rodríguez Holguín, S.; Corral, M. Working memory over a six-year period in young binge drinkers. *Alcohol* **2017**, *61*, 17–23. [CrossRef]
32. Reiter, A.M.F.; Deserno, L.; Wilbertz, T.; Heinze, H.-J.; Schlagenhauf, F. Risk Factors for Addiction and Their Association with Model-Based Behavioral Control. *Front. Behav. Neurosci.* **2016**, *10*, 26. [CrossRef]
33. Nebe, S.; Kroemer, N.B.; Schad, D.J.; Bernhardt, N.; Sebold, M.; Müller, D.K.; Scholl, L.; Kuitunen-Paul, S.; Heinz, A.; Rapp, M.A.; et al. No association of goal-directed and habitual control with alcohol consumption in young adults. *Addict. Biol.* **2018**, *23*, 379–393. [CrossRef] [PubMed]
34. World Health Organization. *AUDIT: The Alcohol Use Disorders Identification Test: Guidelines for Use in Primary Care/Thomas F. Babor ... [et al.]*, 2nd ed.; Available online: https://apps.who.int/iris/handle/10665/67205 (accessed on 30 March 2020).
35. Opitz, A.; Beste, C.; Stock, A.-K. Alcohol Hangover Differentially Modulates the Processing of Relevant and Irrelevant Information. *J. Clin. Med.* **2020**, *9*, 778. [CrossRef] [PubMed]
36. Widmark, E.M.P. *Die Theoretischen Grundlagen und Die Praktische Verwendbarkeit der Gerichtlich-medizinischen Alkoholbestimmung*; Urban und Schwarzenberg: Berlin, Germany, 1932.
37. Watson, P.E.; Watson, I.D.; Batt, R.D. Total body water volumes for adult male and females estimated from simple anthropometric measurements. *Am. J. Clin. Nutr.* **1980**, *33*, 27–39. [CrossRef] [PubMed]
38. Rohsenow, D.J.; Marlatt, G.A. The balanced placebo design: Methodological considerations. *Addict. Behav.* **1981**, *6*, 107–122. [CrossRef]
39. Verster, J.C.; Stephens, R.; Penning, R.; Rohsenow, D.; McGeary, J.; Levy, D.; McKinney, A.; Finnigan, F.; Piasecki, T.M.; Adan, A.; et al. The alcohol hangover research group consensus statement on best practice in alcohol hangover research. *Curr. Drug Abus. Rev.* **2010**, *3*, 116–126. [CrossRef]
40. McKinney, A.; Coyle, K.; Penning, R.; Verster, J.C. Next day effects of naturalistic alcohol consumption on tasks of attention. *Hum. Psychopharmacol.* **2012**, *27*, 587–594. [CrossRef]
41. Howland, J.; Rohsenow, D.J.; Greece, J.A.; Littlefield, C.A.; Almeida, A.; Heeren, T.; Winter, M.; Bliss, C.A.; Hunt, S.; Hermos, J. The effects of binge drinking on college students' next-day academic test-taking performance and mood state. *Addiction* **2010**, *105*, 655–665. [CrossRef]
42. Swift, R.; Davidson, D. Alcohol hangover: Mechanisms and mediators. *Alcohol Health Res. World* **1998**, *22*, 54–60.
43. Chapman, L.F. Experimental induction of hangover. *Q. J. Stud. Alcohol* **1970**, *5* (Suppl. S5), 67–86.
44. Rohsenow, D.J.; Howland, J.; Arnedt, J.T.; Almeida, A.B.; Greece, J.; Minsky, S.; Kempler, C.S.; Sales, S. Intoxication with bourbon versus vodka: Effects on hangover, sleep, and next-day neurocognitive performance in young adults. *Alcohol. Clin. Exp. Res.* **2010**, *34*, 509–518. [CrossRef]
45. McKinney, A.; Coyle, K. Next day effects of a normal night's drinking on memory and psychomotor performance. *Alcohol Alcohol.* **2004**, *39*, 509–513. [CrossRef] [PubMed]
46. Rohsenow, D.J.; Howland, J.; Winter, M.; Bliss, C.A.; Littlefield, C.A.; Heeren, T.C.; Calise, T.V. Hangover sensitivity after controlled alcohol administration as predictor of post-college drinking. *J. Abnorm. Psychol.* **2012**, *121*, 270–275. [CrossRef] [PubMed]
47. Hogewoning, A.; Van de Loo, A.; Mackus, M.; Raasveld, S.J.; De Zeeuw, R.; Bosma, E.R.; Bouwmeester, N.H.; Brookhuis, K.A.; Garssen, J.; Verster, J.C. Characteristics of social drinkers with and without a hangover after heavy alcohol consumption. *Subst. Abus. Rehabil.* **2016**, *7*, 161–167. [CrossRef] [PubMed]
48. van Schrojenstein Lantman, M.; Mackus, M.; Roth, T.; Verster, J.C. Total sleep time, alcohol consumption, and the duration and severity of alcohol hangover. *Nat. Sci. Sleep* **2017**, *9*, 181–186. [CrossRef] [PubMed]
49. Devenney, L.E.; Coyle, K.B.; Roth, T.; Verster, J.C. Sleep after Heavy Alcohol Consumption and Physical Activity Levels during Alcohol Hangover. *J. Clin. Med.* **2019**, *8*, 752. [CrossRef]
50. van Schrojenstein Lantman, M.; Mackus, M.; van de Loo, A.J.A.E.; Verster, J.C. The impact of alcohol hangover symptoms on cognitive and physical functioning, and mood. *Hum. Psychopharmacol.* **2017**, *32*, e2623. [CrossRef]
51. Rummery, G.A.; Niranjan, M. *On-Line Q-Learning Using Connectionist Systems*; Cambridge University Engineering Department: Cambridge, UK, 1994; Volume 166.

52. Eppinger, B.; Walter, M.; Li, S.-C. Electrophysiological correlates reflect the integration of model-based and model-free decision information. *Cogn. Affect. Behav. Neurosci.* **2017**, *17*, 1–16. [CrossRef]
53. Gershman, S.J. Empirical priors for reinforcement learning models. *J. Math. Psychol.* **2016**, *71*, 1–6. [CrossRef]
54. Jarosz, A.F.; Wiley, J. What Are the Odds? A Practical Guide to Computing and Reporting Bayes Factors. *J. Probl. Solving* **2014**, *7*, 2–9. [CrossRef]
55. Obst, E.; Schad, D.J.; Huys, Q.J.; Sebold, M.; Nebe, S.; Sommer, C.; Smolka, M.N.; Zimmermann, U.S. Drunk decisions: Alcohol shifts choice from habitual towards goal-directed control in adolescent intermediate-risk drinkers. *J. Psychopharmacol.* **2018**, *32*, 855–866. [CrossRef]
56. Stock, A.-K.; Schulz, T.; Lenhardt, M.; Blaszkewicz, M.; Beste, C. High-dose alcohol intoxication differentially modulates cognitive subprocesses involved in response inhibition. *Addict. Biol.* **2014**. [CrossRef] [PubMed]
57. Stock, A.-K.; Riegler, L.; Chmielewski, W.X.; Beste, C. Paradox effects of binge drinking on response inhibition processes depending on mental workload. *Arch. Toxicol.* **2016**, *90*, 1429–1436. [CrossRef] [PubMed]
58. Stock, A.-K.; Blaszkewicz, M.; Beste, C. Effects of binge drinking on action cascading processes: An EEG study. *Arch. Toxicol.* **2014**, *88*, 475–488. [CrossRef]
59. Verster, J.C.; van de Loo, A.J.A.E.; Benson, S.; Scholey, A.; Stock, A.-K. The Assessment of Overall Hangover Severity. *J. Clin. Med.* **2020**, *9*, 786. [CrossRef]
60. Zwosta, K.; Ruge, H.; Goschke, T.; Wolfensteller, U. Habit strength is predicted by activity dynamics in goal-directed brain systems during training. *Neuroimage* **2018**, *165*, 125–137. [CrossRef]
61. Vaghi, M.M.; Cardinal, R.N.; Apergis-Schoute, A.M.; Fineberg, N.A.; Sule, A.; Robbins, T.W. Action-Outcome Knowledge Dissociates From Behavior in Obsessive-Compulsive Disorder Following Contingency Degradation. *Biol. Psychiatry Cogn. Neurosci. Neuroimaging* **2019**, *4*, 200–209. [CrossRef]
62. Schad, D.J.; Jünger, E.; Sebold, M.; Garbusow, M.; Bernhardt, N.; Javadi, A.-H.; Zimmermann, U.S.; Smolka, M.N.; Heinz, A.; Rapp, M.A.; et al. Processing speed enhances model-based over model-free reinforcement learning in the presence of high working memory functioning. *Front. Psychol.* **2014**, *5*, 1450. [CrossRef]
63. Deserno, L.; Wilbertz, T.; Reiter, A.; Horstmann, A.; Neumann, J.; Villringer, A.; Heinze, H.-J.; Schlagenhauf, F. Lateral prefrontal model-based signatures are reduced in healthy individuals with high trait impulsivity. *Transl. Psychiatry* **2015**, *5*, e659. [CrossRef]
64. Kool, W.; Gershman, S.J.; Cushman, F.A. Cost-benefit arbitration between multiple reinforcement learning systems. *Psychol. Sci.* **2017**, *28*, 1321–1333. [CrossRef]
65. Bolenz, F.; Kool, W.; Reiter, A.M.; Eppinger, B. Metacontrol of decision-making strategies in human aging. *Elife* **2019**, *8*, e49154. [CrossRef]
66. Bensmann, W.; Kayali, Ö.F.; Beste, C.; Stock, A.-K. Young frequent binge drinkers show no behavioral deficits in inhibitory control and cognitive flexibility. *Prog. Neuropsychopharmacol. Biol. Psychiatry* **2019**, *93*, 93–101. [CrossRef]
67. Patzelt, E.H.; Kool, W.; Millner, A.J.; Gershman, S.J. Incentives Boost Model-Based Control Across a Range of Severity on Several Psychiatric Constructs. *Biol. Psychiatry* **2019**, *85*, 425–433. [CrossRef]
68. Erol, A.; Karpyak, V.M. Sex and gender-related differences in alcohol use and its consequences: Contemporary knowledge and future research considerations. *Drug Alcohol Depend.* **2015**, *156*, 1–13. [CrossRef]
69. Prat, G.; Adan, A.; Sánchez-Turet, M. Alcohol hangover: A critical review of explanatory factors. *Hum. Psychopharmacol.* **2009**, *24*, 259–267. [CrossRef]
70. van Lawick van Pabst, A.E.; Devenney, L.E.; Verster, J.C. Sex Differences in the Presence and Severity of Alcohol Hangover Symptoms. *J. Clin. Med.* **2019**, *8*, 867. [CrossRef]
71. Vatsalya, V.; Stangl, B.L.; Schmidt, V.Y.; Ramchandani, V.A. Characterization of hangover following intravenous alcohol exposure in social drinkers: Methodological and clinical implications. *Addict. Biol.* **2018**, *23*, 493–502. [CrossRef]
72. Nixon, S.J.; Prather, R.; Lewis, B. Sex differences in alcohol-related neurobehavioral consequences. *Handb. Clin. Neurol.* **2014**, *125*, 253–272. [CrossRef]
73. Otto, A.R.; Raio, C.M.; Chiang, A.; Phelps, E.A.; Daw, N.D. Working-memory capacity protects model-based learning from stress. *Proc. Natl. Acad. Sci. USA* **2013**, *110*, 20941–20746. [CrossRef]
74. Worthy, D.A.; Cooper, J.A.; Byrne, K.A.; Gorlick, M.A.; Maddox, W.T. State-based versus reward-based motivation in younger and older adults. *Cogn. Affect Behav. Neurosci.* **2014**, *14*, 1208–1220. [CrossRef]

75. Eppinger, B.; Walter, M.; Heekeren, H.R.; Li, S.-C. Of goals and habits: Age-related and individual differences in goal-directed decision-making. *Front. Neurosci.* **2013**, *7*, 253. [CrossRef]
76. Tolstrup, J.S.; Stephens, R.; Grønbaek, M. Does the severity of hangovers decline with age? Survey of the incidence of hangover in different age groups. *Alcohol. Clin. Exp. Res.* **2014**, *38*, 466–470. [CrossRef] [PubMed]
77. Thumin, F.; Wims, E. The perception of the common cold, and other ailments and discomforts, as related to age. *Int. J. Aging Hum. Dev.* **1975**, *6*, 43–49. [CrossRef] [PubMed]
78. Shenhav, A.; Botvinick, M.M.; Cohen, J.D. The expected value of control: An integrative theory of anterior cingulate cortex function. *Neuron* **2013**, *79*, 217–240. [CrossRef] [PubMed]

© 2020 by the authors. Licensee MDPI, Basel, Switzerland. This article is an open access article distributed under the terms and conditions of the Creative Commons Attribution (CC BY) license (http://creativecommons.org/licenses/by/4.0/).

Article

Sleep after Heavy Alcohol Consumption and Physical Activity Levels during Alcohol Hangover

Lydia E. Devenney [1], Kieran B. Coyle [1], Thomas Roth [2] and Joris C. Verster [3,4,*]

1. School of Psychology, Life and Health sciences, Ulster University, BT52 1SA Londonderry, Northern Ireland; Devenney-l2@ulster.ac.uk (L.E.D.); kb.coyle@ulster.ac.uk (K.B.C.)
2. Sleep Disorders and Research Center, Henry Ford Health System, Detroit, MI 48202, USA; troth@hfhs.org
3. Division of Pharmacology, Utrecht University, 3584CG Utrecht, The Netherlands
4. Centre for Human Psychopharmacology, Swinburne University, Vic 3122 Melbourne, Australia
* Correspondence: j.c.verster@uu.nl; Tel.: +31-30-253-6909

Received: 2 April 2019; Accepted: 25 May 2019; Published: 27 May 2019

Abstract: Alcohol consumption can negatively affect sleep quality. The current study examined the impact of an evening of alcohol consumption on sleep, and next day activity levels and alcohol hangover. $n = 25$ healthy social drinkers participated in a naturalistic study, consisting of an alcohol and alcohol-free test day. On both days, a GENEactiv watch recorded sleep and wake, and corresponding activity levels. In addition, subjective assessments of sleep duration and quality were made, and hangover severity, and the amount of consumed alcoholic beverages were assessed. Alcohol consumption was also assessed in real-time during the drinking session, using smartphone technology. The results confirmed, by using both objective and subjective assessments, that consuming a large amount of alcohol has a negative impact on sleep, including a significant reduction in objective sleep efficiency and significantly lower self-reported sleep quality. Activity levels during the hangover day were significantly reduced compared to the alcohol-free control day. Of note, next-morning retrospective alcohol consumption assessments underestimated real-time beverage recordings. In conclusion, heavy alcohol consumption impairs sleep quality, which is associated with increased next day hangover severity and reduced activity levels. The outcome of this study underlines that, in addition to retrospectively reported data, real-time objective assessments are needed to fully understand the effects of heavy drinking.

Keywords: sleep; daytime activity; alcohol; hangover

---

1. Introduction

Alcohol hangover refers to the combination of mental and physical symptoms, experienced the day after a night of heavy drinking, starting when blood alcohol concentration (BAC) approaches zero [1]. Several factors may aggravate hangover severity and corresponding performance impairment, and one of them is the quality and duration of sleep after a heavy drinking session. Both hangover and sleep disturbances have shown to significantly impair potentially dangerous daily activities such as driving a car [2,3]. While people often report falling asleep immediately after alcohol consumption [4], the quality of sleep is often disturbed by the over production of glutamine [5]. Glutamine is a natural stimulant and alcohol produces both stimulant and sedative effects [6]. The stimulating effects of alcohol are thought to be associated with rising BACs (while drinking), whereas the sedative effects are associated with already high BAC levels [7]. The stimulating effects are linked to the activation of dopamine release in the brain's 'reward circuitry' [6]. During alcohol consumption glutamine production is suppressed, and when alcohol leaves the body, the body then attempts to recover lost levels of glutamine. The increased glutamine levels after consumption has ceased and is referred to as glutamine rebound [8]. Roehrs et al. [9] found that when glutamine rebound occurs, increased waking and light sleeping was observed during the second half of the sleep period. On a normal night rapid eye movement (REM) and non-REM sleep periods alternate throughout the night with an average of six to seven cycles, however, after an evening of drinking this is reduced to two to three cycles [10]. It is therefore vital to further examine the relationship between alcohol consumption, sleep, and the alcohol hangover. Up to now, several studies have addressed this issue, and the collected evidence comes from either retrospective self-report or real-time assessments such as polysomnography.

1.1. Self-Report

Most evidence on the association between alcohol hangover and sleep comes from self-report, either gathered in clinical studies or via (retrospective) surveys. These revealed that drinking time often goes at the expense of total sleep time, and that alcohol has a detrimental effect on sleep quality. For example, in a controlled study, Finnigan et al. [11] observed that subjects fell asleep faster after alcohol consumption and reported reduced next-day alertness. McKinney and Coyle [12] examined alcohol hangover effects and sleep in 48 social drinkers. Applying a naturalistic study design, the researchers did not interfere with drinking behavior and no restrictions were placed on the subjects sleep behavior. Similar to Finnigan et al., McKinney and Coyle found that sleep was disrupted after alcohol consumption and next-day fatigue was significantly increased. After alcohol consumption, sleep was qualified as less satisfying, refreshing, and restful. Further, subjects went to bed significantly later when compared to the alcohol-free day, resulting in a significantly reduced total sleep time (TST). Moreover, with higher amounts of alcohol intake, sleep onset latency (time of falling asleep—time to bed; SOL) further reduced. Similar findings were reported by Hogewoning et al. [13] who's naturalistic study revealed that drinking time goes at the expense of TST and that time-to-bed is significantly delayed by more than 1.5 h after alcohol consumption compared to an alcohol-free evening.

Rohsenow et al. [14] examined powerplant performance in $n = 61$ merchant marine cadets the day following an evening of alcohol administration to achieve a BAC of 0.11%. Results were compared to an alcohol-free control test day. After an 8h period of supervised sleep, subjects reported significantly improved sleep quality in the alcohol condition. This unexpected finding may be explained by the fact that after alcohol consumption subjects reported significantly reduced sleep latency until sleep onset. Powerplant performance was not impaired in the hangover state.

Van Schrojenstein Lantman et al. [15] conducted a survey among 578 Dutch University students examining the impact of TST on the presence and severity of their past months latest alcohol hangover. Subjects who consumed more alcohol slept significantly longer. A positive correlation was found between TST and the duration of the alcohol hangover state. However, at the same time, prolonged TST was associated with significantly reduced overall hangover severity. Thus, reduced TST was associated with more severe hangover complaints. In a second survey by van Schrojenstein Lantman et al. [16], 335 adults reported that sleep quality was significantly worse after their latest alcohol consumption session that resulted in a hangover, and that daytime sleepiness was significantly increased compared to a regular alcohol-free day.

With regard to daytime activity, several studies revealed self-reports of increased apathy and hangover symptoms suggesting reduced activity during alcohol hangover [17,18].

1.2. Real-Time Assessments

In 2010, Rohsenow et al. applied polysomnography to examine sleep in relation to alcohol hangover in $n = 95$ social drinkers [19]. In a double-blind study, sleep was assessed after alcohol administration to achieve a BAC of 0.11% and an alcohol-free control day. Alcohol significantly decreased sleep efficiency and rapid eye movement sleep, and next-day self-reported sleepiness was significantly increased during hangover. Significantly worse hangovers were reported by subjects with reduced sleep efficiency and shorter TST. When hangover severity increased, less time was spent in rapid eye movement sleep.

Earlier polysomnography studies with lower alcohol dosages revealed similar effects on sleep [9,20,21]. Alcohol significantly reduced sleep latency and the time spent in REM sleep. In the first half of the night, alcohol significantly increased the time spent in deep sleep (stage 3 and 4), while in the second half of the night, time spent in stage 1 sleep (drowsy light sleep) was significantly increased. The observations confirmed previous findings that after alcohol consumption people fall asleep quicker, spent less time in REM sleep in the first 4 h of sleeping [9]. The next 4 h, i.e., the second half of the night, sleep is more disturbed and fragmented, often characterized by multiple awakenings and increased time spent in Stage 1 sleep. Roehrs et al. [9] conducted a Multiple Sleep Latency Test (MSLT) the day following alcohol consumption (peak BAC 0.08%) or placebo. The assessments showed that throughout the post-alcohol day subjects were sleepier, as evidenced by the fact that they fell asleep significantly faster when compared to the alcohol-free day.

More recently, Wilkinson et al. [22] applied actigraphy to a study with ten healthy subjects without sleep disturbances. Subjects continuously wore an actigraph, starting three nights before a day of alcohol consumption until 4 days thereafter. In the two days before the alcohol challenge, TST was on average 8.0 h and no naps were recorded. On the test day, at 9AM, alcohol was administered in a controlled laboratory setting to achieve a peak BAC of approximately 0.14%. Sleep behavior on the day and subsequent night were examined and next day (24 h after the start of alcohol consumption), subjects completed the Acute Hangover Scale [23]. Seven out of 10 subjects took an unscheduled afternoon nap, on average 8.7h after drinking, which lasted 0.6h. The authors further analyzed the data separately for those who napped ($n = 7$) and those who did not have a nap ($n = 3$) after alcohol consumption. The analysis revealed that the groups did not significantly differ on TST or hangover severity. Limitations of the study include its small sample size, and that alcohol was administered at 9AM in the morning. Therefore, it is unclear to what extend this study mimics real-life drinking and the 'normal' hangover experience.

To our knowledge, physical activity levels during the hangover state have not been investigated previously. Additionally, real-time assessments of sleep and alcohol consumption are usually not conducted in hangover research. However, emerging research [24,25], provides a foundation in Ecological Momentary Assessment (EMA), which can be used to collect real time data. Here, EMAs were used to collect alcohol consumption measures every morning for 4–14 days [24] and one week [25]. Analysis of EMA data showed that subjects exceeded the threshold for binge drinking on drinking occasions [24]. The analysis also revealed more severe hangovers in adolescents (15–19 years) than adult heavy drinkers (21 or over) [24]. Moreover, severe hangover symptoms predicted less alcohol consumption on that particular day [24]. Using EMA messaging, Riordan et al. [25] showed that messages relating to short-term and long-term health and social consequences reduced alcohol consumption in female subjects but not males during orientation (freshers) week and semester 1. These studies [24,25] demonstrate the potential use of EMAs in both research and clinical interventions.

Building on this, smartphone technology has been implemented to collect hourly intoxication ratings, and alcohol and water consumption. In addition the current study was conducted to examine sleep after an evening of heavy alcohol consumption and its relationship to next day hangover severity and physical activity. In order to closely mimic a real life drinking experience, the study had a naturalistic design in which the researchers monitored but did not intervene with alcohol consumption or other activities and behaviors, nor did the researchers control time-to-bed or wake up time. Thus, it provides a unique amalgamation of both the laboratory and naturalistic approach through the use of objective real time measures in a natural environment.

Considering the literature presented above, it is predicted that alcohol consumption may not be accurately reported following a night of heavy drinking. It is also anticipated that sleep time will occur later in the evenings where alcohol is consumed. Finally, it is predicted that sleep efficiency and TST will be reduced during a hangover and participants will engage in less demanding physical activities during a hangover.

2. Methods

Twenty-eight healthy social drinkers (students of Ulster University) were recruited to participate in this naturalistic study. Participants were excluded for head injury, pharmaceutical treatment, pregnancy, and previous treatment for alcohol or drug abuse. Social drinking status, i.e., not being alcohol dependent was verified by self-report, and by completion of the Short Michigan Alcohol Screening Test (SMAST) [26]. Participants with a SMAST score greater than three were excluded from participation in the study. Ethical approval for this study was obtained from the ethics committee at Ulster University. All subjects provided written informed consent, and the study was conducted in accordance with the "Code of Ethics and Conduct" of the British Psychological Society (2009).

2.1. Design

The study comprised an evening of alcohol consumption and an alcohol-free (control) test day. Both experimental testing days occurred in free-living conditions whereby subjects did not have study or work commitments, or mandatory training to attend. Using a naturalistic study design, subjects consumed alcohol at a venue of their own choice, and the type and quantity of alcohol and activities during the evening were not controlled by the researchers in order to closely mimic real-life drinking occasions [6]. The investigators did not interfere with the participant's activities and behavior.

2.2. GENEactiv Accelerometer Assessments of Sleep and Activity

On each test day, participants were asked to wear, on their non-dominant hand, a GENEActiv accelerometer [27,28] to objectively assess activity levels and sleep. The GENEActiv accelerometer continuously records activity, environmental temperature and light exposure. The watch could not be operated by the participants, nor did they have access to the data collected. The device allows for raw data to be transferred wirelessly in real time and saved as an open source or csv. The data then can be analyzed in statistical packages such as SPSS and R [27,28]. Esliger et al. [29] has validated and calibrated the GENEActiv accelerometer using Metabolic Equivalent of Tasks (METs) and Signal Vector Magnitudes (SVM, magnitude of watch movement). METs represent the energy costs of physical activity [30]. One MET refers to an individual's resting metabolic rate and can be calculated by dividing the volume of oxygen (VO_2) used during the activity by 3.5 (1 MET = 3.5 mL O_2/kg/min) [21,22]. The outcome intensity levels, categorized by Esliger et al. [29], and included in this study were: Sedentary (<1.5, METs), light (1.5–3.99 METs), moderate (4.00–6.99 METs), and vigorous (7+ METs) activity. The corresponding cut off points were set at 386 SVM (sedentary to light), 542 SVM (light to moderate), and 1811 SVM (moderate to vigorous). Outcome measures included the percentage of time spent in sedentary, light, moderate and vigorous activity from waking up to midnight, and total METs spent on the hangover and control day were calculated. Continuous measurements of activity level allowed calculation of time to bed, time of falling asleep, wake up time, TST, sleep efficiency (i.e., the ratio of total sleep time and the time spent in bed), and number and the median duration of nightly awakenings/activity. Using this data, sleep onset latency (time of falling asleep—time to bed) was computed.

2.3. Self-Reported Sleep (Next Morning)

Self-reported assessments of sleep comprised questions on time to bed, time of falling asleep, wake up time, sleep onset latency, and total sleep time. Using this information sleep efficiency was computed. To evaluate last night's sleep, subjects rated several aspects of their sleep quality on a 7-point bi-polar scale ranging from extremely, quite, and slightly, around a midpoint of four (neither). Sleep quality was assessments by six bipolar ratings including good-bad, satisfying-not satisfying, restful-not restful, refreshing-not refreshing, and light-deep. This scale has previously been implemented successfully in hangover research [12] and was completed each test day in the morning.

2.4. Assessments of Alcohol Consumption (Real-Time and Retrospective)

On the day before the drinking session, a Droidsurvey/iSurvey app was installed on subjects' smartphones, and they registered with the program (Harvest Your Data). Participants were identified through coded usernames and responses were recorded and synced to the researcher's account online. When offline, the application stored data until the device went online. On the drinking occasion, real-time smartphone assessments of alcohol consumption were made. The app required touch screen responses to four short questions and took approximately 1 min to complete. Participants were asked to set reminders on their alarm to complete the app once hourly throughout their drinking episode. One of the questions assessed the amount of alcoholic beverages consumed during the past hour.

The number of units of alcoholic beverages that was consumed last night was also assessed the following morning. To help recall and calculate the amount of beverages consumed pictures were shown with the drinks and corresponding standardized UK units that included wine, beer, alcopops, and shots of spirits (mixers).

2.5. Assessments of Hangover Severity (Next Morning)

In the morning on each test day, the Acute Hangover Scale [23] was completed to measure the severity of 9 symptoms, including hangover, thirsty, tired, headache, dizziness/faintness, loss of appetite, stomach ache, nausea, and heart racing. Each item could be scored on a Likert scale ranging from 0 to 7, with the anchors none (0), mild (1), moderate (4), and incapacitating (7). The mean of the item scores represents overall hangover severity, with higher scores representing more severe hangovers.

2.6. Statistical Analysis

Statistical analysis was conducted with SPSS, version 24. Mean (SD) were computed for each variable. Results from the hangover day and control day were compared using paired sample *t*-tests, and in case the data was not normally distributed the nonparametric Mann-Whitney U test was applied. Differences were considered significant if $p < 0.05$. The relationship between objective and subjective sleep assessments, and other variables was investigated by computing Spearman's Rank correlation coefficients, using difference scores (alcohol—control day). Correlations were considered significant if $p < 0.05$.

3. Results

Three participants did not attend the testing sessions, and as a result, 25 participants completed both testing sessions. On the experimental day, participants' retrospectively reported a mean (SD) of 8.0 beverages (SD = 2.7), and their mean (SD) hangover severity was 2.2 (0.9). Descriptive statistics of the study sample are summarized in Table 1.

Results from the real time data collection of alcoholic drinks consumption revealed that a mean (SD) of 11.4 (3.8) beverages were consumed. However, participants reported a mean (SD) of 8.0 beverages (SD = 2.7) the following day. While the Pearson's product-moment correlation revealed a significant association between alcohol consumption reported next day and in real time ($r = 0.57$, $p < 0.01$), a paired samples T-test revealed a significant difference in real time and next day reports of alcohol consumption ($t(22) = -5.133$, $p = 0.0001$).

Table 1. Demographics and alcohol consumption characteristics.

	Mean (SD)
Demographics	
n	25
Male/Female	12/13
Age (years)	26.0 (7.1)
Age of first drink (years)	14.9 (1.7)
Usual Total sleep time (TST) (h:min)	6:33 (1:55)
Alcohol consumption on study night	
Reported units of alcohol consumed (real-time)	11.4 (3.8)
Reported units of alcohol consumed (retrospective)	8.0 (2.7)
Start time drinking (h:min)	20:48 (3:47)
Stop time drinking (h:min)	01:17 (1:12)
Duration of alcohol consumption (min)	269 (149.2)
Consumed more alcohol than planned (Yes/No)	10/15
Mean (SD) hangover severity	2.2 (0.9)

3.1. GENEActiv Sleep Assessments

A summary of all GENEactiv sleep assessments in the hangover and control condition is given in Table 2.

Table 2. GENEactiv sleep assessments.

Sleep Outcomes	Alcohol Day	Control Day	
	Mean (SD)	Mean (SD)	*p*-Value
Sleep start time	02:41 (1:17)	00:41 (1.16)	0.00 *
Wake-up time	9:46 (1:37)	8.56 (1:53)	0.07
Time in bed (h:min)	9:27 (2:46)	9:22 (2:14)	0.85
Total sleep time (TST) (h:min)	6:34 (3:45)	7:59 (4:42)	0.16
Sleep efficiency (%)	69.0 (16.7)	80.0 (15.2)	0.04 *
Number of nightly activity periods	8.4 (5.5)	8.0 (6.1)	0.81
Median duration nightly activity (min)	27.5 (59.9)	33.5 (79.0)	0.46

Significant differences ($p < 0.05$) between the alcohol and control test day are indicated by *.

The GENEactiv assessments further revealed that sleep efficiency was significantly worse after alcohol consumption ($p = 0.04$). Time to bed was significantly later on the alcohol and the control day. Wake up time did not significantly differ between the alcohol and control day. Interestingly, TST after alcohol consumption did not significantly differ from the alcohol-free control day.

An example of the visual output and summary data is presented in Figure 1. It can be seen in Figure 1 that there is a delay in sleep time after alcohol consumption. Of note, drinking occurred on Friday evening.

The subject in this example went to bed 1 h and 17 min later after alcohol consumption when compared to the alcohol-free control night. In the control condition, 64% of participants went to bed at or after midnight and this was the case for 100% of participants on the alcohol test day. In line with this, TST was about 1.5 h shorter on the alcohol test day, confirming that usually drinking time goes at the expense of sleeping time. However, given the large variability in TST between subjects, the difference in TST between the alcohol and control test days did not reach statistical significance ($p = 0.16$).

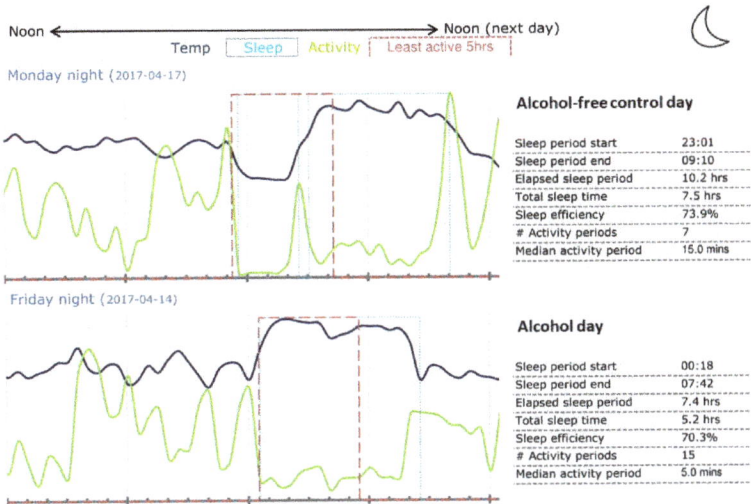

Figure 1. Visual output and summary data provided by the GENEActiv.

3.2. Self-Reported Sleep

Self-reported sleep outcomes are summarized in Table 3. Ratings of sleep quality revealed that subjective quality of sleep was significantly worse after consuming alcohol than on the control day ($p = 0.046$), as well as significantly less restful ($p = 0.001$) and less refreshing ($p = 0.01$). Sleep was also rated more satisfying when participants did not consume alcohol, although the difference with the alcohol test day did not reach significance ($p = 0.07$).

Table 3. Self-reported sleep outcomes.

	Alcohol Day	Control Day	
	Mean (SD)	Mean (SD)	p-Value
Start time sleeping	02:28 (1:14)	00:23 (1:11)	0.00 *
Sleep onset latency (min)	29 (41)	49 (62)	0.22
Total sleep time (TST)	06:40 (1:53)	7:01 (1:50)	0.44
Wake-up time	09:00 (2:25)	8:25 (1:14)	0.00 *
Sleep efficiency (%)	94.5% (17.0%)	91.6% (8.6%)	0.55
Sleep quality [1]			
Good-Bad	3.7 (1.6)	2.8 (1.3)	0.046 *
Satisfying-Not Satisfying	4.0 (1.6)	3.4 (1.3)	0.07
Refreshing-Not Refreshing	4.2 (1.4)	3.1 (1.2)	0.01 *
Restful-Not Restful	4.6 (1.3)	3.4 (1.4)	0.01 *
Light-Deep Sleep	5.0 (1.7)	5.0 (1.2)	0.92

Significant differences ($p < 0.05$) between the alcohol and control day are indicated by *. [1] Higher scores represent poorer sleep quality.

3.3. Correspondence between Objective and Subjective Sleep Assessments

A Spearman's R correlation analysis (using alcohol—control difference scores) comparing objective and subjective sleep measures revealed a significant negative relationship for TST ($r = -0.41$, $p = 0.04$). In addition, positive correlations were found between objective and subjective time of sleep onset ($r = 0.71$, $p < 0.001$) and wake-up time ($r = 0.46$, $p = 0.02$). There was no significant correlation between objective and subjective sleep efficiency, and paired *t*-tests revealed that this outcome significantly differed between the assessment methods ($p = 0.03$). Other sleep outcomes did not significantly differ between subjective and objective assessments.

3.4. Correspondence between Sleep Assessments, Alcohol Consumption, and Hangover Severity

Spearman's R correlation revealed that the total number of units of alcohol consumed and mean hangover severity did not significantly correlate with any of the sleep outcomes.

3.5. Physical Activity

An example of physical activity assessment on the hangover day and the control day is given in Figure 2. It is evident from Figure 2 that activity levels were reduced on the hangover day. Most time was spent in the sedentary activity mode. In this example, moderate activity levels were seen on the control day, which were absent on the hangover day. It can be hypothesized that the absence of moderate activity levels are associated with the large reduction in total sleep time (i.e., more than 2 h in this subject) the night before the hangover day.

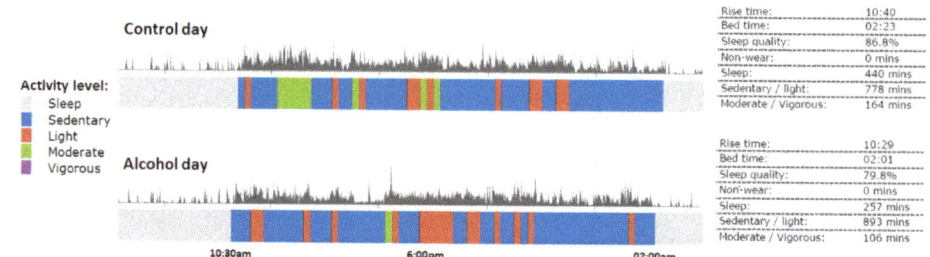

Figure 2. Physical activity levels on the alcohol and control day.

The overall results on physical activity for the hangover and control day are summarized in Figure 3. First, the percentage of time spent on vigorous activity was significantly less ($p = 0.03$) on the hangover day (2.8%) compared to the control day (10.0%). Second, a significantly ($p = 0.01$) higher percentage of time was spent in a sedentary manner on the hangover day (63.6%) compared to the control day (50.8%). Third, no significant difference between the hangover and control day was observed for percentage of time spent on light activity ($p = 0.86$) and moderate activity ($p = 0.09$).

Figure 4 summarizes the amount of MET.minutes for different physical activity levels on the hangover day and control day. Total energy spent on the hangover day (1870 METs) was lower compared to the control day (2279 METs). However, the difference did not reach statistical significance ($p = 0.37$), presumably due to the large standard deviations observed on both the hangover and control day (SD = 1487 and SD = 1549, respectively). Wilcoxon Signed-Ranks analyses revealed no significant differences between hangover and control day for sedentary ($p = 0.43$), light ($p = 0.62$) or moderate ($p = 0.06$) MET.minutes. Thus, although a significantly less percentages of time was spent on sedentary activity on the hangover day compared to the control day, the level of energy spent at this level did not significantly differ between the test days. On the control day, the mean vigorous MET.minutes was significantly higher compared to the hangover day ($p = 0.02$).

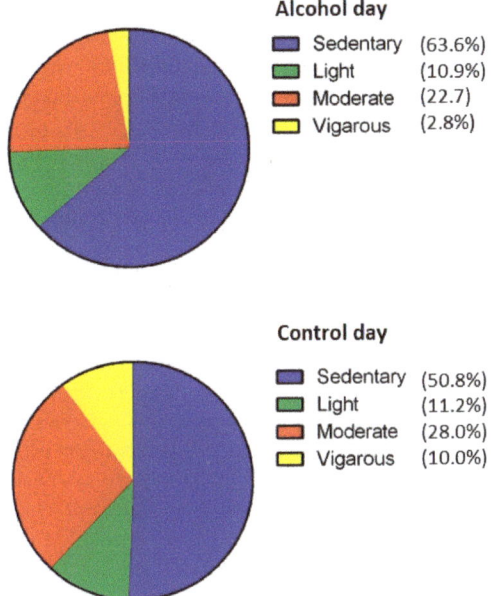

Figure 3. Percentage spent engaging in sedentary, light, moderate and vigorous activity on the alcohol day and control day.

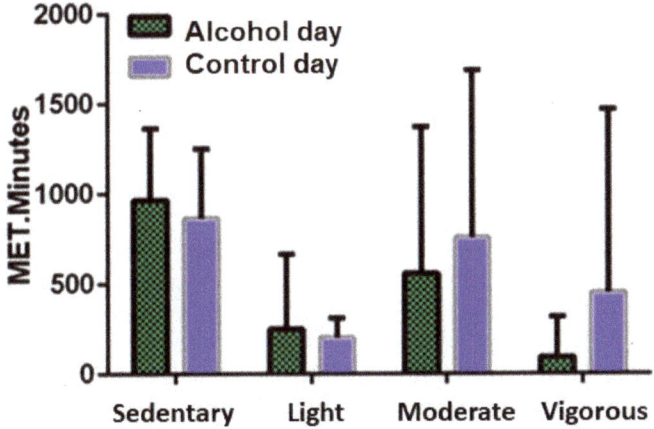

Figure 4. MET.minutes for different physical activity levels on the alcohol day and control day.

3.6. Sleep and Drinking Variables Associated with Percentual Changes in Activity

A Spearman's R correlation analysis was computed to compare the differences scores of percentages spent at the 4 activity levels (hangover-control day) with the total units of alcohol consumed, hangover severity, and sleep outcomes. Significant positive correlations were found between total units of alcohol consumed and differences in the percentage of light activity ($r=0.45$, $p = 0.02$). No significant association was found between percentual changes of activity levels and mean hangover severity.

With regards to objective sleep measures, differences in percentages of sedentary activity positively correlated with wake-up time ($r = 0.40$, $p = 0.049$) as well as TST ($r = 0.53$, $p = 0.01$). Moreover MET.minutes of moderate activity negatively correlated with wake-up time ($r = -0.44$, $p = 0.03$).

However, no other objective sleep measures correlated with MET.minute measurements. In terms of subjective sleep, the percentage of sedentary activity positively correlated with sleep onset latency ($r = 0.52$, $p = 0.01$) and moderate activity level MET.Minutes negatively correlated with sleep onset latency ($r = -0.53$, $p = 0.01$). Subjective measures of sleep quality revealed no significant differences relating to the percentage of activity levels or MET.minutes.

4. Discussion

Using both objective and subjective assessments, this study confirmed that consuming a large amount of alcohol has a negative impact on sleep, including a significantly reduced objective sleep efficiency and significantly lower subjective sleep quality. The study further revealed that next-day activity levels are significantly reduced during the alcohol hangover.

In relation to previous research, these results are in line with previous investigations of sleep and hangover [5,6]. Also in the current study, time to bed was significantly delayed by more than 1 h. However, in contrast to previous research [12,13,15], the difference in TST between the alcohol and control day in our study did not reach statistical significance, and in contrast to previous research [8,12] no significant association was found between objective TST and hangover severity. An explanation for the absence of significant effects may be the fact that there was great variability in TST between the subjects in this sample. Second, similar to previous findings by Rohsenow et al. [19], in the current study, sleep efficiency was significantly reduced after alcohol consumption. Whereas Rohsenow et al. [12] found a significant negative correlation between sleep efficiency and alcohol hangover, this correlation was not significant in the current study. In line with previous research [12,16], but not all [13], sleep quality was significantly poorer after alcohol consumption. Finally, several studies reported a reduced SOL after alcohol consumption [9,11,12,14]. While in the current study, subjective SOL was 20 min shorter after alcohol, the difference with the control day did not reach statistical significance. A correlation between alcohol consumption and hangover severity was not found. This observation supports recent findings [15] but contrasts with others [16]. It should be noted that the current study had less power than other studies that had a larger numbers of participants, so non-significant results can also be a result of low power for those particular analyses.

4.1. Objective Versus Subjective Assessments

The objective and subjective sleep assessments were generally in agreement with each other. However, some discrepancies were also noted. Most notably, while after consuming alcohol objective TST was reduced by more than 1 h, the difference in subjective TST was much smaller (<30 min). A similar discrepancy was found for objective and subjective assessments of sleep efficiency. The reason for these differences in subjective and objective assessments are unclear, but it may be related that subjects are unaware of the number and duration of nightly awakenings.

One could argue that these differences in outcomes underline the need for including real-time assessments to complement subjective sleep reports in clinical studies. Objective sleep may differ from perceived sleep and this discrepancy is not captured by relying solely on self-report. In this context, previous research in other areas has shown that subjects are sometimes unaware of performance or mood changes. For example, subjects were unaware of impairment in on-road driving tests after administering pharmacological treatment, while in contrast to this perception the objective assessments demonstrated that their actual driving performance showed clinically relevant impairment [31]. Thus, relying solely on patient perceptions of mood and impairment may therefore be dangerous in real life (e.g., they may decide to drive a car while they are not fit to drive) and bias clinical trial outcomes. For future research it is therefore recommended that, when possible, subjective ratings are complemented by objective assessments.

4.2. Recall Bias

When using a naturalistic study design, accurate capturing of alcohol consumption data is critical, as researchers are not present during the drinking session. It is therefore important to discuss the possible limitations of the methodologies used in this study.

While it should be taken into account that both being intoxicated and being hungover can result in inaccurate answering, this study revealed that retrospective reporting resulted in significant under-reporting of actual alcohol consumption. Retrospective self-report measures have long since been implemented in data collection. In the case of alcohol consumption this involves using retrospective memory to recall the type and number of alcoholic beverages consumed [32], as well as applying one's ability to comprehend the question being asked, make decisions about the accuracy of the information recalled, and format an answer [33]. It is however unlikely that these factors may have played a role in the current student sample. Further, participants may be reluctant to disclose information about the number of drinks consumed the previous night. Additionally, this is unlikely to be an issue, as both research data and study participation was treated anonymously. The timing and place of data collection can however have had a significant impact on what is reported by study participants [34]. For example, diminished retrospective ability to recall the amounts of drinks consumed may be an issue, especially the morning after a night's drinking when subjects suffer from an alcohol hangover. Research has shown that under these circumstances memory may be impaired [35,36]. Given this, accurately reporting alcohol consumption the day following a night's drinking remains challenging, and is an issue inherent to conducting alcohol research. Furthermore, periods of memory loss while a person is intoxicated (e.g., blackouts or lapses of attention) can hinder both accurate retrospective recall [37] and real-time reporting.

The use of smart phone and wearable technologies has garnered increasing attention in the fields of addiction research, as they offer the possibility of real time data collection at times people use alcohol and/or drugs in a natural setting and they also offer a more accurate assessment of alcohol consumption over time [38–41]. While real time data may be inaccurate because subjects are intoxicated, this data collection technique is not affected by recall bias.

The integration of mobile phone technology to real time alcohol consumption data has been considered for some time [38]. While mobile phones were not as robust at the time, research concluded that it was a feasible alternative to paper and pencil self-monitoring [38]. Nonetheless, since then, few other studies have implemented smartphone technology to capture real-time alcohol consumption data and compare this to retrospective assessments [41–43]. Monk, Heim, and Price [41] applied smartphone technologies to investigate real-time alcohol consumption. Their application was designed to give hourly prompts to participants to select the context and number of drinks consumed. In line with our findings, their study confirmed that participants significantly under-report alcohol consumption when assessed by retrospective self-report measures. A difference of almost four drinks was reported on daily alcohol consumption (8.5 in real-time versus 4.2 retrospective). Taken together, these findings underline the importance of using real-time assessments when accurate real word evidence is needed.

4.3. Daytime Activity Levels

To our knowledge, this is the first study that directly compared objective assessments of levels of daytime activity during the hangover state and an alcohol-free control day. The assessments revealed that daytime activity levels were significantly reduced during alcohol hangover. These findings confirm previous self-reports of increased apathy and reduced alertness during the hangover state [16,17], which can be regarded as indications of reduced activity during alcohol hangover. While the GENEactiv watch objectively assessed the levels and duration of daytime activity, in the current study it was not assessed in what type of activity the subjects were actually engaged. Future research should address which specific activities are affected, delayed or fully skipped, and which other activities are conducted in the same manner as on an alcohol-free day. Also, motives for possible behavioral and activity level changes are largely unknown. Previous research has shown that potentially dangerous activities such as driving a car are significantly impaired during the hangover state [2]. It would be interesting to examine whether or not these types of behaviors are avoided or delayed during the hangover state, and how these behavioral changes and corresponding decision making are motivated by drinkers.

4.4. Limitations and Objectives for Future Research

The study has several limitations that should be addressed. First of all, there were great inter-individual differences in the data, which prevented several differences to reach statistical significance. While the study was adequately powered to demonstrate relevant differences between the hang over and control day, one should take into account that the hangover is a very personal experience that may vary from occasion to occasion. Hence, after consuming the same amount of alcohol, within a subject mood and cognitive effects and the impact on sleep may differ from drinking to drinking occasion. Therefore, future research should capture data from more than one drinking night to get a better overall view of the impact of alcohol consumption on sleep and subsequent hangover state. Second, not all assessments made by the GENEactiv watch were equally informative. For example, the standard output for nightly awakenings was the median duration of these awakenings. It would be more useful to have data on the frequency and duration of each individual nightly awakening. Unfortunately, with the current GENEactiv set-up, we could not recover this data. Third, while the assessments capture physical activity it would also be interested to collect objective data on mental activity during the alcohol hangover state. Electroencephalogram (EEG) and functional magnetic resonance imaging (fMRI) studies should be conducted to provide more insight on this topic. Up to now, only one small pilot study used fMRI to examine brain activity during alcohol hangover [44]. The authors reported that, in the hangover condition, subjects' task performance was not significantly affected. However, maintenance of accurate performance level required compensatory increased activity of prefrontal and temporal brain structures. The authors concluded that correctly performing tasks in the hangover state requires significantly more mental effort compared to performing them on a normal non-drinking day. The interaction between physical and mental activity, and the mediating role of motivation and effort to accomplish tasks during hangover is a largely unexplored, an important topic for future research.

5. Conclusions

The current study showed that sleep duration and quality is significantly negatively affected after alcohol consumption. Furthermore, during the hangover state, activity levels are significantly reduced. Our study advocates for implementing both objective and subjective assessments, and combine both real-time and retrospective measures to provide a more accurate view of a heavy drinking session, and its aftereffects.

Author Contributions: Conceptualization, L.E.D., J.C.V., T.R. and K.B.C.; Methodology, L.E.D.; Formal Analysis, L.E.D. and J.C.V.; Investigation, L.E.D.; Writing—Original Draft Preparation, L.E.D. and J.C.V.; Writing-Review & Editing, K.B.C. and T.R. All authors approved the final manuscript.

Acknowledgments: We wish to thank Noel Brick and the GENEactiv team for providing information on the GENEactiv watch data collection methodology, and members of the Alcohol Hangover Research Group for their suggestions on the study design.

Conflicts of Interest: T.R. has received grants/research support from Aventis, Cephalon, Glaxo Smith Kline, Neurocrine, Pfizer, Sanofi, Schering-Plough, Sepracor, Somaxon, Syrex, Takeda, TransOral, Wyeth and Xenoport and has acted as a consultant for Abbott, Acadia, Acoglix, Actelion, Alchemers, Alza, Ancil, Arena, Astra Zeneca, Aventis, AVER, BMS, BTG, Cephalon, Cypress, Dove, Elan, Eli Lilly, Evotec, Forest, Glaxo Smith Kline, Hypnion, Impax, Intec, Intra-Cellular, Jazz, Johnson & Johnson, King, Lundbeck, McNeil, Medici Nova, Merck, Neurim, Neurocrine, Neurogen, Novartis, Orexo, Organon, Prestwick, Procter-Gamble, Pfizer, Purdue, Resteva, Roche, Sanofi, Schering-Plough, Sepracor, Servier, Shire, Somaxon, Syrex, Takeda, TransOral, Vanda, Vivometrics, Wyeth, Yamanuchi, and Xenoport. J.C.V. has received grants/research support from the Dutch Ministry of Infrastructure and the Environment, Janssen, Nutricia, Red Bull, Sequential, and Takeda, and has acted as a consultant for the Canadian Beverage Association, Centraal Bureau Drogisterijbedrijven, Clinilabs, Coleman Frost, Danone, Deenox, Eisai, Janssen, Jazz, More Labs, Purdue, Red Bull, Sanofi-Aventis, Sen-Jam Pharmaceutical, Sepracor, Takeda,

Toast!, Transcept, Trimbos Institute, Vital Beverages, and ZBiotics. The other authors have no potential conflicts of interest to disclose.

References

1. Van Schrojenstein Lantman, M.; Mackus, M.; van de Loo, A.J.A.E.; Verster, J.C. Development of a definition for the alcohol hangover: Consumer descriptions and expert consensus. *Curr. Drug Abuse Rev.* **2016**, *9*, 148–154. [CrossRef] [PubMed]
2. Verster, J.C.; Bervoets, A.C.; de Klerk, S.; Vreman, R.A.; Olivier, B.; Roth, T.; Brookhuis, K.A. Effects of alcohol hangover on simulated highway driving performance. *Psychopharmacology* **2014**, *231*, 2999–3008. [CrossRef] [PubMed]
3. Jongen, S.; Perrier, J.; Vuurman, E.F.; Ramaekers, J.G.; Vermeeren, A. Sensitivity and validity of psychometric tests for assessing driving impairment: Effects of sleep deprivation. *PLoS ONE* **2015**, *10*, e0117045. [CrossRef]
4. Ebrahim, I.O.; Shapiro, C.M.; Williams, A.J.; Fenwick, P.B. Alcohol and sleep I: effects on normal sleep. *Alcohol. Clin. Exp. Res.* **2013**, *37*, 539–549. [CrossRef]
5. Bajaj, L.; Singh, R. Alcohol hangover-its effects on human body. *J. Addict. Clin. Res.* **2018**, *2*, 14–16. [CrossRef]
6. Hendler, R.A.; Ramchandani, V.A.; Gilman, J.; Hommer, D.W. Stimulant and sedative effects of alcohol. In *Behavioral Neurobiology of Alcohol Addiction*; Springer: Heidelberg, Germany, 2011; pp. 489–509.
7. Martin, C.S.; Earleywine, M.; Musty, R.E.; Perrine, M.W.; Swift, R.M. Development and validation of the biphasic alcohol effects scale. *Alcohol. Clin. Exp. Res.* **1993**, *17*, 140–146. [CrossRef] [PubMed]
8. Simpson, C.W.; Resch, G.E.; Millington, W.R.; Myers, R.D. Glycyl-L-glutamine injected centrally suppresses alcohol drinking in P rats. *Alcohol* **1998**, *16*, 101–107. [CrossRef]
9. Roehrs, T.; Yoon, J.; Roth, T. Nocturnal and next-day effects of ethanol and basal level of sleepiness. *Hum. Psychopharmacol. Clin. Exp.* **1991**, *6*, 307–311. [CrossRef]
10. Roehrs, T.; Roth, T. Sleep, sleepiness, and alcohol use. *Alcohol Res. Health* **2001**, *25*, 101–109. [PubMed]
11. Finnigan, F.; Hammersley, R.; Cooper, T. An examination of next-day hangover effects after a 100 mg/100 mL dose of alcohol in heavy social drinkers. *Addiction* **1998**, *93*, 1829–1838. [CrossRef] [PubMed]
12. McKinney, A.; Coyle, K. Alcohol hangover effects on measures of affect the morning after a normal night's drinking. *Alcohol Alcohol.* **2006**, *41*, 54–60. [CrossRef]
13. Hogewoning, A.; Van de Loo, A.J.A.E.; Mackus, M.; Raasveld, S.J.; De Zeeuw, R.; Bosma, E.R.; Bouwmeester, N.H.; Brookhuis, K.A.; Garssen, J.; Verster, J.C. Characteristics of social drinkers with and without a hangover after heavy alcohol consumption. *Subst. Abuse Rehabil.* **2016**, *7*, 161–167. [CrossRef]
14. Rohsenow, D.J.; Howland, J.; Minsky, S.J.; Arnedt, J.T. Effects of heavy drinking by maritime academy cadets on hangover, perceived sleep, and next-day ship power plant operation. *J. Stud. Alcohol* **2006**, *67*, 406–415. [CrossRef]
15. Van Schrojenstein Lantman, M.; Mackus, M.; Roth, T.; Verster, J.C. Total sleep time, alcohol consumption and the duration and severity of alcohol hangover. *Nat. Sci. Sleep* **2017**, *9*, 181–186. [CrossRef]
16. Van Schrojenstein Lantman, M.; Roth, T.; Roehrs, T.; Verster, J.C. Alcohol hangover, sleep quality, and daytime sleepiness. *Sleep Vigil.* **2017**, *1*, 37–41. [CrossRef]
17. Penning, R.; McKinney, A.; Verster, J.C. Alcohol hangover symptoms and their contribution to overall hangover severity. *Alcohol Alcohol.* **2012**, *47*, 248–252. [CrossRef]
18. Van Schrojenstein Lantman, M.; Mackus, M.; van de Loo, A.J.A.E.; Verster, J.C. The impact of alcohol hangover symptoms on cognitive and physical functioning, and mood. *Hum. Psychopharmacol.* **2017**, *32*. [CrossRef]
19. Rohsenow, D.J.; Howland, J.; Arnedt, J.T.; Almeida, A.B.; Greece, J.; Minsky, S.; Kempler, C.S.; Sales, S. Intoxication with bourbon versus vodka: Effects on hangover, sleep, and next-day neurocognitive performance in young adults. *Alcohol Clin. Exp. Res.* **2010**, *34*, 509–518. [CrossRef]
20. Roehrs, T.; Papineau, K.; Rosenthal, L.; Roth, T. Ethanol as a hypnotic in insomniacs: Self administration and effects on sleep and mood. *Neuropsychopharmacol* **1999**, *20*, 279–286. [CrossRef]
21. Feige, B.; Gann, H.; Brueck, R.; Hornyak, M.; Litsch, S.; Hohagen, F.; Riemann, D. Effects of alcohol on polysomnographically recorded sleep in healthy subjects. *Alcohol Clin. Exp. Res.* **2006**, *30*, 1527–1537. [CrossRef]
22. Wilkinson, A.N.; Afshar, M.; Ali, O.; Bhatti, W.; Hasday, J.D.; Netzer, G.; Verceles, A.C. Effects of binge alcohol consumption on sleep and inflammation in healthy volunteers. *J. Int. Med. Res.* **2018**. [CrossRef]

23. Rohsenow, D.J.; Howland, J.; Minsky, S.J.; Greece, J.; Almeida, A.; Roehrs, T.A. The acute hangover scale: A new measure of immediate hangover symptoms. *Addict. Behav.* **2007**, *32*, 1314–1320. [CrossRef] [PubMed]
24. Huntley, G.; Treloar, H.; Blanchard, A.; Monti, P.M.; Carey, K.B.; Rohsenow, D.J.; Miranda, R., Jr. An event-level investigation of hangovers' relationship to age and drinking. *Exp. Clin. Psychopharmacol.* **2015**, *23*, 314. [CrossRef] [PubMed]
25. Riordan, B.C.; Conner, T.S.; Flett, J.A.; Scarf, D. A brief orientation week ecological momentary intervention to reduce university student alcohol consumption. *J. Stud. Alcohol Drugs* **2015**, *76*, 525–529. [CrossRef]
26. Selzer, M.L.; Vinokur, A.; van Rooijen, L. A self-administered Short Michigan Alcoholism Screening Test (SMAST). *J. Stud. Alcohol* **1975**, *36*, 117–126. [CrossRef]
27. ActivInsights. The GENEActiv Product Range. Available online: https://www.activinsights.com/products/geneactiv (accessed on 1 August 2018).
28. GENEActiv. Open Platform. Available online: https://open.geneactiv.org/ (accessed on 1 August 2018).
29. Esliger, D.W.; Rowlands, A.V.; Hurst, T.L.; Catt, M.; Murray, P.; Eston, R.G. Validation of the GENEA Accelerometer. *Med. Sci. Sports Exerc.* **2011**, *43*, 1085–1093. [CrossRef]
30. Jette, M.; Sidney, K.; Blümchen, G. Metabolic equivalents (METS) in exercise testing, exercise prescription, and evaluation of functional capacity. *Clin. Cardiol.* **1990**, *13*, 555–565. [CrossRef] [PubMed]
31. Verster, J.C.; Roth, T. Drivers can poorly predict their own driving impairment: A comparison between measurements of subjective and objective driving quality. *Psychopharmacol* **2012**, *219*, 775–781. [CrossRef]
32. Baldwin, W. Information no one else knows: The value of self-report. In *The Science of Self-Report*; Psychology Press: Hove, UK, 1999; pp. 15–20.
33. Jobe, J.B.; Herrmann, D.J. Implications of models of survey cognition for memory theory. *Basic Appl. Mem. Res.* **1996**, *2*, 193–205.
34. Godden, D.R.; Baddeley, A.D. Context-dependent memory in two natural environments: On land and underwater. *Br. J. Psychol.* **1975**, *66*, 325–331. [CrossRef]
35. McKinney, A.; Coyle, K. Next day effects of a normal night's drinking on memory and psychomotor performance. *Alcohol Alcohol.* **2004**, *39*, 509–513. [CrossRef]
36. Verster, J.C.; Van Duin, D.; Volkerts, E.R.; Schreuder, A.H.C.M.L.; Verbaten, M.N. Alcohol hangover effects on memory functioning and vigilance performance after an evening of binge drinking. *Neuropsychopharmacol* **2003**, *28*, 740–746. [CrossRef]
37. White, A.M. What happened? Alcohol, memory blackouts, and the brain. *Alcohol Res. Health* **2003**, *27*, 186–197.
38. Verster, J.C.; Tiplady, B.; McKinney, A. Mobile technology and naturalistic study designs in addiction research. *Curr. Drug Abuse Rev.* **2012**, *5*, 169–171. [CrossRef]
39. Dulin, P.L.; Alvarado, C.E.; Fitterling, J.M.; Gonzalez, V.M. Comparisons of alcohol consumption by timeline follow back vs. smartphone-based daily interviews. *Addict. Res. Theory* **2017**, *25*, 195–200. [CrossRef]
40. Collins, R.L.; Kashdan, T.B.; Gollnisch, G. The feasibility of using cellular phones to collect ecological momentary assessment data: Application to alcohol consumption. *Exp. Clin. Psychopharmacol.* **2003**, *11*, 73. [CrossRef]
41. Monk, R.L.; Heim, D.; Qureshi, A.; Price, A. I have no clue what I drunk last night using Smartphone technology to compare in-vivo and retrospective self-reports of alcohol consumption. *PLoS ONE* **2015**, *10*, e0126209. [CrossRef]
42. Luczak, S.E.; Rosen, I.G.; Wall, T.L. Development of a real-time repeated-measures assessment protocol to capture change over the course of a drinking episode. *Alcohol Alcohol.* **2015**, *50*, 180–187. [CrossRef]
43. Krenek, M.; Lyons, R.; Simpson, T.L. Degree of correspondence between daily monitoring and retrospective recall of alcohol use among men and women with comorbid AUD and PTSD. *Am. J. Addict.* **2016**, *25*, 145–151. [CrossRef]
44. Howland, J.; Rohsenow, D.J.; McGeary, J.E.; Streeter, C.; Verster, J.C. Proceedings of the 2010 symposium on hangover and other residual alcohol effects: Predictors and consequences. *Open Addict. J.* **2010**, *3*, 131–132. [CrossRef]

© 2019 by the authors. Licensee MDPI, Basel, Switzerland. This article is an open access article distributed under the terms and conditions of the Creative Commons Attribution (CC BY) license (http://creativecommons.org/licenses/by/4.0/).

Article

Relationship between Alcohol Hangover and Physical Endurance Performance: Walking the Samaria Gorge

Joris C Verster [1,2,3], Aikaterini Anogeianaki [1], Darren Kruisselbrink [4], Chris Alford [5] and Ann-Kathrin Stock [6,*]

[1] Division of Pharmacology, Utrecht Institute for Pharmaceutical Sciences (UIPS), Utrecht University, 3584CG Utrecht, The Netherlands; j.c.verster@uu.nl (J.C.V.); kanogeianaki@gmail.com (A.A.)
[2] Institute for Risk Assessment Sciences (IRAS), Utrecht University, 3584CM Utrecht, The Netherlands
[3] Centre for Human Psychopharmacology, Swinburne University, Melbourne, VIC 3122, Australia
[4] Centre of Lifestyle Studies, Acadia University, Wolfville, NS B4P 2R6, Canada; darren.kruisselbrink@acadiau.ca
[5] Psychological Sciences Research Group, University of the West of England, Bristol BS16 1QY, UK; chris.alford@uwe.ac.uk
[6] Cognitive Neurophysiology Department of Child and Adolescent Psychiatry, Faculty of Medicine of the TU Dresden, University of Dresden, D-01307 Dresden, Germany
* Correspondence: Ann-Kathrin.Stock@uniklinikum-dresden.de

Received: 13 December 2019; Accepted: 30 December 2019; Published: 31 December 2019

Abstract: Alcohol hangover is a potentially debilitating state. Several studies have demonstrated that it does not seem to impair strength or short-term endurance, but its effects on continuous exercise performance/long-term endurance have never been investigated. Therefore, the aim of the current study was to assess hiking performance of participants who walked the 15.8 km Samaria Gorge in Crete, Greece. Participants completed a survey in the morning before walking the Gorge, and in the afternoon after completion of the walk. Demographics, data on previous evening alcohol consumption, sleep, hangover symptoms, and walking performance were assessed. Data from $N = 299$ participants with a mean (SD) age of 38.9 (11.0) years were analyzed. $N = 223$ participants (74.6%) consumed alcohol the evening before walking the Samaria Gorge, and $N = 176$ (78.9%) of those reported a hangover. They consumed a mean (SD) of 3.0 (1.8) alcoholic drinks (10 g alcohol each) with a corresponding next-morning hangover severity of 4.6 (2.4) on a 0–10 scale. Participants with a hangover reported feeling significantly more exhausted after the walk compared to participants with no hangover. The groups did not significantly differ in duration of the walk, and the number and duration of breaks. Overall hangover severity, assessed either before, during, or after walking the Samaria Gorge was not significantly correlated with any walking outcome. In conclusion, hungover participants experienced significantly more exhaustion when performing physical activity at the same level as non-hungover participants.

Keywords: physical performance; hiking; walking; water consumption; alcohol; hangover; sleep; Samaria Gorge

1. Introduction

The alcohol hangover is defined as the combination of mental and physical symptoms experienced the day after a single episode of heavy drinking, starting when the blood alcohol concentration (BAC) approaches zero [1]. A growing body of evidence shows that cognitive functioning and mood are negatively affected in the hangover state [2,3], which may result in impaired daily activities such as job performance [4,5], riding a bicycle [6], or driving a car [7–9].

Much less research has been devoted to the possible impact of the alcohol hangover on physical performance. Van Schrojenstein Lantman et al. [10] investigated the impact of individual hangover symptoms on physical performance. Participants reported that being tired, sleepiness, headache, nausea, and weakness had the greatest impact on their physical performance in the hangover state. However, the few studies that have investigated physical performance in the hangover state to date have failed to find significant effects. In 30 male students, Nelson et al. [11] examined various physical performance activities, including grip strength, a 45-s bicycle ergometer test assessing the maximum number of revolutions against 10 Lb resistance, softball throwing, a vertical jump, and push-ups. Performance in the hangover state did not differ significantly from performance on the control day for any of the tests. Karvinen et al. [12] examined physical performance in 30 firemen and policemen on grip strength, back lift, vertical jump test and a 5-min bicycle ergometer test on a hangover and control day. In this study, also, no significant performance differences were found between the hangover and control day. In a more recent study, Kruisselbrink et al. [13] examined physical activity in the hangover state in 12 females. They performed a 6-min submaximal treadmill run at 6 miles per hour, followed by a run to exhaustion once the treadmill had been raised to a 7% grade. In addition, a grip strength test was conducted. Again, no significant performance difference was observed during the hangover state vs. control.

Devenney et al. [14] assessed physical activity levels during the hangover state. Compared to an alcohol-free control day, activity levels were significantly reduced in the hangover state. In particular, the percentage of time spent on vigorous activity was significantly reduced and a significantly higher percentage of time was spent in a sedentary manner on the hangover day. No significant differences between the hangover and control day were found for time spent on light and moderate activity. Unfortunately, the type of activities participants engaged in was not recorded in this study. Nevertheless, the data show that the activity level is reduced to less demanding levels during the hangover state. Based on this observation, it is tempting to hypothesize that participants may be less capable of performing physical activity at vigorous levels during the hangover state, but more research is needed to verify this.

It should be noted that physical performance in these studies was assessed with tests of relatively short duration and can often be categorized as assessing strength (e.g., grip strength) or balance (e.g., standing on one foot) or short-term endurance (e.g., a bicycle test). As of now, there are however no studies that have examined long-term endurance, i.e., physical performance effects over a longer time, such as a 10 km walk, long-distance running, playing a football match, or skiing. Furthermore, the sample sizes of these studies (12–30 subjects) were relatively small, which may have contributed to observing nonsignificant findings. Therefore, the aim of the current study was to investigate the effects of alcohol hangover on endurance performance within a naturalistic setting. To this extent, walking performance was examined in participants walking the 15.8 km Samaria Gorge, in Crete, Greece.

2. Materials and Methods

The study was conducted in the summer of 2019, at Crete, Greece. Participants who booked a tour to walk the Samaria Gorge were invited to participate in the study. They were recruited on the bus drive towards the entrance of the Samaria Gorge. Participants met the inclusion criteria if they were 18–65 years old and could understand and write English. A researcher was present to help clarify any questions that arose while completing the surveys. The study was conducted by researchers from Utrecht University. The survey was anonymous and participants did not receive an incentive for completing the surveys. The Ethics Committee of the Faculty of Social and Behavioral Sciences of Utrecht University granted ethical approval (approval code FETC17-061). $N = 307$ participants provided informed consent to take part in the study.

2.1. The Walk

Data were collected on 9 different day-excursions to the Samaria Gorge between 23 July and 11 August 2019. Weather data for these days at the Samaria Gorge were obtained from www.worldweatheronline.com. On all days except one (partly cloudy), the weather was described as 'sunny'.

No rainfall was recorded on any of the excursions, the average humidity was 56% with a pressure of 1009 mb. The average temperature rose from 27.7 °C at 21:00 to 29.2 °C at 12:00, 28.9 °C at 15:00 and 27.2 °C at 18:00. A paired samples t-test revealed no significant temperature difference between individual test days and the mean 9-day temperature ($p = 0.988$).

Figure 1A schematizes the day tour of the participants. They were picked up at their accommodation between 06:15 and 06:45 in the morning. Participants had booked their day-excursion to the Samaria Gorge either from Chania ($n = 2$ excursions), Rethymnon ($n = 4$ excursions) or Herakleion ($n = 3$ excursions). Depending on their city of residence, they arrived at the Samaria Gorge after a bus drive of approximately 2 to 3 h: Chania (08:35), Rethymnon (09:00) or Herakleion (09:15). From Hora Sfakia, all busses returned to the participants' accommodations in the afternoon at 18:00. Figure 1B shows the several resting locations where participants could drink water during the walk. These are simple tap points, as there are no inhabited villages or stores from which to buy refreshments in the Samaria Gorge.

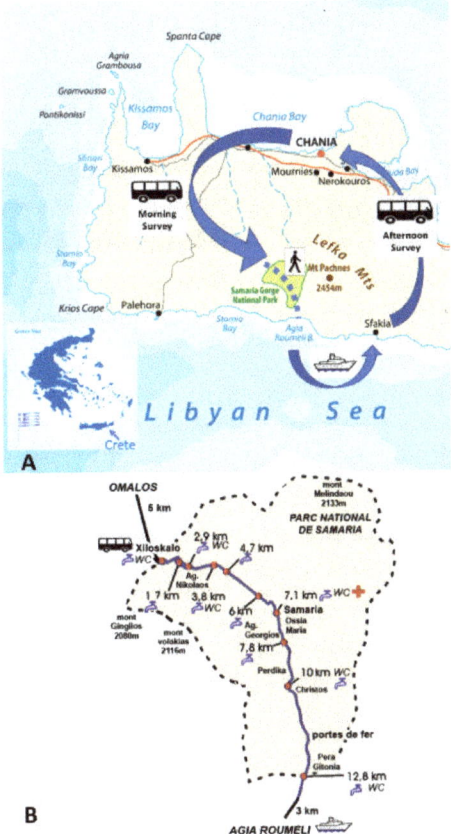

Figure 1. Schematic representation of the day tour and Samaria Gorge walk. (**A**) is adapted and used with permission from www.depositphotos.com. (**B**) is adapted and used with permission from Psarakis Travel Agency (www.psarakistravel.gr).

The Samaria Gorge walk is 15.8 km long. It starts in Xiloskalo at an altitude of 1230 m and ends in Agia Roumeli, a small village at the Libyan Sea. Agia Roumeli can only be reached by walking the Samaria Gorge or by boat. The actual walk through the Samaria Gorge is about 12.8 km long, and after the end of the gorge, it takes another 3 km of walking to reach Agia Roumeli. Hence, participants walk 15.8 km in total, starting at an altitude of 1230 m, slowly descending to sea level in Agia Roumeli. The first part of the walk is 3.8 km long with a 580 m descent, followed by 3.7 km with a 310 m descent. Thereafter, the decent is less steep (3.6 km with a decent of 170 m). The last 4.7 km has a decent of 170 m towards sea level. An impression of the Samaria Gorge walk is given in Figure 2.

Figure 2. Impressions of the Samaria Gorge walk. (**A,B**) Over the first kilometers, participants descend a narrow switchback path, which is partly bordered by a wooden handrail. A large part of the walk is trough rough terrain, i.e., the slippery stones of a dried-up riverbed. In winter, the riverbed is filled with water, but in summer the river has mostly dried up. Participants cross the riverbed several times, sometimes aided by wooden improvised bridges. (**C,D**) there is no paved road but participants have to follow a trail through a rocky valley. Samaria is the main resting place for most participants, as this uninhabited tiny village contains benches in the shade, a water tap point and a medical post. Through the last part of the walk, the cliffs become higher and narrower. (**E**) At its narrowest point, called the "Gates", the Samaria Gorge is approximately 4 m wide, while the cliffs rise over 300 m high. (**F**) Thereafter, the final fairly flat part of the walk ends in Agia Roumeli at the seaside. Participants can relax, eat and drink in the taverns, and recover from the walk. From Agia Roumeli, a ferry brings the participants to Hora Sfakia in about 1 h. From there, coaches leave at 18:00 to bring the participants back to their accommodation. Figures adapted and used with permission from www.depositphotos.com.

2.2. Data Collection

Two surveys were conducted. The first survey was completed in the bus, before walking the Samaria Gorge. Demographic information was collected. Participants provided information on the past evening's alcohol consumption. They reported the number of alcoholic drinks consumed (European sizes were provided as examples, which contain 10 g alcohol each) as well as the start and stop time of drinking. Together with information on sex and weight, this allowed for the calculation of their estimated blood alcohol concentration (BAC) using a modified Widmark formula [15]. Being drunk/intoxicated was rated on an 11-point scale ranging from 0 (absent) to 10 (extreme) [16,17]. The number of smoked cigarettes was recorded as well, as previous research has shown an interaction between smoking and hangover [18]. To capture possible effects of sleep [14,19,20], total sleep time was recorded as well as the (consciously perceived) number of nightly awakenings. Sleep quality was rated on an 11-point scale ranging from 0 (very poor) to 10 (excellent) [21,22].

Physical activity levels were assessed with the IPAQ (International Physical Activity Questionnaire)—short form [23]. This measure assesses the intensity and duration of physical activity and sitting time that people do as part of their daily lives. The outcomes are presented in Metabolic Equivalent Task (MET)-minutes. MET minutes represent the amount of energy spent when engaged in a physical activity. One MET is what you expend when you are at rest. When calculating the MET-minutes per week for the different activity levels, the duration of activity is multiplied with 3.3 (walking), 4 (moderate physical activity) or 8 (vigorous physical activity). The assessments were made for usual physical activity at home, and for physical activity during their holiday in Crete.

Overall hangover severity before starting the walk was rated on a single item 11-point scale ranging from 0 (absent) to 10 (extreme) [24]. Using the same 11-point scale, the severity of a number of individual symptoms that are often related to hangover was also assessed. These included fatigue (being tired), sleepiness, thirst, headache, nausea, loss of appetite, dizziness, stomach pain, heart racing, weakness, anxiety, depression, tension/stress, and anger/hostility. The past year's immune fitness was assessed with the Immune Status Questionnaire (ISQ) [25]. The ISQ consists of 7 items, including 'common cold', 'diarrhea', 'sudden high fever', 'headache', 'muscle and joint pain', 'skin problems (e.g., acne and eczema)' and 'coughing'. The items are scored on a 5-point Likert scale stating how often the participants experienced these complaints during the past year, including 'never', 'sometimes', 'regularly', 'often', and '(almost) always'. The overall ISQ score ranges from 0 (poor) to 10 (excellent), with higher scores indicating a better immune fitness. Current immune fitness (in Crete) was assessed using the 1-item perceived immune functioning scale [26,27]. In a similar way, current 'physical fitness' and 'mental fitness' were assessed. The scores on these scales ranged from 0 (very bad) to 10 (very good). Finally, participants estimated the amount of effort it would take them to walk the Samaria Gorge on an 11-point scale ranging from 0 (absolutely no effort) to 10 (extreme effort).

The surveys were collected by the researcher once completed. When the participants arrived at the entrance of the Samaria Gorge and started their walk, the bus returned by road and drove towards Hora Sfakia, a small seaside city where the participants were picked up at 18:00 to return by bus to their accommodation. The second survey was completed at the end of the day (between 18:00 and 19:00), during the bus drive to their accommodation. In the second survey, participants reported how they experienced walking the Samaria Gorge. Data were collected on the total duration of the walk, the number and duration of breaks, and the amount of water consumed while walking. Participants rated the amount of effort it took them to walk the Samaria Gorge on an 11-point scale ranging from 0 (absolutely no effort) to 10 (extreme effort). Their level of exhaustion was assessed on an 11-point scale ranging from 0 (absolutely not) to 10 (extremely exhausted). Using the scale described above, overall hangover severity was assessed for two additional time frames: (1) retrospectively for how they were during the walk, and (2) in real-time after the walk (when travelling on the bus to their accommodation). Of note, anxiety, depression, tension/stress, and anger/hostility were omitted from the 'during the walk' assessment. Finally, the number of alcoholic drinks consumed and cigarettes smoked after walking the Samaria Gorge were recorded, and subjective intoxication was rated as described above [16,17].

2.3. Statistical Analysis

All surveys were anonymous. However, because date of birth and sex were entered on both the morning and afternoon survey, they could be matched with each other. Data from participants who used drugs, or who did not the complete the questions on previous evening alcohol consumption were omitted from the statistical analysis. Statistical analyses were conducted with SPSS (IBM Corp. Released 2013. IBM SPSS Statistics for Windows, Version 25.0. Armonk, NY, USA: IBM Corp.). Mean and standard deviation (SD) were computed for each variable. Most data in the study did not follow a normal distribution. Therefore, nonparametric statistics were used to analyze the data. As a nonparametric analog of analysis of variance (ANOVA), the Kruskal–Wallis test was used to compare outcomes of the 'no alcohol', 'no hangover', and 'hangover' groups. If significant, Independent Samples Mann–Whitney U tests (the nonparametric analog of independent t-tests) were used to make paired comparisons between two individual groups (e.g., compare the 'hangover' versus 'no hangover' group).

To compare assessments made within subjects (e.g., before versus after walking), Related Samples Wilcoxon Signed Rank tests were conducted (the nonparametric analog of paired t-tests). Spearman's rho correlations were used to compute correlations. Results were considered statistically significant if $p < 0.05$. In case of multiple related comparison (i.e., hangover symptoms), a Bonferroni's correction was applied to account for multiplicity. Linear stepwise regression analyses (for which independent variables do not need to be normally distributed or continuous) were conducted to determine which of the individual variables assessed in this study were significant predictors of walking outcomes.

3. Results

3.1. Demographics and Physical Activity Levels

$N = 307$ participants provided consent to participate in the study. One subject used illicit drugs and her data was excluded from the analysis. $N = 7$ other participants did not report on previous evening alcohol consumption, and these incomplete datasets and were also excluded from the analyses. Data for the remaining $N = 299$ participants were included in the analyses. Except for $N = 6$ participants from Canada (2%), all participants originated from 17 European countries. Most participants came from Germany (14.0%), the UK (12.5%), The Netherlands (11.4%), Poland (7.0%), Belgium (6.7%), and Greece (6.1%). $N = 223$ participants (74.6%) consumed alcohol the evening before walking the Samaria Gorge, and $N = 176$ (78.9%) of those reported a hangover score greater than zero in the morning. Their demographics are summarized in Table 1.

Table 1. Demographics and physical activity levels of participants with and without a hangover.

Demographics	Overall	No Alcohol	Alcohol, No Hangover	Alcohol, Hangover	p-Value
N	299	76	47	176	
Age (years)	38.9 (11.0)	39.4 (12.1)	37.5 (11.4)	39.0 (10.3)	0.506
BMI (kg/m^2)	23.9 (2.4)	23.5 (2.4)	24.4 (2.7)	24.0 (2.4)	0.061
ISQ	8.6 (1.9)	8.8 (1.9)	7.6 (2.5) ‡	8.7 (1.5)	0.006
General health	7.4 (1.1)	7.5 (1.0)	7.9 (1.2)	7.3 (1.0)	0.005
Weekly alcohol consumption at home	4.8 (3.6)	2.0 (1.7)	4.5 (4.0) ‡	5.9 (3.3) *X	0.000
Physical activity at home					
Vigorous (MET-min/w)	961 (1728)	879 (1937)	1351 (1757) ‡	883 (1617) *	0.035
Moderate (MET-min/w)	1869 (2470)	1928 (3048)	2004 (3868)	1806 (1551)	0.100
Walking (MET-min/w)	3298 (2240)	3511 (1913)	2937 (3725) ‡	3305 (1810) *	0.004
Total (MET-min/w)	5972 (4231)	6124 (4130)	6201 (6008)	5843 (3666)	0.637
Time spent sitting (min)	385 (119)	404 (121)	382 (150)	377 (108)	0.490
On holiday in Crete					
Vigorous (MET-min/w)	708 (1432)	823 (1722)	892 (1406)	611 (1299)	0.258
Moderate (MET-min/w)	1588 (1531)	1665 (1766)	1319 (1981)	1627 (1273) *	0.019
Walking (MET-min/w)	3409 (2361)	4099 (2964)	3319 (2953) ‡	3141 (1788) X	0.028
Total (MET-min/w)	5588 (3937)	6498 (4569)	5536 (4979)	5219 (4274)	0.155
Time spent sitting (min)	353 (111)	355 (114)	291 (117) ‡	369 (102)	0.000
Evening before					
Number of cigarettes smoked	3.7 (5.9)	0.8 (2.8)	1.3 (3.1) ‡	5.6 (6.6) * X	0.000
Total sleep time (h)	6.1 (0.8)	6.3 (0.8)	6.2 (1.2)	5.9 (0.6) * X	0.000
Number of nightly awakenings	1.1 (1.0)	0.9 (1.0)	1.1 (1.2)	1.3 (0.9) X	0.006
Sleep quality	6.1 (1.9)	7.2 (1.9)	7.1 (1.6)	5.4 (1.6) * X	0.000

p-value from the Kruskal–Wallis test is shown, comparing the outcomes of the three groups. If the group effect was significant ($p < 0.05$), Independent Samples Mann–Whitney U tests were conducted to investigate paired comparisons between the individual groups. Significant differences ($p < 0.0001$) between the 'hangover' group and 'no hangover' group are indicated by *. Significant differences ($p < 0.0001$) between the 'no hangover' group and 'no alcohol' group are indicated by ‡. Significant differences ($p < 0.0001$) between the 'hangover' group and 'no alcohol' group are indicated by X. Abbreviations: BMI = body mass index, ISQ = immune status questionnaire, MET = metabolic equivalent of task, min = minutes, /w = per week.

Table 1 shows that at home, participants in the 'hangover' group consumed significantly more alcohol ($p < 0.0001$) compared to the 'no hangover' group and 'no alcohol' group. The evening before the walk, participants with a hangover smoked significantly more cigarettes ($p < 0.0001$), and reported a significantly reduced sleep quality ($p < 0.0001$) and sleep duration ($p = 0.022$) than the 'no hangover' group. Although some differences between the intensity levels were observed, total physical activity at home and in Greece did not significantly differ between the groups.

3.2. Drinking Characteristics and Hangover Symptom Severity

Alcohol consumption characteristics the evening before their Samaria Gorge excursion for drinkers with and without a hangover on the day of their walk are summarized in Table 2.

Table 2. Drinking characteristics.

	No Alcohol	No Hangover	Hangover	p-Value
Evening before walking				
Number of alcoholic drinks	0.0 (0.0)	0.9 (1.6)	3.0 (1.8)	0.000 *
Subjective intoxication	—	1.3 (1.9)	4.6 (2.4)	0.000 *
Start time drinking (h.min)	—	19.10 (2.4)	17.40 (1.8)	0.000 *
Stop time drinking (h.min)	—	21.45 (1.9)	20.13 (1.9)	0.000 *
Duration of drinking (h)	—	2.6 (2.4)	2.5 (1.5)	0.157
Estimated BAC (%)	—	0.02 (0.03)	0.03 (0.03)	0.001 *
Test day				
Overall hangover severity before walking	—	0.0 (0.0)	4.6 (2.1)	0.000 *
Overall hangover severity during walking	—	1.1 (1.8)	3.4 (2.3)	0.000 *
Overall hangover severity after walking	—	1.1 (2.1)	3.5 (2.4)	0.000 *
Afternoon after walking				
Number of alcoholic drinks	0.4 (0.9)	1.0 (1.7)	1.0 (1.3)	0.112
Subjective intoxication	0.6 (1.6)	0.9 (2.1)	1.5 (2.0)	0.010 *
Number of cigarettes smoked	0.4 (1.5)	1.0 (2.3)	3.4 (4.5)	0.000 *

An Independent Samples Mann–Whitney U test was conducted to compare the 'no hangover' and 'hangover' group. Significant differences between the 'hangover' and 'no hangover' group ($p < 0.05$) are indicated by *.

Table 2 shows that, the evening before their excursion, participants in the 'hangover' group consumed on average 3.0 alcoholic drinks (European size, 10 g alcohol each). Although this seems a low amount of alcohol in comparison to, for example, student samples [24], it equates to half their usual weekly alcohol consumption (see Table 1). Overall hangover severity declined during the day. Participants were allocated to the 'no hangover' group when overall hangover severity was rated zero in the morning. However, some of these participants did report positive hangover scores during or after the walk. Hence, the mean hangover severity is not zero in the 'no hangover' group for these assessments. Severity scores for individual hangover symptoms, rated before, during (retrospectively), and after walking the Samaria Gorge are summarized in Table 3.

In the morning, hangover symptom severity scores were usually significantly higher in the 'hangover' group compared to the 'no hangover' group. During and after the walk, no significant differences were found between the 'hangover' group and 'no hangover' group.

The hangover group experienced a significant reduction in severity scores during the walk for sleepiness, headache, nausea, dizziness, and stomach pain. No significant differences were observed between individual symptom severity assessment 'during' or 'after' the walk, indicating that the experience of symptom severity had plateaued during the walk.

Although often not considered in hangover research, various symptoms experienced during the hangover state are also experienced by participants who reported having no hangover or consumed no alcohol at all, thus reporting overall hangover severity scores of zero. Although their severity scores are usually low and fairly constant, ratings for fatigue, thirst, and weakness significantly increased during the walk in both the 'no alcohol' and 'no hangover' group ($p < 0.0001$). Compared to the morning assessments, there was also a significant increase in heart racing severity during the walk in the 'no hangover' group. Again, no significant differences between 'during' and 'after' walking assessments were observed in these groups.

Table 3. Hangover and symptom severity before, during and after walking the Samaria Gorge.

Group	No Alcohol N = 76			No Hangover (in the Morning) N = 47			Hangover (in the Morning) N = 176		
Symptom	Before	During	After	Before	During	After	Before	During	After
Fatigue	3.6 (2.0)	5.0 (2.4) bd	5.3 (2.3)	2.6 (1.9)	5.9 (2.7) bd	6.1 (2.5) ba	6.1 (1.7) *, X	6.3 (1.6) X	6.3 (1.7) X
Sleepiness	4.0 (2.5)	3.6 (2.5)	4.1 (2.4)	2.5 (2.3)	3.7 (3.1)	4.2 (3.0)	6.5 (2.0) *, X	4.8 (2.6) bd	5.3 (2.3) X, ba
Thirst	2.1 (2.3)	3.8 (2.8) bd	4.1 (2.9)	2.3 (2.2)	5.1 (2.8) bd	5.1 (2.5) ba	5.6 (2.8) *, X	5.1 (2.6)	5.2 (2.6)
Headache	0.7 (1.7)	0.6 (1.4)	1.0 (2.1)	0.4 (0.9)	1.6 (2.6)	0.8 (1.4)	2.2 (3.0) X	0.9 (1.7) bd	1.1 (2.0) ba
Nausea	0.3 (0.7)	0.3 (0.9)	0.3 (1.1)	0.1 (0.3)	0.5 (1.2)	0.3 (0.9)	0.8 (1.6)	0.4 (0.9) bd	0.4 (1.3)
Dizziness	0.4 (1.2)	0.4 (1.0)	0.7 (1.5)	1.0 (1.8)	0.6 (1.2)	0.9 (1.7)	1.3 (1.7) *, X	0.6 (1.2) bd	0.9 (1.6)
Stomach pain	0.6 (1.4)	0.4 (1.1)	0.7 (1.5)	0.3 (0.8)	0.6 (1.1)	0.8 (1.5)	2.1 (2.2) *, X	0.9 (1.5) bd	1.0 (1.4) ba
Heart racing	0.7 (1.6)	1.5 (2.0)	1.2 (2.0)	0.3 (0.7)	2.8 (2.7) bd	2.6 (2.3) ba, ‡	2.3 (2.1) *, X	2.1 (2.0)	1.9 (1.7) X
Weakness	0.6 (1.2)	2.2 (2.2) bd	2.0 (2.1)	0.5 (0.8)	2.8 (2.1) bd	2.6 (2.5)	2.3 (1.9) X	2.9 (1.9)	2.4 (1.9)
Loss of appetite	1.0 (1.9)	NA	0.6 (1.6)	1.4 (1.8)	NA	1.1 (1.6)	1.0 (1.7)	NA	0.5 (1.2)
Anxiety	0.6 (1.1)	NA	0.5 (1.3)	0.7 (1.6)	NA	0.2 (0.6)	1.2 (1.6)	NA	0.8 (1.2)
Depression	0.1 (0.5)	NA	0.2 (0.8)	0.3 (1.0)	NA	0.0 (0.3)	0.3 (1.0)	NA	0.1 (0.5)
Tension, stress	0.6 (1.5)	NA	0.6 (1.7)	0.7 (1.5)	NA	0.5 (1.6)	0.4 (1.0)	NA	0.4 (0.9)
Anger, hostility	0.2 (0.7)	NA	0.3 (0.7)	0.6 (1.5)	NA	0.4 (1.0)	0.3 (0.9)	NA	0.3 (0.7)

Mean (SD) severity scores are shown. A Bonferroni's correction ($p < 0.0001$) was applied to account for multiple comparisons. Related Samples Wilcoxon Signed Rank tests were conducted to investigate pairwise comparisons between the timepoints of assessment within each group. Significant differences between 'before walking' and 'after walking' assessments ($p < 0.0001$) are indicated with 'ba'. Significant differences between 'before walking' and 'during walking' assessments ($p < 0.0001$) are indicated with 'bd'. No significant differences between 'during walking' and 'after walking' assessments ($p < 0.0001$) were observed. Between-group comparisons were made with a Kruskal–Wallis test. If the group effect was significant ($p < 0.001$), Independent Samples Mann–Whitney U tests were conducted to investigate paired comparisons between the individual groups. Significant differences ($p < 0.0001$) between the 'hangover' group and 'no hangover' group are indicated by *. Significant differences ($p < 0.0001$) between the 'no hangover' group and 'no alcohol' group are indicated by ‡. Significant differences ($p < 0.0001$) between the 'hangover' group and 'no alcohol' group are indicated by X. Abbreviation: NA = not assessed.

Table 3 further summarizes alcohol consumption and smoking data that occurred after the completion of walking the Samaria Gorge. Across all groups, alcohol consumption, smoking, and subjective intoxication levels were relatively low.

As this alcohol consumption and smoking occurred after the walk while waiting for the bus to return to their apartments/hotel, they have no impact on the dependent variables in this study (i.e., the walking outcomes). Their effects on hangover severity assessed after the walk are described in Section 3.7.

3.3. Walking Performance

Table 4 summarizes the walking outcomes and fitness measures of the three groups.

Participants in the 'hangover' group felt significantly more exhausted after walking the Samaria Gorge compared to participants in the 'no hangover' group ($p = 0.004$). The latter was anticipated by participants of the 'hangover' group, as they rated the expected effort to complete the walk as significantly higher than the 'no hangover' group ($p < 0.0001$). However, actual effort scores did not significantly differ between the 'hangover' group and 'no hangover' group ($p = 0.136$). Also, participants with a hangover consumed significantly more water compared to participants with no hangover (300 mL more on average, $p = 0.043$) and those who consumed no alcohol (400 mL more on average).

Interestingly, the immune fitness rating before walking was significantly better in the 'no hangover' group ($p < 0.0001$) compared to the 'hangover' group and the 'no alcohol' group ($p < 0.0001$). In the 'hangover' group, the immune fitness rating did not significantly correlate with overall hangover severity ($r = 0.101$, $p = 0.184$).

Table 4. Performance assessments related to walking the Samaria Gorge.

	No Alcohol	Alcohol, no Hangover	Alcohol, Hangover	p-Value
Assessed before walking				
Immune fitness	7.5 (1.3)	8.3 (1.5) ‡	7.3 (1.2) *	0.000
Mental fitness	8.6 (1.3)	8.6 (1.2)	8.5 (1.2)	0.628
Physical fitness	6.9 (1.5)	7.4 (1.8)	6.8 (1.4)	0.057
Expected effort to walk the Gorge	8.0 (1.9)	6.6 (1.9) ‡	8.4 (1.4)	0.000
Assessed after walking				
Immune fitness	6.9 (1.2)	6.8 (1.7)	6.3 (1.4) X	0.003
Mental fitness	7.9 (1.4)	7.8 (1.7)	7.7 (1.5)	0.636
Physical fitness	6.8 (1.3)	6.4 (1.8)	6.1 (1.4) X	0.004
Effort to walk the Gorge	8.1 (1.4)	7.9 (1.7)	8.3 (1.5)	0.135
Exhaustion	8.3 (1.4)	7.8 (2.2)	8.8 (1.4) * X	0.002
Walking time (hours)	6.1 (0.7)	6.1 (0.9)	6.0 (0.6)	0.734
Number of breaks	4.3 (1.7)	4.4 (1.9)	4.0 (1.1)	0.666
Total duration of breaks (minutes)	24.3 (12.5)	26.7 (19.0)	25.6 (8.7)	0.227
Water consumed during the walk (liters)	2.3 (0.9)	2.4 (0.8)	2.7 (0.7) * X	0.001

p-value from the Kruskal–Wallis test is shown, comparing the outcomes of the three groups. If the group effect was significant ($p < 0.05$), Independent Samples Mann–Whitney U tests were conducted to investigate paired comparisons between the individual groups. Significant differences ($p < 0.0001$) between the 'hangover' group and 'no hangover' group are indicated by *. Significant differences ($p < 0.0001$) between the 'no hangover' group and 'no alcohol' group are indicated by ‡. Significant differences ($p < 0.0001$) between the 'hangover' group and 'no alcohol' group are indicated by X.

3.4. Correlates of Walking Performance

Overall hangover severity, assessed either before, during (in retrospect), or after walking the Samaria Gorge was not significantly correlated with any walking outcome.

Linear stepwise regression models were computed, including sex, age, BMI, weekly alcohol consumption, Total MET/week home and Crete, time spent sitting home and Crete, ISQ, general health rating, immune/menta/physical fitness in morning, number of cigarettes smoked, overall hangover severity (but NOT all individual hangover symptoms) in the morning, total sleep time, nightly awakenings, sleep quality, all drinking outcomes, and group ('no alcohol', 'no hangover', 'hangover'). The models are summarized in Table 5.

3.5. Sex Differences

Of the participants with a hangover, $N = 105$ were men and $N = 71$ were women. Only a few sex differences were observed. With regards to demographics, women had a significantly lower BMI than men (22.8 ± 1.9 vs. 24.9 ± 2.3 kg/m^2, $p < 0.0001$), spent less time on walking at home (2918 ± 1585 vs. 3566 ± 1911 MET-minutes, $p = 0.035$), reported less physical activity (total METs per week) at home (5014 ± 2974 vs. 6390 ± 3981 MET-minutes, $p = 0.037$), and reported a significantly lower weekly alcohol consumption at home (3.9 ± 1.8 vs. 7.4 ± 3.4 alcoholic drinks, $p < 0.0001$). Women's reported past year immune fitness was significantly higher than men's (9.1 ± 1.4 vs. 8.5 ± 1.6, $p = 0.004$).

The evening before the walk, women smoked significantly fewer cigarettes (4.3 ± 5.8 vs. 6.6 ± 6.9 cigarettes, $p = 0.026$), and consumed significantly less alcohol (2.3 ± 1.3 vs. 3.5 ± 1.9 alcoholic drinks, $p < 0.0001$), but over a significantly shorter time (2.1 ± 1.5 vs. 2.8 ± 1.3 h, $p < 0.0001$), resulting in a significantly higher estimated BAC in women (0.036% ± 0.03% vs. 0.028% ± 0.03%, $p = 0.006$). Subjective intoxication during drinking (4.2 ± 2.3 vs. 4.9 ± 2.5, $p = 0.051$) and next morning hangover severity (4.2 ± 2.2 vs. 4.8 ± 2.1, $p = 0.063$) did not significantly differ between men and women. Nonetheless, hangover severity was significantly lower in women than in men, as assessed retrospectively when thinking back on their experience during the walk (2.7 ± 2.1 vs. 3.8 ± 2.4, $p = 0.001$) and after the walk (2.7 ± 2.1 vs. 4.0 ± 2.4, $p < 0.0001$). During the walk, women consumed significantly less water than men (2.3 ± 0.6 vs. 2.9 ± 0.5 L, $p < 0.0001$), but no significant sex differences were found on any walking outcome.

Table 5. Correlates of walking performance.

	Predictive Validity of the Model	Contribution of Individual Variables
Exhaustion	11.0%	Physical fitness rate before walking (6.3%) Group (2.6%) Physical activity (total METs) in Crete (2.1%)
Effort	29.6%	Physical activity (total METs) in Crete (14.5%) Physical fitness rate before walking (6.7%) Number of cigarettes smoked (3.4%) Group (1.9%) Number of alcoholic drinks consumed (1.6%) BMI (1.5%)
Duration of the walk	29.2%	Physical fitness rate before walking (11.0%) Physical activity (total METs) in Crete (5.1%) BMI (3.7%) Past year's immune status (ISQ) (3.7%) Usual weekly alcohol consumption (3.0%) Time spent sitting on a week day in Crete (2.7%)
Number of breaks	19.9%	Time spent sitting on a week day at home (7.9%) BMI (3.7%) Physical activity (total METs) in Crete (3.5%) Sex (2.9%) Group (1.9%)
Total duration of breaks	17.5%	Usual weekly alcohol consumption (6.5%) BMI (6.2%) Duration of drinking evening before (4.8%)
Amount of water consumed	21.1%	Sex (15.1%) Physical activity (total METs) in Crete (7.2%) Past year's immune status (ISQ) (5.2%) BMI (3.2%) Physical fitness rate before walking (1.4%)

The regression models show that physical activity (total METs) in Crete, the physical fitness rate before walking, and BMI are important factors that predict walking outcomes. Albeit modest, 'group' membership ('no alcohol', 'no hangover', 'hangover') significantly predicted exhaustion after the walk, effort to perform the walk, and the number of breaks. Sex had the largest impact on water consumption.

3.6. The Association of Water Consumption and Hangover Severity

In the 'hangover' group, the amount of consumed water was significantly correlated with the duration of the walk ($r = 0.160$, $p = 0.033$), the number of breaks ($r = 0.234$, $p = 0.002$), the total duration of the breaks ($r = 0.173$, $p = 0.022$), and the level of reported exhaustion after the walk ($r = 0.161$, $p = 0.034$). Water consumption was not significantly correlated with the amount of effort to complete the walk ($r = 0.016$, $p = 0.831$). No significant correlations were found between the amount of consumed water and overall hangover severity, or other individual hangover symptom scores that were assessed either before, during or after walking the Samaria Gorge. It was further investigated whether the changes in severity scores were associated with the amount of water consumed during the walk. A difference score (Δ, afternoon assessment − morning assessment) was calculated for overall hangover severity and correlated with water intake while walking the Samaria Gorge. The amount of water consumed did not significantly correlate with Δ overall hangover severity ($r = 0.066$, $p = 0.388$) (see Figure 3). The hangover severity difference score between the assessment during walking and the assessment before walking also did not significantly correlate with the amount of water consumed ($r = 0.103$, $p = 0.174$).

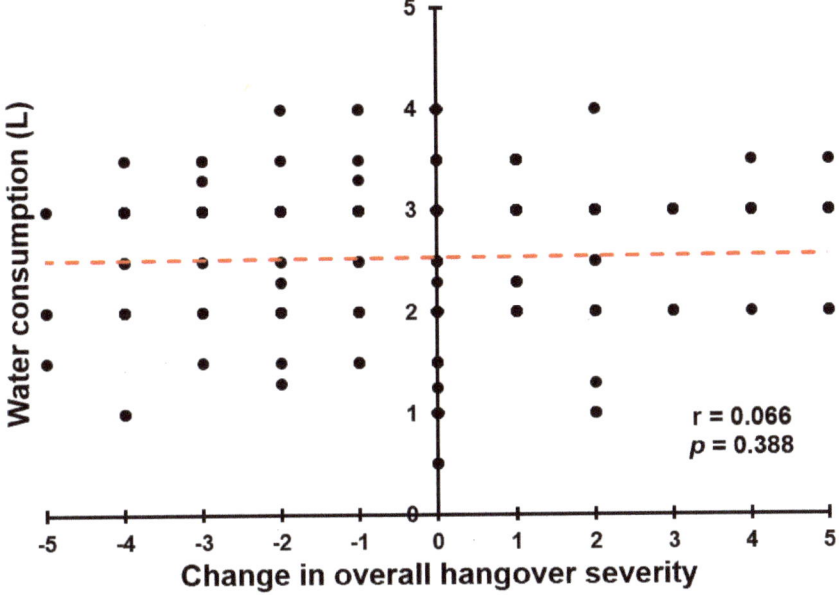

Figure 3. Relationship between water consumption and change in overall hangover severity. A difference score (Δ) was calculated for overall hangover severity, by deducting the morning severity rating from the afternoon severity rating. Red dotted line represents the Spearman's rho correlation, which was not significant.

3.7. The Association of Alcohol Consumption and Smoking after Walking and Hangover Severity

Alcohol consumption after the walk was low across all groups, averaging about one alcoholic drink. However, a consumption range of zero to eight drinks was observed. Similarly, the reported number of cigarettes smoked after the walk varied between zero and 17 cigarettes. For the subgroup that had a hangover in the morning, a difference score (Δ, afternoon assessment − during walking assessment) was calculated for overall hangover severity and correlated with the amount of alcohol consumed and number of cigarettes smoked after walking the Samaria Gorge. A significant correlation was found with the amount of alcohol consumed ($r = 0.329$, $p < 0.0001$), suggesting that alcohol consumption in the afternoon was associated with increased hangover severity (See Figure 4A). The correlation with number of cigarettes smoked after the walk was not significant ($r = 0.119$, $p = 0.125$). Those who consumed alcohol ($N = 99$) had a significant increase in hangover severity (Δ, afternoon assessment − during walking assessment), mean (SD) + 0.4 (1.9), while subjects that did not consume alcohol ($N = 75$) showed a further decrease in hangover severity after the walk, mean (SD) − 0.3 (1.6) (See Figure 4B). The difference observed change scores of the groups was statistically significant ($p = 0.005$).

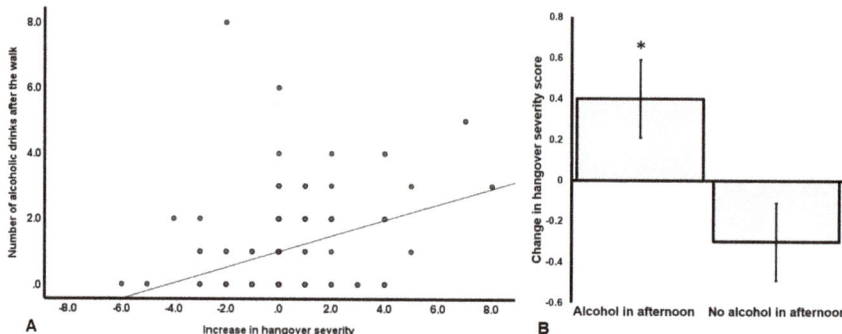

Figure 4. The relationship of alcohol consumption and hangover severity. (**A**) shows the Spearman's correlation between alcohol consumption and the change in hangover severity (Δ, afternoon assessment − during walking assessment). A positive correlation was found suggesting that higher levels of alcohol consumption during hangover (after the walk) were associated with experiencing more severe hangovers. (**B**) shows the change in hangover severity scores (Δ, afternoon assessment − during walking assessment) for subjects that consumed alcohol after the walk versus subjects that did not consume alcohol after walking the Samaria Gorge. An increase in hangover severity was observed in those who consumed alcohol after the walk, whereas a decrease in hangover severity was observed in subjects who did not consume alcohol after walking the Samaria Gorge. Error bars represent standard errors. The significant difference ($p = 0.005$) between drinkers and those who did not consume alcohol after the walk is indicated by *.

When interpreting the data in Figure 4, it should be taken into account that the observed absolute differences are small on a hangover scale ranging from 0 to 10. For the two groups individually, the difference in change scores (hangover severity assessed after walking minus before walking) did not reach statistical significance ($p = 0.056$ and $p = 0.082$ for those who did and did not consume alcohol, respectively). The two groups did not differ in the amount of water consumed during the walk ($p = 0.199$), nor on any of the demographic variables assessed in this study.

4. Discussion

The data showed that walking the Samaria Gorge while having a hangover was associated with significantly higher exhaustion scores when compared to participants with no hangover or who consumed no alcohol. However, no significant differences between the groups were found in the duration of the walk, or number and duration of breaks. The amount of water consumed was significantly correlated with the duration of the walk, the number of breaks, the total duration of the breaks, and the level of reported exhaustion after the walk. Further, participants with a hangover consumed significantly more water during the walk. However, neither hangover severity, nor its reduction during the day, were significantly correlated with the amount of water consumed (see Figure 3). The regression analyses revealed that the observed differences in walking outcomes between participants with and without a hangover are related to differences in levels of physical activity in Crete, reported physical fitness, and BMI. The regression analyses revealed that in this sample, sleep outcomes had no relevant impact on walking performance. Furthermore, overall hangover severity did not significantly correlate with any walking outcome and was not a significant predictor in the regression analyses, even though group membership was. The latter suggests that the relative severity of a hangover is not important with regard to endurance performance, but the fact whether or not you experience a hangover per se is a significant determinant.

As walking the Samaria Gorge is an activity that most people do together with others (e.g., in the current study, several families and groups of friends participated), it is understandable that these individuals all have the same total walking time, number and duration of breaks. Indeed, all three groups completed the walk in approximately the same time, with the same total duration of breaks, providing a useful between groups control. It is therefore understandable that participants with a hangover reported that more effort was needed to perform at this same physical exercise level and reported to be significantly more exhausted than participants who reported no hangover or who had consumed no alcohol.

Several factors may have influenced our findings. First, this was a naturalistic study, in which the researchers did not intervene with the activities of the participants, nor did they receive any instructions [28]. Participants received no instructions on how to walk the Gorge. The results could have been different if, for example, participants were instructed to walk the Gorge as quickly as possible, or when water intake was restricted to a pre-set volume. Participants were also not aware that they would need to retrospectively rate their hangover symptoms experienced during the walk. Had they known, they might have monitored their subjective experience more closely which might have produced more accurate ratings.

Second, this was a non-student sample of middle-aged adults. As a consequence, this sample consumed much less alcohol in the evening before the walk, as compared to the amounts of alcohol usually seen in hangover studies conducted in student populations [24]. Despite their lower estimated BAC levels, a considerable number of participants reported having a hangover in the morning. Since their weekly alcohol consumption was relatively low, the increase in alcohol consumption seen on holiday in Crete (i.e., about half their usual weekly alcohol intake at home consumed at a single drinking occasion in Crete) may have been sufficient to elicit a hangover, despite the fact that the estimated BAC (0.03%) was well below the levels commonly reported in student samples. This observation supports observations in other studies that lower estimated BAC levels can also elicit a hangover, and that the previously suggested lower limit of 0.11% [29] should be abandoned [30].

Third, participants may have anticipated the fact that they had to complete the Samaria Gorge and, as a result, adjusted their drinking and sleep behavior. Perhaps they also practiced for the walk. Although we did not directly assess this with our surveys, the data suggests that overall weekly physical activity levels did not significantly differ between the groups. In contrast, sleep quality was rated as significantly poorer among participants with a hangover. Moreover, a substantial number of participants (8.4%) consumed more alcohol in the evening before the walk than they normally did in a full regular week at home, and on average drinkers consumed about half their usual weekly alcohol intake on this single study occasion in Crete. The latter suggests that the knowledge of having to walk the Samaria Gorge was not a motivation for participants to moderate their alcohol consumption on the evening before the walk. However, the 'no hangover' group consumed less alcohol than the 'hangover' group. This may suggest that this group did moderate their alcohol consumption in anticipation that they had to walk the Samaria Gorge.

Interestingly, the 'no hangover' group happened to give significantly higher ratings of their immune fitness and physical fitness, as compared to the two other groups. Previous studies have reported a relationship between perceived immune fitness and hangover susceptibility [31], but not with hangover severity [32]. In the current study, hangover severity scores also did not significantly correlate with the immune fitness rating in the 'hangover' group. There is anecdotal evidence suggesting that water consumption brings hangover relief. Indeed, the 'hangover' group consumed significantly more water than participants from the 'no hangover' group, and more than those who consumed 'no alcohol'. However, the observed reduction in hangover severity during the day was not significantly correlated with the amount of water consumed during the walk. Previous research showed that consumption of water before going to bed did not affect next-day hangover severity [33]. It is important to note that causality cannot be concluded from these correlational analyses. The absence of a significant correlation can either imply that consumption of water is not effective in the relief of hangovers, or that having a hangover does not stimulate individuals to consume more water. We also did not found evidence for the 'hair of the dog' effect, i.e., the hypothesis that hangover severity will attenuate by consuming alcohol [34,35]. Instead, the results suggest the opposite, as hangover severity scores increased in those who consumed alcohol in the afternoon and reduced in those who did not consume alcohol. However, these results must be interpreted with caution as, although not very likely, we did not record possible alcohol consumption during the walk. Furthermore, we did not record water and soft drink consumption after the walk. Therefore, future controlled trials are necessary to determine the impact of water and alcohol consumption on the presence and severity of alcohol hangover.

Finally, hangover symptoms were also reported by participants with no hangover, and participants who did not consume any alcohol the evening before walking the Samaria Gorge. The reported symptoms are common complaints, such as sleepiness and thirst, that can be experienced by anyone, including without any specific intervention or event that could explain their occurrence (e.g., alcohol consumption). Some symptoms may be related to sleep loss, whereas other symptoms may be related to the temperature in the Gorge (e.g., the rise in severity scores of thirst while walking), or the physical demands to perform the walk (e.g., increases in scores of weakness during the walk). This points to the importance of using overall hangover severity ratings instead of composite ratings of individual hangover symptoms, as it appears that participants in the 'no hangover' group and 'no alcohol' group did not attribute their experience of individual symptoms such as sleepiness and thirst to alcohol hangover since their overall hangover severity rating was zero. This finding underlines the importance of including an overall hangover severity score in research design instead of relying on composite symptom scores to calculate an overall hangover score, which is done when using the three most frequently used hangover scales [36–38].

5. Conclusions

In conclusion, this study suggests that a significant hangover may be experienced after relatively low levels of alcohol consumption for this older non-student group. In addition, physical endurance performance was associated with experiencing significantly more exhaustion during the hangover state. Future research should investigate possible hangover effects on other physical activities, such as short-term anaerobic performance (e.g., a 100 m sprint or power lifting) or other forms of long term aerobic exercise (e.g., swimming laps or running long distance).

Author Contributions: Conceptualization and design of the study, A.A., D.K., and J.C.V., data acquisition, A.A.; data analysis, J.C.V.; data interpretation, A.A., J.C.V., D.K., C.A., and A.-K.S.; writing original draft, J.C.V., C.A., D.K.; all authors critically reviewed the manuscript for important intellectual content and approved the final version. All authors have read and agreed to the published version of the manuscript.

Funding: This research received open access funding by the Publication Fund of the TU Dresden.

Conflicts of Interest: C.A. has undertaken sponsored research, or provided consultancy, for a number of companies and organizations including Airbus Group Industries, Astra, British Aerospace/BAeSystems, Civil Aviation Authority, Duphar, FarmItalia Carlo Erba, Ford Motor Company, ICI, Innovate UK, Janssen, LERS Synthélabo, Lilly, Lorex/Searle, UK Ministry of Defense, Quest International, Red Bull GmbH, Rhone-Poulenc Rorer, and Sanofi Aventis. Over the past 3 years, J.C.V. has received grants/research support from the Dutch Ministry of Infrastructure and the Environment, Janssen Research and Development, and Sequential, and has acted as a consultant/advisor for Clinilabs, More Labs, Red Bull, Sen-Jam Pharmaceutical, Toast!, and ZBiotics. The other authors have no potential conflicts of interest to disclose.

References

1. Van Schrojenstein Lantman, M.; van de Loo, A.J.; Mackus, M.; Verster, J.C. Development of a definition for the alcohol hangover: Consumer descriptions and expert consensus. *Curr. Drug Abuse Rev.* **2016**, *9*, 148–154. [CrossRef] [PubMed]
2. Gunn, C.; Mackus, M.; Griffin, C.; Munafò, M.R.; Adams, S. A systematic review of the next-day effects of heavy alcohol consumption on cognitive performance. *Addiction* **2018**, *113*, 2182–2193. [CrossRef] [PubMed]
3. McKinney, A. A review of the next day effects of alcohol on subjective mood ratings. *Curr. Drug Abuse Rev.* **2010**, *3*, 88–91. [CrossRef] [PubMed]
4. Frone, M.R. Employee psychoactive substance involvement: Historical context, key findings, and future directions. *Annu. Rev. Organ. Psychol. Organ. Behav.* **2019**, *6*, 273–297. [CrossRef]

5. Bhattacharya, A. *Financial Headache. The Cost of Workplace Hangovers and Intoxication to the UK Economy*; Institute of Alcohol Studies: London, UK, 2019.
6. Hartung, B.; Schwender, H.; Mindiashvili, N.; Ritz-Timme, S.; Malczyk, A.; Daldrup, T. The effect of alcohol hangover on the ability to ride a bicycle. *Int. J. Legal Med.* **2015**, *129*, 751–758. [CrossRef]
7. Verster, J.C. Alcohol hangover effects on driving and flying. *Int. J. Disabil. Hum. Dev.* **2007**, *6*, 361–367. [CrossRef]
8. Verster, J.C.; Bervoets, A.C.; de Klerk, S.; Vreman, R.A.; Olivier, B.; Roth, T.; Brookhuis, K.A. Effects of alcohol hangover on simulated highway driving performance. *Psychopharmacology* **2014**, *231*, 2999–3008. [CrossRef]
9. Verster, J.C.; van der Maarel, M.; McKinney, A.; Olivier, B.; de Haan, L. Driving during alcohol hangover among Dutch professional truck drivers. *Traffic Inj. Prev.* **2014**, *15*, 434–438. [CrossRef]
10. Van Schrojenstein Lantman, M.; Mackus, M.; van de Loo, A.J.A.E.; Verster, J.C. The impact of alcohol hangover symptoms on cognitive and physical functioning, and mood. *Hum. Psychopharm.* **2017**, *32*, e2623. [CrossRef]
11. Nelson, D.O. Effects of ethyl alcohol on the performance of a selection of gross motor tests. *Res. Q.* **1959**, *30*, 312–320.
12. Karvinen, E.; Miettinen, M.; Ahlman, K. Physical performance during hangover. *Q. J. Stud. Alcohol.* **1962**, *23*, 208–215. [PubMed]
13. Kruisselbrink, L.D.; Martin, K.L.; Megeney, M.; Fowles, J.R.; Murphy, R.J. Physical and psychomotor functioning of females the morning after consuming low to moderate quantities of beer. *J. Stud. Alcohol.* **2006**, *67*, 416–420. [CrossRef] [PubMed]
14. Devenney, L.E.; Coyle, K.B.; Roth, T.; Verster, J.C. Sleep after heavy alcohol consumption and physical activity levels during alcohol hangover. *J. Clin. Med.* **2019**, *8*, E752. [CrossRef] [PubMed]
15. Watson, P.E.; Watson, I.D.; Batt, R.D. Prediction of blood alcohol concentrations in human subjects. Updating the Widmark Equation. *J. Stud. Alcohol. Drugs* **1981**, *42*, 547–556. [CrossRef] [PubMed]
16. Verster, J.C.; Benjaminsen, J.M.E.; van Lanen, J.H.M.; van Stavel, N.M.D.; Olivier, B. Effects of mixing alcohol with energy drink on objective and subjective intoxication: Results from a Dutch on-premise study. *Psychopharmacology* **2015**, *232*, 835–842. [CrossRef]
17. Van de Loo, A.J.A.E.; van Andel, N.; van Gelder, C.A.G.H.; Janssen, B.S.G.; Titulaer, J.; Jansen, J.; Verster, J.C. The effects of alcohol mixed with energy drink (AMED) on subjective intoxication and alertness: Results from a double-blind placebo-controlled clinical trial. *Hum. Psychopharmacol.* **2016**, *31*, 200–205. [CrossRef] [PubMed]
18. Jackson, K.M.; Rohsenow, D.J.; Piasecki, T.M.; Howland, J.; Richardson, A.E. Role of tobacco smoking in hangover symptoms among university students. *J. Stud. Alcohol. Drugs* **2013**, *74*, 41–49. [CrossRef]
19. Van Schrojenstein Lantman, M.; Mackus, M.; Roth, T.; Verster, J.C. Total sleep time, alcohol consumption and the duration and severity of alcohol hangover. *Nat. Sci. Sleep* **2017**, *9*, 181–186. [CrossRef]
20. Van Schrojenstein Lantman, M.; Roth, T.; Roehrs, T.; Verster, J.C. Alcohol hangover, sleep quality, and daytime sleepiness. *Sleep Vigil.* **2017**, *1*, 37–41. [CrossRef]
21. Donners, A.A.M.T.; Tromp, M.D.P.; Garssen, J.; Roth, T.; Verster, J.C. Perceived immune status and sleep: A survey among Dutch students. *Sleep Disord.* **2015**, *2015*, 721607. [CrossRef]
22. Abdulahad, S.; Huls, H.; Balikji, S.; van de Loo, A.J.A.E.; Roth, T.; Verster, J.C. Irritable bowel syndrome, immune fitness and insomnia: Results from an online survey among people with sleep complaints. *Sleep Vigil.* **2019**, *3*, 121–129. [CrossRef]
23. Craig, C.L.; Marshall, A.L.; Sjöström, M.; Bauman, A.E.; Booth, M.L.; Ainsworth, B.E.; Pratt, M.; Ekelund, U.; Yngve, A.; Sallis, J.F.; et al. International physical activity questionnaire: 12-country reliability and validity. *Med. Sci. Sports Exerc.* **2003**, *35*, 1381–1395. [CrossRef] [PubMed]
24. Hogewoning, A.; van de Loo, A.J.A.E.; Mackus, M.; Raasveld, S.J.; de Zeeuw, R.; Bosma, E.R.; Bouwmeester, N.H.; Brookhuis, K.A.; Garssen, J.; Verster, J.C. Characteristics of social drinkers with and without a hangover after heavy alcohol consumption. *Subst. Abuse Rehab.* **2016**, *7*, 161–167. [CrossRef] [PubMed]
25. Wilod Versprille, L.J.F.; van de Loo, A.J.A.E.; Mackus, M.; Arnoldy, L.; Sulzer, T.A.L.; Vermeulen, S.A.; Abdulahad, S.; Huls, H.; Baars, T.; Kraneveld, A.D.; et al. Development and validation of the Immune Status Questionnaire (ISQ). *Int. J. Environ. Res. Public Health* **2019**, *16*, 4743. [CrossRef]

26. Van Schrojenstein Lantman, M.; Otten, L.S.; Mackus, M.; de Kruijff, D.; van de Loo, A.J.A.E.; Kraneveld, A.D.; Garssen, J.; Verster, J.C. Mental resilience, perceived immune functioning, and health. *J. Multidiscip. Healthc.* **2017**, *10*, 107–112. [CrossRef]
27. Mackus, M.; de Kruijff, D.; Otten, L.S.; Kraneveld, A.D.; Garssen, J.; Verster, J.C. Differential gender effects in the relationship between perceived immune functioning and autism spectrum disorder scores. *Int. J. Environ. Res. Public Health* **2017**, *14*, 409. [CrossRef]
28. Verster, J.C.; van de Loo, A.J.A.E.; Adams, S.; Stock, A.-K.; Benson, S.; Alford, C.; Scholey, A.; Bruce, G. Naturalistic study design in alcohol hangover research: Advantages, limitations, and solutions. *J. Clin. Med.* **2019**, *8*, 2160. [CrossRef]
29. Verster, J.C.; Stephens, R.; Penning, R.; Rohsenow, D.; McGeary, J.; Levy, D.; McKinney, A.; Finnigan, F.; Piasecki, T.M.; Adan, A.; et al. The Alcohol Hangover Research Group consensus statement on best practice in alcohol hangover research. *Curr. Drug Abuse Rev.* **2010**, *3*, 116–127. [CrossRef]
30. Verster, J.C.; Kruisselbrink, L.D.; Slot, K.A.; Anogeianaki, A.; Adams, S.; Alford, C.; Arnoldy, L.; Ayre, E.; Balikji, S.; Benson, S.; et al. Sensitivity to experiencing alcohol hangovers: Reconsideration of the 0.11% blood alcohol concentration (BAC) threshold for having a hangover. *J. Clin. Med.* **2020**, *9*, 179.
31. Van de Loo, A.J.A.E.; Mackus, M.; van Schrojenstein Lantman, M.; Kraneveld, A.D.; Garssen, J.; Scholey, A.; Verster, J.C. Susceptibility to alcohol hangovers: The association with self-reported immune status. *Int. J. Environ. Res. Public Health* **2018**, *15*, 1286. [CrossRef]
32. Van de Loo, A.J.A.E.; van Schrojenstein Lantman, M.; Mackus, M.; Scholey, A.; Verster, J.C. Impact of mental resilience and perceived immune functioning on the severity of alcohol hangover. *BMC Res. Notes* **2018**, *11*, 526. [CrossRef] [PubMed]
33. Kösem, Z.; van de Loo, A.J.A.E.; Fernstrand, A.M.; Garssen, J.; Verster, J.C. The impact of consuming food or drinking water on alcohol hangover. *Eur. Neuropsychopharm.* **2015**, *25*, S604. [CrossRef]
34. Verster, J.C. The "hair of the dog": A useful hangover remedy or a predictor of future problem drinking? *Curr. Drug Abuse Rev.* **2009**, *2*, 1–4. [CrossRef] [PubMed]
35. Hunt-Carter, E.E.; Slutske, W.S.; Piasecki, T.M. Characteristics and correlates to relieve hangover in a college sample. *Alcohol. Clin. Exp. Res.* **2005**, *29*, 152A.
36. Penning, R.; McKinney, A.; Bus, L.D.; Olivier, B.; Slot, K.; Verster, J.C. Measurement of alcohol hangover severity: Development of the alcohol hangover severity scale (AHSS). *Psychopharmacology* **2013**, *225*, 803–810. [CrossRef] [PubMed]
37. Slutske, W.S.; Piasecki, T.M.; Hunt-Carter, E.E. Development and initial validation of the Hangover Symptoms Scale: Prevalence and correlate of hangover symptoms in college students. *Alcohol. Clin. Exp. Res.* **2003**, *27*, 1442–1450. [CrossRef]
38. Rohsenow, D.J.; Howland, J.; Minsky, S.J.; Greece, J.; Almeida, A.; Roehrs, T.A. The acute hangover scale: A new measure of immediate hangover symptoms. *Addict. Behav.* **2007**, *32*, 1314–1320. [CrossRef]

 © 2019 by the authors. Licensee MDPI, Basel, Switzerland. This article is an open access article distributed under the terms and conditions of the Creative Commons Attribution (CC BY) license (http://creativecommons.org/licenses/by/4.0/).

Article

The Effects of SJP-001 on Alcohol Hangover Severity: A Pilot Study

Joris C Verster [1,2,3,*], **Thomas A Dahl** [4], **Andrew Scholey** [3] **and Jacqueline M Iversen** [4]

[1] Division of Pharmacology, Utrecht Institute for Pharmaceutical Sciences (UIPS), Utrecht University, 3584CG Utrecht, The Netherlands
[2] Institute for Risk Assessment Sciences (IRAS), Utrecht University, 3584CM Utrecht, The Netherlands
[3] Centre for Human Psychopharmacology, Swinburne University, Melbourne VIC 3122, Australia; andrew@scholeylab.com
[4] Sen-Jam Pharmaceutical, 223 Wall St., #130, Huntington, NY 11743, USA; tadahl@sen-jam.com (T.A.D.); jackie@sen-jam.com (J.M.I.)
* Correspondence: j.c.verster@uu.nl; Tel.: +31-302-536-909

Received: 14 February 2020; Accepted: 29 March 2020; Published: 31 March 2020

Abstract: Background. Despite a clear market need and many hangover products available, currently there is no hangover treatment that is supported by substantial scientific evidence demonstrating its efficacy and safety. A pilot study was conducted to investigate the effects of a potential new hangover treatment, SJP-001, and its constituents (220 mg naproxen and 60 mg fexofenadine) on hangover severity. Methods. $N = 13$ healthy social drinkers (36.3 ± 8.9 years old) participated in a double-blind, factorial design, cross-over study. On each test day, they consumed their own choice of alcohol up to a self-reported level sufficient to elicit a next-day hangover. Treatments were administered prior to onset of drinking. Next morning, hangover severity was assessed with the Acute Hangover Scale (AHS). Subjects were included in the efficacy analysis only if they reported a hangover after placebo. Results. $N = 5$ subjects (60% male, 35.2 ± 9.0 years old) were included in the analysis. They consumed a mean (SD) of 4.6 ± 1.1 units of alcohol and had an average peak breath alcohol concentration (BrAC) of 0.065% across conditions. Compared to placebo, SJP-001 significantly improved the AHS overall hangover severity score (0.8 ± 0.3 versus 1.5 ± 0.9, $p = 0.042$). Compared to placebo, SJP-001 also reduced scores on the individual item 'hangover', although the observed improvement (−1.6) did not reach statistical significance ($p = 0.102$). The differences from placebo after naproxen alone and fexofenadine alone were not statistically significant. SJP-001 also improved scores for the individual hangover symptoms tired, thirsty, headache, dizziness, nausea, and loss of appetite, but these effects did not reach statistical significance. Discussion. Compared to placebo, SJP-001 significantly reduced overall hangover severity. The effects of SJP-001 should be further examined in a double-blind, placebo-controlled trial with a larger sample size and controlled administration of sufficient amounts of alcohol to provoke a more substantial alcohol hangover.

Keywords: alcohol; hangover; treatment; prevention; SJP-001; naproxen; fexofenadine

1. Introduction

The alcohol hangover is defined as "the combination of negative mental and physical symptoms which can be experienced after a single episode of alcohol consumption, starting when blood alcohol concentration (BAC) approaches zero" [1,2]. Recent research suggests that there is no BAC threshold for producing hangover, but that hangovers may occur at all BAC levels, and are most likely elicited following consumption of more alcohol than usual on a drinking occasion [3,4].

Alcohol hangovers are typically characterized by a combination of symptoms, affecting mood, cognition, and physical functioning [5–8]. These negatively impact daily activities including, but not restricted to, job performance [9] and driving [10–12]. In the USA, the annual economic costs of alcohol hangover in terms of absenteeism and presenteeism have been estimated at $173 billion annually [13]. A recent UK study rated the annual economic costs of hangover at 4 billion GBP [14]. Unsurprisingly, consumers have expressed a clear need for an effective hangover treatment [15]. However, to date, little research has been devoted to the development of effective and safe hangover treatments, and currently there is no evidence-based hangover treatment [16–18]. The pathology of alcohol hangover is not yet elucidated [19–21], although alcohol metabolism and the immune response to alcohol consumption are current research foci [22–25].

The immune system likely plays a role in the development of alcohol hangover, with significant immune reactions (e.g., changes in blood cytokine levels and C-reactive protein) being associated with heavy alcohol consumption and subsequent hangovers [23,26–30]. In addition, alcohol and acetaldehyde liberate histamine from mast cells and depress histamine elimination by inhibiting diamine oxidase [31]. Independently of alcohol consumption, prostaglandin and histamine release contributes to inflammation, pain (including headache), and fatigue [32–34]. These are all symptoms of alcohol hangover [5,6]. Thus, it is hypothesized that an intervention that prevents the release of prostaglandin and histamine may serve as an effective treatment to prevent alcohol hangover.

SJP-001 has been developed as such a potential new treatment for prevention of alcohol hangover. SJP-001 is a combination of two over-the-counter (OTC) oral generic drugs (i.e., a nonsteroidal anti-inflammatory drug (220 mg naproxen) and an H_1-antagonist (60 mg fexofenadine)). Both naproxen and fexofenadine are individually marketed in USA as over-the-counter (OTC) drugs [35,36]. The anti-inflammatory properties of naproxen are well documented [37]. Although prescribed primarily for its H_1 antagonist activity [38], fexofenadine also exhibits some anti-inflammatory properties by modulating the release of a variety of proinflammatory mediators. For example, in a study to evaluate the immunomodulatory properties of antihistamines, it was found that fexofenadine downregulated IL 4–induced production of IL 5 and suppressed IL 12–induced secretion of IFN γ [39].

Given this background, a pilot study was conducted to determine if SJP-001 can reduce hangover severity when taken prior to drinking in a sample of healthy social drinkers.

2. Methods

The study followed a randomized, double-blind, placebo-controlled, factorial, cross-over design. In addition to a screening day, there were four study visits, separated by a washout period of three to nine days. The study was sponsored by Sen-Jam Pharmaceutical (JMI Capital Group) and conducted by Clinilabs, Inc. in accordance with the guidelines of the Declaration of Helsinki and its latest amendments. Ethics approval was obtained from the Chesapeake Institutional Review Board (Study number: Pro00016219) and written informed consent was obtained from all subjects. Data from the study are on file (Sen-Jam Pharmaceutical) and available upon reasonable request.

2.1. Subjects

$N = 16$ self-reported moderate drinkers that previously experienced alcohol hangovers were recruited via online advertisement. Screening procedures included a brief physical examination including height and weight, collection of demographic information, medical history and vital signs, review of prior and concomitant medications, and a urine drug and pregnancy screen, and a breathalyzer alcohol test. To be included, subjects had to be nonsmoking men or women between 25 and 65 years old, have a body mass index (BMI) between 19 and 32 kg/m^2, have a regular, habitual bedtime between 21:30 and 24:00 h, and have a good general health as determined by a thorough medical history and physical examination including vital signs, conducted by the study physician. Subjects had to be self-reported moderate drinkers of alcohol, which was approximated with a breath alcohol concentration (BrAC) of 0.04%–0.11% on usual drinking occasions (corresponding to 2 to 5 or 3 to 7 alcoholic drinks for a 70 kg female and male, respectively, over a 2 to 3 h period). Subjects were included if they reported that this amount of alcohol usually resulted in a next-day hangover. Subjects were excluded if they reported acute illness within 14 days prior to screening visit, experienced an allergic reaction or upper respiratory tract infection within 7 days of screening visit, had been vaccinated within 7 days of screening visit, had a history of clinically significant allergies (except for untreated, asymptomatic, seasonal allergies at time of dosing), hematological, renal, endocrine, pulmonary, gastrointestinal, cardiovascular, hepatic, or neurological disease, cancer or diabetes, or psychiatric illness, previous or current Substance-Related Disorder as defined by DSM-5, self-reported usual consumption of more than 14 units of alcohol per week, recent (within one month) or current use of tobacco or nicotine products. In addition, subjects were excluded if medical examination revealed a clinically significant, unstable medical illness, a positive alcohol breathalyzer or urine drug screen test (including cocaine, THC-marijuana, opiates, amphetamines, methamphetamine, phencyclidine, benzodiazepines, barbiturates, methadone, MDMA-ecstasy, oxycodone, and propoxyphene) which was provided at screening, when subjects had a blood pressure >140/90 mm/Hg or heart rate >100 bpm. Women who were pregnant or breastfeeding, or had a positive urine pregnancy test at screening were also excluded. Subjects taking any prescription or OTC oral pain medication(s) or antihistamine drug, or who previously experienced an allergic reaction or adverse event associated with aspirin, nonsteroidal anti-inflammatory drugs (NSAID), or antihistamine usage were excluded. Finally, subjects were excluded if they were unwilling to forgo caffeine consumption with or following dinner on each treatment night or who were unwilling to comply with study restrictions for prohibited medications/foods throughout study participation.

2.2. Procedures

During the four test days, each participant arrived at the clinical unit at approximately 5 pm. On the first test day, subjects were randomized to one of four treatment sequences according to Latin Square assignment. On test days, after general health assessments, including a urine drug and pregnancy test and a breathalyzer test to ensure a BrAC of zero, subjects were served a standardized dinner. Alcohol was available with dinner. Subjects received an oral dose of either SJP-001 (220 mg naproxen and 60 mg fexofenadine), naproxen alone (220 mg), fexofenadine alone (60 mg), or placebo. Treatments consisted of two oral capsules administered 30 min (first capsule) and 15 min (second capsule) before the start of alcohol consumption, taken with approximately 240 mL of water. Treatments were self-administered by subjects under the supervision of study personnel.

Subjects remained together in a lounge room during dinner and thereafter for the evening. They were permitted to socialize, read, and watch television. Study staff were present to monitor alcohol and food consumption and general behavior. Foods high in histamine such as red meat, lamb, and aged cheese were not served. Subjects were provided a variety of alcoholic beverages from which to choose including red and white wine, various types of beer, champagne, and liquors. Water was provided, along with other nonalcoholic beverages as mixers. They were instructed to consume the amount of alcohol (and nonalcoholic drinks including water) that had in the past resulted in a hangover, and consume the same type and amount of alcohol on each test day. For example, if a subject reported at screening that consuming four glasses of red wine in a 3 h period had previously resulted in a hangover, the target for that subject was to consume four glasses of red wine at each visit. Subjects could mix different types of alcohol and were allowed to drink at their own pace; all alcohol consumption was completed within a maximum 3 h period. Light snacks (e.g., pretzels, potato chips, nuts) were provided during the alcohol consumption period. At the end of alcohol consumption, a breathalyzer test was conducted to assess BAC. Subjects stayed overnight in the study center. They were required to go to bed at their usual habitual bedtime. Subjects slept in a private room with a bed, night table, writing table, and chair, and were instructed to remain in bed for 8 h with the lights turned off. Subjects were allowed to deviate from these instructions to use the restroom during the night. Study staff made rounds during the night to monitor subjects' safety. Next morning, if necessary, subjects were awakened 8 h after their bedtime. Within 10–20 min after awakening, and prior to breakfast or the consumption of any coffee or other caffeine-containing beverages, a breathalyzer test and brief neurological assessments (tests examining walking and heel-to-toe walking, and the Romberg test to evaluate balance) were conducted. A postsleep questionnaire was completed, including questions regarding total sleep time, sleep onset latency, number of nightly awakenings, time awake while in bed, and a rating of sleep quality on a scale from 1 (poor) to 10 (excellent). Next-morning hangover severity was assessed with the Acute Hangover Scale [40]. Thereafter, subjects received breakfast and the test day was ended.

2.3. Assessment of Hangover Severity

The Acute Hangover Scale [40] consists of nine items including 'hangover', 'thirsty', 'tired', 'headache', 'dizziness/faintness', 'loss of appetite', 'stomachache', 'nausea', and 'heart racing', which could be rated on a scale ranging from 0 to 7. The anchors of the Likert-type scale were 'none' (score of 0), 'mild' (score of 1), 'moderate' (score of 4), and 'incapacitating' (score of 7). Overall hangover severity was computed by calculating the average score across the AHS nine items. In the interest of safety, subjects with hangover symptoms were confined to the study center at the discretion of the clinician until symptoms had improved.

2.4. Statistical Analysis

Subjects were included in the statistical analysis only if they reported a hangover on the placebo test day. The reasons for this were twofold. Firstly, the absence of a hangover in the placebo condition implies that the subject had not complied with the instructions to drink to levels which would typically produce hangover. Secondly, we wished to evaluate if the combination of naproxen and fexofenadine in SJP-001 was superior to either compound alone. This required us to compare effects on a hangover in the placebo group (note that the absence of a hangover in any of the nonplacebo arms could theoretically be attributable to efficacy of the treatment so such cases could not be excluded). Statistical analyses were conducted with SPSS (IBM Corp. Released 2013. IBM SPSS Statistics for Windows, Version 25.0, IBM Corp. Armonk, NY, USA). Mean and standard deviation (SD) were computed for all variables. The primary outcome measure of the study was the average AHS score. Secondary outcomes were the individual AHS items. Overall hangover severity and individual symptom ratings after SJP-001 and placebo were compared applying a nonparametric Related-Samples Wilcoxon Signed Rank test.

3. Results

Of the $N = 16$ subjects that were screened, $N = 13$ met all inclusion and exclusion criteria and participated in the study. Seven subjects who reported no hangover on the placebo test day were excluded from the statistical analysis (see previous section). Another subject was excluded due to significant sleep difficulties in the clinical setting. $N = 5$ subjects were included in the final dataset. For one of these subjects, the last test day (naproxen) was discontinued due to noncompliance with study procedures during the drinking session. The demographic data of the included subjects are summarized in Table 1.

Table 1. Demographics, morphometrics, and drinking characteristics of the sample.

	Overall	Subject 7	Subject 8	Subject 9	Subject 12	Subject 16
Sex (male/female)	3/2	Male	Male	Male	Female	Female
Age (years)	35.2 (9.0)	27	47	33	27	42
Weight (lb)	170.8 (13.7)	169.4	192.6	172.8	162.4	156.6
Height (in)	67.4 (2.6)	67.5	68.9	70.7	65.5	64.2
BMI (kg/m^2)	26.4 (1.5)	26.1	28.5	24.3	26.6	26.7
Habitual bedtime [1] (h:min)	23:12 (0:34)	00:00	23:30	23:00	22:30	23:00
Preferred alcohol type [2]	-	Rum	Mixed [3]	Mixed [4]	Rum	Vodka
Units alcohol for hangover [5]	4.6 (1.1)	3	6	5	4	5

Mean and standard deviation (SD) are shown for the overall sample and five individual subjects. [1]: the individual habitual bedtime was used for each test day. [2]: these drinks were also consumed on each test day. [3]: subject consumed white wine (Chardonnay), red wine (Pinot Noir), and tequila on a usual drinking occasion. [4]: subject consumed Scotch whiskey, beer, and red wine on a usual drinking occasion. [5]: this amount of alcohol units was also consumed on each test day. Abbreviation: BMI = body mass index.

In the study, subjects consumed on average 4.6 alcoholic drinks and had an average BrAC of 0.065% at the end of the drinking session. BrACs did not statistically differ between conditions (see Table 2). After the drinking session, subjects slept in the clinical research unit. Table 2 summarizes the sleep outcomes for each treatment condition. No significant differences were found between SJP-001 and placebo.

Table 2. Alcohol consumption and sleep outcomes.

	SJP-001	Fexofenadine	Naproxen	Placebo
Units of alcohol consumed	4.6 (1.1)	4.6 (1.1)	4.6 (1.1)	4.6 (1.1)
BrAC (3h) (%)	0.064 (0.034)	0.068 (0.026)	0.053 (0.030)	0.072 (0.038)
Total sleep time (min)	412.0 (45.9)	377.0 (66.3)	406.0 (59.3)	421.0 (44.2)
Number of nightly awakenings	1.6 (1.1)	2.0 (1.4)	1.5 (1.9)	1.6 (1.1)
Sleep onset latency (min)	34.0 (18.5)	52.0 (30.9) *	45.0 (17.3)	12.2 (11.3)
Time awake while in bed (min)	29.4 (30.7)	34.2 (38.8)	13.0 (12.9)	36.0 (38.7)
Sleep quality	7.2 (1.5)	4.8 (3.6)	8.3 (1.0)	7.8 (1.6)

Mean and standard deviation (SD) are shown for alcohol consumption and sleep outcomes in the four treatment conditions. Significant differences ($p < 0.05$) between the treatments and placebo are indicated by *. Abbreviation: BrAC = breath alcohol concentration.

Mean ± SD AHS scores were mild and equaled 1.5 ± 0.9 after placebo, 0.8 ± 0.3 after SJP-001, 1.0 ± 0.7 after fexofenadine, and 0.7 ± 0.7 after naproxen. Scores on individual hangover symptoms are listed in Table 3. Compared to placebo, SJP-001 significantly improved overall hangover severity ($p = 0.042$), whereas the differences from placebo after naproxen ($p = 0.066$) and fexofenadine ($p = 0.345$) were not statistically significant (See Figure 1A).

Table 3. Mean Acute Hangover Scale and individual hangover symptom scores.

Symptoms	SJP-001	Fexofenadine	Naproxen	Placebo
Hangover	0.8 (0.8)	1.6 (1.5)	0.8 (0.5)	2.4 (1.3)
Thirsty	2.4 (1.8)	1.6 (1.1)	1.5 (2.4)	2.8 (2.4)
Tired	2.2 (1.9)	2.4 (2.1)	1.8 (2.1)	2.8 (1.6)
Headache	0.4 (0.5)	1.6 (1.5)	0.3 (0.5)	1.6 (2.5)
Dizziness/faintness	0.0 (0.0)	0.2 (0.4)	1.0 (2.0)	1.0 (1.2)
Loss of appetite	0.4 (0.5)	1.0 (1.7)	0.3 (0.5)	1.0 (0.7)
Stomachache	0.8 (1.3)	0.4 (0.9)	0.3 (0.5)	0.6 (0.5)
Nausea	0.4 (0.9)	0.2 (0.4)	0.3 (0.5)	1.0 (1.7)
Heart racing	0.4 (0.5)	0.2 (0.4)	0.0 (0.0)	0.2 (0.4)
Mean AHS score	0.8 (0.3) *	1.0 (0.7)	0.7 (0.7)	1.5 (0.9)

Mean (SD) are shown for the AHS and its individual items for the four treatment conditions. Abbreviation: AHS = Acute Hangover Scale. Significant differences ($p < 0.05$) between the treatments and placebo are indicated by *.

Also, compared to placebo, SJP-001 reduced scores on the individual item 'hangover', although the difference of −1.6 did not reach statistical significance in this small sample ($p = 0.102$) (see Figure 1B). SJP-001 also improved scores for other individual hangover symptoms including tired, thirsty, headache, dizziness/faintness, nausea, and loss of appetite (none of the differences between SJP-001 and placebo reached statistical significance). After fexofenadine alone and naproxen alone, none of the individual item scores differed significantly from placebo.

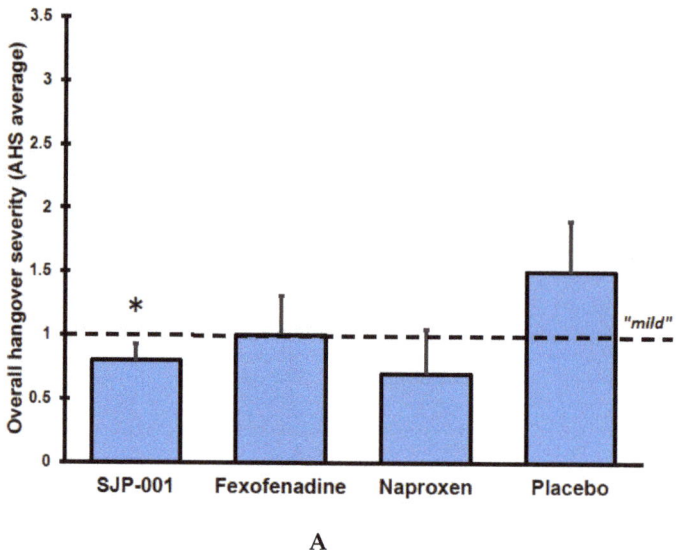

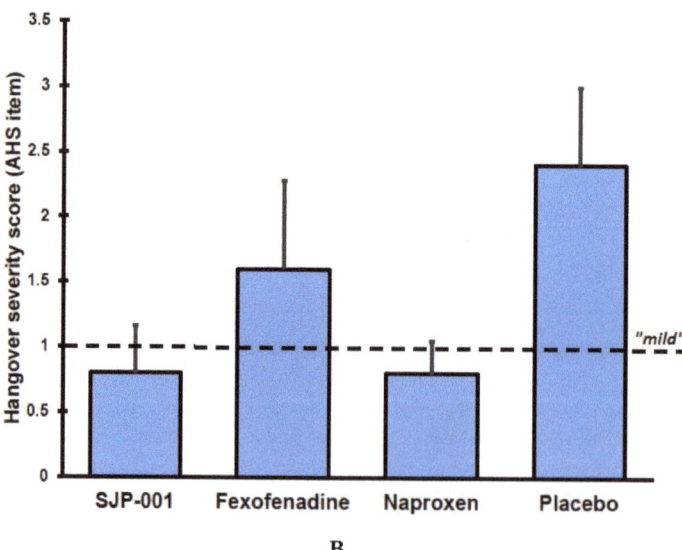

Figure 1. Overall hangover severity by group. (**A**) shows the average AHS scores; (**B**) shows the scores on the individual AHS item 'hangover'. Mean scores with standard errors are shown. Abbreviation: AHS = Acute Hangover Scale. Significant differences ($p < 0.05$) between the treatments and placebo are indicated by *.

4. Discussion

The findings of the current pilot study suggest that SJP-001 may be effective in the prevention of alcohol hangover. However, the observations should be interpreted with caution, as this pilot study had a small sample size (albeit using a cross-over design). The observed significant reduction in overall hangover severity, assessed both via the average AHS score and the individual item "hangover", justifies further investigation of the properties of SJP-001 in future randomized controlled trials (RCTs) with an adequate sample size and improved methodology.

It has been argued that when conducting pilot studies with a small sample size, one should focus on descriptive statistics and estimation rather than formal hypothesis testing to infer whether findings are clinically relevant or not [41–43]. To further evaluate descriptive data of the current study, difference scores in hangover severity between SJP-001 and placebo were calculated. These mean (SD) difference scores equal −0.64 (0.62) for the AHS overall hangover severity score and −1.6 (1.67) for the one-item hangover severity score. Applying the anchor-based approach [44], both the average AHS and one-item hangover score are between the anchors 'mild' and 'moderate' after placebo which were relevantly improved towards scores between the anchors 'none' and 'mild' after SJP-001. When applying the one-half SD criterion as a benchmark [45], the minimal clinically important difference (MCID) should be greater than −0.31 on the AHS and greater than −0.84 for the one-item hangover severity score. Thus, both the AHS (−0.64 > −0.31) and the one-item hangover severity difference score (−1.6 > 0.84) can be considered as clinically relevant reductions in hangover severity. However, to infer clinical relevance, the single-item hangover severity score is likely to be more accurate with regard to clinical relevance than the AHS aggregate symptoms score [46]. Taken together, evaluating descriptive data further supports the decision to conduct adequately powered RCTs to enable more definitive conclusions regarding the efficacy of SJP-001.

Turning to individual symptoms, it may be argued that the effects of SJP-001 are relatively similar to those observed when naproxen alone is administered. Importantly, for the AHS scores, the difference between SJP-001 and placebo (but not naproxen and placebo) is statistically significant. This is likely due to the fact that, although the mean AHS scores of SJP-001 and naproxen are similar, there is a considerable difference in the corresponding standard deviation of SJP-001 and naproxen (0.3 versus 0.7, respectively). One could argue that the lower variance after SJP-001 in contrast to naproxen alone is a result of the naproxen–fexofenadine combination resulting in a more stable, less variable treatment effect due to the anti-inflammatory properties of fexofenadine. Also, the achieved BrAC after SJP-001 was higher than after naproxen (0.064% versus 0.053% BrAC, respectively), which may have resulted in an enhanced treatment effect after naproxen compared to that of SJP-001. Had equal BrACs been achieved, the effect of SJP-001 may have been superior to naproxen alone. Alternatively, it may be that the observed difference in standard deviation reflects the small sample size of this pilot study, and will not be replicated in a larger, well-powered RCT. Future research should investigate these hypotheses.

In addition to assessing initial evidence regarding the effectiveness of SJP-001 in preventing hangover, an important goal of this pilot study was to verify whether the current study design is suitable for large sample, double-blind RCTs. From the current pilot study, several important lessons were learned regarding limitations to the study design.

Firstly, the sample size of this pilot study was small. Future RCTs should be adequately powered. An important reason for this small study size was that of the $N = 13$ subjects that completed the study, only $N = 5$ were suitable to be included in the data analysis. Although subjects consumed the self-reported amount of alcohol that was sufficient to induce hangover, seven subjects reported to have no hangover in the placebo condition. In future RCTs, the amount of alcohol to provoke a hangover should be determined in more detail to prevent such a large dropout. Currently, a hangover sensitivity scale is in development to aid this process. This approach will increase the likelihood that a hangover will be present on each test day.

The exclusion of subjects who had no hangover following placebo only was necessary to allow meaningful statistical comparison across the groups. It is possible that excluding individuals not experiencing hangover in the placebo condition may have somehow increased the probability of a significant outcome favoring SJP-001. If this were the case, however, one might also expect significant effects when comparing the naproxen-only and fexofenadine-only arms with placebo. To verify this, we conducted two additional analyses, in which we compared naproxen-only versus placebo and fexofenadine-only versus placebo, including only subjects with a placebo score greater than zero. Again, no significant differences between the treatments were found for the AHS and one-item overall hangover severity score. Therefore, we believe that excluding subjects without a hangover in the placebo condition did not affect the probability of obtaining a significant outcome favoring SJP-001.

A second issue in this pilot study is that subjects could consume their preferred types of alcohol. Although the type(s) of drink(s) were the same on each test day, eliminating intraindividual differences, this does introduce interindividual variability. Additionally, alcoholic drinks have variable amounts of congeners and histamines, which may impact study outcomes. A high congener content has been shown to negatively impact hangover severity [47]. One obvious solution to this is to administer one standard type of alcoholic drink to all subjects. On the other hand, this may not adequately reflect participants' usual drinking behavior. Achieving a balance between experimental control and ecological validity is a challenge for hangover research [48].

One subject was excluded because of experiencing significant sleep difficulties in the clinical research unit. Sleeping in a new (and especially clinical) environment may require adaptation. The current study did not include a sleep rehearsal night during screening to identify possible 'first-night effects' or other sleep disturbances [49,50]. It is advisable to include a sleep rehearsal night in future RCTs to screen for subjects with possible sleep disturbances. Further, the current study used subjective measures of sleep. Subjective assessments of sleep do not always correspond to objective sleep measures [51], such as polysomnography or actigraphy. Future RCTs could usefully include objective sleep measures, for example, by using mobile technology such as the GENEactiv device [51].

Finally, the assessment of alcohol hangover depends solely on self-report. The most direct method to assess hangover severity is a single question allowing subjects to rate their hangover severity, for example, a 0 to 7 score as used for the AHS [40], or a 0 to 10 rating scale ranging from absent to extreme [52]. Since hangover may differ qualitatively from person to person, a single question is more likely to accurately reflect overall hangover severity than an aggregate score of variable hangover symptoms, such as that utilized with current hangover scales [40,53,54]. Aggregate symptom scores may be lower than one-item overall hangover severity scores as the scales may include low-frequency/low-severity symptoms and omit high-frequency/high-severity symptoms [46]. Notably in this pilot study, the single-item hangover score was higher, and better differentiated SJP-001 from placebo than the AHS aggregate symptom score. Relying on self-report to investigate hangovers is necessary, as there is currently no known objective criteria or biomarker for hangover severity. It is, however, important to include biomarkers of immune function in future RCTs, to assess the immune response to alcohol consumption in the placebo condition, and to infer whether SJP-001 is capable of reducing this effect. Such data will be important to provide insight into the mechanism(s) of action of SJP-001.

In conclusion, the data from this pilot study suggest that SJP-001 is effective in preventing alcohol hangover. However, the sample size of this study was very small. Therefore, the effects of SJP-001 should be further evaluated in RCTs with a larger sample size.

Author Contributions: Conceptualization, T.A.D. and J.M.I.; methodology, T.A.D. and J.M.I.; formal analysis, J.C.V.; interpretation of data, J.C.V., A.S., T.A.D. and J.M.I.; writing—original draft preparation, J.C.V.; writing—review and editing, J.C.V., A.S., T.A.D. and J.M.I.; funding acquisition, J.M.I. All authors have read and agreed to the published version of the manuscript.

Funding: This study was funded by Sen-Jam Pharmaceutical and conducted by Clinilabs.

Conflicts of Interest: Over the past 36 months, A.S. has held research grants from Abbott Nutrition, Arla Foods, Bayer, BioRevive, DuPont, Fonterra, Kemin Foods, Nestlé, Nutricia-Danone, Verdure Sciences. He has acted as a consultant/expert advisor to Bayer, Danone, Naturex, Nestlé, Pfizer, Sanofi, Sen-Jam Pharmaceutical, and has received travel/hospitality/speaker fees from Bayer, Sanofi, and Verdure Sciences. Over the past 36 months, J.C.V. has held grants from the Dutch Ministry of Infrastructure and the Environment, Janssen, Nutricia, and Sequential, and acted as a consultant/expert advisor to Clinilabs, More Labs, Red Bull, Sen-Jam Pharmaceutical, Toast!, and ZBiotics. T.A.D. is partner and Head of Product Development and Regulatory Affairs of Sen-Jam Pharmaceutical. J.M.I. is founder and Head of Clinical Development of Sen-Jam Pharmaceutical. This independent study was conducted by Clinilabs, without involvement of the authors and the funder. The funder was involved in the design of the study, interpretation of data, writing of the manuscript, and in the decision to publish the results.

References

1. Van Schrojenstein Lantman, M.; van de Loo, A.J.; Mackus, M.; Verster, J.C. Development of a definition for the alcohol hangover: Consumer descriptions and expert consensus. *Curr. Drug Abus. Rev.* **2016**, *9*, 148–154. [CrossRef]
2. Verster, J.C.; van de Loo, A.J.A.E.; Benson, S.; Scholey, A.; Stock, A.-K. Updating the definition of the alcohol hangover. *J. Clin. Med.* **2020**, *9*, 823. [CrossRef] [PubMed]
3. Verster, J.C.; Kruisselbrink, L.D.; Slot, K.A.; Anogeianaki, A.; Adams, S.; Alford, C.; Arnoldy, L.; Ayre, E.; Balikji, S.; Benson, S.; et al. Sensitivity to experiencing alcohol hangovers: Reconsideration of the 0.11% blood alcohol concentration (BAC) threshold for having a hangover. *J. Clin. Med.* **2020**, *9*, 179. [CrossRef] [PubMed]
4. Verster, J.C.; Kruisselbrink, L.D.; Anogeianaki, A.; Alford, C.; Stock, A.K. Relationship of alcohol hangover and physical endurance performance: Walking the Samaria Gorge. *J. Clin. Med.* **2020**, *9*, 114. [CrossRef] [PubMed]
5. Penning, R.; McKinney, A.; Verster, J.C. Alcohol hangover symptoms and their contribution to overall hangover severity. *Alcohol Alcohol.* **2012**, *47*, 248–252. [CrossRef]
6. Van Schrojenstein Lantman, M.; Mackus, M.; van de Loo, A.J.A.E.; Verster, J.C. The impact of alcohol hangover symptoms on cognitive and physical functioning, and mood. *Hum. Psychopharmacol.* **2017**, *32*, e2623. [CrossRef]
7. Gunn, C.; Mackus, M.; Griffin, C.; Munafò, M.R.; Adams, S. A systematic review of the next-day effects of heavy alcohol consumption on cognitive performance. *Addiction* **2018**, *113*, 2182–2193. [CrossRef]
8. McKinney, A. A review of the next day effects of alcohol on subjective mood ratings. *Curr. Drug Abus. Rev.* **2010**, *3*, 88–91. [CrossRef]
9. Frone, M.R. Employee psychoactive substance involvement: Historical context, key findings, and future directions. *Annu. Rev. Organ. Psychol. Organ. Behav.* **2019**, *6*, 273–297. [CrossRef]
10. Verster, J.C. Alcohol hangover effects on driving and flying. *Int. J. Disabil. Hum. Dev.* **2007**, *6*, 361–367. [CrossRef]
11. Verster, J.C.; Bervoets, A.C.; de Klerk, S.; Vreman, R.A.; Olivier, B.; Roth, T.; Brookhuis, K.A. Effects of alcohol hangover on simulated highway driving performance. *Psychopharmacology* **2014**, *231*, 2999–3008. [CrossRef] [PubMed]
12. Verster, J.C.; van der Maarel, M.; McKinney, A.; Olivier, B.; de Haan, L. Driving during alcohol hangover among Dutch professional truck drivers. *Traffic Inj. Prev.* **2014**, *15*, 434–438. [CrossRef] [PubMed]
13. Sacks, J.J.; Gonzales, K.R.; Bouchery, E.E.; Tomedi, L.E.; Brewer, R.D. 2010 National and State Costs of Excessive Alcohol Consumption. *Am. J. Prev. Med.* **2015**, *49*, e73–e79. [CrossRef] [PubMed]
14. Bhattacharya, A. *Financial Headache: The Cost of Workplace Hangovers and Intoxication to the UK Economy*; Institute of Alcohol Studies: London, UK, 2019.
15. Mackus, M.; van Schrojenstein Lantman, M.; van de Loo, A.J.A.E.; Nutt, D.J.; Verster, J.C. An effective hangover treatment: Friend or foe? *Drug Sci. Policy Law* **2017**. [CrossRef]
16. Pittler, M.H.; Verster, J.C.; Ernst, E. Interventions for preventing or treating alcohol hangover: Systematic review of randomized trials. *Br. Med. J.* **2005**, *331*, 1515–1518. [CrossRef] [PubMed]
17. Verster, J.C.; Penning, R. Treatment and prevention of alcohol hangover. *Curr. Drug Abus. Rev.* **2010**, *3*, 103–109. [CrossRef] [PubMed]
18. Jayawardena, R.; Thejani, T.; Ranasinghe, P.; Fernando, D.; Verster, J.C. Interventions for treatment and/or prevention of alcohol hangover: Systematic review. *Hum. Psychopharmacol.* **2017**, *32*, e2600. [CrossRef]
19. Penning, R.; van Nuland, M.; Fliervoet, L.A.L.; Olivier, B.; Verster, J.C. The pathology of alcohol hangover. *Curr. Drug Abus. Rev.* **2010**, *3*, 68–75. [CrossRef]

20. Tipple, C.T.; Benson, S.; Scholey, A. A review of the physiological factors associated with alcohol hangover. *Curr. Drug Abus. Rev.* **2016**, *9*, 93–98. [CrossRef]
21. Palmer, E.; Tyacke, R.; Sastre, M.; Lingford-Hughes, A.; Nutt, D.; Ward, R.J. Alcohol Hangover: Underlying Biochemical, Inflammatory and Neurochemical Mechanisms. *Alcohol Alcohol.* **2019**, *54*, 196–203. [CrossRef]
22. Van de Loo, A.J.A.E.; Mackus, M.; Korte-Bouws, G.A.H.; Brookhuis, K.A.; Garssen, J.; Verster, J.C. Urine ethanol concentration and alcohol hangover severity. *Psychopharmacology* **2017**, *234*, 73–77. [CrossRef] [PubMed]
23. Van de Loo, A.J.A.E.; Knipping, K.; Mackus, M.; Kraneveld, A.D.; Garssen, J.; Scholey, A.; Bruce, G.; Verster, J.C. Differential effects on acute saliva cytokine response following alcohol consumption and alcohol hangover: Preliminary results from two independent studies. *Alcohol. Clin. Exp. Res.* **2018**, *42*, 20A.
24. Mackus, M.; van Schrojenstein Lantman, M.; van de Loo, A.J.A.E.; Brookhuis, K.A.; Kraneveld, A.D.; Garssen, J.; Verster, J.C. Alcohol metabolism in hangover sensitive versus hangover resistant social drinkers. *Drug Alcohol Depend.* **2018**, *185*, 351–355. [CrossRef] [PubMed]
25. Wiese, J.; McPherson, S.; Odden, M.C.; Shlipak, M.G. Effect of Opuntia ficus indica on symptoms of the alcohol hangover. *Arch. Intern. Med.* **2004**, *164*, 1334–1340. [CrossRef]
26. Parantainen, J. Prostaglandins in alcohol intolerance and hangover. *Drug Alcohol Depend.* **1983**, *11*, 239–248. [CrossRef]
27. Kaivola, S.; Parantainen, J.; Osterman, T.; Timonen, H. Hangover headache and prostaglandins: Prophylactic treatment with tolfenamic acid. *Cephalalgia* **1983**, *3*, 31–36. [CrossRef]
28. Kim, D.J.; Kim, W.; Yoon, S.J.; Choi, B.M.; Kim, J.S.; Go, H.J.; Kim, Y.K.; Jeong, J. Effects of alcohol hangover on cytokine production in healthy subjects. *Alcohol* **2003**, *31*, 167–170. [CrossRef]
29. Mandrekar, P.; Catalano, D.; Girouard, L.; Szabo, G. Human monocyte IL-10 production is increased by acute ethanol treatment. *Cytokine* **1996**, *8*, 567–577. [CrossRef]
30. Szabo, G.; Mandrekar, P.; Girouard, L.; Catalano, D. Regulation of human monocyte functions by acute ethanol treatment: Decreased tumor necrosis factor-alpha, interleukin-1 beta and elevated interleukin-10, and transforming growth factor-beta production. *Alcohol. Clin. Exp. Res.* **1996**, *20*, 900–907. [CrossRef]
31. Zimatkin, S.M.; Anichtchik, O.V. Alcohol-histamine interactions. *Alcohol Alcohol.* **1999**, *34*, 141–147. [CrossRef]
32. Antonova, M.; Wienecke, T.; Olesen, J.; Ashina, M. Prostaglandins in migraine: Update. *Curr. Opin. Neurol.* **2013**, *26*, 269–275. [CrossRef] [PubMed]
33. Harmer, I.M.; Harris, K.E. Observations on the vascular reactions in man in response to histamine. *Heart* **1926**, *13*, 381–394.
34. Worm, J.; Falkenberg, K.; Olesen, J. Histamine and migraine revisited: Mechanisms and possible drug targets. *J. Headache Pain* **2019**, *20*, 30. [CrossRef] [PubMed]
35. Food and Drug Administration (FDA). Naproxyn (naproxen) Prescribing Information. Available online: https://www.accessdata.fda.gov/drugsatfda_docs/label/2017/017581s113,018164s063,020067s020lbl.pdf (accessed on 10 February 2020).
36. Food and Drug Administration (FDA). Allegra (Fexofenadine Hydrochloride) Prescribing Information. Available online: https://www.accessdata.fda.gov/drugsatfda_docs/label/2003/20786se8-014,20872se8-011,20625se8-012_allegra_lbl.pdf (accessed on 10 February 2020).
37. Brogden, R.N.; Heel, R.C.; Speight, T.M.; Avery, G.S. Naproxen up to date: A review of its pharmacological properties and therapeutic efficacy and use in rheumatic diseases and pain states. *Drugs* **1979**, *18*, 241–277. [CrossRef] [PubMed]
38. Amichai, B.; Grunwald, M.H.; Brenner, L. Fexofenadine hydrochloride–a new anti-histaminic drug. *IMAJ* **2001**, *3*, 207–209. [PubMed]
39. Ashenager, M.S.; Grgela, T.; Aragane, Y.; Kawada, A. Inhibition of cytokine-induced expression of T-cell cytokines by antihistamines. *J. Investig. Allergol. Clin. Immunol.* **2007**, *17*, 20–26.
40. Rohsenow, D.J.; Howland, J.; Minsky, S.J.; Greece, J.; Almeida, A.; Roehrs, T.A. The acute hangover scale: A new measure of immediate hangover symptoms. *Addict. Behav.* **2007**, *32*, 1314–1320. [CrossRef]
41. Lancaster, G.A.; Dodd, S.; Williamson, P.R. Design and analysis of pilot studies: Recommendations for good practice. *J. Eval. Clin. Pract.* **2004**, *10*, 307–312. [CrossRef]
42. Thabane, L.; Ma, J.; Chu, R.; Cheng, J.; Ismaila, A.; Rios, L.P.; Robson, R.; Thabane, M.; Giangregorio, L.; Goldsmith, C.H. A tutorial on pilot studies: The what, why and how. *BMC Med. Res. Methodol.* **2010**, *10*, 1–10. [CrossRef]

43. Lee, E.C.; Whitehead, A.L.; Jacques, R.M.; Julious, S.A. The statistical interpretation of pilot trials: Should significance thresholds be reconsidered? *BMC Med. Res. Methodol.* **2014**, *14*, 41. [CrossRef]
44. Norman, G.R.; Sloan, J.A.; Wyrwich, K.W. Interpretation of changes in health-related quality of life: The remarkable universality of half a standard deviation. *Med. Care* **2003**, *41*, 582–592. [CrossRef] [PubMed]
45. McGlothlin, A.E.; Lewis, R.J. Minimal clinically important difference: Defining what really matters to patients. *JAMA* **2014**, *312*, 1342–1343. [CrossRef] [PubMed]
46. Verster, J.C.; van de Loo, A.J.A.E.; Benson, S.; Scholey, A.; Stock, A.-K. The assessment of overall hangover severity. *J. Clin. Med.* **2020**, *9*, 786. [CrossRef]
47. Rohsenow, D.J.; Howland, J. The role of beverage congeners in hangover and other residual effects of alcohol intoxication: A review. *Curr. Drug Abus. Rev.* **2010**, *3*, 76–79. [CrossRef]
48. Verster, J.C.; van de Loo, A.J.A.E.; Adams, S.; Stock, A.-K.; Benson, S.; Alford, C.; Scholey, A.; Bruce, G. Advantages and limitations of naturalistic study designs and their implementation in alcohol hangover research. *J. Clin. Med.* **2019**, *8*, 2160. [CrossRef]
49. Webb, W.B.; Campbell, S.S. The first night effect revisited with age as variable. *Waking Sleep.* **1979**, *3*, 319–324.
50. Le Bon, O.; Staner, L.; Hoffmann, G.; Dramaix, M.; San Sebastian, I.; Murphy, J.R.; Kentos, M.; Pelc, I.; Linkowski, P. The first-night effect may last more than one night. *J. Psychiatr. Res.* **2001**, *35*, 165–172. [CrossRef]
51. Devenney, L.E.; Coyle, K.B.; Roth, T.; Verster, J.C. Sleep after heavy alcohol consumption and physical activity levels during alcohol hangover. *J. Clin. Med.* **2019**, *8*, 752. [CrossRef]
52. Hogewoning, A.; van de Loo, A.J.A.E.; Mackus, M.; Raasveld, S.J.; de Zeeuw, R.; Bosma, E.R.; Bouwmeester, N.H.; Brookhuis, K.A.; Garssen, J.; Verster, J.C. Characteristics of social drinkers with and without a hangover after heavy alcohol consumption. *Subst. Abus. Rehabil.* **2016**, *7*, 161–167. [CrossRef]
53. Penning, R.; McKinney, A.; Bus, L.D.; Olivier, B.; Slot, K.; Verster, J.C. Measurement of alcohol hangover severity: Development of the alcohol hangover severity scale (AHSS). *Psychopharmacology* **2013**, *225*, 803–810. [CrossRef] [PubMed]
54. Slutske, W.S.; Piasecki, T.M.; Hunt-Carter, E.E. Development and initial validation of the Hangover Symptoms Scale: Prevalence and correlates of hangover symptoms in college students. *Alcohol. Clin. Exp. Res.* **2003**, *27*, 1442–1450. [CrossRef] [PubMed]

© 2020 by the authors. Licensee MDPI, Basel, Switzerland. This article is an open access article distributed under the terms and conditions of the Creative Commons Attribution (CC BY) license (http://creativecommons.org/licenses/by/4.0/).

Article

Effects of Rapid Recovery on Alcohol Hangover Severity: A Double-Blind, Placebo-Controlled, Randomized, Balanced Crossover Trial

Andrew Scholey [1], Elizabeth Ayre [1], Ann-Kathrin Stock [2], Joris C Verster [1,3] and Sarah Benson [1,*]

[1] Centre for Human Psychopharmacology, Swinburne University, Melbourne, VIC 3122, Australia; andrew@scholeylab.com (A.S.); besayre24@gmail.com (E.A.); j.c.verster@uu.nl (J.C.V.)
[2] Cognitive Neurophysiology, Department of Child and Adolescent Psychiatry, Faculty of Medicine, TU Dresden, Fetscherstr. 74, 01307 Dresden, Germany; Ann-Kathrin.Stock@uniklinikum-dresden.de
[3] Division of Pharmacology, Utrecht Institute for Pharmaceutical Sciences (UIPS), Utrecht University, 3584CG Utrecht, The Netherlands
* Correspondence: sarahmichellebenson@gmail.com; Tel.: +44-(3)921-452-12

Received: 31 May 2020; Accepted: 3 July 2020; Published: 9 July 2020

Abstract: The aim of this study was to evaluate the efficacy of putative hangover treatment, Rapid Recovery, in mitigating alcohol hangover (AH) symptom severity. Using a double-blind, randomized, placebo-controlled, balanced crossover design, 20 participants attended the laboratory for two evenings of alcohol consumption, each followed by morning assessments of AH severity. Participants were administered Rapid Recovery and placebo on separate visits. In the first testing visit, participants self-administered alcoholic beverages of their choice, to a maximum of 1.3 g/kg alcohol. Drinking patterns were recorded and replicated in the second evening testing visit. In the morning visits, AH severity was assessed using questionnaires measuring AH symptom severity and sleep quality, computerized assessments of cognitive functioning as well as levels of blood biomarkers of liver function (gamma-glutamyl transferase (GGT)) and inflammation (high-sensitive C-reactive protein (hs-CRP)). There were no differences in the blood alcohol concentrations (BAC) obtained in the Rapid Recovery (mean = 0.096%) and placebo (mean = 0.097%) conditions. Participants reported significantly greater sleep problems in the Rapid Recovery compared to placebo condition, although this difference was no longer significant following Bonferroni's correction. There were no other significant differences between Rapid Recovery and placebo. These data suggest that Rapid Recovery has no significant effect on alcohol hangover nor on associated biomarkers.

Keywords: hangover; alcohol; hangover treatment; inflammation; liver function

1. Introduction

Alcohol hangover (AH) is defined as the combination of negative mental and physical symptoms which can be experienced after a single episode of alcohol consumption, starting when blood alcohol concentration (BAC) approaches zero [1,2]. It is characterized by a general state of malaise and a range of physical and psychological symptoms including headache [3,4], fatigue [5], nausea [4] and reduced cognitive functioning [6–8]. These symptoms negatively impact daily activities such as driving [9,10], job performance [11,12] and studying [4].

AH is pervasive, affecting 75% of all social drinkers [13]. As well as subjective effects it contributes to significant economic costs. It is estimated that, due to associated absenteeism and presenteeism, AH costs the UK economy between £1.2 billion and £1.4 billion per year [14] (i.e., approximately US $1.5 to $1.7 billion) and the Australian economy over AUS $3 billion annually [15] (i.e., approximately US $1.8 billion). Assessing the full cost, beyond absenteeism and presenteeism, it has been estimated that AH costs the American economy some US $179 billion per year [16].

The physiological causes of AH are largely unknown. Analyses of blood, saliva and urine samples indicate that concentrations of various hormones, electrolytes, free fatty acids, triglycerides, lactate, ketone bodies, cortisol, glucose and biomarkers of dehydration do not appear to correlate with hangover symptom severity [17,18]. In a recent review of biological factors that contribute to AH, Palmer et al. [19] concluded that alcohol metabolites, inflammatory factors, neurotransmitter alterations and mitochondrial dysfunction are the most likely contributors to AH severity.

Alcohol is predominately broken down in the liver, where it is metabolized by alcohol dehydrogenase (ADH) to acetaldehyde, which is then itself metabolized by aldehyde dehydrogenase (ALDH) to acetate. Acetate is then broken down into water and carbon dioxide for elimination. Acetaldehyde is rapidly metabolized by most individuals so that blood acetaldehyde levels typically remain low and it is unlikely to be present during AH. Nevertheless, acetaldehyde is highly toxic. It can cause tissue damage [20,21] and its presence in the body has been associated with hangover-like symptoms, including nausea, sweating, rapid pulse and headache [22,23]. It has been argued that increased acetaldehyde concentration and its long-lasting effects contribute to the presence of hangover symptoms [24,25]. However, the one human study to assess the effects of blood acetaldehyde levels on AH severity failed to find any evidence for a correlation between peak acetaldehyde concentration and hangover severity [26]. However, this one study does not provide sufficient evidence to exclude the possibility of an association between acetaldehyde and AH severity.

Evidence collected in several animal [27–30], human [31] and in vitro studies [32,33] indicate severe effects of ethanol on inflammatory processes. Inflammatory responses can also result in a variety of hangover-related symptoms, including nausea, vomiting, headache, negative mood and cognitive impairment [34,35]. Several studies have demonstrated evidence for elevated cytokine levels during hangover [31,36–38]. Another marker of inflammation, C-reactive protein (CRP), has been reported in two studies to correlate with AH severity [25,39], while another study has failed to find an association between CRP levels and AH [40]. However, the reliability of one of the studies that reported an association [25] is questionable as the assays that were used to measure CRP had limited detection sensitivity. This resulted in almost one-quarter of the data being outside the detection limit. Since this study, the development of highly sensitive assays to measure CRP levels have enabled more accurate measurement of CRP.

The current lack of understanding of the pathology of AH has hindered the development of an effective hangover treatment. Despite this there remains a high consumer demand [41] and many currently available products are advertised as mitigating AH severity. Yet there is no hangover treatment on the market with robust evidence for efficacy. Proposed treatments that have been investigated in human research showed either no effect, or minimal and differential reduction in the presence or severity of some but not other hangover symptoms [42].

The treatment of AH is further complicated by individual variation in hangover symptom frequency and severity [43,44], of which, genetic variations contribute about 40%–45% [45]. The influence of genetic variations on AH is evident when considering the efficacy of Korean pear juice, which has been shown to effectively reduce certain AH symptoms according to aldehyde dehydrogenase (ALDH) genotype. Specifically, it is effective in carriers of the ALDH2*1/*1 and ALDH2*1/*2 alleles, while being ineffective in the ALDH2*2/*2 genotype [46]. Variations in aldehyde dehydrogenase genes are also responsible for alcohol-induced flush reactions that are evident in 36% of people descended from East Asia [47]. Symptoms include flushing of the face, neck and shoulders, along with symptoms commonly associated with hangover, including headache and nausea, which are caused by elevated circulating levels of acetaldehyde [48]. Individuals who experience alcohol-induced flush reactions also display greater susceptibility to AH [49] and sensitivity to AH symptom severity [50], adding further support for the important role of acetaldehyde in AH. However, it should also be noted that Lee et al. [44] investigated the effects of Korean pear in a sample of 14 healthy male-only Asian subjects. Therefore, more research is needed to confirm these findings in groups of non-Asian descent men and women.

The aim of this investigation was to examine the effects of Rapid Recovery on AH symptom severity, inflammation, sleep quality and cognitive functioning. Rapid Recovery is an oral capsule that contains the amino acid L-cysteine and B and C group vitamins. It is proposed by the manufacturers that these ingredients will improve acetaldehyde metabolism and reduce oxidative stress. L-cysteine plays a role in reversing oxidization in the liver, with animal research showing that L-cysteine accelerates the breakdown and reduces the accumulation of acetaldehyde [51]. Another rodent study found that the administration of L-cysteine combined with vitamins B-1 and C reduced mortality caused by acetaldehyde poisoning [52]. In the current study, we tested the hypothesis that Rapid Recovery would reduce AH severity in social drinkers. A number of relevant biomarkers were co-monitored.

2. Experimental Section

2.1. Method

This study was conducted in accordance with the Declaration of Helsinki and was approved by the Swinburne University Human Research Ethics Committee (SUHREC, 2018/275). This study was registered with the Australian New Zealand Clinical Trials Registry (ANZCTR, ACTRN12618001996257).

2.2. Design

This study was a semi-naturalistic, randomized, double-blind, placebo-controlled, crossover clinical trial. The laboratory was set-up to simulate a bar-like environment and participants consumed alcoholic drinks of their choice and at their own pace, to a maximum of 1.3 g/kg alcohol. Participants were administered either placebo or active treatment over two testing visits.

2.3. Participants

Twenty-three participants who were healthy, aged 21–50 years old and regularly experienced hangovers were enrolled in the study. Three participants withdrew at the first morning visit, two withdrew due to illness and one failed to meet the eligibility requirement of a (BAC of 0.00% at the morning visit. The final sample consisted of 20 participants (65% female) with a mean age of 30.30 years (range 25–43 years old).

All participants were free of any current or history of drug or alcohol abuse, medically treated liver or renal impairment, pregnancy or breast feeding in females, and current use of any medication that could potentially affect the outcome of the study.

2.4. Measures

2.4.1. Breath Alcohol Concentration (BAC)

BAC was measured at the beginning of each testing visit to ensure a reading of 0.00%. In the evening testing visits, BAC was measured approximately 20-min after the final alcoholic drink. BAC was collected using a regularly calibrated Lion Alcolmeter SD400PA.

2.4.2. Assessment of Hangover Severity

Overall hangover severity was measured using a single one-item rating and severity of 23 hangover symptoms were rated on an 11-point Likert scale ranging from 0 to 10, with higher scores indicating a more severe hangover. The 23 items were derived from the Alcohol Hangover Severity Scale, the Hangover Symptoms Scale and the Acute Hangover Scale [53,54]. This composite scale has been successfully implemented in previous hangover research [55].

2.4.3. Sleep Quality Assessments

Self-reported assessments of sleep quality comprised of the Groningen Sleep Quality Scale (GSQS) [56] and the Karolinska Sleepiness Scale (KSS) [57], which measured sleep quality during the previous night and current sleepiness, respectively. The GSQS comprises of 15 sleep complaints requiring a "yes" or "no" response indicating whether they had been experienced during the previous night's sleep. Scores range from 0 to 14, (the first item is not scored) with higher scores indicating poorer sleep quality. The KSS requires participants to indicate their level of fatigue in the last five minutes on a single-item using a nine-point Likert scale. Higher scores indicate greater levels of sleepiness. These scales have been implemented successfully in previous hangover research [9,55].

2.4.4. Assessment of Biomarkers for Inflammation and Liver Function

High-sensitivity C-reactive protein (hs-CRP) tests were used to measure inflammation and gamma-glutamyl transferase (GGT) tests were used to measure liver function.

2.4.5. Assessment of Cognitive Performance

Cognitive performance was measured using the following tests available on the Vienna Test System (Schuhfried GmbH, Moedling, Austria). This test system assesses cognitive functioning that influence driving ability. The entire battery required approximately 15–20 min to complete.

Reaction Test (RT)

This test measures reaction time and motor time in response to optical and acoustic signals [58]. Participants were asked to place and leave their index finger on a pressure-sensitive key (i.e., rest key). Using the same index finger, participants were required to react as quickly as possible to the signals by pressing a target key before retiring their finger to the rest key. Performance was measured according to mean reaction time, mean motor time and number of correct reactions.

Determination Test (DT)

This test assesses reactive stress tolerance, divided attention and mental flexibility [59]. Participants are presented with various visual and auditory stimuli and are required to respond the stimuli by pressing corresponding response buttons with either their hands or feet, using the response panel and foot pedals of the Vienna Test System. Performance was assessed according to reaction time, number of correct responses, number of errors and number of missed responses.

Adaptive Tachistoscopic Traffic Perception Test (ATAVT)

This test assesses visual observation skills, visual orientation ability, speed of perception and skills in obtaining a traffic overview [60]. Images of traffic situations appeared briefly on a computer screen and the participant was asked to state what was in each image, by choosing from five answer options; motor vehicle, road sign, traffic light, pedestrian and bicycle. Performance was measured according to reaction time and the number of errors made.

2.4.6. Perceived Treatment Order

Awareness of the allocated condition order (active-placebo or placebo-active) was measured at the end of the trial. Participants were asked which treatment (active or placebo) they believed they had received on the first and second testing visit.

2.5. Procedure

Prior to undergoing any testing procedures, participants provided written informed consent and were assessed for eligibility. Participants then underwent training and practice in completing the RT, DT and ATAVT tasks, and provided a baseline blood sample for hs-CRP and GGT analyses.

All testing visits were held in the laboratory, with intoxication visits held between 17:00 and 00:30, and hangover visits held the following morning between 7:00 and 11:00. The two evening visits were held within 7–14 days of one another. During the evening visits, the laboratory was set-up to mimic a bar and background music was played while participants socialized with one another.

Participants were advised to avoid alcohol for 24-h prior to the intoxication visits, food and drink (other than water) for 2-h prior to all testing visits and alcohol, drugs, food and caffeine between the evening and morning visits. At the beginning of each evening visit, participants were provided with a meal, the type and quantity of food consumed in the first visit was recorded and replicated in the second evening visit. Participants were then instructed to freely consume the drink type(s) of their choice (of wine, cider, beer, spirits), to a maximum of 1.3 g/kg alcohol. The time that each drink was started and finished was recorded and drinking behavior was replicated in the second evening visit. Participants were administered the first dose of the study treatment with their final drink and were provided the second dose to self-administer upon their first awakening the following morning. The study treatment was either placebo (corn flour) or Rapid Recovery (L-cysteine, thiamine, pyridoxine and ascorbic acid). The contents of the study treatments were controlled by a laboratory independent of the manufacturer.

Participants returned to the laboratory the following morning where they were initially breathalyzed to ensure a BAC reading of 0.00%. Once deemed eligible, participants were able to commence the testing procedures.

2.6. Statistics and Analyses

Statistical analyses were conducted using SPSS, Version 25 (IBM Corp, Armonk, NY, USA). All variables were analyzed using paired sample t-tests comparing Rapid Recovery with placebo. The sleep quality assessments and Hs-CRP and GGT levels were correlated with overall hangover severity. Lastly, a chi-square test was used to determine whether there was a significant difference between correct and incorrect perceived treatment order.

In order to further investigate whether the obtained data was more in favor of the null hypothesis (H0, i.e., the assumption of no differences between the active and placebo condition) or more in favor of the alternative hypothesis (H1, i.e., the assumption of differences between the active and placebo condition), add-on Bayesian statistics were conducted using the standard settings of SPSS for the respective tests. Based on the cutoffs suggested by Wagenmakers, et al. [61] the Bayes factor (BF) of 1 does not provide evidence for either hypothesis. Larger BF values provide stronger evidence for the H0 (compared to the H1), while smaller BF values provide stronger evidence for the H1 (compared to the H0), given the obtained data. Specifically, values 1–3 (1/3–1) are seen as anecdotal evidence for the H0, values 3–10 (1/10–1/3) are seen as substantial evidence for the H0, values of 10–30 (1/30–1/10) are seen as strong evidence for the H0, values of 30–100 (1/100–1/30) are seen as very strong evidence for the H0, and values of >100 (<1/100) are seen as extreme evidence for the H0.

3. Results

3.1. BAC Levels

BAC levels obtained in the active (mean = 0.096%, sd = 0.023) and placebo (mean = 0.097%, sd = 0.028) conditions did not significantly differ ($t(19) = 0.507$, $p = 0.618$). Add-on Bayesian analyses provided substantial evidence for the null hypothesis (BF = 5.183), showing that BAC concentrations did indeed not differ between conditions.

3.2. Hangover Symptom Severity

The only hangover symptom to significantly differ according to testing condition was "sleep problems" $t(19) = 2.10$, $p = 0.049$, with more severe sleep problems in the active (mean = 2.59, sd = 2.86) compared to placebo (mean = 1.63, sd = 1.75) condition. Following Bonferroni's correction, this difference was no longer significant. Add-on Bayesian analyses for the non-significant effects revealed that in most cases, the obtained BF provided substantial evidence for the H0, as indicated by BF values between 3 and 10. For the other factors ("reduced appetite", "sweating", "heart beating" and "vomiting"), the Bayesian analyses still provided anecdotal evidence in favor of the H0, as indicated by BF values between 1 and 3. Taken together, all of these findings support the assumption that none of the investigated measures improved during the active condition. Hangover symptom severity scores can be found in Table 1, below.

Table 1. Hangover symptom severity scores (means and standard deviations) in the Rapid Recovery and placebo conditions. Descriptive data is given in the left columns, while the p value obtained from paired samples t-tests and the Bayes factor (BF) value obtained in case of non-significant differences (i.e., p values < 0.05) are provided in the right columns.

Item	Placebo M (SD)	Rapid Recovery M (SD)	p Value	BF Value
Single-Item Severity Scale				
'How severe is your hangover?'	3.18 (2.69)	3.22 (2.07)	0.962	5.856
Hangover Symptom Composite Scale				
Concentration problems	5.65 (2.40)	5.43 (1.92)	0.748	5.570
Thirst	4.94 (2.51)	5.12 (1.35)	0.775	5.630
Tiredness	4.78 (2.35)	4.83 (2.43)	0.957	5.854
Sleepiness	4.53 (2.340)	4.66 (2.50)	0.874	5.790
Headache	3.31 (3.07)	3.31 (2.51)	0.996	5.862
Apathy	2.90 (2.64)	2.79 (2.22)	0.856	5.768
Clumsiness	2.65 (2.01)	2.66 (2.11)	0.994	5.862
Weakness	2.54 (2.63)	2.71 (2.07)	0.835	5.737
Sensitivity to light	2.26 (2.54)	2.38 (2.06)	0.811	5.698
Nausea	1.77 (1.72)	2.37 (2.47)	0.365	3.904
Sleep problems	1.63 (1.75)	2.59 (2.86)	0.049 *	/
Reduced appetite	1.61 (1.80)	2.57 (2.75)	0.219	2.774
Dizziness	1.53 (1.54)	2.05 (1.97)	0.360	3.870
Stomach pain	1.37 (2.38)	1.35 (2.08)	0.977	5.860
Shaking, shivering	1.23 (1.66)	0.95 (1.31)	0.407	4.169
Anxiety	1.18 (1.54)	1.07 (1.20)	0.668	5.350
Confusion	1.17 (1.54)	0.98 (0.95)	0.569	4.991
Regret	1.05 (1.68)	0.77 (1.00)	0.429	4.298
Sweating	0.93 (1.08)	1.40 (1.68)	0.238	2.940
Heart beating	0.90 (1.31)	1.41 (1.76)	0.234	2.903
Depression	0.73 (0.88)	0.75 (0.97)	0.935	5.843
Heart racing	0.67 (0.95)	0.84 (0.93)	0.247	3.016
Vomiting	0.39 (0.53)	0.89 (1.70)	0.193	2.529

Note: M: Mean; SD: Standard deviation; *: $p < 0.05$.

3.3. Sleep Quality and Cognitive Performance

There were no significant differences between the Rapid Recovery and placebo conditions on the Groningen Sleep Quality Scale (GSQ), Karolinska Sleepiness Scale (KSS), reaction test (RT), determination test (DT) and adaptive tachistoscopic traffic perception test (ATAVT). The mean scores and standard deviations are displayed in Table 2, below. Self-rated overall hangover severity significantly correlated with GSQ ($r = 0.552$, $p = 0.012$) and KSS ($r = 0.764$, $p \leq 0.001$) scores in the placebo condition. The same result was found in the treatment condition, with overall hangover severity scores significantly correlating with GSQ ($r = 0.638$, $p = 0.002$) and KSS ($r = 0.762$, $p < 0.001$) scores.

Table 2. Sleep quality and cognitive performance scores (means and standard deviations) in the Rapid Recovery and placebo conditions. Descriptive data is given in the left columns, including Groningen Sleep Quality Scale (GSQS), Karolinska Sleepiness Scale (KSS), reaction test (RT), determination test (DT) and adaptive tachistoscopic traffic perception test (ATAVT) while the p value obtained from paired samples t-tests and the BF value obtained in case of non-significant differences (i.e., p values < 0.05) are provided in the right columns.

Item	Rapid Recovery M (SD)	Placebo M (SD)	p-Value	BF Value
Sleep Quality				
GSQ	4.10 (3.54)	3.35 (3.01)	0.429	4.300
KSS	5.00 (2.15)	4.85 (2.06)	0.845	5.753
RT				
Reaction time (milliseconds)	426.79 (57.23)	415.37 (79.40)	0.297	3.341
Motor time (milliseconds)	152.84 (37.55)	156.89 (34.64)	0.584	4.788
Number of correct reactions	15.95 (0.23)	16.00 (0.00)	0.331	3.580
DT				
Reaction time (milliseconds)	676.80 (61.20)	668.90 (62.35)	0.544	4.770
Number of correct responses	283.05 (33.58)	289.32 (27.03)	0.367	3.826
Number of errors	21.42 (11.76)	22.21 (11.54)	0.681	5.265
Number of missed responses	14.21 (8.34)	13.95 (6.93)	0.823	5.586
ATAVT				
Reaction time (seconds)	8.98 (1.50)	9.03 (1.47)	0.893	5.675
Number of errors	5.95 (2.37)	6.00 (3.79)	0.956	5.718

Note: M: Mean; SD: Standard deviation.

3.4. Levels of Biomarkers for Inflammation and Liver Function

There were no significant differences in hs-CRP and GGT levels in the Rapid Recovery compared to placebo condition (all $p \geq 0.376$). Bayesian add-on analyses further provided substantial evidence for the H0 (all BF ≥ 3.587), thus demonstrating that both measures did not differ across conditions. Furthermore, hs-CRP and GGT levels did not significantly correlate with self-rated overall hangover severity in either of the testing conditions (all $p \geq 0.286$). For both conditions, add-on Bayesian analyses provided substantial evidence for the lack of correlation between hs-CRP and overall hangover ratings (all BF ≥ 3.000) and for the lack of correlation between GGT and overall hangover (all BF ≥ 3.008). The hs-CRP and GGT levels can be found in Table 3, below.

Table 3. High sensitivity C-reactive protein (hs-CRP) and gamma-glutamyl transpeptidase (GGT) levels (means and standard deviations) at baseline and in the Rapid Recovery and placebo conditions. Descriptive data is given in the left columns, while the p value obtained from paired samples t-tests comparing the placebo and active condition, as well as the BF value obtained in case of non-significant differences (i.e., p values < 0.05) are provided in the right columns ($N = 16$).

	Baseline M (SD)	Rapid Recovery M (SD)	Placebo M (SD)	p-Value	BF Value
Hs-CRP (mg/L)	1.78 (2.86)	1.49 (2.28)	1.43 (2.37)	0.813	5.150
GGT (U/L)	27.56 (14.39)	28.31 (15.17)	27.13 (13.85)	0.376	3.587

Note: M: Mean; SD: Standard deviation.

3.5. Percieved Treatment Order and Adverse Events

A total of 60% of the participants guessed the correct condition order, indicating adequate blinding, $\chi^2 = 0.80$, $p = 0.371$.

There were no reported adverse events associated with Rapid Recovery.

4. Discussion

The current study assessed the effects of Rapid Recovery on hangover symptom severity. The hypothesis that Rapid Recovery would reduce AH severity was not supported. There were no significant differences between placebo and Rapid Recovery on self-rated overall hangover severity, sleep quality, CRP and GGT levels and cognitive performance. Furthermore, Bayesian add-on analyses provided credible evidence that the assumption of a null effect was more likely (than the assumption of non-significant/residual differences), given the obtained data. Of the 23 hangover symptoms that were assessed, the only significant difference between Rapid Recovery and placebo was found on the symptom of "sleep problems", which was worse following Rapid Recovery administration. The ineffectiveness of Rapid Recovery to reduce AH severity may indicate that administration of l-cysteine combined with B and C vitamins does not improve acetaldehyde metabolism, or that acetaldehyde is not responsible for AH severity.

The results of this study further support the relationship between poor sleep quality and hangover severity [8,62–64], with significant and positive correlations between AH severity, and poor previous night's sleep quality and current sleepiness. The current study failed to provide any evidence for a correlation between AH severity and CRP or GGT levels, which remained within normal ranges at each testing timepoint. Currently, the evidence for an association between AH severity and CRP is mixed, with some studies [25,39] demonstrating support for an association, while one other study [40], consistent with the findings of the current study, failed to find significant correlation between AH severity and CRP. While GGT is a reliable biomarker of liver damage caused by chronic heavy drinking [65–67], previous research has indicated that GGT levels are not associated with AH susceptibility [49] and, consistent with the findings of this study, are not necessarily elevated during AH [39]. Although it was not demonstrated in the current study, compelling evidence indicates an impairing effect of AH on immune functioning, and more reliable markers of this may include, but are not limited to, interleukin (IL)-6, IL-10, IL-12 and tumor necrosis factor (TNF)-α [37,68,69].

This investigation utilized a novel, controlled and ecologically valid methodology, which was found to successfully induce AH. Participants obtained mean BACs of 0.096% and 0.097% in the Rapid Recovery and placebo conditions, respectively, levels beyond that required to induce a hangover [70]. By enabling participants to self-administer alcohol within a controlled laboratory setting, we were able to overcome several commonly occurring methodological issues within the area of AH research. While methods of alcohol dosing used in previous laboratory studies assessing AH have been criticized for not mimicking real-life drinking behaviors, naturalistic studies have been criticized for lacking experimental control, and relying on self-reported alcohol intake to calculate estimated BAC [4,71–73]. The current study used methodology which combined the advantages of naturalistic approaches (i.e., participants drinking alcohol of their choice in a social setting) with those of laboratory studies (i.e., a controlled environment, objective measures of BAC and other biomarkers, veracity of treatment administration).

There were several limitations in this study. Firstly, we allowed participants to consume their preferred type of alcohol to ensure drinking behaviors replicated real-life drinking. Although drinking behaviors were consistent across the two testing visits, which eliminated intraindividual differences, this introduced interindividual variability. Alcoholic drinks contain various concentrations of congeners, which have been found to increase AH severity [17,74]. On the other hand, the fact that each individual's session was matched, somewhat, mitigates against this influencing our results. Furthermore, it is possible that the effectiveness of Rapid Recovery is dependent on individual factors, for example, genetic variations or tolerance to alcohol. This was evident in literature on the effectiveness of Korean pear juice in treating certain AH symptoms in particular genetic subgroups but not others [46]. The sample size of this study was too small to allow meaningful subgroup analysis. Lastly, we did not assess hangover symptoms following alcohol abstinence because, although interesting, this was not necessary for the aim of this study, i.e., comparing Rapid Recovery and placebo. As such, we are unable to determine the severity of AH obtained in this study. While mean BAC levels are beyond those deemed required to induce a hangover [70], the mean overall hangover severity score is relatively low. Recent evidence indicates that BAC may not be the most appropriate predictor of AH severity, which is better predicted by levels of subjective intoxication and increased alcohol consumption compared to usual [70]. While these factors were not assessed in the current study, future AH research should aim to include measures of subjective intoxication and typical alcohol intake.

In conclusion, the findings from this study suggest that the administration of Rapid Recovery does not mitigate AH severity, and Hs-CRP and GGT levels are not associated with AH. Further research is required to assess the impact of an effective hangover treatment on alcohol consumption and determine whether it would encourage excessive drinking. Importantly, an effective hangover treatment would not mitigate all adverse factors associated with heavy drinking, such as chronic disease and injury. The development of an effective hangover treatment is currently hindered by a lack of understanding of the pathology of AH. As such, future research should continue to assess the pathology of AH to enable the development of treatments that target key mechanisms involved in the AH.

Author Contributions: Conceptualization, S.B. and A.S.; methodology, S.B. and A.S.; formal analysis, S.B. and A.S.; investigation, S.B., A.S. and E.A.; data curation, S.B. and A.S.; writing—original draft preparation, S.B. and A.S.; writing—review and editing, S.B., A.S., E.A., J.C.V., and A.-K.S.; supervision, S.B. and A.S.; project administration, S.B., A.S. and E.A.; funding acquisition, S.B. and A.S. All authors have read and agreed to the published version of the manuscript.

Funding: This research was funded by Phoenix Pharmaceutical.

Conflicts of Interest: Over the past 36 months, A.S. has held research grants from Abbott Nutrition, Arla Foods, Bayer, BioRevive, DuPont, Fonterra, Kemin Foods, Nestlé, Nutricia-Danone, and Verdure Sciences. He has acted as a consultant/expert advisor to Bayer, Danone, Naturex, Nestlé, Pfizer, Sanofi, Sen-Jam Pharmaceutical, and has received travel/hospitality/speaker fees from Bayer, Sanofi, and Verdure Sciences. Over the past 36 months, J.C.V. has held grants from Janssen, Nutricia, and Sequential, and acted as a consultant/expert advisor to Clinilabs, More Labs, Red Bull, Sen-Jam Pharmaceutical, Toast!, and ZBiotics. S.B. has received funding from Red Bull GmbH, Kemin Foods, Sanofi Aventis, Phoenix Pharmaceutical, BioRevive, Australian Government Innovations Scheme and GlaxoSmithKline. A.K.S. has received funding from Daimler and Benz.A.J.A.E.V.D.L. has no conflicts of interest to declare. The funders had no role in the design of the study; in the collection, analyses, or interpretation of data; in the writing of the manuscript, or in the decision to publish the results.

References

1. Verster, J.C.; Scholey, A.; van de Loo, A.J.; Benson, S.; Stock, A.-K. Updating the definition of the alcohol hangover. *J. Clin. Med.* **2020**, *9*, 823. [CrossRef]
2. van Schrojenstein Lantman, M.; van de Loo, A.J.A.E.; Mackus, M.; Verster, J.C. Development of a definition for the alcohol hangover: Consumer descriptions and expert consensus. *Curr. Drug Abuse Rev.* **2016**, *9*, 148–154. [CrossRef] [PubMed]
3. Rohsenow, D.J.; Howland, J.; Minsky, S.J.; Greece, J.; Almeida, A.; Roehrs, T.A. The Acute Hangover Scale: A new measure of immediate hangover symptoms. *Addict. Behav.* **2007**, *32*, 1314–1320. [CrossRef]
4. Verster, J.C.; Stephens, R.; Penning, R.; Rohsenow, D.; McGeary, J.; Levy, D.; McKinney, A.; Finnigan, F.; Piasecki, T.M.; Adan, A.; et al. The alcohol hangover research group consensus statement on best practice in alcohol hangover research. *Curr. Drug Abuse Rev.* **2010**, *3*, 116–126. [CrossRef] [PubMed]
5. Mc Kinney, A.; Coyle, K. Alcohol hangover effects on measures of affect the morning after a normal night's drinking. *Alcohol Alcohol.* **2006**, *41*, 54–60. [CrossRef] [PubMed]
6. Benson, S.; Ayre, E.; Garrisson, H.; Wetherell, M.A.; Verster, J.C.; Scholey, A. Alcohol Hangover and Multitasking: Effects on Mood, Cognitive Performance, Stress Reactivity, and Perceived Effort. *J. Clin. Med.* **2020**, *9*, 1154. [CrossRef]
7. Gunn, C.; Mackus, M.; Griffin, C.; Munafò, M.R.; Adams, S. A systematic review of the next-day effects of heavy alcohol consumption on cognitive performance. *Addiction* **2018**, *113*, 2182–2193. [CrossRef] [PubMed]
8. Scholey, A.; Benson, S.; Kaufman, J.; Terpstra, C.; Ayre, E.; Verster, J.C.; Allen, C.; Devilly, G.J. Effects of alcohol hangover on cognitive performance: Findings from a field/internet mixed methodology study. *J. Clin. Med.* **2019**, *8*, 440. [CrossRef]
9. Verster, J.C.; Bervoets, A.C.; de Klerk, S.; Vreman, R.A.; Olivier, B.; Roth, T.; Brookhuis, K.A. Effects of alcohol hangover on simulated highway driving performance. *Psychopharmacology* **2014**, *231*, 2999–3008. [CrossRef]

10. Alford, C.; Broom, C.; Carver, H.; Johnson, S.J.; Lands, S.; Reece, R.; Verster, J.C. The Impact of Alcohol Hangover on Simulated Driving Performance During a 'Commute to Work'—Zero and Residual Alcohol Effects Compared. *J. Clin. Med.* **2020**, *9*, 1435. [CrossRef]
11. Frone, M.R. Employee psychoactive substance involvement: Historical context, key findings, and future directions. *Ann. Rev. Organ. Psychol. Organ. Behav.* **2019**, *6*, 273–297. [CrossRef]
12. Moore, R.S.; Ames, G.M.; Duke, M.R.; Cunradi, C.B. Food service employee alcohol use, hangovers and norms during and after work hours. *J. Subst. Use* **2012**, *17*, 269–276. [CrossRef] [PubMed]
13. Howland, J.; Rohsenow, D.J.; Allensworth-Davies, D.; Greece, J.; Almeida, A.; Minsky, S.J.; Arnedt, J.T.; Hermos, J. The incidence and severity of hangover the morning after moderate alcohol intoxication. *Addiction* **2008**, *103*, 758–765. [CrossRef] [PubMed]
14. Bhattacharya, A. *Financial Headache: The Cost of Workplace Hangovers and Intoxication to the UK Economy*; Institute of Alcohol Studies: London, UK, 2019.
15. Roche, A.; Pidd, K.; Kostadinov, V. Alcohol- and drug-related absenteeism: A costly problem. *Aust. N. Z. J. Public Health* **2016**, *40*, 236–238. [CrossRef]
16. Bouchery, E.E.; Harwood, H.J.; Sacks, J.J.; Simon, C.J.; Brewer, R.D. Economic costs of excessive alcohol consumption in the US, 2006. *Am. J. Prev. Med.* **2011**, *41*, 516–524. [CrossRef] [PubMed]
17. Penning, R.; van Nuland, M.; Fliervoet, L.A.L.; Olivier, B.; Verster, J.C. The pathology of alcohol hangover. *Curr. Drug Abuse Rev.* **2010**, *3*, 68–75. [CrossRef]
18. Tipple, C.; Benson, S.; Scholey, A. A review of the physiological factors associated with alcohol hangover. *Curr. Drug Abuse Rev.* **2016**, *9*, 93–98. [CrossRef]
19. Palmer, E.; Tyacke, R.; Sastre, M.; Lingford-Hughes, A.; Nutt, D.; Ward, R.J. Alcohol Hangover: Underlying Biochemical, Inflammatory and Neurochemical Mechanisms. *Alcohol Alcohol.* **2019**, *54*, 196–203. [CrossRef]
20. Powers, S.K.; Hamilton, K. Antioxidants and exercise. *Clin. Sports Med.* **1999**, *18*, 525–536. [CrossRef]
21. Yan, T.; Zhao, Y.; Zhang, X. Acetaldehyde induces cytotoxicity of SH-SY5Y cells via inhibition of Akt activation and induction of oxidative stress. *Oxidative Med. Cell. Longev.* **2016**, *4512309*. [CrossRef]
22. Khan, M.A.; Jensen, K.; Krogh, H. Alcohol-Induced hangover; A double-blind comparison of pyritinol and placebo in preventing hangover symptoms. *Q. J. Stud. Alcohol* **1973**, *34*, 1195–1201. [CrossRef] [PubMed]
23. Baluci, C.; Saliba, C.; Gutierrez, G.; Collie, A.; Agius, C. Cognitive effects of acute alcohol consumption are reduced by TEX-OE pre-conditioning. *J. Psychopharmacol.* **2005**, A25.
24. Kaivola, S.; Parantainen, J.; Österman, T.; Timonen, H. Hangover headache and prostaglandins: Prophylactic treatment with tolfenamic acid. *Cephalalgia* **1983**, *3*, 31–36. [CrossRef] [PubMed]
25. Wiese, J.; McPherson, S.; Odden, M.C.; Shlipak, M.G. Effect of Opuntia ficus indica on symptoms of the alcohol hangover. *Arch. Intern. Med.* **2004**, *164*, 1334–1340. [CrossRef] [PubMed]
26. Ylikahri, R.; Huttunen, M.; Eriksson, C.; Nikklä, E. Metabolic studies on the pathogenesis of hangover. *Eur. J. Clin. Investig.* **1974**, *4*, 93–100. [CrossRef]
27. Vallés, S.L.; Blanco, A.M.; Pascual, M.; Guerri, C. Chronic ethanol treatment enhances inflammatory mediators and cell death in the brain and in astrocytes. *Brain Pathol.* **2004**, *14*, 365–371. [CrossRef]
28. Tiwari, V.; Kuhad, A.; Chopra, K. Suppression of neuro-inflammatory signaling cascade by tocotrienol can prevent chronic alcohol-induced cognitive dysfunction in rats. *Behav. Brain Res.* **2009**, *203*, 296–303. [CrossRef]
29. Kishore, R.; Hill, J.R.; McMullen, M.R.; Frenkel, J.; Nagy, L.E. ERK1/2 and Egr-1 contribute to increased TNF-α production in rat Kupffer cells after chronic ethanol feeding. *Am. J. Physiol.-Gastrointest. Liver Physiol.* **2002**, *282*, G6–G15. [CrossRef]
30. Pascual, M.; Montesinos, J.; Marcos, M.; Torres, J.L.; Costa-Alba, P.; García-García, F.; Laso, F.J.; Guerri, C. Gender differences in the inflammatory cytokine and chemokine profiles induced by binge ethanol drinking in adolescence. *Addict. Biol.* **2017**, *22*, 1829–1841. [CrossRef]
31. Neupane, S.P.; Skulberg, A.; Skulberg, K.R.; Aass, H.C.D.; Bramness, J.G. Cytokine changes following acute ethanol intoxication in healthy men: A crossover study. *Mediat. Inflamm.* **2016**, *3758590*. [CrossRef]
32. Davis, R.L.; Syapin, P.J. Ethanol increases nuclear factor-κB activity in human astroglial cells. *Neurosci. Lett.* **2004**, *371*, 128–132. [CrossRef] [PubMed]
33. Szabo, G.; Mandrekar, P.; Oak, S.; Mayerle, J. Effect of ethanol on inflammatory responses. *Pancreatology* **2007**, *7*, 115–123. [CrossRef] [PubMed]

34. Harrison, N.A.; Brydon, L.; Walker, C.; Gray, M.A.; Steptoe, A.; Critchley, H.D. Inflammation causes mood changes through alterations in subgenual cingulate activity and mesolimbic connectivity. *Biol. Psychiatr.* **2009**, *66*, 407–414. [CrossRef] [PubMed]
35. Dantzer, R. Cytokine-Induced sickness behaviour: A neuroimmune response to activation of innate immunity. *Eur. J. Pharmacol.* **2004**, *500*, 399–411. [CrossRef]
36. Gonzalez-Quintela, A.; Dominguez-Santalla, M.; Perez, L.; Vidal, C.; Lojo, S.; Barrio, E. Influence of acute alcohol intake and alcohol withdrawal on circulating levels of IL-6, IL-8, IL-10 and IL-12. *Cytokine* **2000**, *12*, 1437–1440. [CrossRef]
37. Kim, D.-J.; Kim, W.; Yoon, S.-J.; Choi, B.-M.; Kim, J.-S.; Go, H.J.; Kim, Y.-K.; Jeong, J. Effects of alcohol hangover on cytokine production in healthy subjects. *Alcohol* **2003**, *31*, 167–170. [CrossRef]
38. Van de Loo, A.; Slot, K.; Kleinjan, M.; Knipping, K.; Garssen, J.; Verster, J. Time-Dependent changes in saliva cytokine concentrations during alcohol hangover: A comparison of two naturalistic studies. *Alcohol. Clin. Exp. Res.* **2016**, *40*, 95A.
39. Bang, J.S.; Chung, Y.H.; Chung, S.J.; Lee, H.S.; Song, E.H.; Shin, Y.K.; Lee, Y.J.; Kim, H.-C.; Nam, Y.; Yeong, J.H. Clinical effect of a polysaccharide-rich extract of Acanthopanax senticosus on alcohol hangover. *Die Pharm.-An. Int. J. Pharm. Sci.* **2015**, *70*, 269–273.
40. George, A.; Udani, J.K.; Yusof, A. Effects of Phyllanthus amarus PHYLLPROTM leaves on hangover symptoms: A randomized, double-blind, placebo-controlled crossover study. *Pharm. Biol.* **2019**, *57*, 145–153. [CrossRef]
41. Mackus, M.; van Schrojenstein Lantman, M.; van de Loo, A.J.A.E.; Nutt, D.; Verster, J.C. An effective hangover treatment: Friend or foe? *Drug Sci. Policy Law* **2017**, *3*, 2050324517741038. [CrossRef]
42. Verster, J.C.; Penning, R. Treatment and prevention of alcohol hangover. *Curr. Drug Abuse Rev.* **2010**, *3*, 103–109. [CrossRef] [PubMed]
43. Verster, J.C.; van Schrojenstein Lantman, M.; Mackus, M.; van de Loo, A.J.; Garssen, J.; Scholey, A. Differences in the Temporal Typology of Alcohol Hangover. *Alcohol. Clin. Exp. Res.* **2018**, *42*, 691–697. [CrossRef] [PubMed]
44. Verster, J.C.; Slot, K.A.; Arnoldy, L.; van Lawick van Pabst, A.E.; van de Loo, A.J.; Benson, S.; Scholey, A. The Association between Alcohol Hangover Frequency and Severity: Evidence for Reverse Tolerance? *J. Clin. Med.* **2019**, *8*, 1520. [CrossRef] [PubMed]
45. Slutske, W.S.; Piasecki, T.M.; Nathanson, L.; Statham, D.J.; Martin, N.G. Genetic influences on alcohol-related hangover. *Addiction* **2014**, *109*, 2027–2034. [CrossRef] [PubMed]
46. Lee, H.-S.; Isse, T.; Kawamoto, T.; Baik, H.W.; Park, J.Y.; Yang, M. Effect of Korean pear (*Pyruspyrifolia* cv. Shingo) juice on hangover severity following alcohol consumption. *Food Chem. Toxicol.* **2013**, *58*, 101–106. [CrossRef]
47. Brooks, P.J.; Enoch, M.-A.; Goldman, D.; Li, T.-K.; Yokoyama, A. The alcohol flushing response: An unrecognized risk factor for esophageal cancer from alcohol consumption. *PLoS Med.* **2009**, *6*. [CrossRef]
48. Harada, S.; Agarwal, D.; Goedde, H.; Tagaki, S.; Ishikawa, B. Possible protective role against alcoholism for aldehyde dehydrogenase isozyme deficiency in Japan. *Lancet* **1982**, *320*, 827. [CrossRef]
49. Yokoyama, M.; Yokoyama, A.; Yokoyama, T.; Funazu, K.; Hamana, G.; Kondo, S.; Yamashita, T.; Nakamura, H. Hangover susceptibility in relation to aldehyde dehydrogenase-2 genotype, alcohol flushing, and mean corpuscular volume in Japanese workers. *Alcohol. Clin. Exp. Res.* **2005**, *29*, 1165–1171. [CrossRef] [PubMed]
50. Yokoyama, M.; Suzuki, N.; Yokoyama, T.; Yokoyama, A.; Funazu, K.; Shimizu, T.; Shibata, M. Interactions between migraine and tension-type headache and alcohol drinking, alcohol flushing, and hangover in Japanese. *J. Headache Pain* **2012**, *13*, 137–145. [CrossRef] [PubMed]
51. Donohue, T.M., Jr.; Tuma, D.J.; Sorrell, M.F. Acetaldehyde adducts with proteins: Binding of [14C] acetaldehyde to serum albumin. *Arch. Biochem. Biophys.* **1983**, *220*, 239–246. [CrossRef]
52. Sprince, H.; Parker, C.M.; Smith, G.G.; Gonzales, L.J. Protection against acetaldehyde toxicity in the rat by L-cysteine, thiamin and L-2-methylthiazolidine-4-carboxylic acid. *Agents Actions* **1974**, *4*, 125–130. [CrossRef] [PubMed]
53. Penning, R.; McKinney, A.; Bus, L.D.; Olivier, B.; Slot, K.; Verster, J.C. Measurement of alcohol hangover severity: Development of the Alcohol Hangover Severity Scale (AHSS). *Psychopharmacology* **2013**, *225*, 803–810. [CrossRef] [PubMed]

54. Slutske, W.S.; Piasecki, T.M.; Hunt-Carter, E.E. Development and initial validation of the Hangover Symptoms Scale: Prevalence and correlates of hangover symptoms in college students. *Alcohol. Clin. Exp. Res.* **2003**, *27*, 1442–1450. [CrossRef]
55. Hogewoning, A.; Van de Loo, A.; Mackus, M.; Raasveld, S.; De Zeeuw, R.; Bosma, E.; Bouwmeester, N.; Brookhuis, K.; Garssen, J.; Verster, J. Characteristics of social drinkers with and without a hangover after heavy alcohol consumption. *Subst. Abuse Rehabil.* **2016**, *7*, 161. [CrossRef] [PubMed]
56. van der Meulen, W.M.-H.; Wijnberg, J.; Hollander, J.; De Diana, I.; Van den Hoofdakker, R. Measurement of subjective sleep quality. In Proceedings of the Amsterdam: Fifth European Sleep Congress of the European Sleep Research Society, Amsterdam, The Netherlands, 2–5 September 1980; p. 98.
57. Åkerstedt, T.; Gillberg, M. Subjective and objective sleepiness in the active individual. *Int. J. Neurosci.* **1990**, *52*, 29–37. [CrossRef] [PubMed]
58. Dinges, D.F.; Powell, J.W. Microcomputer analyses of performance on a portable, simple visual RT task during sustained operations. *Behav. Res. Methods Instrum. Comput.* **1985**, *17*, 652–655. [CrossRef]
59. Neuwirth, W.; Benesch, M. *Manual DT: Determination Test*, version 33.00; SCHUHFRIED GmbH: Mödling, Austria, 2007.
60. Schuhfried, G. *Manual ATAVT: The Adaptive Tachistoscopic Traffic*; SCHUHFRIED GmbH: Mödling, Austria, 2009.
61. Wagenmakers, E.-J.; Wetzels, R.; Borsboom, D.; Van Der Maas, H.L. Why psychologists must change the way they analyze their data: The case of psi: Comment on Bem. *J. Pers. Soc. Psychol.* **2011**, *100*, 426–432.
62. van Schrojenstein Lantman, M.; Roth, T.; Roehrs, T.; Verster, J.C. Alcohol hangover, sleep quality, and daytime sleepiness. *Sleep Vigil.* **2017**, *1*, 37–41. [CrossRef]
63. van Schrojenstein Lantman, M.; Mackus, M.; Roth, T.; Verster, J.C. Total sleep time, alcohol consumption, and the duration and severity of alcohol hangover. *Nat. Sci. Sleep* **2017**, *9*, 181. [CrossRef]
64. Devenney, L.E.; Coyle, K.B.; Roth, T.; Verster, J.C. Sleep after heavy alcohol consumption and physical activity levels during alcohol hangover. *J. Clin. Med.* **2019**, *8*, 752. [CrossRef]
65. Alatalo, P.; Koivisto, H.; Puukka, K.; Hietala, J.; Anttila, P.; Bloigu, R.; Niemelä, O. Biomarkers of liver status in heavy drinkers, moderate drinkers and abstainers. *Alcohol Alcohol.* **2009**, *44*, 199–203. [CrossRef] [PubMed]
66. Conigrave, K.M.; Degenhardt, L.J.; Whitfield, J.B.; Saunders, J.B.; Helander, A.; Tabakoff, B.; Group, W.I.S. CDT, GGT, and AST as markers of alcohol use: The WHO/ISBRA collaborative project. *Alcohol. Clin. Exp. Res.* **2002**, *26*, 332–339. [CrossRef] [PubMed]
67. Conigrave, K.M.; Davies, P.; Haber, P.; Whitfield, J.B. Traditional markers of excessive alcohol use. *Addiction* **2003**, *98*, 31–43. [CrossRef] [PubMed]
68. Palmer, E.O.; Arnoldy, L.; Ayre, E.; Benson, S.; Balikji, S.; Bruce, G.; Chen, F.; van Lawick van Pabst, A.E.; van de Loo, A.J.A.E.; O'Neill, S.; et al. In Proceedings of the 11th Alcohol Hangover Research Group Meeting, Nadi, Fiji. *Proceedings* **2020**, *43*, 1. [CrossRef]
69. Raasveld, S.; Hogewoning, A.; Van de Loo, A.; De Zeeuw, R.; Bosma, E.R.; Bouwmeester, N.; Lukkes, M.; Brookhuis, K.; Knipping, K.; Garssen, J. Cytokine concentrations after heavy alcohol consumption in people with and without a hangover. *Eur. Neuropsychopharmacol.* **2015**, *25*, 228. [CrossRef]
70. Verster, J.C.; Kruisselbrink, L.D.; Slot, K.A.; Anogeianaki, A.; Adams, S.; Alford, C.; Arnoldy, L.; Ayre, E.; Balikji, S.; Benson, S.; et al. Sensitivity to experiencing alcohol hangovers: Reconsideration of the 0.11% Blood Alcohol Concentration (BAC) threshold for having a hangover. *J. Clin. Med.* **2020**, *9*, 179. [CrossRef]
71. Stephens, R.; Grange, J.A.; Jones, K.; Owen, L. A critical analysis of alcohol hangover research methodology for surveys or studies of effects on cognition. *Psychopharmacology* **2014**, *231*, 2223–2236. [CrossRef]
72. Prat, G.; Adan, A.; Pérez-Pàmies, M.; Sànchez-Turet, M. Neurocognitive effects of alcohol hangover. *Addict. Behav.* **2008**, *33*, 15–23. [CrossRef]
73. Verster, J.C.; van de Loo, A.J.; Adams, S.; Stock, A.-K.; Benson, S.; Scholey, A.; Alford, C.; Bruce, G. Advantages and Limitations of Naturalistic Study Designs and Their Implementation in Alcohol Hangover Research. *J. Clin. Med.* **2019**, *8*, 2160. [CrossRef]
74. J. Rohsenow, D.; Howland, J. The role of beverage congeners in hangover and other residual effects of alcohol intoxication: A review. *Curr. Drug Abuse Rev.* **2010**, *3*, 76–79. [CrossRef]

© 2020 by the authors. Licensee MDPI, Basel, Switzerland. This article is an open access article distributed under the terms and conditions of the Creative Commons Attribution (CC BY) license (http://creativecommons.org/licenses/by/4.0/).

MDPI
St. Alban-Anlage 66
4052 Basel
Switzerland
Tel. +41 61 683 77 34
Fax +41 61 302 89 18
www.mdpi.com

Journal of Clinical Medicine Editorial Office
E-mail: jcm@mdpi.com
www.mdpi.com/journal/jcm

www.ingramcontent.com/pod-product-compliance
Lightning Source LLC
LaVergne TN
LVHW070237100526
838202LV00015B/2142